Ultra Poincaré Chaos and Alpha Labeling

A new approach to chaotic dynamics

Online at: https://doi.org/10.1088/978-0-7503-5162-1

Ultra Poincaré Chaos and Alpha Labeling

A new approach to chaotic dynamics

Marat Akhmet

Department of Mathematics Middle East Technical University, Ankara, Turkey

IOP Publishing, Bristol, UK

ISBN   978-0-7503-5162-1 (ebook)
ISBN   978-0-7503-5160-7 (print)
ISBN   978-0-7503-5163-8 (myPrint)
ISBN   978-0-7503-5161-4 (mobi)

DOI   10.1088/978-0-7503-5162-1

Version: 20241101

IOP ebooks

British Library Cataloguing-in-Publication Data: A catalogue record for this book is available from the British Library.

Published by IOP Publishing, wholly owned by The Institute of Physics, London

IOP Publishing, No.2 The Distillery, Glassfields, Avon Street, Bristol, BS2 0GR, UK

US Office: IOP Publishing, Inc., 190 North Independence Mall West, Suite 601, Philadelphia, PA 19106, USA

To the memory of my beloved mother, Sultanova Railya

Contents

Part IV   Differential equations with ultra Poincaré chaos

5   Alpha unpredictable differential equations   5-1

Part V   Numerical alpha unpredictability

6   Ultra Poincaré chaos numerically   6-1

Preface

Recently, we started two novel insights in the theory of dynamics: ultra Poincaré chaos (formerly Poincaré chaos [1–3, 7, 9, 13, 19]) which couples with alpha unpredictability (previously known as unpredictability [4, 5, 9, 10]) and alpha labeling (formerly referred to as labeling [4, 7, 19]). The first one is a fundamental characteristic of complexity. The second is a universal tool to discover and arrange complexity in dynamics and geometries. The manuscript takes a powerful approach to studying complexities through mathematical methods, consequently shaping circumstances for deep comprehension of the world in motion. Additionally, we are confident that alpha labeling will increase the clearness of the geometrical and topological vision of the world. The continuum of numbers can be complimented and substituted with alpha labeling, which will help observe and utilize more effectively sophisticated geometrical objects. We suggest the reader consider the space of alpha labels as a new mathematical structure. Together with numbers and sets, it will become a leading instrument for research of any mathematical complexity. Meanwhile, the investigation is only an initial step in discovering the complicated physical and geometrical world around us. Thus, results on the global analysis of time and space [35, 44] can be continued by studying the micro-world, which is full of discontinuities and complexities. Researchers are still investigating sophisticated dynamics of differential and even discrete deterministic equations at the embryonic level. At the same time, a study has begun on the complexities of perfectly discontinuous problems in a non-random fashion. Possibly, the idea that the reconciliation of quantum mechanics and the theory of relativity will happen based on sophisticated space-time geometries will be fueled by the present research, in particular. If revolutionary discoveries in physics caused the appearance of such mathematical structures as sets, vectors, and functional spaces, the theory of complexities demands the introduction of adequate fundamentals. There is strong confidence that chaos as a dynamic characteristic in our presentation can be indicated in many random processes. Thus, our results predict chaotic dynamics to be common in deterministic and stochastic processes.

Alpha labeling is an operator acting on a domain of dynamics and mapping it to an abstract similarity set. The operator has to be determined such that the motions of the original dynamics agree with those of alpha labels. If the map is one-to-one, then it is isometric. There dynamics may exist; first of all, stochastic processes, which request one-to-several and even one-to-infinity labeling such that alpha label spaces are not metric. Our suggestions are more effective than the conservative symbolic counterpart and any former Markov partition approach [29–32, 45, 53, 56]. One can see that in applications to the logistic equation, Baker transformations, Smale horseshoe maps, Bernoulli schemes, and Markov chains. Verification of chaos for fractals is another advantage of our method. Alpha labeling was pioneered for stochastic processes with finite state spaces, and analysis of arbitrary stochastic processes can also be accomplished using the labeling. Generally, the theory of stochastic differential equations can be extended in light of present suggestions. In other words, it can be developed in a new way. The decimal number system is universal for the world reflection. Alpha labeling is very analogous to that.

Moreover, it is a generalization of numbers because it makes us confident that the study is in the flow of Pythagorean and Al-Khwarizmi doctrines. It can be realized for ordinary, partial, functional differential, discrete equations, and hybrid systems. Alpha labels are a new effective mathematical structure because the method suggests three principal novelties: the distance function for alpha spaces is not a metric, in general, since the label operator is not bijection; the diameter condition for the abstraction of continuity in initial data; and the separation condition, which implies alpha unpredictability and sensitivity.

In this book, we present a range of new research techniques that have the potential to revolutionize the study of complex processes and sophisticated geometries. These techniques, including the numerical method for chaos indication, *the sequential test*, which is also known as *test for alpha unpredictability* or *ultra Poincaré chaos test*, offer a fresh perspective and can provide valuable insights by observing convergence and divergence in time sequences. This potential for transformation and advancement in our field is inspiring and promising.

While working on new types of recurrence, we were surprised to discover that, despite extensive literature, there is a lack of numerical examples and simulations for both the functions and solutions of differential equations. Meanwhile, the demands of industries, particularly in fields such as neuroscience, artificial intelligence, and other modern domains, require numerical representations of motions that are already well supported by theoretical frameworks. Our research addresses these challenges comprehensively, as we are the first to construct samples of Poisson stable and unpredictable functions using solutions to the logistic equation. We have also determined these functions randomly through realizations of Bernoulli schemes and Markov chains. In this way, we have successfully extended dynamical and stochastic methods of functions construction to meet new circumstances.

Our research also introduces the concept of "delta synchronization," a type of dynamics effective for recognizing various synchronized models. This type of dynamics, which we demonstrate in our work, would become a robust approach suitable for generalized, identical models and other types of synchronization that cannot be clarified. This is a significant contribution to the field, as it offers a new perspective on the study of synchronization in various systems. Besides this, *asymptotic visualization* of recurrent motions, such as alpha unpredictable, Poisson stable, and almost periodic, has been introduced and realized in many of our projects. Another is the *method of included intervals* for confirmation of alpha unpredictability and Poisson stability in differential and discrete equations dynamics. The *kappa property* is newly suggested to approve alpha unpredictability of compartmental periodic functions. The new methods witness that the boundaries of nonlinear dynamics can be extended to new positions, where universalism and realism of science will be strengthened significantly.

Our manuscript's methods and results benefit senior researchers, engineers, and a wide range of specialists, including those working in artificial neural networks, machine and deep learning, computer sciences, quantum computers, and applied and pure mathematics. This broad applicability underscores the value and relevance of our research to the wider academic community and the potential impact it can have on various fields.

One must confirm that the novelties do not have limits in applications if one proceeds to create mathematics like modern research. Nevertheless, if modified, one can only be confident that all mathematical structures will be needed. Each structure follows the demands of science and occurs because of them, and alpha labeling is no exception to this tradition.

Complexity in science is a last-century phenomenon and understood variously in *chaos, fractals, turbulence, random processes, and physical–chemical and algorithmic complexity* [33, 34, 37, 39, 40, 46, 51]. These researches will be much more productive with alpha labeling. Curiously, the research of complexity has not produced a structure adequate to the challenge.

To prevent misunderstanding with the regard to novelty, we should mention our former publication [4] in IOP. The present book is a continuation of the last one, but not only so. Much of the manuscript consists of recent contributions to the theory of recurrent, alpha unpredictable, and Poisson stable continuous and discontinuous functions, which have also been researched as solutions of quasi-linear differential and discrete equations. The dynamics of classical models are provided by focusing on a more transparent and profound presentation of known results. This, in turn, provides convenient circumstances for further extensions of complexity research. Such studies create optimism that the dynamic structures of old models can be elaborated by alpha labeling. This time, abstract hyperbolic chaos is verified for the Smale Horseshoe map and Baker multi-folded transformation. The reader is invited to compare what is done in this book with the results known through traditional methods. The approach of alpha labeling is constructive in exploring dynamics in fractals. Chaos was arranged in those with self-similarities. The episodic results for the Cantor set and Sierpinski carpet are replaced now with the universal method research [7]. Surprisingly, we have explored how to arrange chaotic dynamics in sets, which are not fractals but Euclidean cubes, and to all those that are topologically related to them, and even finite sets. The concept of abstract fractals was introduced [19], which can be helpful for modern science as a whole. It is much more accessible now to construct chaotic models. Presently, we suggest structuring a state space by alpha labeling such that similarity mapping precisely imitates the motion under research, and it is not a conjugacy or even semiconjugacy. This is one of the main novelties of the approach, which has yet to be considered in the literature. A sufficient background for complexity in terms of similarity dynamics can be found in our research. It is built with clearly formulated and easy-to-verify diameter and separation conditions, which are significant inventions for the study. This is another reason alpha labeling is principally advantageous and deserves a book.

The cornerstone of the recently introduced chaotic dynamics is the *alpha unpredictable point*, a Poisson stable point specification. Remember that the point is an ultimate type of oscillation in the Poincaré–Birkhoff theory [26, 43, 51, 54]. Thus, our suggestions prolong the line of recurrent motions to a new frontier such that the research joins with chaos investigation. It is important to emphasize Lorenz chaos relies entirely on the collective behavior of orbits, such as sensitivity, dense cycles, and transitivity [34, 39]. Oppositely, ultra Poincaré chaos [9, 10] is based on the alpha unpredictability concept, in which complexity is focused uniquely on a single orbit motion. By simulating a solution, we can conclude not only on chaos presence but also that the motion is alpha unpredictable [18]. This is, from one side, principally new for chaos study and, from another side, leads us back to the roots of classical dynamics, which considers a single stable motion as the main object of discussion [43, 54]. We believe the novel motion will make chaos research more beneficial for theoretical studies and applications. Initial discoveries have already been made in papers [1, 2], where the perception of random processes has been deepening through the novelties invented for our research. In the present book, the reader will find many exciting models revisited. One can formulate that *alpha unpredictability implies chaos* to illuminate the alpha unpredictability concept. As we predicted, the results have been utilized for topological dynamics [41, 42, 55], confirming the generality of our theory and its prospects.

Building on the previous discussion, it is important to clarify why the term "ultra Poincaré chaos is appropriate for the phenomenon investigated in our research. Chaos serves as a direct extension of the recurrence theorem, initially proposed by the eminent French mathematician for non-periodic motions, which also underpins ergodicity. Specifically, alpha unpredictability can be regarded as a distinct continuation of what was referred to as Poisson stability in [52]. As a result, our findings are particularly well suited as candidates for this "ultra'" characteristic within the framework of Poincaré's research. They not only align closely with the founder's ideas but also offer valuable insights into his understanding of complexity.

The fundamental definitions and theorems concerning alpha labeling have been significantly elaborated in chapter 2. Alpha unpredictable functions are assigned new constructive properties. The arguments' structure of compartmental alpha unpredictable functions follow the ideas for quasi-periodicity in [27, 28, 36], but on an essentially advanced level, when dependence on the part of several variables on diagonal is chaotic, while on others' it is non-chaotic. One can say that chaos is involved in functions' behavior. The proposals can initiate intensive studies of alpha unpredictable functions since more characteristics are involved in the discussion. The first results can be found in the book. Another significant part of the study is alpha unpredictable functions determined randomly. Methodically, they rely entirely on alpha labeling. Confirmations for the actuality of the contribution issue one can find in fundamental results by B. Ornstein [47–50] on isomorphism between deterministic and nondeterministic processes and insights of Nicolis and Prigogine [46] on the importance of individual stochastic motion for chaos investigation. Thus, we push for further theory extension by considering alpha unpredictable solutions of stochastic differential equations with Markov chains for inputs and coefficients. It is easily predictable that outputs with the same qualitative characteristics are expected. Consequently, new theorems of existence and stability for alpha unpredictable solutions are essential to the manuscript.

This work would not have been possible without the collaboration of Ejaily Al-Ejaily from Libya, Madina Tleubergenova, Roza Sejilova, Akylbek Zhamanshin,

Zahira Nugaeva, Nauryzbay Aviltay, Mukhtar Zhassuzak from Kazakhstan, Duygu Arugaslan-Cincin, Gulbahar Erim, Kagan Baskan and Cihan Yesil from Türkiye, Witesyavwirwa Vianney Kambale from Austria, and Astrit Tola from Albania. Special thanks to Sanzhar Hodzhaev from Uzbekistan for the picture of Al-Khwarizmi. My heartfelt thanks go to colleagues with whom the results have been discussed visiting their universities, Prof. Michal Fechkan from Slovakia, Prof. Kyandoghere Kyamakya from Austria, Prof. Chuandong Li, Prof. Gang Tiang, Prof. Lei Zhang, Dr. He Zhang from China, Prof. Murathan Dauylbayev, Prof. Yedilkhan Amirgaliyev from Kazakhstan, Prof. Valentin Panchenko from Australia, Prof. Durdimurod Durdiev, Dr. Umid Durdiev and Dr. Zavqiddin Bozorov from Uzbekistan, Prof. Mukhtar Kirane from France, Prof. Angel Manuel Ramos del Olmo from Spain, and Prof. Dilmurat Tursunov from Kyrgyzstan.

I am deeply grateful to Ashley Gasque Stec, Erika Radzvilaite, Senior Commissioning Editors and Bethany Hext, e-books Coordinator of IOP Publishing, for their invitation, professionalism, and permanent collaboration on the book's preparation.

References

[1] Akhmet M 2019 Domain-structured chaos in discrete random processes ArXiv e-prints (1912.10478)

[2] Akhmet M 2022 Unpredictablity in Markov chains *Carpathian J. Math.* **38** 13–9

[3] Akhmet M and Alejaily E M 2019 Domain-structured chaos in a Hopfield neural network *Int. J. Bifurcation Chaos* **29** 195–205

[4] Akhmet M 2021 *Domain Structured Dynamics* (Bristol: IOP Publishing)

[5] Akhmet M 2020 *Almost Periodicity, Chaos and Asymptotic Equivalence* (Berlin: Springer)

[6] Akhmet M and Alejaily E M 2019 Finite dimensional space chaotification *Kazakh Math. J.* **19** 21–6

[7] Akhmet M and Alejaily E M 2021 Abstract similarity, chaos and fractals *Discrete Continuous Dyn. Syst. Ser.* B **26** 2479–97

[8] Akhmet M, Arugaslan Cincin D, Tleubergenova M and Nugayeva Z 2021 Unpredictable? oscillations for Hopfield-type neural networks with delayed and advanced arguments *Math. Open Access* **9** 1–19

[9] Akhmet M and Fen M O 2016 Poincaré chaos and unpredictable?functions *Commun. Nonlinear Sci. Numer. Simul.* **48** 85–94

[10] Akhmet M and Fen M O 2016 Unpredictable?points and chaos *Commun. Nonlinear Sci. Numer. Simul.* **40** 1–5

[11] Akhmet M and Fen M O 2017 Existence of unpredictable?solutions and chaos *Turk. J. Math.* **41** 254–66

[12] Akhmet M and Fen M O 2018 Non-autonomous equations with unpredictable?solutions *Commun. Nonlinear Sci. Numer. Simul.* **59** 657–70

[13] Akhmet M, Fen M O and Alejaily E M 2018 *Dynamics motivated by Sierpinski fractals* ArXiv e-prints

[14] Akhmet M, Fen M O, Tleubergenova M and Zhamanshin A A 2019 Unpredictable? solutions of linear differential and discrete equations *Turk. J. Math.* **43** 2377–89

[15] Akhmet M, Fen M O, Tleubergenova M and Nugayeva Z 2020 Unpredictable?solutions of linear impulsive systems *Math. Open Access* **8** 1–16
[16] Akhmet M, Fen M O, Tleubergenova M and Zhamanshin A 2019 Poincare chaos for a hyperbolic quasilinear system *Miskolc Math. Notes* **20** 33–44
[17] Akhmet M, Fen M O, Tleubergenova M and Zhamanshin A 2019 Unpredictable?solutions of linear differential and discrete equations *Turk. J. Math.* **43** 2377–89
[18] Akhmet M, Fen M O and Tola A 2023 A numerical analysis of Poincaré chaos *Discontinuity, Nonlinearity Complexity* **12** 183–195
[19] Akhmet M and Alejaily E M 2021 Abstract fractals *Discontinuity, Nonlinearity Complexity* **10** 135–42
[20] Akhmet M, Seilova R D, Tleubergenova M and Zhamanshin A 2020 Shunting inhibitory cellular neural networks with strongly unpredictable?oscillations *Commun. Nonlinear Sci. Numer. Simulat.* **89** 105287
[21] Akhmet M, Tleubergenova M and Nugayeva Z 2020 Strongly unpredictable?oscillations of Hopfield-type neural networks *Math. Open Access* **8** 1–14
[22] Akhmet M, Tleubergenova M and Zhamanshin A 2020 Quasilinear differential equations with strongly unpredictable?solutions *Carpathian J. Math.* **36** 341–9
[23] Akhmet M, Tleubergenova M and Zhamanshin A 2019 Neural networks with Poincare chaos *ACM Int. Conf. Proc. Series* pp 34–7
[24] Akhmet M, Tleubergenova M and Zhamanshin A 2020 Inertial neural networks with unpredictable?oscillations *Math. Open Access* **8** 1–11
[25] Akhmet M and Tola A 2020 Unpredictable?strings *Kazakh Math. J* **20** 16–22
[26] Birkhoff G D 1991 *Dynamical Systems* (Providence, RI: Colloquium Publications)
[27] Bohl P 1900 Ueber Einige Differentialgleichungen Allgemeinen Charakters, Welche in der Mechanik Anwendbar Sind *PhD Thesis* Dorpat
[28] Bohl P 1906 Über eine differentialgleichung der strungstheorie *Grelles J.* **131** 268–321
[29] Bowen R 1970 Markov partitions and minimal sets for axiom A diffeomorphisms *Am. J. Math.* **92** 907–18
[30] Bowen R 1970 Markov partitions for axiom a diffeomorphisms *Am. J. Math.* **92** 725–47
[31] Bowen R 1973 Symbolic dynamics for hyperbolic flows *Am. J. Math.* **95** 429–40
[32] Bunimovich L and Ya Sinai 1980 Markov partitions for dispersed billiards *Commun. Math. Phys.* **78** 247–80
[33] Chaitin G 1975 Randomness and mathematical proof *Sci. Am.* **232** 47–52
[34] Devaney R L 1987 *An Introduction to Chaotic Dynamical Systems* (Menlo Park, CA: Addison-Wesley)
[35] Hawking S 1994 *Black Holes and Baby Universes* (London: Bantam Press)
[36] Esclangon E 1904 *Les Fonctions Quasi-Periodiques* (Paris: Gauthier-Villars)
[37] Kahane J P 1985 Sur le chaos multiplicatif *Ann. Sci. Math. Québ.* **9** 105–50
[38] Lasota A and Mackey M C 1985 *Probabilistic Properties of Deterministic Systems* (Cambridge: Cambridge University Press)
[39] Li T Y and Yorke J A 1975 Period three implies chaos *Am. Math. Mon.* **82** 985–92
[40] Mandelbrot B B 1975 *Les Objets Fractals: Forme, Hasard, et Dimension* (Paris: Flammarion)
[41] Mahajan A, Thakur R and Das R 2024 Sensitivity and unpredictability in semiflows on topological spaces *Commun. Nonlinear Sci. Numer. Simulat.* **133** 107949
[42] Miller A 2019 Unpredictable?points and stronger versions of Ruelle–Takens and Auslander–Yorke chaos *Topology Appl.* **253** 7–16

[43] Nemytskii V V and Stepanov V V 1960 *Qualitative Theory of Differential equations* (Princeton, NJ: Princeton University Press)

[44] Penrose R and Hawking S 1996 *The Nature of Space and Time* (Princeton, NJ: Princeton University Press)

[45] Pesin Y and Weiss H 1996 On the dimension of deterministic and random cantor-like sets, symbolic dynamics, and the Eckmann–Ruelle conjecture *Commun. Math. Phys.* **182** 105–53

[46] Nicolis G and Prigojin I 1989 *Exploring Complexity* (New York: Freeman)

[47] Ornstein D 1970 Bernoulli shifts with the same entropy are isomorphic *Adv. Math.* **4** 337–52

[48] Ornstein D 1995 In what sense can a deterministic system be random? *Chaos Solitons Fractals* **59** 139–41

[49] Ornstein D 2004 Kolmogorov, random processes, and Newtonian dynamics *Russ. Math. Sur.* **59** 313–7

[50] Ornstein D and Weiss B 1998 On the Bernoulli nature of systems with some hyperbolic structure *Ergod. Theor. Dynam. Syst.* **18** 139–41

[51] Poincaré H 1957 *New Methods of Celestial Mechanics Volumes I–III* (New York: Dover)

[52] Poincaré H 1890 Sur le probléme des trois corps et les équations de la dynamique *Acta Math.* **13** 1–270

[53] Schuster H G and Just W 2005 *Deterministic Chaos: An Introduction* (Weinheim: Wiley-VCH)

[54] Sell G R 1971 Topological dynamics and ordinary *Differential Equations* (London: Van Nostrand Reinhold Company)

[55] Thakur R and Das R 2020 Strongly Ruelle–Takens, strongly Auslander–Yorke and Poincaré chaos on semiflows *Commun. Nonlinear Sci. Numer. Simulat.* **81** 105018

[56] Wiggins S 1988 *Global Bifurcation and Chaos, Analytical Methods* (New York: Springer)

Author biography

Marat Akhmet

Dr. Marat Akhmet currently holds the position of Professor of Mathematics at Middle East Technical University in Ankara, Türkiye. He completed his Bachelor's degree in Mathematics at Aktobe State University in Kazakhstan and earned his Doctorate in Differential equations and Mathematical Physics from Kyiv State University in Ukraine.

Dr. Akhmet has received notable accolades, including the Science Prize from TUBITAK (Türkiye, 2015) for his contributions to scientific research and the Prof. C. S. Hsu Award (UC Berkeley, 2021) for his exceptional outcomes in the study of non-linear analysis. He has been invited as a plenary speaker at 17 international conferences across the UK, Kazakhstan, Türkiye, the USA, France, and Italy, and has organized over 20 scientific meetings on both national and international levels. With a deep-rooted passion for research and teaching, Dr. Akhmet has consistently been recognized with the "Best Performance Award" at Middle East Technical University throughout his 20-year tenure. He has mentored 20 Ph.D. students, many of whom now contribute to research institutes and universities in Türkiye, Kazakhstan, and Libya.

His expertise is evident through his eight published books, which have made significant contributions to the field. These include titles such as *Principles of Discontinuous Dynamical Systems* (Springer, 2010), *Non-Linear Hybrid Continuous Discrete-Time Models* (Atlantis Press, 2011), *Neural Networks with Discontinuous Impact Activations* (Springer, 2014), *Replication of Chaos in Neural Networks, Economics, and Physics* (Springer, 2015), *Bifurcation in Autonomous and Nonautonomous Differential equations with Discontinuities* (Springer, 2017), *Dynamics with Chaos and Fractals* (Springer, 2020), *Almost Periodicity, Chaos and Asymptotical Equivalence* (Springer, 2020), and *Domain Structured Dynamics: Unpredictability, Chaos, Randomness, Fractals, Differential equations, and Neural Networks* (IOP, 2021). These publications reflect his comprehensive understanding of the subject and serve as essential resources for both students and researchers.

Dr. Akhmet has made significant contributions to the foundational theory of discontinuous, almost periodic functions and the solutions of impulsive systems. He developed the *B*-equivalence method for systems characterized by variable moments of impulses, proving particularly effective in studying models with parameter-dependent discontinuities. His recent work explores sophisticated bifurcation problems and discontinuous dynamics featuring grazing phenomena, particularly relevant in mechanics. He has proposed a novel approach centered on singularities in recent papers, revealing how various effects can compensate for one another to yield regular dynamics.

Dr. Akhmet has also made notable progress in analyzing differential equations with piecewise constant arguments. His innovative method of using equivalent integral equations has replaced the traditional reduction to discrete equations. This incorporation of classical results from operator theory has led to a new classification of generalized dependence on piecewise constant arguments. Additionally, he successfully solved the Second Peskin conjecture concerning integrate-and-fire biological oscillators, formalizing the firing connection to achieve synchronization, thus paving the way for advancements related to the First Peskin conjecture. Numerous intriguing synchronization problems for biological oscillators and neural networks remain ripe for exploration through his methodologies.

The asymptotic equivalence of equations serves as a powerful qualitative tool, allowing Dr Akhmet to derive refined conditions for equivalence through special transformations of linear systems, significantly expanding the class of equivalent equations. His research elucidates that fractals are not only the product of iterative construction but can also represent states and orbits, a notion he first demonstrated in the literature. Consequently, the aesthetic appeal of fractals can now be appreciated along the time axis as well.

Chaos remains an enticing subject of study, yet its relation to the qualitative theory of differential equations has been largely underexplored. Dr. Akhmet has proposed a replication of chaos method, central to which is an input-output mechanism that preserves known chaotic characteristics under specific conditions aligned with differential and discrete equations. While this mechanism has been implemented for quasi-linear systems, it holds substantial potential for other non-linear analytical methods.

Over the past decade, Dr. Akhmet has introduced two pivotal concepts within dynamics: ultra Poincaré chaos and alpha labeling. The first is closely tied to alpha unpredictability, a core aspect of chaotic behavior, while the second serves as a universal tool for analyzing and organizing complexity across dynamics, algorithms, and geometries. These innovations have proven fruitful for uncovering and constructing chaos in various models, laying the groundwork for new types of complex functions. He has also developed numerical indicators for chaotic proc-esses, ensuring effective synchronization for models where traditional strategies fall short.

Part I

Introduction

Ultra Poincaré Chaos and Alpha Labeling
A new approach to chaotic dynamics
Marat Akhmet

Chapter 1

Historical and philosophical overview

I say unto you: one must still have
chaos in oneself to be able to give birth to a dancing star.
I say unto you: you still have chaos in yourselves.
—Friedrich Nietzsche, Thus Spoke Zarathustra

This book introduces novel concepts related to dynamical systems and the complex-
ities that accompany them. It encourages and provides opportunities for a renewed
exploration of the hierarchy of mathematical structures, the foundations of dynamics.
This chapter offers a historical and philosophical perspective, delving into topics such
as ultra Poincaré chaos, alpha labeling, and alpha unpredictability, all of which have
been recently discussed in the author's publications. The chapter also presents a
fascinating range of dynamical and chaos uncertainties that serve as criteria for chaos.
Additionally, it proposes elements for the classification of chaos to enhance effective-
ness in this domain.

1.1 Outline

Dynamics on alpha labels and correspondingly *chaos by alpha labeling* are designed
for universal analysis of models in science. It is helpful to prove chaos in fractals and
random processes and to extend chaos in hyperbolic sets. We believe it will provide
essential light on turbulence comprehension [69, 99] and related statistical theories of
Gaussian multiplicative chaos [77]. The basis for dynamics on alpha labels is the
abstract similarity set, which is also said to be the *set of alpha labels* or *alpha set*, in
what follows. The alpha sets help construct abstract fractals (alpha fractals), which
can be a theoretical base for new sophisticated geometries. Next, the abstract
similarity map is determined in the set, and correspondingly, we call it the *alpha*

map. The set and map with *alpha distance* make *alpha dynamics* (formerly, abstract similarity dynamics). Investigations of our research start with properly assigned alpha labels. That is, with an appropriate choice of labels for states of dynamics. The alpha labeling formally is an operator, determined on a set of states and maps points of the domain to alpha labels and has to be determined such that trajectories of the original dynamics agree with orbits in the alpha set. If alpha labeling is one-to-one, then the transformation is isometric. For some dynamics, first of all, random processes, we apply one-to-several and even one-to-infinity alpha labeling, and consequently, the label space is not a metric one. We call it *alpha space*. Dynamics on alpha labels is more effective than conservative symbolic counterpart and any Markov partition approach [34–36, 38, 80, 86, 97]. It is seen in applications of our research to the logistic equation, Baker and Smale horseshoe maps, Bernoulli processes, and Markov chains. Proof of chaos for fractals is another benefit of alpha labeling. The dynamics have been initially explored for random processes with finite state spaces. However, we believe that many other stochastic processes can also be analyzed using this method. Furthermore, the theory of stochastic differential equations can be expanded based on the current suggestions. The decimal number system serves as a universal reflection of the world. Similarly, alpha labeling parallels this concept, reinforcing our confidence that the study aligns with the principles of Pythagoras and Al-Khwarizmi. In the future, this approach can be broadly applied to ordinary, partial, functional differential discrete equations, and hybrid systems.

Alpha labeling is an effective approach for exploring both discrete and continuous motions, as well as complex geometries. Unlike symbolic dynamics and coding, which are more concrete, our alpha labeling framework introduces more abstract concepts such as alpha distance, diameter, and separation conditions. In this work, we present a comprehensive overview of findings related to this labeling technique. We revisit logistic, Baker, and Smale horseshoe maps to illustrate the advantages of our method. Our analysis provides clear insights into continuity and sensitivity within the dynamics of ultra Poincaré Li–Yorke, and Devaney chaos. This technique proves valuable for studying fractals that exhibit self-similarity. Additionally, we have obtained significant confirmations of alpha unpredictability in random processes characterized by finite state spaces, Bernoulli schemes, and Markov chains, indicating a broader applicability of alpha unpredictability across various stochastic processes. We anticipate that the first and second laws for alpha unpredictable strings, as detailed in the book, will also influence developments in probability theory. Overall, alpha sets are expected to assume a role analogous to that of numbers and sets, particularly in the context of the mathematics of complexity.

For those unfamiliar with the theory, chaos is often illustrated by the butterfly effect, a concept popularized by E Lorenz to depict this phenomenon. However, professionals working with dynamical systems might face some confusion, as continuity in initial conditions coexists with sensitivity. Therefore, it is essential to formalize both continuity and sensitivity to identify new forms of chaotic dynamics. We propose utilizing diameter and separation conditions for alpha labeled processes, as the first assures continuity and the second provides sensitivity. This allows for a novel and universal abstraction of the dynamical characteristics found in Lorenz chaos.

The cornerstone of recently introduced dynamics is *alpha unpredictable point*, which is a specification of the Poisson stable point, which is known as the ultimate type of oscillations in the Poincaré–Birkhoff theory [32, 73, 82, 87]. Thus, our research prolongs the line of recurrent motions to a new frontier such that the line meets chaos investigation. It is important to emphasize that Devaney's chaos relies entirely on the collective characteristics of orbits, such as sensitivity, dense cycles, and transitivity [45, 64]. Oppositely, ultra Poincaré chaos [17, 18] is based on the alpha unpredictability concept, which is focused uniquely on a single motion. Moreover, by simulating a solution, we can conclude not only on chaos presence but also that the motion is alpha unpredictable [20, 21, 25]. That is, from one side, it is principally new for chaos study, and from another, it leads us back to the roots of classical dynamics, which considers motion the main object of discussion. What we are doing is like that in theory for a single electron or atom. In our opinion, the novel motion will make chaos research more beneficial inside the theoretical study and for global applications in science and industry. First discoveries in this direction have already been described in papers [2, 7, 11], where we have deepened the perception of random processes through alpha labeling. The reader will find several attractive old models revisited in the present book. To give more light on the concept, let us say, similarly to the title in [64], that *alpha unpredictability implies chaos*. Though the unpredictability is of much weaker recurrence than periodicity, the results in [18, 64] make two boundary points of the spectrum such that all other types of recurrence, quasi-periodicity, almost periodicity, and others can be found inside of it. We continue to look for what one calls the order in chaos.

Researching stochastic dynamics and probability means finding the process's necessary features. Deterministic chaos formally is a list of conditions that must exist to conclude that the phenomenon is present in a model. Consequently, one can search for some ingredients of chaos in random dynamics and must be confident that we do not abolish randomness in the study. This sort of research would be more accessible if conditions for chaos were more fit to what was waiting, and we are working on the problem by introducing various ways of investigating chaos. The results of our activity are in this book. For example, it is shown that alpha unpredictability can be seen in Bernoulli schemes and Markov chains. Relations of chaos and random processes are the focus of many researches [61, 76]. It should be emphasized that ultra Poincaré chaos specifically may play a most important role in the direction, as opposed to conventional types of chaos.

Scientists who work in the field might be busy with several interesting questions. They are: why is the amount of chaos research results still not as large as that of its antithesis, non-chaotic dynamics; is it true that the theory of chaos already exists such as of quantum mechanics, for instance; if the research is just a part of dynamical systems theory (P Holmes guessed that this is true [54]); why total investigations of chaos did not begin immediately from the research of H Poincaré, G Birkhoff, and H Hilmy, even though E Lorenz respectfully made himself a reference on their research [65]; if the word *chaos* reflects the studies adequately or if it is a metaphor, and is far from reality; is the phrase 'order in chaos' confusing, and would it be better to say 'chaos is order'; has the word *chaos* been suitably promotional or has it instead impeded the progress of science; do we need new mathematical structures in addition to numbers, sets, topological and metric spaces for the studies; how far should we trust numerical experiments or experimental data; and can one accept the realizations of random processes as orbits of chaotic motions. These and many other questions are under consideration in the manuscript.

For deterministic processes, our research demonstrates how alpha labeling is beneficial for constructing Poisson stable and alpha unpredictable functions. The applications of the method for deterministic and stochastic processes are different, and this is also under discussion in the present paper. Alpha unpredictability is a new type of recurrence, and its role as a unique property is emphasized in the research of differential equations. Thus, a remarkably new area of qualitative analysis has been created, shaping a significant part of the theory. Finally, we demonstrate the productivity of the concept for random processes. The appearance of alpha unpredictability in Bernoulli schemes is completed with the first and second laws of large alpha unpredictable strings. The sequences are observed in Markov chains with finite state space, continuous and discrete time. Thus, the sentence *alpha unpredictability implies chaos* does not fully describe the range of our proposals, since this property can be much more productive in future developments (figure 1.1).

Symbolic dynamics is the research instrument closest to alpha labeling [34–36, 38, 80, 86, 97]. The method of investigation by projection on orbits formed by Bernoulli shifts along infinite sequences of symbols is the most powerful tool for studying chaos nowadays. Despite numerous merits of the approach and results on the way, one can see, taking into account our proposals, that there is a strong deficiency in the 'symbolic' road of research, since for any discussed problem one has to establish the projection on spaces of sequences, which is an auxiliary and additional mapping. Finding it and, most importantly, proving topological equivalency between spaces is

Figure 1.1. Pythagoras (570 BC–490 BC) shaped numbers as the global mathematical structure (Google).

challenging. This search is an art similar to Lyapunov functions construction in many difficult problems of discovering model stability. Presently, we suggest structuring a state space by alpha labeling such that mapping alpha labels precisely reflects the motion of elements under alpha labeling. The operator is not conjugative or semiconjugative, since one of the two sets is not a metric, but an alpha label space. This is one of the main novelties of the approach. A sufficient background for complexity in terms of similarity dynamics can be found in our research. It is built of clearly formulated and easy to verify diameter and separation conditions, a significant novelty in our study. This is another reason alpha labeling is principally different from previous methods and deserves the book. The dynamics method on alpha labels will be developed further, and many other fruitful features of the fundamentals will be proposed to produce many universal and interesting discoveries. One can easily see that we describe sophisticated motions transparently, without reflecting them in symbolic dynamics. We promise this is an easy way of chaos learning, and it will be seen theoretically and in applications soon. Surprisingly, it is possible to simplify proof of chaos in classical models. Moreover, the method is universally applicable for proving chaos in fractals. This is predictable since self-similarity is the basis of classical legendary fractals construction. Furthermore, we extract the dynamical essence of hyperbolic sets by using chaos through alpha labels. In our recent research this is done for a clear vision of chaos in several models of the area. This time, we consider the Smale Horseshoe and Baker multi-folded transformation and suggest considering Anosov automorphisms, Plykin attractors, and other similar constructions for prospective analysis. In chapters 7 and 8 we will discuss the benefits of nearing chaotic and stochastic motion concepts. The chapters will discuss necessary features of alpha dynamics in such processes as Bernoulli schemes and Markov chains. Finally, we are suggesting scientific use of a new concept of modular chaos, when motions are whirling among various alpha sets. This generates new opportunities to investigate unbounded motions and random processes with continuous and discrete time.

Finally, we make a philosophical remark. We believe that the alpha labels concept is in the stream of Pythagorean doctrine since the scientists declared the unlimited power of numbers. 'The so-called Pythagoreans, who were the first to take up mathematics, not only advanced this subject, but were obsessed with it, they fancied that principles of mathematics were principles of all things' Aristotle, Metaphysics, 1–5, c. 350 BC. The number ten was regarded as perfect [39]. We also compare the concept with ideas of Al-Khwarizmi (780–850), born in Horezm, Uzbekistan, who made practical applications of decimals. The great scientist considered numbers to be a way of describing the natural world, which consequently significantly increased the universality of numbers. For us, it is essential that decimals can be alpha labeled, and therefore alpha labels are the immediate inheritance of Al-Khwarizmi's research [28].

1.2 A new mathematical structure

Alpha labeling is the central topic of this manuscript and can be regarded as a mathematical structure, similar to numbers, sets, and metric and topological spaces. Some of these structures are so fundamental that they have been revered as paradisiacal, while others are so ubiquitous that we use them daily, often without realizing we are engaging with highly sophisticated products of human intellect. These structures are potentially far more significant than mere collections of theorems, definitions, and theories that rely on them. To visually convey their importance, imagine these structures as vast oceans, in which modern mathematical knowledge floats and takes form, much like tankers, cruise ships, boats, and fishing trawlers navigating through oceanic waters.

The structures do not have limits in applications if one proceeds to create mathematical information like modern research. However, one can only be confident that a structure will be needed in the future if it is modified. It is traditional for each new structure to follow the demands of research and develop accordingly, and alpha labeling is no exception. The study of *complexity* in science, the phenomenon of the 20th century, which is understood in various presentations: *chaos, fractals, turbulence, random processes, physical–chemical and algorithmic complexity* [41, 45, 64, 66, 75, 77, 82], can be much more productively researched if considered in the structure of alpha labels, in our opinion. Surprisingly, the great challenge, such as complexity research, has not yet given birth to a new structure suitable to the importance of the problems. To illustrate our perception of the alpha label's potential role, figure 1.2 is presented, where the ship *Complexity* is placed in the Sea of Alpha Labels. In addition to alpha labeling, this book also delves into the application of abstract similarity in the research of chaos, fractals, and random processes. While the first topic pertains to sophisticated dynamics, the second explores complexity in geometry, and the third involves the analysis of stochastic processes. This approach brings deterministic and nondeterministic dynamics together on the same platform through their functional representation of trajectories.

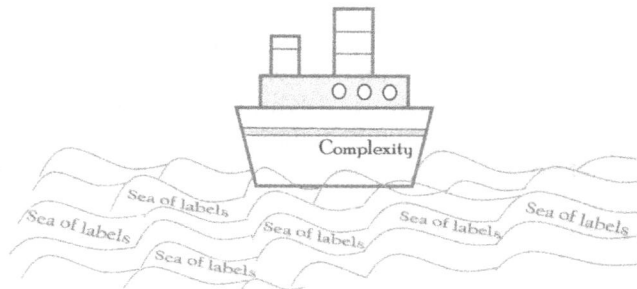

Figure 1.2. The ship *Complexity* is navigating through the Sea of Alpha Labels to achieve new and exciting milestones ahead.

Figure 1.3. Ships *Chaos*, *Fractals*, and *Random Processes* are in the Sea of Alpha Labels.

To provide a global perspective on the book's content, imagine ships named *Chaos*, *Fractals*, and *Random Processes* sailing on the Sea of Alpha Labels, as depicted in figure 1.3.

1.3 Alpha labels in spiraled hierarchy of mathematical structures

It is accurate to say that the doctrine of numbers is deeply connected to Pythagoras of Samos, an Ionian Greek philosopher who lived from 570 to 495 BC. He was among the first to emphasize the unique importance of numerology, believing that numbers form the foundation of the Universe itself. For example, he illustrated the significance of the number three through the concept of the triad—end, middle, and beginning—which he considered fundamental to everything. While some philosophers posited that water, fire, Earth, and air combined to form the world, Pythagoras pioneered mathematical physics by using arithmetic ratios to measure the quantitative relationships of music through the length of strings. Notably, he managed to study the *invisible* phenomenon of sound mathematically. W. Heisenberg recognized this numerical analysis of music as one of the most significant advances in human science.

Al-Khwarizmi (780–850) was among the first to recognize the significance and universality of decimals and their fractions for precise astronomical observations (figure 1.4). He introduced Hindu-Arabic numerals and their arithmetic to the Western world, earning recognition as one of the greatest mathematicians in history.

The analysis of the more intricate properties of numbers demanded a new type of structure—sets. German mathematician R. Dedekind (1831–1916) made pioneering strides in this area by defining infinite sets as 'similar to a proper part of itself' through the concept of similarity. Dedekind completed the construction of real numbers as a mathematical structure. As such, Pythagoras, Al-Khwarizmi, and R. Dedekind are regarded as the key architects of the powerful role that numbers hold in modern mathematics. Dedekind's work laid the groundwork for the research of G. Cantor (1845–1918), who is celebrated as the founder of set theory. This development has enhanced the application of numbers in science. Remarkably, sets underpin many other structures such as metric and topological spaces.

Figure 1.4. Alpha labels are a direct evolution of decimals introduced by Al-Khwarizmi (born c. 780–died c. 850).

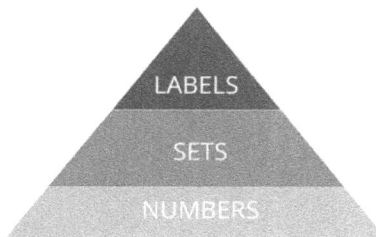

Figure 1.5. The three fundamental structures needed for complexity research.

In figure 1.5, 'sets' are depicted as being above 'numbers,' symbolizing that sets support numbers, even as numbers are crucial to the concept of sets and follow them in both theory and application. Sets exist on a higher level of abstraction, underscoring their foundational role in modern mathematics.

The alpha labels are on the highest level of abstraction in the triad. Further in the manuscript, it will be seen that they are, in many senses, similar to real numbers. If one accepts the development of concepts in the spiraling road, as philosophers accept it, then understand that the alpha labels would be better seen over the numbers directly, not through sets as in figure 1.5, since numbers can be alpha labeled. Their decimal presentation is one of the best examples of alpha labeling.

Alpha labels occupy the highest level of abstraction in this hierarchy. Further on in the manuscript, it will become apparent that alpha labels bear many similarities to real numbers. If we adopt the philosophical viewpoint that concepts develop along a spiraling road, it would make sense to position alpha labels directly above numbers

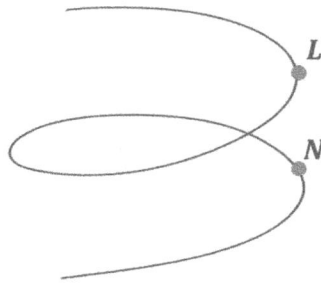

Figure 1.6. Alpha labels (*L*) are strictly over numbers (*N*) on the spiral of concepts.

rather than sets, as shown in figure 1.5, given that numbers can be alpha labeled. The decimal representation of numbers is one of the best examples of alpha labeling. To emphasize the strong relationship between numbers and alpha labels, consider figure 1.6, where they are placed one above the other in the spiral. This illustrates that alpha labels can be understood as a transformative evolution of numbers within the creation of mathematics.

We believe that the developing theory of alpha sets will become significant not only in the study of complexity but also in the theories of numbers, sets, functional analysis, operators, probability, dynamical systems, and differential equations. Our expectations for alpha labeling extend to its playing a fundamental and universal role in mathematics.

P. Holmes highlights the necessity for new structures, stating: 'The ideas of dynamical-systems theory have also led to several new methods of data analysis; dimension computations, Lyapunov exponents, phase-space reconstruction, spectra of dimensions, etc, although much remains to be learned about the applicability and validity of such methods' [54]. Another concern is the current trend of simplifying many chaotic models for investigation. We propose adapting alpha labeling to break free from this tradition and provide a universal foundation for methods.

Furthermore, as noted in [54], there is a 'doubt that strange attractors are quite as prevalent, or important.' We believe that the geometry of chaotic domains is not necessarily fractal-like. Dynamics often dominate over geometry, a notion that aligns with ideas in [75].

1.4 Universality and minimalism of alpha labeling and unpredictability

Remarkably, uncountable real numbers are constructed entirely through the finite set of digits 0, 1, 2, 3, 4, 5, 6, 7, 8, 9. When equipped with distance and order relations, they form a universal structure for describing, exploring, and exploiting various aspects of the world, particularly in industry. This universality, combined with a minimal and finite set of elements, highlights a fascinating aspect of numbers. It is this very universality and minimalism that we aim to replicate in the creation of new mathematical structures, such as alpha sets and dynamics. Moreover, the structure of alpha sets aligns with these principles. Alpha labels use indices derived from a finite alphabet, adhering to simple relationships and straightforward mappings. Special conditions are introduced for alpha sets that guarantee properties of continuity and alpha unpredictability (sensitivity) for the alpha map. With these minimal means, one can address problems of complexity and harness the power of mathematics. Numbers themselves can be considered as alpha labeled, demonstrating that alpha values inherit the beneficial properties of numbers. In this sense, numbers are a particular case of alpha labels, carrying forward their universal applicability but at a new level of abstraction.

The practical applications of alpha sets and labeling are vast and varied. These concepts are not just theoretical constructs but tools that can be deployed in the investigation of deterministic dynamics, theoretical analysis in geometry, probability, stochastic dynamics, and many other fields. The design of alpha labeling is specifically tailored for research in areas such as turbulence, probability, neuroscience, artificial intelligence, quantum computing, physicochemical processes, and algorithmic complexity. This extensive range of applications underscores the relevance and potential of these concepts in future research endeavors.

There is strong perception that finding and generating a partition united with one-to-one correspondence between states of dynamics and sequences of symbols make research of original processes utterly equivalent to that of symbolic dynamics. When partition is finite, the equivalence is not correct, but the coincidence of all dynamical properties is true only if alpha labeling is applied for dynamics, due to that mappings agree, and alpha labels are assigned to each point. Another benefit of our approach is that alpha labeling being flexible as it is not necessarily one-to-one. One can apply one-to-several and one-to-infinity alpha labels to discover chaotic ingredients in finite space dynamics and random processes. In particular, if both a set under labeling and an alpha set are metric spaces, we are dealing with semiconjugacy.

The epoch of discovering chaos in already existing models started with H. Poincaré's three-body problem [82] and Y. Ueda's [93, 94] analysis of the Duffing equation. Nowadays, scientists are searching for sophisticated dynamics in traditional models, which is a great activity. A model is universal and consequently correct if it is connected to chaos. That is, it either produces the dynamics in its present state or with new values of parameters or through extension of the model in a certain way. We accept the application and reconsideration of traditional models for chaos production as a fact of minimalism and, consequently, for alpha unpredictability and labeling.

The minimalism of alpha labeling will be understood later, in the future, when other ways of chaos definitions maximally apply the structure and dynamics on it. That is, alpha sets inherit the merits of numbers, as ancient geometrical constructions are enveloped by the theoretical frames of Pythagoras, Al-Khwarizmi, R. Descartes, and R. Dedekind. At the crossroads of the last two centuries, H. Poincaré presented a significant challenge to mathematics: to serve the needs of physics by embracing the theories of relativity and quantum mechanics. Today, we are tasked with developing mathematical approaches to explore chaos, complex algorithms, intricate geometrical structures, random processes, and brain activity—collectively, the domain of complexity. Alpha labels are anticipated to provide a universal framework for this endeavor. They are designed to address global research challenges and to facilitate a deeper understanding of complexity phenomena. While our focus will be on chaos, fractals, and stochastic processes, we believe that this framework can be valuable across various scientific and industrial fields.

The fundamental understanding that any locally in time regular and stable motion of industry or real-world is unstable and chaotic if considered in time globally, is a compelling argument for the use of alpha labeling. This concept underscores the need for a universal structure like alpha labeling to help us make sense of the complexity and chaos that pervades our world.

1.5 Principle of excessiveness: the Universe approximates science

The principle we are about to explore is not merely a theoretical idea but one with practical implications across all fundamental mathematical concepts. It suggests that, in many instances, our understanding exceeds the immediate requirements of applications. For example, the concept of asymptotic stability for a model is often discussed with time extending to infinity, even though the model may not actually be in use for such prolonged periods. Consequently, this principle holds direct relevance and applicability in various scientific fields.

Fractals are defined at infinitely small scales; however, the Planck units [84] render this smallness impractical, at least for deterministic purposes. Fields such as mechanics, chemistry, and economics utilize derivatives and integrals, whose definitions rely on arbitrarily small segments of space and time. Yet, these are again constrained by the Planck units. We refer to this phenomenon in science as the *principle of excessiveness*. This principle can be defended in mathematics due to potential requirements since the duration of an experiment remains unknown. Therefore, models must be adaptable to any timescale, ready for use at any moment. Often, we cannot determine how deeply a biological or geological analysis should penetrate, yet we must be prepared for any possibility and avoid rigid boundaries in our modeling and mathematical assumptions.

Real numbers serve as a prime example of excessiveness, as for any application, the corresponding approximation has finite accuracy. Following this paradigm, we consider alpha labeling to be an excessive tool, one we believe can be exceptionally powerful for addressing future challenges. This principle is essential for analyzing fractals and chaos, as Lorenz sensitivity or proximality in Li–Yorke chaos cannot be simplified into approximations. This holds true for ultra Poincaré chaos as well, given the convergence conditions [18].

From a philosophical standpoint, reflect on Plato's world of ideas [43]: in the aspects of the world that we have come to understand, it is not science that merely approximates nature; instead, the world aligns more closely with the concepts put forth by science. Consequently, the principles of quantum mechanics are not in conflict with the fundamentals of calculus.

Ultimately, the principle of uncertainty acts as the antithesis to excessiveness, providing a necessary complement in our attempts to reflect the Universe through scientific models.

1.6 Potential and actual uncertainty in dynamics

In this part of the chapter, we shall discuss the problem of uncertainty in dynamics. Of course, the discussion will relate to the principle of uncertainty by W. Heisenberg (1901–76). One has to start by saying that in 1927, the German physicist published a paper on the *principle of uncertainty*, which says that the more precisely determined position of a particle badly affects the estimation of speed for it. Later, the principle of limits existence for experimental knowledge of dynamical characteristics was indicated in biology, history, economics, and so on. Let us clarify why two kinds of the phenomenon, *actual* and *potential* uncertainties, are considered in our research. Remember that physics and mathematics have *two sorts of infinity*. The first one is *potential infinity*. For instance, it is connected with the possibility to halve a section of real numbers infinitely many times. Another is *actual infinity*, which is for a set of points in an interval of real numbers, for instance. Similarly, two types of dynamical uncertainty are suggested in the discussion.

Uncertainty is not just a theoretical concept, but a pervasive reality in our world. In our exploration, we will anchor this concept in the practical realm of differential equations, which form the bedrock of dynamics. The principle of uncertainty takes on a particularly intriguing form when applied to the analysis of dynamics using the theory of differential equations. This application allows us to delve into the details of the principle, making it more tangible and relatable. We will also discuss the principle of W. Heisenberg, which can be seen as the ultimate limit case of another generic phenomenon, the potential principle of uncertainty. To bring these ideas to life, we will use a differential equation as our illustrative tool.

$$\frac{dx}{dt} = f(t, x). \tag{1.1}$$

To discuss the principle, we will consider motions by solutions for differential equations since they are about *derivatives* and *positions*. Let us consider the differential equation (1.1) and, consequently, function $f(t, x)$ at the right-hand-side, which uniquely determines the equation and *derivative*, and the *position* is the value of a solution at a fixed moment. In other words, we consider a process P as a union of derivative D and state S, which are given in coordinates. Thus, symbolically, the paradigm of uncertainty can be present as the sum $P = D + S$. The process P cannot be observed precisely, despite that the components D and S can be separately determined clearly, but not together and simultaneously. Remember that the derivative equals function $f(t, x)$. Specifically, $f(t, x)$ depends on a parameter, say, μ, such that $f(t, x, \mu)$. We know that this dependence is not so clear, which is one reason for the potential principle of uncertainty. In other words, if one is to observe the derivative D of process P, changes of function $f(t, x)$ must be monitored. Thus, one should follow the change of differential equation, that is, D, and the position S simultaneously. Monitoring both is an impossible real-time task since we should verify that they must satisfy the equation (1.1). We suggest calling the phenomenon *potential principle of uncertainty*. It does not relate only to differential equations. It generally says that considering a process P, one can observe precisely only derivative D or state S, but not both of them simultaneously, since they must also agree in a relation, let us say, (1.1). Moreover, this is the task of permanent agreement between the two for all moments in $[0, T]$. If the value of T collapses to zero, we obtain the principle of quantum mechanics as an ultimate case, in particular. If one looks for a metaphor, one can find that the relation between two types of uncertainty is similar to that of potential and actual infinity. We suggest discussing the potential principle of uncertainty and accepting that of W. Heisenberg as a sort of actual principle.

Let us provide one more metaphor for uncertainty. If the potential principle can be compared with the arc tan function, the actual principle is like the Heaviside one. We accept potential uncertainty for discussion, though it does not provide specific quantitative characteristics. The comparison with the quantum mechanic's principle clearly illustrates that the deal is not in a differential equation since the zero-length duration of time makes a differential equation impossible. Correspondingly, the potential principle of uncertainty is fundamentally true for any dynamics, and it is an abstract concept. Remember that we do not know the physical reason for Heisenberg's principle but accept it. Analogously, although we do not know a physical, biological, or any other reason for the potential principle of uncertainty, to emphasize the generality, one can say that the principle is an abstract concept and not connected directly to differential equations. The observations that have been made prove that the limits of quantum mechanics are standard for any dynamics as a phenomenon. Moreover, theoretically, one can extend details of the Schrdinger equation, let us say the wave function, for other dynamics, analogously. We hope the subsequent scientific developments will give more arguments and analogies for the extension, inspiring further exploration and discussion in this fascinating field.

There are numerous observations of processes in very different areas of science, which the idea of W. Heisenberg has not directly influenced. However, they are inevitable for the mainstream of progress. One of them, which is similar to what we have discussed above, is of Yu N. Harari. 'This is one of the distinguishing marks of history as an academic discipline—the better you know a particular historical period, the *harder* it becomes to explain why harder things happened one way and not another. Those with only a superficial knowledge of a certain period tend to focus only on the possibility that was eventually realized. They offer a just-so story to explain with hindsight why that outcome was inevitable. Those more deeply informed about the period are much more cognizant of the roads not taken' [55].

Thus, there exist two complementary concepts: potential and actual uncertainties. The first one is not less generic than the second one, and this can be a source of open problems for dynamics, which is the main object of science. Actual uncertainty, being the ultimate case of its potential counterpart, can provide reasons for more deep research, and vice versa. Knowledge of potential uncertainty may be a cause for research of actual uncertainty.

1.7 Actual uncertainty of chaos

We have now formulated two fundamental principles, which can help to clarify appearances of reality in science. They are of potential and actual uncertainties. Let us discuss whether chaos research must be connected with one of the two. Consider first deterministic Lorenz chaos. The source of uncertainty is sensitivity, which is the main ingredient of irregular dynamics. We will discuss it in a way adapted for the research and common for different types of chaos. A dynamics $f(q, t)$ is sensitive at a point p provided that there exists a positive number e, sequences of points p_n and moments s_n, $n = 1, 2, \ldots$, such that $p_n \to p$ as $n \to \infty$, and distance between $f(p, s_n)$ and $f(p_n, s_n)$ is larger than e for all n. There is the condition, which is not mentioned in studies of chaos, the continuous dependence on initial states, and we are strongly confident that without the continuity there is no chaos in its classical version since the dependence implies that sequence s_n must increase to infinity. That is, the divergence is hidden at *infinity*, and this is on one side of the uncertainty since to learn sensitivity we have to look for points p_n arbitrarily close to point p, but this makes less observable moments, where the divergence happens. Choosing points near the initial point is another side of uncertainty. Oppositely, if one tries to determine large *moments* of the divergence, precision for the location of corresponding points p_n decreases, and vice versa. The principle of actual uncertainty for chaos is based on sensitivity.

Now, let us consider ultra Poincaré chaos for the same reason of uncertainty. The alpha unpredictable point is the origin of this chaos [17–19]. A point p of a dynamics is unpredictable if there exists a positive number e, sequences t_n and s_n increasing to infinity, such that $p_n = f(t_n, p)$ converges to p as n tends to infinity, and points $f(t_n + s_n, p)$ are on distance larger than e from $f(s_n, p)$ for all n. Again, we are in a position where chaos can not be observed precisely in experiments since the closer approximation points $f(t_n, p)$ are observed, and the near to infinity are moments of divergence s_n. Conversely, if one determines a large moment s_n, then finding corresponding t_n is a difficult task. That is, the more precisely determined s_n means less accuracy for p_n. Thus, we are busy with uncertainty. Since it relates to dynamics, we can accept that one more case of the principle is under discussion. Moreover, we say that it is actual uncertainty since it is formulated in terms of ultra Poincaré and Lorenz chaos. We invite readers to accept it for chaos as an actual uncertainty since the sequences of points p_n, and moments t_n, s_n, are not less theoretical than the Heisenberg quantities.

Now, we are near to completing the discussion on uncertainty for chaos and suggesting numerical characteristics like those for quantum uncertainty. It is proposed that for individual dynamics, positive constants h_α and h_β exist such that

$$h_\alpha \leqslant s_n d(p_n, p) \leqslant h_\beta, \tag{1.2}$$

where d is the distance function. By considering the formula, one can see that placing p_n close to p makes s_n uncertain at infinity and vice versa. The inequalities (1.2) can also be accepted as formalization for alpha unpredictability. The notations for Lorenz and ultra Poincaré lines have been chosen to make inequalities easily readable for both. The principle difference is that sensitivity is designed for collective characteristics while alpha unpredictability is for an individual motion. Precisely, $p_n = f(t_n, p)$ for alpha unpredictability, and is independent on p for sensitivity.

The discussions above give more arguments that deterministic chaos and quantum mechanics are related to the world's reality. Moreover, we believe that supported by the actual principle of chaos, one can find analogs of the wave functions and Schrödinger equation for the deterministic phenomenon. It seems there must be other relations between space and time sequences. The separation number e can be said to be the constant of uncertainty for ultra Poincaré and Lorenz chaos since the smallness of error for the initial point does not affect the separation size.

As we have already proven the ultra Poincaré chaos for the Bernoulli scheme and Markov chains [4, 7, 24], this is a way to consider uncertainty for stochastic dynamics.

What is the philosophical conclusion for our discussion? The more uncertainties there are for dynamics, the better. An uncertainty is the main argument that a theoretical phenomenon deserves to be researched since it is about the unity of opposites. So it is good that the uncertainty for chaos has been shown. By the way, it is true also for abstract similarity dynamics and, consequently, for chaos by alpha labeling.

1.8 Monads spiral up to abstract similarity

Self-similarity is fundamental in modern science, with origins dating back to the 17th century when Gottfried Leibniz introduced recursive self-similarity [100]. The concept reemerged in the 19th century when Karl Weierstrass presented, in 1872, a function that is continuous everywhere but differentiable nowhere, demonstrating a self-similar curve.

The phenomenon is particularly significant in mathematics—especially in geometric, metric, and topological spaces [50, 59, 66, 72]—is defined, in simple terms, as a property of objects whose parts, at all scales, are similar to the whole. Statistically self-similar sets exhibit the same statistical properties across multiple scales. Our recent research [12] introduces abstract similarity, a property of alpha sets, extending the concept beyond geometry to a more universal and abstract domain. We further discuss the implications and applications of similarity in various articles and books [4–7, 22, 24].

This concept permeates various fields, including wavelets, fractals, and graph systems [60], turbulence, and statistics [69, 77]. The idea of spatial self-similar sets was first mathematically defined by Moran in 1946 [72], and his construction has since been extensively studied and expanded. Research has generalized Moran's approach using symbolic spaces to develop different classes of Moran-like geometric constructions [79, 80, 91], leading to the Cantor-like set. Exciting definitions of self-similarity, focusing on dimension and measure, are discussed in numerous papers [30, 48, 50, 51, 56, 59, 62, 74, 90].

Our manuscript considers sets in Euclidean space, defining a self-similar set as a union of its images under an alpha-map. This research extends the concept of self-similarity to metric and topological spaces, hence the term *abstract self-similarity*. Our method does not rely on special functions; instead, the *similarity map* or *alpha map* naturally arises from the alpha set structure, serving as a generalization of the Bernoulli shift on symbolic sequences. Unlike classical approaches, our primary focus is on distance rather than dimension and measure. Nonetheless, our approach may prove valuable for extending results in abstract self-similarity from previous works [30, 48, 50, 51, 56, 59, 62, 74, 90]. We anticipate that abstract self-similarity will pave new paths in mathematics and be a useful tool in other fields such as harmonic analysis, discrete mathematics, probability, and operator algebras [60].

Any universal concept has deep historical roots, and alpha labeling is no exception. The following discussion elaborates on its development.

The renowned German mathematician and philosopher Gottfried Leibniz (1646–1716) provided profound insights into the world of similarities, a concept now known as *recursive self-similarity* [37, 100]. In the manuscript *Monadology*, Leibniz illustrates the concept vividly: 'Each portion of matter may be conceived as a garden full of plants and like a pond full of fishes. However, each branch of every plant, each member of every animal, each drop of its liquid parts is also some such garden or pond.' Leibniz saw the Monad as the basic unit of existence, with each Monad containing another world within it, reflecting a self-similar recursive structure.

Thus, Leibniz perceived the world as a self-similar universe. Subsequent mathematical research has refined the understanding of self-similarity [50, 59]. Our current study aligns with this philosophical universality, potentially bringing the study of similarity back to the philosophical ideas of the 17th century. To illustrate our approach to the study of similarity, consider the diagram in figure 1.7, which encapsulates over three centuries of scientific history.

The similarity, which was established in *Monadology*, is caused by the absence of interactions and the presence of inserted synchronization, and this arranges the identity and excludes any 'deformations.' It was commented in [85] by B. Russel: 'Leibniz held that every monad mirrors the Universe, not because the Universe affects it, but because God has given it a nature which spontaneously produces this result. There is a 'pre-established harmony' between the changes in one monad and those in another, which produces the semblance of interaction. This is an extension of the two clocks, which strike at the same moment because each keeps perfect time.

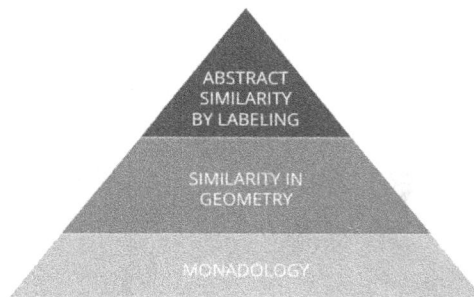

Figure 1.7. The three doctrines of similarity fundamental for complexity research.

Leibniz has an infinite number of clocks, all arranged by the Creator to strike at the same instant, not because they affect each other, but because each is a perfectly accurate mechanism. To those who thought the pre-established harmony odd, Leibniz pointed out what admirable evidence it afforded of the existence of God' [85].

To provide more comprehension of monads in physical sense, let us cite B. Russel again, 'Space, as it appears to the senses, and as it is assumed in physics, is not real, but it has a real counterpart, namely the arrangement of the monads in a three-dimensional order according to the point of view from which they mirror the world. Each monad sees the world in a certain perspective peculiar to itself; in this sense, we can speak, somewhat loosely, of the monad as having a spatial position' [85].

Another interesting observation by the author is that 'Monads form a hierarchy, in which some are superior to others in the clearness and distinctness with which they mirror the Universe. In all, perception has some degree of confusion, but the amount of confusion varies according to the dignity of the monad concerned' [85]. The remark about hierarchy in our circumstances can be interpreted as the degree of similarity for alpha sets when it is observed theoretically, but in reality, one can not see them at all, and assume the hierarchy could be seen if alpha labeling were to be completed. From another point of view, a hierarchy presence in a model may be an instrument for alpha labeling because of rectangles in an alpha set model.

One can not ignore Russel's final remark on the philosophy by G. Leibniz: 'What I, for my part, think best in his theory of monads is his two kinds of space, one subjective, in the perceptions of each monad, and one objective, consisting of the assemblage of points of view of the various monads. This, I believe, is still useful in relating perception to physics' [85]. In this sentence, one can find the dialectic vision of integrity of a part and a whole, which must be remembered considering the alpha labeling theory.

The concept of similarity has been articulated not only in philosophy but also in literature, as demonstrated by Jonathan Swift (1667–1745) in his work *Gulliver's Travels* [92], where he created the intriguing Lilliputians and Giants. Earlier, Pieter Bruegel the Elder (1525–69) presented this notion in his renowned painting *Hunters in the Snow*, showcasing similarity at four different scales through the images of men in the three ponds, and the hill illustrated in figure 1.8.

Figure 1.8. Hunters in the Snow, Pieter Bruegel the Elder, 1565, Kunsthistorisches Museum, Vienna.

Historically, the idea of self-similarity in science began to emerge in 1872, when Karl Weierstrass introduced a function that is continuous everywhere yet differentiable nowhere. The graph of the Weierstrass function became a prime example of a self-similar curve. Subsequently, G. Cantor, who was influenced by Weierstrass's lectures, constructed subsets known as Cantor sets. Two notable contemporaries, the German mathematician Felix Klein (1849–1925) and the French scientist Henri Poincaré (1854–1912), contributed to the notion of self-similarity by introducing a category referred to as self-inverse fractals.

Following the ideas of Henri Poincaré, Swedish mathematician Helge von Koch (1870–1924) defined a self-similar figure known as the Koch snowflake in 1904. Other self-similar sets are Sierpinski gasket and carpet, which were introduced in 1916. Paul Lévy (1886–1971) was a French mathematician who significantly contributed to stochastic processes, random walks, and probability. Related to the study, he expanded on self-similar sets we now call the Lévy C curves. Later, two French mathematicians, Pierre Fatou (1878–1929) and Gaston Julia (1893–1978), studied independently iterations of rational functions in the complex plane, which were visualized by computers in the research of Benoit Mandelbrot (1924–2010). The term *fractal* was coined by Mandelbrot in 1975 [66] to describe specific geometrical structures that exhibit self-similarity. Since then, this word has been employed to denote all the sets mentioned above, and the field became known as *fractal geometry*. Consequently, the concept is axiomatically linked to self-similarity, which is applied to an acceptable definition of fractals. They can be defined as sets that display self-similarity at all scales. However, Mandelbrot defines a fractal as a set whose Hausdorff dimension is strictly larger than its topological one [67, 68]. To sum up, self-similarity and dimension are the most essential features of fractals. The traditional connection between them is that self-similarity became the easiest way to determine sets that have fractional dimensions [44]. Thus, for the following discussion, we have recursive self-similarity by Leibniz and the phenomenon in geometry. Moreover, we establish spaces with trans-functions of Hutchinson and Falconer, determined by iteration function systems of Barnsley, and fractals. Besides classical approaches, we shall consider alpha sets, and abstract fractals. It is of great interest to observe the relations between them and each other, which are seen in figure 1.7. The middle part concerns Barnsley's iterated function systems as the most concrete constructive instrument for self-similar sets. Despite theoretical solid generality, others in [51, 59] can not be used in dynamics research so much and consider problems of measure and dimension. Nevertheless, the result by Hutchinson should be stressed since there is a unique set invariant concerning each family of contraction maps [59].

In the triad of figure 1.7, only Leibniz's theory and the abstract similarity are for a universal world description. Our doctrine is based on abstract similarity and can be considered for each event, which can be alpha labeled. We have agreed that abstract similarity is valid for real numbers and finite-dimensional cubes. Consequently, it is possible to see the similarity everywhere where numbers and vectors can be seen. The alpha labeling can also be realized for topological spaces. Then, the universality of numbers implies the same is true for abstract similarity. Thus, we claim that the concept of G. Leibniz has been 'parameterized' by alpha labeling. In other words, we have brought the philosophical depth of the German genius to a mathematical level. Similarly, the decimals of Al-Khwarizmi finally parameterized the geometrical vision of Greek scientists. The diagram of figure 1.9 shows the connection of two concepts for similarity. It demonstrates that we believe in a new stage of world research, which is ultimately sophisticated. In particular, alpha labeling [4, 19] can exploit strongly iterated functions systems [31]. Moreover, the metric, topological, and measure achievements in similarity [50, 59] can be realized for irregular dynamics if the suggestions of alpha labeling theory are applied. Thus, we clarify that abstract self-similarity is strongly related to fractals, if they are self-similar geometrically or defined through dimension. A specific relation of fractals and alpha labeling is seen in the diagram of figure 1.10, where abstract self-similarity is proved, but geometrically the set is not self-similar. Duffing equation dynamics have modified the carpet. We can also confirm that the set is of fractal dimensions, but it relies on our discussion in [8].

Let us consider the state similarity determined by trans-function [50, 51, 59] and self-similarity based on topological equivalence to the symbolic dynamics [72], and iterated functions system [31]. If one compares them with monads and abstract similarity, they are very specific since they relate only to geometrical objects.

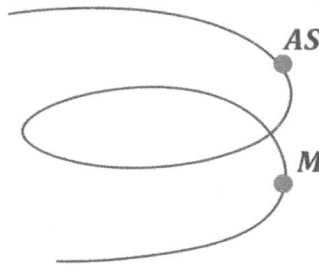

Figure 1.9. The abstract similarity (*AS*) is positioned above that of monads (*M*) within the evolutionary spiral of concepts.

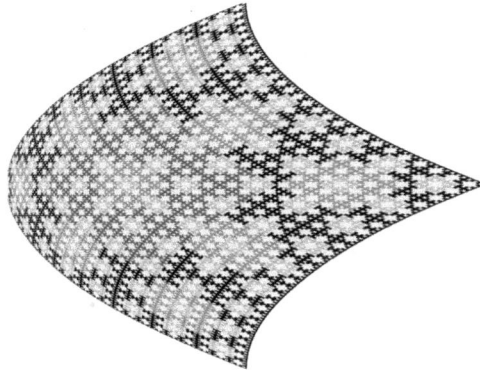

Figure 1.10. Sierpinski carpet modified by the Duffing equation. Reproduced from [19], with permission from Springer Nature.

Nevertheless, iterated function systems by Barnsley can be beneficial in applying alpha labeling to many complexity-related problems if amalgamated with our theory. There are two principal open problems: arranging alpha labeling for a mathematical problem and finding corresponding iteration functions for chaotic or regular process alpha labeled and subdued to alpha-map. To say that similarities in geometry are not universal would be a mistake since one can say the same about Euclidean space. Hence, many mathematical problems can be solved by trans-functions or iterated function systems. The deficiency of geometrical similarity methods is focused either on linear functions-transforms of J. Hutchinson [59] or complex-valued polynomials [50]. However, the alpha labeling itself provides limitless theoretical opportunities.

Here, the approach connected to sets determined by iteration of functions can be named the 'method of algorithms.' The similitudes of Hutchinson or iterated functions systems can determine them. Why do we suggest using the concept of algorithms? The word is also suitable for a verbal description of the doctrine by Leibniz: iterated mappings in abstract similarity sets. If one agrees with the term, it can be claimed that alpha labeling can be strongly connected to algorithms. They are also related to the definition and properties of specific geometrical objects as fractals. Corresponding interesting researches are done in papers [51, 59]. Thus, the focus of alpha labeling applications should be the triad: alpha label sets, algorithms, and fractals. Another essential object for research is the random processes, which are already the focus of self-similarity studies.

Let us return to observing figure 1.10. It pictures a set such that its similarity can not be proven by the conservative methods, but can by alpha labeling. That is abstract similarity. We started with the original Sierpinski carpet to obtain the set in figure 1.10. The set is alpha labeled such that points are considered as elements of an abstract similarity set. Next, they are used as initial points of a Duffing equation dynamics [4, 8, 19]. The endpoints of all trajectories with the common finite moments, which start in the carpet, set the result in the figure. Points inherit the alpha labels of initial members, and the properties for abstract similarity are valid because of continuous dependence on the initial data. Lipschitz condition can be verified by detailed analysis of the equivalent integral equation. Thus, alpha labeling is an instrument of self-similarity verification. It is evident that the spatial methods can not provide the conclusion.

Let us make final remarks. Alpha labeling is a universal mathematical structure for the research of self-similarity, while methods based on spatial analysis are for specific cases. It is a general abstract concept, which can not be reduced to the conservative geometrical perception of similarity. Our theory is in strong relation to the original philosophical ideas of G. Leibniz. As the fundamental argument for this, we can apply alpha labeling for intervals of real, and subsequently complex numbers and all other structures based on them. Consequently, everything related to intervals of real numbers admits abstract self-similarity. Only by subduing these immensely huge regions of real-world to alpha labeling algorithms can one say that the ideas of G. Leibniz are universal. Nevertheless, the existing methods for spatial similarity are effective instruments for the alpha labeling of sets. It is already proven for the iterated function systems in our research. More deep connections exist between alpha labeling and spatial functional methods. Hutchinson's results [59] prove that for each set of contraction maps, there exists an invariant set that should be an issue for the subsequent investigations in the direction.

1.9 Alpha labeling for more chaos and fractals

Our research findings [4, 12], which are novel and groundbreaking, confirm that phase portraits of chaotic dynamics reveal that the dimensional space occupied by an orbit and its closure is not less than that of the model. This is true for all chaotic orbits within the closure, leading to the intriguing conclusion that chaos is more prevalent than regular motions. Furthermore, our research challenges the notion that chaotic dynamics are confined to fractals. It demonstrates the efficacy of alpha labeling for organizing chaos on all self-similar fractals and those topologically related to them. Additionally, we have characterized chaos on Euclidean cubes and their topologically equivalent sets, proving that density in the cube points can initiate chaotic motions. Moreover, we have established the presence of ultra Poincaré chaos in Bernoulli schemes and Markov chains. This strongly suggests that chaos, as we present it, can manifest in many random processes. Consequently, our results indicate that chaotic dynamics are generic in both deterministic and stochastic processes. The universality of chaos is inherently linked to abstract similarity. Our dimension paradigm for chaos implies that fractals should also be considered through the lens of alpha labeling. Indeed, the concept of abstract fractals is as generic as similarity and chaos. We can arrange the structure of abstract fractals in topological objects such as finite-dimensional cubes and various other figures. Another significant application of alpha labeling is the consideration of dynamically appearing fractals, which are modified by differential and discrete equations and validated in time-space sets.

Multiple researchers have noted strong relationships between chaos and fractal geometry, evident, for instance, in the phase portraits of the Lorenz, Rössler, or

Chua systems. Furthermore, Fatou–Julia iterations are used to construct Julia and Mandelbrot sets, where neighboring points exhibit distinct behavior, demonstrating sensitivity in fractal structures. Chaos represents the irregularity and divergence of trajectories, while the fractal concept aids in analyzing complex geometric structures. The connection between chaos and fractals becomes clearer when a fractal dimension measures how a trajectory fills phase space, indicating the presence of a strange attractor [71]. Contrary to the conservative approach, our proposals based on alpha labeling apply to both fractal and non-fractal sets.

Significant work on the chaotic nature of fractals has been conducted primarily on specific types, such as totally disconnected fractals [31, 42, 45]. The concept of topological equivalence has been used to prove that these fractal sets are invariant under specific chaotic maps. Relatively few studies have explored chaotic dynamical systems in fractals, with a notable example being the Sierpinski carpet [49], which shows that the dynamical system associated with a shift transformation on the Sierpinski carpet is chaotic in the sense of topological mixing. Our results are particularly relevant; we do not rely on topological conjugacy to prove the presence of chaos in fractal sets. Instead, we consider all basic types of chaos, covering all self-similar fractals, including the Cantor set, Sierpinski carpet, and Koch curve. Furthermore, our approach is applicable to sets in multi-dimensional spaces. For example, the logistic map demonstrates that chaos can extend to higher-dimensional discrete systems [46, 47, 70], necessitating results like the Marotto theorem [70]. By applying the chaotic abstract similarity developed in our research, we have shown that ultra Poincaré Devaney, and Li–Yorke chaos exist in the dynamics of several interconnected and perturbed logistic maps.

Our perspective on chaos prioritizes the domain over the map. Typically, chaotic analysis begins with the description of a map with specific properties like unimodality, hyperbolicity, three-periodicity, or topological conjugacy. The dynamics are then identified, followed by the analysis of the structure of the chaotic set where trajectories are placed. For example, the Li–Yorke map starts with three-periodicity, and only then are the properties of the scrambled set characterized. Similarly, the chaos of unimodal maps with Cantor-like sets and Smale's dynamics are determined through mutual construction of the map and set. In symbolic dynamics, the domain is described as infinite sequences of symbols, and the map is a shift acting on space elements, though the map still takes precedence since sequence properties are defined in relation to the map. In our approach, we first alpha label a domain in metric space by confirming it meets specific conditions suitable for chaos. Then, a similarity map is constructed based on the self-similarity inherent in the domain. Consequently, the map is suitable for abstract self-similar sets and any fractal constructed through self-similarity, thereby proving chaos for these fractals and non-fractal sets.

1.10 Ancient Greek chaos

Chaos, initially appearing as a concept of deity, holds a significant place in Ancient Greek mythology. On http://www.greekmythology.com, we learn that Chaos descends from Ananke and Chronos, two primordial deities. Ananke, the personification of necessity and fate, was depicted holding a spindle and was present when the Universe began, along with her consort, Chronos (time). Ananke and Chronos, with the figure of a serpent, were intertwined and revolved around the primal egg of matter; when they applied force, the primal egg split into the parts of the Universe (Earth, heaven, and sea). In figure 1.11, Chaos is pictured with the All-seeing eye, further emphasizing its divine nature. Ananke was believed to rule over Fate, and being the mother of the Fates, she was the only one who could control their decisions. Thus, if one considers Fate as *Necessity* or a version of *Determinism* and the Chronos is *Time*, then obtain that Chaos is about *Deterministic Dynamics*. Surprisingly, this was already remarked on in ancient myths. The more substantial relation between what was done in Ancient Greece and modern mathematical chaos can be understood if one considers cosmogony: 'Chaos was—most Greek cosmologies tell us—the very first of all, the origin of everything, the empty, unfathomable space at the beginning of Time. But, it was more than just a gaping void—as its name is usually translated from Ancient Greek. Personified as a female, Chaos was the primal feature of the Universe, a shadowy realm of mass and energy from which much of what is powerful (mostly negative and dark) would stem forth in later genealogies.' Source: https://www.greekmythology.com.

The concept of chaos takes on different forms in various mythologies. Buddha's doctrine, for instance, presents a more specific and emotionally negative understanding of chaos, stating that 'Chaos is inherent in all compounded things. Strive on with diligence.' On the other hand, ancient Egyptians had a god named Seth who was associated with chaos, storms, disorder, violence, war, and foreigners.

Figure 1.11. Chaos is pictured with the All-seeing eye: adapted from Magnum Chaos in basilica of Santa Maria Maggiore, Bergamo, by Giovan Francesco Capoferri to a design by Lorenzo Lotto.

These contrasting views of chaos in different mythologies pique our curiosity about the diverse interpretations of this concept.

Thus, only Ancient Greek philosophy and mythology are closest to the universal and emotionless concept of modern chaos. Here, we should remark that the modern theoretical attitude to chaos has lost its *being everywhere* feature, which will be discussed in the following parts of the chapter.

1.11 Alpha unpredictability and alpha labeling decrease formalism of chaos

Loosely speaking, we accept that alpha labeling and unpredictability make chaos presence *less formal* in our life since it provides *more evidences* of chaos in industry and science. Dialectically thinking, we comprehend the decrease of formalism as *becoming* more generic in science and industry than previously. That is more realistic. It does not mean that the theories are not formal; they are, but they help to prove the existence of chaos universally in *the world*. Let us provide a metaphor for this as it is seen in figure 1.12, where, according to our theoretical sources, an alpha unpredictable trajectory is simulated, and each point of the cube is a limit for points of an alpha unpredictable orbit. That is, the cube *entirely* is a domain of chaos. One can imagine that the cube is the Universe, and the trajectory is one of the motions.

In the sense of dynamics, we return research of chaos into depths of differential equations, and any others describing dynamics. The research is not focused on sensitivity or another collective characteristic but alpha unpredictability. That is, an individual property is considered the leading actor in the play. Moreover, asymptotically stable alpha unpredictable solutions of various types of equations are used to indicate chaos presence. That is, irregularity of isolated and asymptotically stable motions became preferable for chaos as it was before: only for regular motions. Individual dynamics is the focus of global science again.

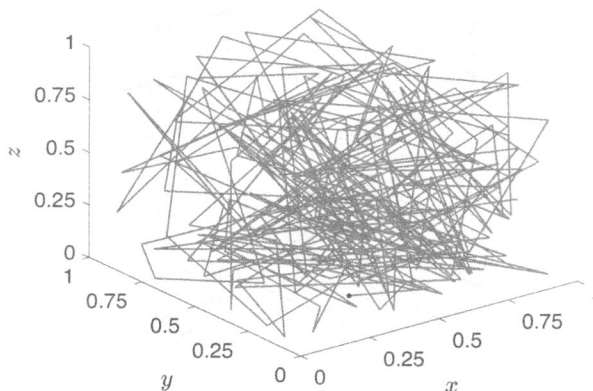

Figure 1.12. Ultra Poincaré chaos on the unit cube.

Possibly, the least formalized comprehension of chaos is the 'butterfly effect,' when the wings of an insect create tiny changes in the atmosphere, which ultimately alter the path of a tornado or delay, accelerate, or even prevent the occurrence of a tornado in another location. One should sincerely confess that the metaphor does not relate to chaos, since the case can not be modeled with any differential equation nowadays and even in the future. Moreover, these effects can be observed even in linear models if the dynamics is unbounded. Nevertheless, it has been a reason for many scientists to give a first glance at the subject. Another source of informal chaos comprehension is a rich collection of phase portraits for chaotic dynamics. They are often considered a strong argument for chaos, even in scientific papers, if any other is not available.

Vital elements of formalism started in the research by E Lorenz in [65], where the sensitivity appeared. T. Li and J. Yorke [63] not only suggested the term chaos but provided their version of chaos formalization about proximality and frequent separation features [64]. Then R. Devaney [45] united the ingredients suggested by Poincaré Birchoff [32] and Lorenz. The most formal descriptions of chaos are positive Lyapunov exponents and Feigenbaum's number [52]. Thus, the phenomenon is formally present through parameters, specific numbers, and sequences, indicating sophisticated dynamics. We have constructed alpha labeling, a powerful tool for chaos demonstration, creation, and alpha unpredictability, which makes the visibility of chaos much more straightforward than before by conservative methods. This is done by the dynamics of the abstract similarity map and by the possibility of observing alpha unpredictable trajectories. Discussions of alpha dynamics support the proofs. Figure 1.13 contains a piece of trajectory for a Duffing equation with a piece-wise constant alpha unpredictable force, and this is a visual example of ultra Poincaré chaos.

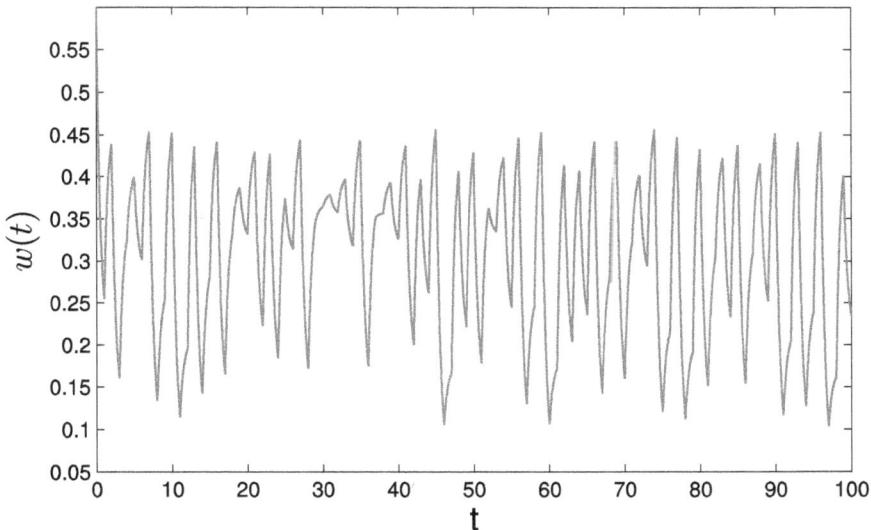

Figure 1.13. Ultra Poncaré chaos in a Duffing equation: adapted from [9].

Concepts introduced in our research are significant contributions to formal attributes of chaos, which is increased in their appearances, and global science for chaos will make them less formal. The alpha unpredictability is a characteristic for a single motion, which makes it possible to analyze chaos by dealing with isolated individual dynamics. That is, we have created, for the first time in history, conditions when one can claim that a motion is chaotic, verifying that it is alpha unpredictable. We are still determining if the reverse is also correct; that is, if any irregular trajectory of chaos is alpha unpredictable. Investigating this is a good task for the future. Nevertheless, there are arguments that this can be true. Our papers have verified that each limited-time realization of Bernoulli schemes or Markov chains with finite state space is an alpha unpredictable function. This is a significant step in formalizing chaos as a universal concept. Amazingly, we can claim now that the realization of a Markov chain in figure 1.14 is a graph for an alpha unpredictable function.

Moreover, a dynamical composition of the chain is another continuous alpha unpredictable function, and its graph is seen in figure 1.15.

Our research has led to realizing alpha unpredictable trajectories in connected non-fractal sets, which bear topological equivalence to finite-dimensional cubes. This discovery, illustrated in figure 1.12, confirms that all self-similar fractals are endowed with chaotic dynamics through alpha labeling. We propose that any time series obtained experimentally can be deemed unpredictable, affirming chaos's universality based on alpha unpredictability.

Another way to reduce the formalization of chaos is alpha labeling with such essential components as abstract similarity sets and maps. The sets consist of alpha labels, which are assigned to be universal for complexity research as real numbers

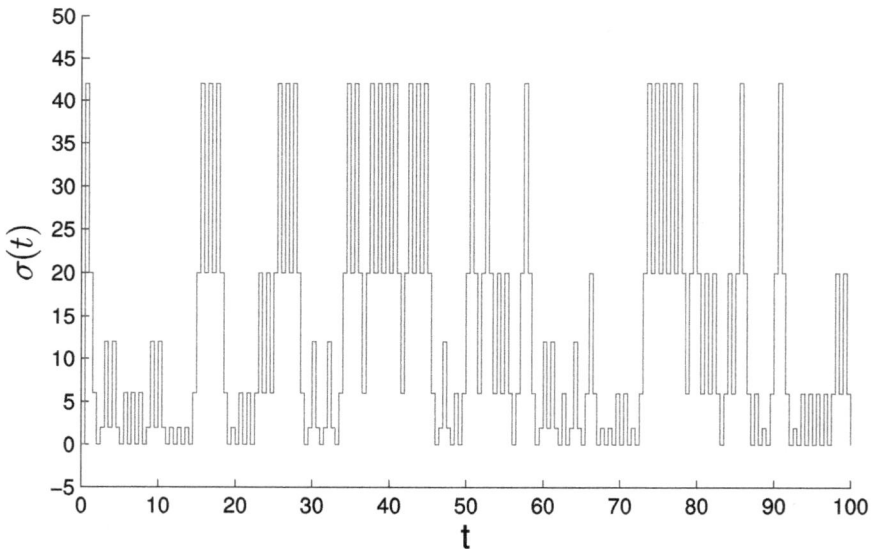

Figure 1.14. Ultra Poncaré chaos in a Markov chain. Reproduced from [10]. CC BY 4.0.

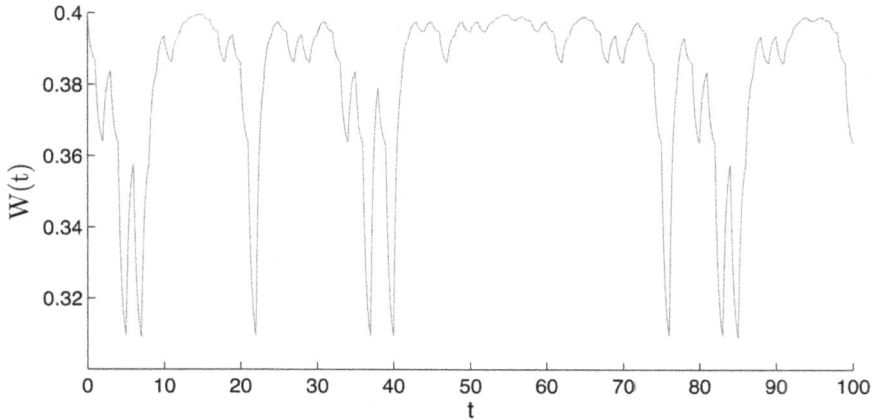

Figure 1.15. A Markovian function as result of a dynamical composition. Reproduced from [10]. CC BY 4.0.

are for science. If alpha labeling happens with the specific diameter and separation properties, chaos can be easily claimed for dynamics. That is, we reduce the task of chaos recognition and chaos creation to alpha labeling of domains. Thus, one can accept that abstract similarity dynamics demonstrates chaos universally. Unsurprisingly, alpha labeling can be easily utilized to check alpha unpredictability and, consequently, ultra Poincaré chaos presence in stochastic processes. Moreover, alpha labeling with proper conditions on the alpha map confirms, besides the ultra Poincaré, also Li–Yorke and Devaney chaos. It is of significant interest if one can, by applying alpha labeling to a new lower level of formalization of Feigenbaums' number, bring bifurcation diagrams and Lyapunov exponents. Next, we expect chaos creation in differential equations by structuring domains through alpha labeling. This is a very grandiose challenge as it can turn research of chaos from toy models to real-world problems.

Reducing formalism in our suggestions also means that we do not distinguish between deterministic and nondeterministic dynamics and statistical and non-statistical descriptions of real-world processes. *The construction of boundaries is formalism.* Moreover, we do not give any meaning for reasons and origins of alpha unpredictable motions-functions. These are the results of randomness or deterministic equations that occur in quantum processes or celestial and economic dynamics. We wish only to find functional characteristics of complexity. Theoretically, we consider processes between infinite past and future. However, this is more than just a theoretical approach. As a result of our discussions, one can make decisions on complexity considering even finite time events. If we observe dynamics that happen on limited time intervals, we include them as episodes of processes with infinite time duration. Random motions are least formal in mathematics and most popular in real-world dynamics. However, we cannot discuss periodicity in processes without mentioning probability. That is, there is no possibility that dynamics can occur in highly probable realizations.

1.12 Mathematical chaos and H. Poincaré

In paper [81], it was realized that transverse homoclinic points would lead to complicated behavior. Subsequently, G. Birkhoff characterized them [32], before in [88, 89] the description was completed. S. Smale proved the final result by applying topological conjugacy to the symbolic dynamics, which is completely avoided in our research.

Smale–Birkhoff homoclinic theorem [54] Let $P\colon \mathbb{R}^2 \to \mathbb{R}^2$ be a diffeomorphism possessing a transversal homoclinic point $\mathbf{q}$ to a hyperbolic saddle point $\mathbf{p}$. Then, for some $N < \infty$, P has a hyperbolic invariant set Λ on which the mth iterate P^m is topologically conjugated to a shift on two symbols.

G. Birkhoff proved many countable periodic points in any neighborhood of a homoclinic point [32], and Smale's construction provides an extension for $\mathbb{R}^n$, and more detailed analysis had been made.

In literature, the studies belonging to H. Poincaré, G. Birkhoff, and S. Smale resulted as Poincaré homoclinic chaos. In what follows, we call the phenomenon issued from the research *H-Poincaré chaos*.

In the same manuscript, the French genius formulated the recurrence theorem [81], which states that specific dynamical systems in continuous time will, after a sufficiently long but finite time, return to a state arbitrarily close to their initial state. Similarly, a state returns precisely to its initial state for discrete time. The final version of the theorem was proved in [40] using the measure theory. What was discovered in [81] are *Poisson stable trajectories*.

As far as we know, the recurrence theorem is not considered as a sort of chaos result. However, everyone who knows the theory basics recognizes that topological transitivity is at least present in the theorem. G. Birkhoff made another contribution in the direction [32] and H. Hilmy [58] who proved that closure of a Poisson stable trajectory, the quasi-minimal set, contains an uncountable set of orbits, which are also Poisson stable and dense in the set. Thus, we can accept that an analog for another property of chaos, the existence of infinite periodic motions dense in the domain, is valid. Only sensitivity was needed to achieve chaos based on the recurrence theorem. Thus, research by H. Poincaré, C. Carathéodory, G. Birkhoff, and H. Hilmy of recurrence had not been completed for chaos since the sensitivity failed. The property was proved by assigning *alpha unpredictability* [4, 17–19] for Poisson stable motions. Thus, the Lorenz line and Poincaré line intersect, and the ultra Poincaré chaos has been introduced. Not to confuse with *H-Poincaré chaos*, we decided, in this paper, to call it *ultra Poincaré chaos*, emphasizing the *alpha unpredictability* property as the reason for the phenomenon. That is, ultra Poincaré chaos happens on the closure of alpha unpredictable trajectory. Now the results by G. Birkhoff and H. Hilmy have obtained new colors. In other words, the recurrence theorem is now extended to chaos, and significant improvements to research initiated by H. Poincaré in [81] have been activated in our investigations.

Over the last several decades, chaos as a dynamic phenomenon has been the focus of mathematical and interdisciplinary investigations. The intensity of the research makes it necessary to discuss delicate questions related to chaos and consider different scientific progress routes.

It is now clear that the research of chaos could be initiated in different ways and can be issued from various mathematical sources. More clearly, we are talking about results by French genius H. Poincaré concerning the famous recurrence theorem and Poisson stability. The reason for pausing in the proceeding line of Poincaré is the invention of sensitivity by E. Lorenz, which stopped the intensive research and has been dominated in the study of complexity for several decades. We guess that switching between lines, or abandoning research of the Poincaré line, happened because the classical theory of dynamical systems was focused on regular types of motion, and mathematicians have been satisfied with the boundaries of the theory lined by authorities. Secondly, advanced definitions and theorems of the dynamical systems theory seemed sophisticated for further continuation to leave the arranged boundaries. Meanwhile, Lorenz's suggestion and its further variations looked simple and attractive. What is important is that sensitivity does not request a deep knowledge of classical dynamical systems theory for applications. That is, the chaos research started as *tabula rasa*, considering the invention of E. Lorenz.

To highlight the significance of our findings, it is worth noting that in [19], we established ultra Poincaré chaos through alpha labeling for Horseshoes. This suggests that alpha unpredictability and sensitivity are valid in both directions of discrete time, thereby illuminating this fundamental result. Furthermore, there is confidence that the proofs of foundational theorems utilizing symbolic dynamics could be simplified using alpha labeling, beginning with the Smale–Birkhoff homoclinic theorem.

In the realm of science, it remains challenging to amend and revise the foundational aspects of existing theories. A historical example is L. Kronecker's opposition to set theory. To contextualize our discussion, we can reference a passage from Yu N. Harari's book [55], where he argues that religion and social relations often serve as barriers to scientific progress. He states, 'Ancient traditions of knowledge admitted only two kinds of ignorance. First, an individual might be ignorant of something important. To obtain the necessary knowledge, all he needed to do was ask somebody wiser. There was no need to discover something that nobody yet knew. For example, if a peasant in some thirteenth-century Yorkshire village wanted to know how the human race originated, he assumed that Christian tradition held the definitive answer. All he had to do was ask the local priest. Second, an entire tradition might be ignorant of unimportant things. By definition, whatever the great gods or the wise people of the past did not bother to tell us was unimportant. For example, if our Yorkshire peasant wanted to know how spiders weave their webs, it was pointless to ask the priest, because there was no answer to this question in any of the Christian Scriptures. That did not mean, however, that Christianity was deficient. Rather, it meant that understanding how spiders weave their webs was unimportant. After all, God knew perfectly well how spiders do it. If this were a vital piece of information, necessary for human prosperity and salvation, God would have included a comprehensive explanation in the Bible' [55].

In our case, the ultimate type of recurrence, Poisson stability, has not been revisited for nearly a century. It is evident that the range of motion types must be extended to strengthen the theory of dynamics for practical applications. It is crucial to avoid an unyielding, almost religious adherence to past results, which may hinder progress. The respect and conservative sentiments surrounding the foundational contributions of Poincaré Birkhoff and others may create barriers to advancing new concepts in dynamics. One might speculate that the various obstacles discussed by Harari about maintaining the status quo also apply to conserving Poincaré's ideas, which have not been adequately developed. Our research aims to address and remove some of these barriers.

1.13 Alpha unpredictability versus sensitivity

Let us start this part of the manuscript with a citation, 'The average person, seeing that we can predict tides pretty well a few months ahead would say, why can't we do the same thing with the atmosphere, it's just a different fluid system, the laws are about as complicated. But I realized that any physical system that behaved non-periodically would be *unpredictable*' [65] (the italic is ours). In his citation, E. Lorenz, the renowned scientist known for unveiling the concept of sensitivity, stated with great precision that physical processes are *unpredictable* if they are non-periodic. However, he was somewhat mistaken, as tidal time series are non-periodic yet exhibit a form of alpha unpredictability, based on contemporary understanding. We believe that, upon recognizing the numerical complexities, Lorenz had two potential paths to explore. One of these paths, sensitivity, was pursued, while the other—alpha unpredictability [4, 17, 18]—was overlooked by Lorenz and others. Our recent papers have since discovered and developed this second approach. In essence, this represented a 'bifurcation' moment for scientific advancement: sensitivity was identified as one branch, while alpha unpredictability remained hidden, resulting in the current state of knowledge. The interplay between the findings and methodologies of Lorenz's work and those of Poincaré will undoubtedly enhance the latter's contributions. In the following sections, we will delve into a more detailed discussion of these two scenarios.

E. Lorenz places an emphasis on sensitivity when he explained deficiency in accomplished simulations. Precisely, it means that for a given motion of dynamics, $f(t, p), f(0, p) = p, t \geqslant 0$, there is an arbitrary near to p point q such that at the moment $\tau > 0$, the distance between $f(\tau, p)$ and $f(\tau, q)$ is not less than e, and the last number is typical for the dynamics, independently of start points. In continuity, the closer the initial values, the larger moment τ of divergence must be.

Consider the scenario for alpha unpredictability. For that, fix a motion, $f(t, p), f(0, p) = p, t \geqslant 0$. It is assumed that p is an accumulation point for an infinite set of points, $f(t_n, p)$, with n naturals, and unbounded sequence t_n. Moreover, for each n there exists a positive moment s_n such that distance between $f(s_n, p)$ and $f(t_n + s_n, p)$ is not less than e, a number common for all points. The existence of sequence s_n and number e is called the *unpredictability* property. For convenience, we shall call the property alpha unpredictability. That is, that unpredictability considered in our former papers starting with [4, 17, 18] now has to be understood as *alpha unpredictability*. We decided to make it different from the former concepts in any other area of science known as unpredictability. The sequence of divergence moments is unbounded; this is valid because of continuity on initial data. Since we are dealing with dynamical systems, the trajectory as an arc is independent of the initial moment, and for each natural n, the orbits of motions $f(t_n + t, p)$ and $f(t, f(t_n, p))$ are identical. The distance between $f(t_n, p)$ and p combined with the constant e guarantees alpha unpredictability. In papers [4, 17, 18] it is proved that the Lorenz sensitivity follows alpha unpredictability. Thus, one can use the famous title for Li–Yorke chaos and confirm that *alpha unpredictability implies chaos* as a result of our studies. Moreover, one can guess that sensitivity and topological transitivity imply alpha unpredictability, but it should be proven. A more detailed discussion of our doctrine can be found in [4, 17–19]. It is rooted in Poisson stability [32, 82], initiated by H. Poincaré. This is why we named the chaos after the great French scientist.

We believe that sensitivity and alpha unpredictability may be the reason for 'erratic' Lorenz numerical simulations. The arguments for research alpha unpredictability are theoretically much higher than those for sensitivity since they request deeper knowledge of qualitative theory for dynamical systems. E. Lorenz's approach is suitable instead for physicists and engineers. At the same time, the content of the Poincaré line is arranged through sophisticated theoretical analysis of dynamical systems and differential equations.

Another interesting question to discuss is, What is a chaotic trajectory? For ultra Poincaré chaos, the question is easy to answer. A chaotic trajectory is, at first, any alpha unpredictable orbit. We consider the closure of the orbit and obtain a quasi-minimal set [73, 87]. As it was shown in [18], dynamics on the set are sensitive. The set is filled by uncountable alpha unpredictable orbits, which are not periodic. So, Lorenz chaos deals with infinitely many unstable regular orbits, but in our case, we are dealing with an uncountable set of alpha unpredictable motions. That is, a new sort of dynamics is discovered, called ultra Poincaré chaos. It is important to remark that besides alpha unpredictable orbits, the dynamics may admit infinitely many unstable periodic motions and quasi-periodic, almost-periodic, and even recurrent or Poisson stable motions, which are not unpredictable. All of them are to be considered chaotic trajectories since they are in the quasi-minimal set and *unstable*. Simply speaking, we admit any trajectory as chaotic if it belongs to the domain of chaos and is not stable. The dynamics of ultra Poincaré chaos are much more sophisticated than that of Lorenz. Similarly to chaos based on alpha unpredictability, one can consider chaotic trajectories of the Lorenz type, which is determined through sensitivity or property relative to it. For example, that of Li–Yorke belongs to the class. There is a trajectory, which is dense in the domain of Lorenz chaos, and it is chaotic. Moreover, the domain is filled with periodic trajectories. They are chaotic because they are *unstable*. It will be a big task in the future to show that a trajectory dense in Lorenz chaos is Poisson stable and unpredictable. Meanwhile, we can provide models known as Devaney or/and Li–Yorke chaotic, and alpha unpredictable motions have been discovered there. For instance, the symbolic dynamics. Thus, working on ultra Poincaré chaos, we define a phase space for other types of chaos, and this is another contribution of our approach to global chaos research. Finally, let us emphasize that no numerical method can determine that a given trajectory is chaotic in Lorenz. However, we have introduced and developed a numerical test for alpha unpredictable motions based on revealing convergence and separation sequences[20, 21]. That is, chaotic trajectories can be recognized individually and synchronized [13–15, 26].

1.14 Ultra Poincaré chaos versus Li–Yorke chaos

Within the realm of sophisticated dynamics, two exceptional types of chaos emerge: ultra Poincaré and Li–Yorke. Unlike other forms of chaos that focus on the divergence of nearby trajectories and the collective behavior of infinitely many motions, these two types uniquely center on a single trajectory to test for chaos. In essence, they proclaim, 'alpha unpredictability implies chaos' and 'period three implies chaos.' This complexity of the two types of chaos is what makes them intriguing and challenging to study. Significantly, alpha unpredictability and periodicity are entirely individual properties of trajectories. One can also mention homoclinic trajectory for the same reason, but it is not self-sufficient to claim chaos. This is why one should compare the two versions of chaos to make the global state of the research more clear through appropriate discussion. Thus, there is a common issue point in the basis for two versions, a single trajectory behavior as chaos initiation. In what follows, we shall provide arguments that ultra Poincaré chaos is universal, while the second one is specific. The period three property is for a narrow class of chaotic models when alpha unpredictability is suitable for universality regardless of mathematical issues.

Li–Yorke research in its original state is formulated only for one-dimensional dynamics. It is described as a Li–Yorke sensitivity based on proximality and frequency separation, which are consequences of Lorenz sensitivity. One can find deep analysis for the subject in paper [27]. Nevertheless, the sensitivity can also be a consequence of the convergence and separation properties of the alpha unpredictability. One can consider Marotto chaos [70] for the same reason.

Ultra Poincaré chaos, which originates from the dynamics of a single motion, has universal applications in mathematical theories, physics, and other fields of science and industry. Its numerical observability and the flexibility to consider motions in any state space, metric or topological, finite or infinite dimensional, make it a valuable tool. By requesting only the existence of two fundamental time sequences for convergence and separation, we can apply this chaos to various dynamics. This practical relevance of the chaos types is what makes them significant in the field. Nevertheless, they may appear in particular cases near alpha unpredictable trajectories. Periodic, almost periodic, and Poisson stable motions can be seen [3, 4, 16]. Generally speaking, existing types of chaos can be considered as particularly relevant concerning ultra Poincaré chaos.

Li–Yorke chaos is a sort of H-Poincaré that is homoclinic chaos. Let us explain our opinion. Remember that period three means existence of three points, a, b and c

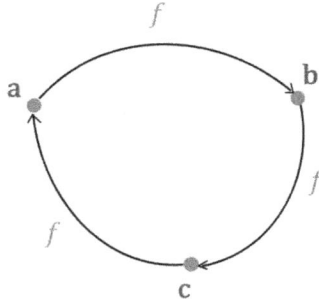

Figure 1.16. Dynamics with three 'homoclinic' points.

for a map f such that $b = f(a)$, $c = f(b)$ and $a = f(c)$. Formally speaking, one can accept b as a homoclinic point concerning points a, and c. The idea of the homoclinic point is seen in the construction of period three. Figure 1.16 shows visual illustrations for our arguments. One can see that a and c can also be considered homoclinic points. We believe that the formality can help to extend the role of H-Poincaré chaos in theory and applications.

In the context of ultra Poincaré chaos, we do not need to adhere strictly to the framework of homoclinic points. Instead, what we propose is a set of more general criteria for analyzing complex dynamics that can be applied to any type of motion. However, the concept of ultra Poincaré chaos does encompass homoclinic chaos when it is extended to include dynamics that consider negative time direction. We have applied this approach to examine the dynamics of the Smale Horseshoe. Overall, homoclinic points can be identified through alpha labeling, as noted in [4].

Marotto chaos cannot be considered a generalization of Li–Yorke chaos, as it represents another instance of homoclinic chaos. In fact, both types of chaos are more closely related to homoclinic chaos than they are to each other. Furthermore, it is important to note that Li–Yorke chaos is primarily associated with one-dimensional dynamics, if not particularly from a topological perspective [78]. A thorough examination of the similarities between the works of Li–Yorke [64], Marotto [70], and the H-Poincaré chaos [97] warrants serious attention. We believe that deeper investigations in this direction could yield more intriguing results. Given that alpha labeling is a key tool for investigating ultra Poincaré chaos, we remain optimistic that this method may be applied in the future to verify Li–Yorke and Marotto snap-repelling dynamics.

1.15 Chaos as a metaphor

Let us cite that 'A metaphor is a figure of speech that, for rhetorical effect, directly refers to one thing by mentioning another. It may provide (or obscure) clarity or identify hidden similarities between two different ideas.' Source: *Wikipedia*.

P. Holmes [54] was the first to explore the term *chaos* as a metaphor for the dynamics being studied. This perspective highlights how real-world processes connect to scientific developments in the mid-twentieth century. Generally, mathematical terminology tends not to provoke strong emotional responses among researchers, largely due to its long-standing usage and familiarity. However, the term chaos stands out because it has only recently emerged in scientific discourse. It is linked to ancient mythological and religious traditions, which is not the case for many other mathematical terms. Consequently, emotions are inevitably intertwined with this area of research. Individuals familiar with Greek tragedies, myths, and the fundamentals of dynamical systems tend to share a specific sentiment regarding the term, while those who are only acquainted with the butterfly effect have different perspectives.

Regarding the *butterfly effect*, one can also provide a portion of criticism for the choice of metaphor. First, remember that even a linear equation may cause sensitivity. Consider, for instance, $x' = 2x$. Two solutions of it, $x_0 e^{2t}$ and $y_0 e^{2t}$, are far from each other with any distance at a suitable moment if only the initial values are the same. So, we have observed that the effect and sensitivity are valid for the 'laminar' dynamics. One can replace the trivial version of the equation with aerodynamic non-chaotic models and obtain sensitivity. A more concrete but not less critical objection to Lorenz's metaphor is that chaos is generally assumed in bounded regions. At the same time, the hurricane or tornado develops in large parts of space, which is not bounded potentially. The first steps in this direction were made in [4, 6], where we apply alpha labeling possibilities to describe unbounded chaos.

We can propose an alternative perspective on the fundamental components of traditional chaos, specifically focusing on sensitivity. This concept, grounded in Lorenz chaos, is viewed as a source of irregularity and is reflected in various assessments, such as positive exponents and period-doubling cascades, which are particularly relevant when examining bounded processes. However, since sensitivity can also be observed in unbounded dynamics, as previously discussed, it leads us to conclude that sensitivity is not a definitive indicator of complex dynamics, especially when it is global but not chaotic. In contrast, alpha unpredictability ensures irregularity even in unbounded processes, and we assert that it should be regarded as a critical or 'internal' condition for dynamics when assessing the characteristics of chaos.

A non-linearity encircled in bounded regions, endowed with sensitivity and extended to infinity in time, affects our feelings differently from regular motions, strongly and deeply connects to art, and gives birth to new, colorful phenomena and emotions. In particular, 'Chaos has created special techniques of using computers and kinds of graphic images, pictures that capture a fantastic and delicate structure underlying complexity' [53]. Lorenz's sensitivity and alpha unpredictability will be heard in music and applied in beautiful symphonies and songs using computers. Even religious-type feelings can be exciting: 'Believers in chaos—and sometimes call themselves believers, converts, or evangelists—speculate about determinism and free will, evolution, and the nature of conscious intelligence. They feel that they are turning back a trend in science toward reductionism, the analysis of systems in terms of their constituent parts: quarks, chromosomes, or neurons. They believe that they are looking for the whole' [53]. One can provide a comment for the citation: as for any belief, there are two sides: buoyant, as a desire which increases the strength and power of scientists; and negative, which makes them passive and even ignorant. Since the influence of metaphors depends on the individual features of scientists, traditions, experiences, erudition, and the ability to imagine, the role of a metaphor may be various, more or less positive or negative.

As is recognized, three scientific revolutions in physics occurred in the last century. They are the theory of relativity, quantum mechanics, and chaos.

It is important to emphasize that *chaos* as a metaphor is rooted in the ancient world, supported by a long history of civilizations. Two others, *relativity* and *quantum*, are *technical* terms, which themselves serve as metaphors instead for modern cultural and scientific generations.

If the choice of *chaos* in [64] can be considered just as a metaphor, then this raises the question of whether it is a good word to choose. Moreover, how helpful is it (does it provide clarity) for science, or is it harmful (does it obscure perception)?

Considering the word's etymology, it might be time to consider the term, *chaos in dynamics*. Is it a good metaphor? It is clear that chaos in dynamics was a new concept in the middle of last century. Moreover, as far as we know, the question has not previously been discussed in the literature, except for several remarks in [54].

A metaphor is helpful to clarify a topic by comparing previously known objects or emphasize a specific feature of an event, for example, in the sentence 'All the world's a stage' by William Shakespeare. We believe that T. Li and J. York considered chaos as a metaphor for the first reason. That is, to say that the phenomenon they introduced is as sophisticated as chaos in its modern sense, and there were no more reasons to discuss. We do not see any specific reason for what is understood as mathematical chaos to be compared with that of the Ancient Greeks. It is rather about modern synonyms such as *anarchy*, *disorganization*, *misrule*, *turmoil*, and others. Here, one can understand why chaos became unpopular for many working on differential and difference equations theories. Many conservative mathematicians, who used to see the strong correspondence between a name and an object, avoid being involved in investigating chaos since they have been confused with this lack of correspondence. Another negative result of the choice is that chaos investigation has not been included as a part of the theory of differential equations research from the beginning. We should remember that the phenomenon under discussion is just about dynamics related to a special differential equation. As we remarked in our articles [4], continuous dependence is a necessary attribute for chaos. The concept is present in particular models, such as the Lorenz system, logistic equation, and Ueda model, without deep qualitative analysis. While in [54], it is mentioned, 'I do not see a 'new science' here, in particular, I do not see that 'chaos theory' even exists as a coherent object, for example, like the quantum and relativity theories.' Thus, we believe many specialists in differential and discrete equations have been repelled away from studying chaos, although it is advantageous. This is what we understand as a negative effect of the metaphor. That is, one can say that word *chaos* has played a negative role in reducing clarity generally, as well in a particular sense considering motions of differential and difference equations. However, many scientists continue the research despite dangers such as 'Attempts to understand (pieces of) the world often start with a metaphor: one studies a simpler system which 'looks like' the real object of interest, but one does not insist on fundamental connections or derivations 'from first principles.' The fact that small sets of differential equations and iterated maps can exhibit complicated behavior has stimulated a lot of such metaphorical studies' [54].

Now, let us consider again the phrases 'order in chaos' or 'chaos and order.' They are nonsense for those busy with complicated phenomena in dynamics for an extended period. Since it is evident that 'chaos' is 'ordered' as much as a stable periodic motion is. The degree of order is much higher than before. It is suitable to say that 'chaos is order' to emphasize knowledge accumulated for the subject. Generally speaking, to work on a theory means to discover order that exists independently of a researcher. There is no reason to ask if 'a theory is order.' Any scientist bringing order to their subject means creating or contributing a theory. In other words, to say 'I am making order within chaos' means 'I am building a theory of chaos.' If one says, 'I am seeking order within chaos,' it is the same truism as saying, 'I am searching for order around asymptotically stable periodicity.' All these show how the phrases are imperfect and, in some sense, harmful.

Nevertheless, chaos investigations provide conditions for flourishing numerical methods in dynamics analysis, especially those motions that last for an infinite time. Remember that we can not refer numerically to non-periodic recurrence or asymptotic properties since they must be observed over an infinite period. Only by definitions, theoretically, can one discuss such phenomena. At the same time, several new methods, such as bifurcation diagrams, Lyapunov exponents, and others, allow us to talk about chaos presence. However, dynamics are assumed to be at infinite time intervals, which is an interesting paradox. Nevertheless, '... although much remains to be learned about the applicability and validity of such methods' [54]. Besides criticizing the lack of theoretical results, P. Holmes refers to the 'tyranny of technique which afflicts many scientists' [54].

The term 'chaos' holds a significant allure, primarily due to the butterfly effect. Its popularity is evident in the numerous films and research papers dedicated to the subject, which significantly boost the impact factors of scientific journals. The interdisciplinary connections that chaos theory fosters are remarkably robust, further enhancing its appeal. In this light, one could argue that the metaphor of chaos is a masterstroke in scientific marketing.

1.16 Does the theory of chaos exist?

An intriguing question arises regarding whether the findings from chaos research can be synthesized into a cohesive *theory*. To address this, we can consider two perspectives. The first perspective comes from P. Holmes, who states, 'I do not see a 'new science' here, in particular, I do not see that 'chaos theory' even exists as a coherent object, for example like the quantum and relativity theories.... However, we do have a loose collection of ideas and methods, many of the latter inherited from 'classical' applied mathematics, which we can add to the scientist's toolbox' [54]. Another important aspect to consider in this discussion is the need for a clear and comprehensive definition of a scientific theory: 'A scientific theory is an explanation of an aspect of the natural world and Universe that has been repeatedly tested and corroborated by the scientific method, using accepted protocols of observation, measurement, and evaluation of results. Where possible, theories are tested under controlled conditions in an experiment' [98].

Considering these two points of view, both are reasonable for chaos theory. The first one is based on the opinion that chaos is a product of mathematical insights and is supported by strong arguments, which we list below. They confirm that the research is still close to the dynamical systems theory and needs to be rigorously proven in many senses if separated. While considering the definition of scientific theory and processes discovered in several areas, such as physics, biology, and mathematics, we observe that the results of chaos can be seen as a theory—and an interdisciplinary one. So, before we start a discussion, let us provide a version of the conclusion. Chaos research has already been shaped as a scientific theory of a new type if one considers it a collection of methods, first of all numerical, protocols and observations, models and experimental processes, which are tested to be chaotic in many areas of science and industry. The theory is much more interdisciplinary than, for example, quantum mechanics if one considers the number of sciences involved in the research and prospect investigations, first of all, in quantum computers and artificial networks. Many possibilities still need to be formulated. From a mathematician's point of view, one can not say that there is a theory of chaos since there are only a couple of chaos cornerstones, such as sensitivity and alpha unpredictability, which are only sometimes consistent with numerical methods and, if proven, only for simplified models.

So, if looking for the existence of chaos theory, we are likely returning to the state of two centuries ago, when mathematicians and physicists (nature-philosophers) were considered the same researchers. Should we talk about complexity instead of chaos if one wants to accept the study as a theory? That is researching chaos as a part of complexity, which indeed is an interdisciplinary subject [75], and admits in focus ergodicity, mathematical chaos, turbulence, quantum mechanics, theory of algorithms, and many other fields. Another interesting observation is about borders between pure and applied mathematics, which can be canceled through chaos investigations. 'There is a great ferment of excitement and activity. The artificial distinction between pure and applied mathematics is weakening. Mathematicians and scientists from different fields are talking to one another. Some are even listening' [54].

We should not admire claims such as 'Where chaos begins, classical science stops' [53]. This may seem true for the present short stage of science development, but globally speaking, it is incorrect. This will be seen better when the roots of chaos investigations are discovered more clearly. Another misunderstanding is caused by the enormous number of publications on the subject, which appear to be only effective investigations. However, after proper unification, the results will be narrowed significantly, and an essential few achievements will be kept.

There is the opinion on chaos universality which is close to that perception of chaos in ancient philosophy and myths: Now that science is looking, chaos seems to be everywhere. A rising column of cigarette smoke breaks into wild swirls. A flag snaps back and forth the wind. A dripping faucet goes from a steady pattern to a random one. Chaos appears in the behavior of the weather, the behavior of an airplane in flight, the behavior of cars clustering on an expressway, and the behavior of oil flowing in underground pipes. No matter the medium, the behavior obeys the same newly discovered laws. That realization has begun to change how business executives make decisions about insurance, how astronomers look at the solar system, and how political theorists talk about the stresses leading to armed conflict' [53]. Appreciating the beauty of the citation, one must not forget that the same sentences can be said about *non-chaotic* motions of the world around, and they will not be less poetic.

The third global remark can be made for the subject's interdisciplinary essence: 'Chaos breaks across the lines that separate scientific disciplines. Because it is a science of the global nature of systems, it has brought together thinkers from fields that had been widely separated' [53]. One must remember that *non-chaotic* dynamics, which has been considered much earlier than chaos, is also very interdisciplinary. Moreover, regular dynamics is still more universal and applicable than chaotic ones. Thus, as soon we consider chaos to be routine dynamics it will become a part of classical dynamical systems theory, significant, but a part.

The most passionate advocates of chaos go so far as to say that the science of twentieth-century *will* be remembered for just three things: relativity, quantum mechanics, and chaos. The last one is supposed to be the third great revolution, which we cannot reject. Besides, chaos eliminates the Laplacian belief in deterministic predictability. However, still, one can agree that it is not as deep an insight as the first two theories. Moreover, remember that the sensitivity was discovered numerically, let us say occasionally. At the same time, quantum mechanics was created on an intense theoretical science level. Although started with the experiment of Michelson-Morley, the theory of relativity is based on the philosophical insights of H. Poincaré [83].

Several remarks could reasonable support the opinion that there are still gaps in the research of chaos. These gaps are arguments against the fact that the analysis is even a part of dynamical systems theory and needs to be developed for it to be that.

The first remark in that sense is about numerical tools for chaos indication: 'The ideas of dynamical-systems theory has also led to several new methods of data analysis; dimension computations, Lyapunov exponents, phase-space reconstruction, spectra of dimensions, etc, although much remains to be learned about the applicability and validity of such methods' [54] Accordingly, '…we should beware of the tyranny of technique which afflicts many scientists: having ignored the work described below for 90 years, a certain type of researcher now sees chaos and strange attractors wherever (s)he looks' [54].

One more warning by P. Holmes is that chaos research still can not be considered part of dynamical systems theory, but this time from the opposite side. This argument is about the theoretical basics of chaos analysis. Precisely, 'Dynamical-systems theory has potentially dangerous side effects in this respect. Certain persons seem to prefer to abandon hard-won, detailed knowledge of problems like turbulence in boundary or shear layers in favor of metaphors, such as coupled map lattices, which have little obvious connection with the underlying physics. In most cases, it remains to relate these to fundamental principles, and so, in turn, the metaphors into models. However, it is certainly true that the notion of universality—topological and metric behavior which is, to some extent, independent of the precise system—confers respectability on some such metaphors and reveals underlying structure similarities across application fields. Unfortunately, none of the famous 'bifurcation to chaos' scenarios–period doubling, quasi-periodic, etc—are sufficiently general that we can entirely dispense with analyses to determine what occurs in specific models' [54].

In several parts of the manuscript [54], P. Holmes expressed skepticism for the theoretical merits of chaos research methods and numerous results. The doubts were expressed approximately three decades ago and are still actual since one cannot see even a discussion of the critique nowadays. There are no efforts to bring the research to a single basement as it requests a theory. What we are suggesting is that alpha labeling and alpha unpredictability can be a starting point to remove doubts and build the *theory of chaos and complexity* in the future. Formally speaking, alpha labeling will play the role of basic mathematical structure for complexity and alpha unpredictability, the most sophisticated motion supported by its functional analysis and numerical tools.

1.17 Ubiquitous chaos and dualism

Why is chaos, as a form of motion, so pervasive? P. Valéry offers an insightful perspective: 'Immobility is a forced and forced transitional state, imposed almost by force, while jumping, precisely measured rhythm, pointe, antreche or dizzying rotations are just a natural way of being. It is precisely this short state that takes us outside of ourselves or takes us away from ourselves and in which we are supported by instability, and stability arises *only by chance*, such a state gives us an idea of a different form of existence in which there are special moments when human abilities reach their limits' [95] (the italic is ours). Thus, the renowned essayist clearly recognized that instability is much more intrinsic to dynamics than stability. In our view, this citation serves as one of the best metaphors for chaos, evident even in dance. It can be acknowledged that this concept extends to all human activities, whether physical or mental.

The development of science itself is unpredictable and chaotic. What it attempts to uncover and predict pales in comparison to the vastness of what actually occurs or could exist in reality. In this sense, chaos is indeed omnipresent, particularly within research processes, which represent merely a small fraction of the Universe's dynamics yet hold immense significance for us.

Before delving into chaotic orbits, it is important to reiterate a core idea: chaos is a pervasive phenomenon in the Universe, and our research into alpha unpredictability and alpha labeling aims to affirm this global presence of chaos. We will present arguments supporting the notion that chaos exists everywhere, one of which is that any natural motion in the Universe can be described as either a finite or infinite arc of a chaotic orbit. We have previously explained the trajectories associated with ultra Poincaré and Lorenz chaos. However, it is evident that as our knowledge advances, new chaotic characteristics will emerge for the benefit of science. We are even now capable of constructing chaotic orbits based on concepts of alpha unpredictability and labeling that have not yet been fully explored.

While some researchers argue that chaos is not a universal phenomenon, a substantial number hold opposing views. Certainly, many scientists remain undecided. To those engaged in the study of chaos, the concept may still seem exotic. Daily life and industrial activities predominantly require a sense of periodicity in motion, which was our initial perception of repeated motion. Ptolemy viewed celestial dynamics as not only periodic but also circular. Along with concepts of equilibrium, these ideas represent the simplest opposites to chaos when considered stable. This perspective tends to foster a conservative attitude among researchers toward unstable motions, even periodic ones.

Experts in differential equations maintain that research should focus exclusively on regular motions, which must be stable and preferably asymptotic. There is ongoing debate regarding whether chaos or regular motions are more prevalent in the Universe. Yet, a prevailing belief is that the presence of chaos is incidental, leading to the impression that it is infrequent in the world. This view is particularly popular among those focused on differential equations, who are accustomed to studying stable oscillations as the foundation of their research.

We are sure those who say that chaos is ubiquitous are correct and believe that it is supported by our several years of research, too. In other words, we believe that any motion and change in the Universe happens chaotically. If an individual motion is observed for a short period, its orbit is part of an alpha unpredictable trajectory. If an orbit were infinitely prolonged, it would be one of the types of chaotic orbits. Moreover, one can say that any process of the real world admits a chaotic 'coordinate.' Thus, chaos appears necessarily everywhere.

The following arguments are that non-chaotic dynamics everywhere are only an illusion. One can accept that periodic or any regular type of motions are results of 'natural' control of chaos since it is known that unstable periodic motions are ingredients of chaos [45, 64]. Moreover, we proved that this is true for quasi-periodic, almost periodic, and Poisson stable motions [23], and they can replace periodicity as the ingredient of chaos for various dynamics. Let us consider the question from the position of a single trajectory. It can be either the orbit, which is dense in the domain of chaos (transitivity) or alpha unpredictable trajectory of ultra Poincaré chaos [4], whose closure is a quasi-minimal set. If one considers that a chaotic set is of fractal dimension, then transitivity and alpha unpredictability can not prove that chaos is ubiquitous—quite the opposite. However, in recent papers [4], alpha labeling demonstrated that a finite-dimensional cube serves as a domain for chaos—that it is a quasi-minimal set. Consequently, it is valid for any bounded domain homeomorphic to cubes. In other words, the opinion by E. Nicolis and I. Prigojin [75] that for a motion to be sophisticated, it is not necessary to happen in complex systems, is accurate. Even dynamics on only two points can be chaotic [4]. To be the domain for ultra Poincaré chaos means that almost every point of the set is alpha unpredictable. Thus, based on our studies, we have one more argument for the density of chaos in the Universe. For unbounded processes, one can suggest results in [6], which can be applied to prove that unbounded sets can be domains for chaos. Another challenging problem is to show that random processes such as Bernoulli schemes or Markov chains can produce alpha unpredictable functions [4, 7, 24]. Since the problem has been solved for stochastic problems with finite state space, one can believe that analogous results are true for many other stochastic processes. Moreover, issuing from abstract similarity dynamics one can show that Barnsley's algorithms can form shapes of leaves and other life pictures. We can solve these problems by alpha labeling since isolated steps can be done randomly, but in summary, one obtains globally deterministic objects. Similarly, randomly falling drops of rain fill a square deterministically. Branches of a tree growing in different directions finally make a plant that deterministically can be predicted as a tree. Random events are finally summarized in the lives of ordinary modern humans. It may be useless to discuss deeply deterministic and nondeterministic backgrounds of the world in our manuscript. Nevertheless, we have produced preliminary contributions for that. We have made efficient instruments—alpha unpredictable point and alpha labeling. Many things have been determined in terms of Poincaré–Birkhoff theory of dynamical systems [32, 82, 87]. Alpha unpredictable orbits, functions, and sequences are introduced as differential and difference equation solutions. We have shown that alpha unpredictable sequences can be infinite realizations for random processes with finite state space. However, alpha unpredictable orbits are trajectories of ultra Poincaré chaos as well as Devaney and Li–Yorke chaos [4]. Thus, we strongly argue that chaos is ubiquitous considering probabilistic processes. The results of the research generate challenging problems. First, we expect that the alpha unpredictability of stochastic processes with a finite space of states will be extended for many, if not all, random processes. The dynamics technique on alpha labels can also be applied to recognize chaos in various ordinary, partial, functional-differential, and discrete equations. The challenging problem is rigorously recognizing chaos in famous models such as the Lorenz system, which is still far from the complete solution.

Searching differential equations, first of all autonomous, with alpha unpredictable or/and chaotic solutions is another challenging problem. Besides, the question appears about how to model the real-world dynamics by differential equations having chaotic solutions. Generally speaking, comprehension of ubiquitous chaos can generate strategic changes in science progress. In particular, it is already seen in social processes: 'History cannot be explained deterministically, and it can not be predicted because it is chaotic. So many forces are at work and their interactions are so complex that tiny variations in the strength of the forces and the way they interact produce huge differences in outcomes' [55]. One can see that the author understands very clearly that social processes are random. We can add that when parameterized and modeled properly, global and local history can also be analyzed in terms of alpha unpredictability and alpha labeling.

Nevertheless, Harari overestimated historical processes considering them as chaotic on the second level '... history is what is called a "level two" chaotic system. Chaotic systems come in two shapes. Level one chaos is chaos that does not react to predictions about it. The weather, for example, is a level-one chaotic system. Though it is influenced by myriad factors, we can build computer models that take more and more of them into consideration and produce better and better weather forecasts. Level two chaos is the chaos that reacts to predictions about it, and therefore can never be predicted accurately. Markets, for example, are a level two chaotic system. What will happen if we develop a computer program that forecasts with percent accuracy the $100 price of oil tomorrow? The price of oil will immediately react to the forecast, which would consequently fail to materialize. If the current price of oil is $90 a barrel, and the infallible computer program predicts that tomorrow it will be $100, traders will rush to buy oil so that they can profit from the predicted price rise. As a result, the price will shoot up to $100 a barrel today rather than tomorrow. Then what will happen tomorrow? Nobody knows. Politics, too, is a second-order chaotic system. Many people criticize Sovietologists for failing to predict the 1989 revolutions and castigate Middle East experts for not anticipating the Arab Spring revolutions of 2011. This is unfair. Revolutions are, by definition, unpredictable. A predictable revolution never erupts' [55].

It is our firm belief that there is no different level for social or historical motions, and the level is only one common for all types of motion because what is suggested as prediction by Harari is just one of 'myriad factors.' That is, the opinion of the author that a prediction and power to realize information are external circumstances is not correct. They carry equal weight in determining historical processes and are endogenous to chaos. In other words, the exceptional (second) type of chaos is a redundant concept, once intellectual activity is treated as a 'coordinate' of process analogous and equal in dynamical sense to all other 'coordinates.' This is, what willingly would be accepted and supported by specialists in neuroscience. Of course, our discussion of what Harari wrote does not reduce the importance of his observations; it just makes chaos more adequate for mathematical research.

For many of us chaos is, first of all, about what is visible, presumably by snapshots of wind turbulence, waterfalls, etc, but chaos's presence has to be considered theoretically, that is, through knowledge of differential equations, mechanics, physics, biology, economics, engineering sciences, and sociology. However, more than knowledge is required. Any realistic motion should be considered globally in terms of time and space. For example, suppose one accepts that the periodicity of a car's wheel dynamics means the absence of chaos. In that case, we object to the motion necessarily stopping at a moment, proving that the motion is not periodic. We claim that the death or birth of a man is an appearance of chaos, not only because they are random events but because they are antitheses of 'regular motion' as a whole.

As it was said by P Valéry, 'A phenomenon that in one world is quite possible, in another—an incredible accident' [95]. That is, the life of an individual seems accidental, 'In any case, the life of an individual is a chain of accidents and more or less accurate answers to random events' [95]. Oppositely, Ananke, the Goddess of Fate, sees it as a predictable chain of necessities. Here, we see a dualism of chaotic and non-chaotic. Any event in the Universe can be considered in both appearances;

it depends on how you look at it, from which world, at which interval of time, and even from points of purpose.

One more citation from Harari is, 'So why study History? Unlike physics or economics, History is not a means for making accurate predictions. We study History not to know the future but to widen our horizons, to understand that our present situation is neither natural nor inevitable and that we consequently have many more possibilities before us than we imagine. For example, studying how Europeans came to dominate Africans enables us to realize that there is nothing natural or inevitable about the racial hierarchy and that the world might well be arranged differently' [55]. A revolution can not be predicted, but it can be avoided. *That is, chaotic dynamics is not predictable but controllable.* As it was said by Immanuel Kant about control, 'Nature even in chaos cannot proceed otherwise than regularly and according to order' [29].

Chaos research by alpha labeling and alpha unpredictability is effective for multidimensional spaces, infinite and finite-dimensional. Indeed, if alpha labeling is realized for a dynamics such that separation and diameter conditions are valid—be sure that ultra Poincaré, Li–Yorke, and Devaney chaos are valid. We believe the method is universal since alpha dynamics can be performed for domains with arbitrary dimensions.

It is curious to see how chaos is considered in different sciences: history, biology, physics, and chemistry. Let us focus on the first: 'One can not predict History, since she is unpredictable. Too many forces are interacting simultaneously' [55]. Considering an engine, we can arrange a stable periodic motion in it. However, are we sure that the engine has no unpredictable parameters? For example, what are a detail's temperature, pressure, and mechanical or chemical characteristics? In other words, when considering an event, we must understand that there are many, if not infinite, coordinates (parameters) related to every single physical event, similar to that in history, and chaos necessarily appears in some of them. From this point of view, we can say that any physical, chemical, or biological process admits chaotic 'coordinates indeed.' Therefore, chaos is present everywhere. Our discussion concluded that chaos is present everywhere, and our suggestions—the abstract similarity dynamics, alpha labeling, and alpha unpredictability—will help to realize this idea effectively in modern science.

1.18 Domain of chaos

Here, we are inclined to dispel the suggestion that sets where chaos happens should also be sophisticated. This is true for the present research, since analysis through alpha labeling unpredictability makes the sets for chaos not only complex, but also primitive. This is close to the ideas of Nicolis and Prigojine [75].

Let us consider the geometry of domains for different types of chaos. That is, sets of space points where chaotic scenarios happen, and whose nature is usually intuitively clear, but theoretically not often specified. To describe the set for Devaney chaos one can use the trajectory, which is dense for chaotic points, and its closure is considered as the domain of chaos. Otherwise, unite all points, which are initial for sensitivity property, then add the trajectory, which is dense in the set to construct the domain. One can also use the third description of the domain as a union of all unstable cycles of the chaos, and take closure of the set. Similarly, the set for Li–Yorke definition consists of the scrambled set united with all cycles of the chaos. It is helpful to remember that a domain can be a fractal for conservative chaos.

Possibly, the most simple description of chaotic domain is for the ultra Poincaré chaos. It is a quasi-minimal set. That is the closure of an alpha unpredictable trajectory. The alpha chaos implies ultra Poincaré, Devaney, and Li–Yorke chaos. Hence, all the above descriptions are suitable for describing the alpha chaos domain. What is most important, the chaos is in the privilege state, since we determine a domain for alpha chaos as the initial step of analysis. That is, we accept as a domain for the chaos the set, which can be alpha labeled. Next, we check if the fundamental diameter and separation conditions are valid. Thus, we have obtained the exceptionally simple algorithm of chaos construction. This implies that we are in circumstances when deep research can be done and unexpected results can be obtained. The alpha chaos can be arranged in any finite-dimensional cube from one side and even in a set of two points. One should be clear that the domain of chaos in the cube is a set dense in it. But the cube itself is a domain for chaos. Since it is the closure of a chaotic trajectory, in this case, we agree that the dimension of alpha chaos in the cube is an integer. That is, the dimension of closure for chaotic domain can be an integer and equal to the dimension of dynamics. Let us remind the reader that the cube is not a fractal, though fractals are considered as very common domains for many chaotic models. The fractals emerge because of the chaos ingredients and continuous maps. It is questionable, if one can prove that the cube is a domain for chaos, without alpha labeling. The map makes the possibility to consider alpha chaos in stochastic processes. We claim that one can organize the chaotic dynamics in any domain, if it is of not less than two distanced points.

References

[1] Addabbo T, Fort A, Rocchi S and Vignoli V 2011 Digitized chaos for pseudo-random number generation in cryptography *Chaos-Based Cryptography: Theory, algorithms and Applications* ed L Kocarev and S Lian Chaos-Based (Berlin, Heidelberg: Springer) pp 67–97

[2] Akhmet M 2019 Domain-structured chaos in discrete random processes ArXiv e-prints

[3] Akhmet M 2020 *Almost Periodicity, Chaos and Asymptotic Equivalence* (Berlin: Springer)

[4] Akhmet M 2021 *Domain Structured Dynamics* (Bristol: IOP Publishing)

[5] Akhmet M 2022 Abstract hyperbolic chaos *Discontinuity Nonlinearity Complexity* **11** 133–8

[6] Akhmet M 2022 Modular chaos for random processes *Discontinuity Nonlinearity Complexity* **11** 191–201

[7] Akhmet M 2022 Unpredictablity in Markov chains *Carpathian J. Math.* **38** 13–9

[8] Akhmet M and Alejaily E M 2019 Generation of fractals as duffing equation orbits *Chaos* **29** 053113

[9] Akhmet M, Tleubergenova M and Zhamanshin A 2023 Unpredictable solutions of duffing type equations with Markov coefficients arXiv (2303.17336)

[10] Akhmet M, Tleubergenova M and Zhamanshin A 2023 Compartmental unpredictable functions *Mathematics* **11** 1069

[11] Akhmet M 2020 A novel deterministic chaos and discrete random processes *ACM Int. Conf. Proc. Series* pp 53–6

[12] Akhmet M and Alejaily E M 2021 Abstract similarity, chaos and fractals *Discrete Continuous Dyn. Syst. Ser.* B **26** 2479–97

[13] Akhmet M, Baskan K and Yesil C 2023 Markovian noise-induced delta synchronization approach for Hindmarsh-Rose model *Chaos Solitons Fractals* **185** 115155

[14] Akhmet M, Baskan K and Yesil C 2022 Delta synchronization of Poincaré chaos in gas discharge-semiconductor systems *Chaos* **32** 083137

[15] Akhmet M, Baskan K and Yesil C 2023 Revealing chaos synchronization below the threshold in coupled Mackey-glass systems *Math. Open Access* **11** 3197

[16] Akhmet M and Fen M O 2016 *Replication of Chaos in Neural Networks, Economics and Physics* (Berlin: Springer)

[17] Akhmet M and Fen M O 2016 Poincaré chaos and unpredictable functions *Commun. Nonlinear Sci. Numer. Simul.* **48** 85–94

[18] Akhmet M and Fen M O 2016 Unpredictable points and chaos *Commun. Nonlinear Sci. Numer. Simul.* **40** 1–5

[19] Akhmet M, Fen M O and Alejaily E M 2020 *Dynamics with Chaos and Fractals* (Berlin: Springer)

[20] Akhmet M, Fen M O and Tola A 2023 A numerical analysis of Poincaré chaos *Discontinuity Nonlinearity Complexity* **12** 183–95

[21] Akhmet M, Fen M O and Tola A 2022 Strange non-chaotic attractors with unpredictable trajectories *J. Vib. Test. Syst. Dyn.* **6** 317–27

[22] Akhmet M and Alejaily E M 2021 Abstract fractals *Discontinuity Nonlinearity Complexity* **10** 135–42

[23] Akhmet M, Seilova R D, Tleubergenova M and Zhamanshin A 2020 Shunting inhibitory cellular neural networks with strongly unpredictable oscillations *Commun. Nonlinear Sci. Numer. Simulat.* **89** 105287

[24] Akhmet M and Tola A 2020 Unpredictable strings *Kazakh Math. J.* **20** 16–22

[25] Akhmet M, Fen M O and Tola A 2019 The sequential test for chaos *arXiv*, (1904.09127)

[26] Akhmet M, Yesil C and Baskan K 2023 Synchronization of chaos in semiconductor gas discharge model with local mean energy approximation *Chaos Solitons Fractals* **167** 113035

[27] Akin E and Kolyada S 2003 Li-Yorke sensitivity *Nonlinearity* **16** 1421–33

[28] Al-Khwarizmi 820 (year of publication) On the calculation with hindu numerals.

[29] Arbib M A 1951 *Critique of Judgment* (New York: Hafner Publishing)

[30] Bandt C and Graf S 1992 Self-similar sets 7. A characterization of self-similar fractals with positive Hausdorff measure *Proceedings of the American Mathematical Society* 995–1001

[31] Barnsley M 1988 *Fractals Everywhere* (New York: Academic)

[32] Birkhoff G D 1991 *Dynamical Systems* (Providence, RI: Colloquium Publications)

[33] Bourbaki N 1950 *Eléments de mathématiques* (Paris: Hermann)

[34] Bowen R 1970 Markov partitions and minimal sets for axiom A diffeomorphisms *Am. J. Math.* **92** 907–18

[35] Bowen R 1970 Markov partitions for axiom a diffeomorphisms *Am. J. Math.* **92** 725–47

[36] Bowen R 1973 Symbolic dynamics for hyperbolic flows *Am. J. Math.* **95** 429–40

[37] Brott B 1998 The form of form: the fold and architecture *Archit. Theory Rev.* **3** 88–111

[38] Bunimovich L and Ya S 1980 Markov partitions for dispersed billiards *Commun. Math. Phys.* **78** 247–80

[39] Burkert W 1972 *Lore and Science in Ancient Pythagoreanism* (Cambridge, MA: Harvard University Press)

[40] Carathéodory C 1919 Über den wiederkehrsatz von Poincaré *Berl. Sitzungsber.* 580–4

[41] Chaitin G 1975 Randomness and mathematical proof *Sci. Am.* **232** 47–52

[42] Chen G and Huang Y 2011 *Chaotic Maps: Dynamics, Fractals and Rapid Fluctuations, Synthesis Lectures on Mathematics and Statistics* (Austin, TX: Morgan and Claypool Publishers)

[43] *The Founders of Western Thought—The Presocratics* ed J M Cooper (Hackett: Indianapolis)

[44] Crownover R M 1995 *Introduction to Fractals and Chaos* (Boston, MA: Jones and Bartlett)

[45] Devaney R L 1987 *An Introduction to Chaotic Dynamical Systems* (Menlo Park, CA: Addison-Wesley)

[46] Diamond P 1976 Chaotic behavior of systems of difference equations *Internat. J. Syst. Sci.* **7** 953–6

[47] Dohtani A 1992 Occurrence of chaos in higher dimensional discrete time systems *SIAM J. Appl. Math.* **52** 1707–21

[48] Edgar G A 1990 *Measure, Topology, and Fractal Geometry* (New York: Springer)

[49] Ercai C 1997 Chaos for the Sierpinski carpet *J. Stat. Phys.* **88** 979–84

[50] Falconer K J 1985 *The Geometry of Fractal Sets* (Cambridge: Cambridge University Press)

[51] Falconer K J 1995 Sub-self-similar sets *Trans. Am. Math. Soc.* **347** 3121–9

[52] Feigenbaum M J 1980 Universal behavior in nonlinear systems *Los Alamos Sci./Summer* **1** 4–27

[53] Gleick J 1987 *Chaos Making a New Science* (New York: Viking Books)

[54] Guckenheimer J and Holmes P J 1997 *Nonlinear Oscillations, Dynamical Systems and Bifurcations of Vector Fields* (New York: Springer)

[55] Harari Y N 2014 *Sapiens: A Brief History of Humankind* (London: Harvill Secker)

[56] Hata M 1985 On the structure of self-similar sets *Japan J. Appl. Math.* **2** 381–414

[57] Hilbert D 1932-1935 *Gesammelte Abhandlungen, 3 volumes* (Berlin: Springer)

[58] Hilmy H 1936 Sur les ensembles quasi-minimaux dans les systèmes dynamiques *IAnn Math.* **37** 899–907

[59] Hutchinson J 1981 Fractals and self-similarity *Indiana Univ. Math. J.* **30** 713–47

[60] Jorgensen P E T 2006 *Analysis and Probability: Wavelets, Signals, Fractals* vol **234** (New York: Springer) Graduate Texts in Mathematics

[61] Lasota A and Mackey M C 1985 *Probabilistic Properties of Deterministic Systems* (Cambridge: Cambridge University Press)

[62] Lau K S, Ngai S M and Rao H 2001 Iterated function systems with overlaps and the self-similar measures *J. London Math. Soc.* **63** 99–115

[63] Li R and Zhou X 2013 A note on chaos in product maps *Turk. J. Math.* **37** 665–75

[64] Li T Y and Yorke J A 1975 Period three implies chaos *Am. Math. Mon.* **82** 985–92

[65] Lorenz E N 1963 Deterministic non-periodic flows *J. Atmos. Sci.* **20** 130–41

[66] Mandelbrot B B 1975 *Les Objets Fractals: Forme, Hasard, et Dimension* (Paris: Flammarion)

[67] Mandelbrot B B 1977 *Fractals: Form, Chance, and Dimension* (San Francisco, CA: W. H. Freeman)

[68] Mandelbrot B B 1983 *The Fractal Geometry of Nature* (New York: Freeman)

[69] Mandelbrot B B 2004 *Fractals and Chaos: The Mandelbrot Set and Beyond* (New York: Springer)

[70] Marotto F R 1978 Snap-back repellers imply chaos in R^n *Math. Anal. Appl.* **63** 199–223

[71] Moon F C 1992 *Chaotic and Fractal Dynamics: An Introduction for Applied Scientists and Engineers* (New York: Wiley)

[72] Moran P A P 1946 Additive functions of intervals and Hausdorff measure *Proc. Cambridge Philosophy* pp 15–23 Soc. 42

[73] Nemytskii V V and Stepanov V V 1960 *Qualitative Theory of Differential equations* (Princeton, NJ: Princeton University Press)

[74] Ngai S M and Wang Y 2001 Hausdorff dimension of overlapping self-similar sets *J. London Math. Soc.* **63** 665–72

[75] Nicolis G and Prigojin I 1989 *Exploring Complexity* (New York: Freeman)

[76] Ornstein D 1970 Bernoulli shifts with the same entropy are isomorphic *Adv. Math.* **4** 337–52

[77] Kahane J P 1985 Sur le chaos multiplicatif *Annales des sciences mathématiques du Québec* **9** 105–50

[78] Kolyada S 2004 Li-yorke sensitivity and other concepts of chaos *Ukrainian Math. J.* **56** 1242–57

[79] Pesin Y 1997 *Dimension Theory in Dynamical Systems: Contemporary Views and Applications* (Chicago, IL: University of Chicago Press)

[80] Pesin Y and Weiss H 1996 On the dimension of deterministic and random cantor-like sets, symbolic dynamics, and the Eckmann–Ruelle conjecture *Commun. Math. Phys.* **182** 105–53

[81] Poincaré H 1890 Sur le probléme des trois corps et les équations de la dynamique *Acta Math.* **13** 1–270

[82] Poincaré H 1957 *New Methods of Celestial Mechanics Volumes I–III* (New York: Dover)

[83] Poincaré H 2001 *The Value of Science (New York: Penquin Random House)* (New York: Penquin Random House)

[84] Rovelli C and Vidotto F 2014 *Covariant Loop Quantum Gravity: An Elementary Introduction to Quantum Gravity and Spinfoam Theory* (Cambridge: Cambridge University Press)

[85] Russel B 1972 *A History of Western Philosophy* (New York: Simon & Schuster)

[86] Schuster H G and Just W 2005 *Deterministic Chaos: An Introduction* (Weinheim: Wiley-VCH)

[87] Sell G R 1971 Topological dynamics and ordinary *Differential Equations* (London: Van Nostrand Reinhold Company)

[88] Smale S 1966 Diffeomorphisms with many periodic points *Differential and Combinatorial Topology: A Symposium in Honor of Marston Morse* (Princeton, NJ: Princeton University Press) pp 63–70

[89] Smale S 1967 Differentiable dynamical systems *Bull. Am. Math. Soc.* **73** 747–817

[90] Spear D W 1992 Measure and self-similarity *Adv. Math.* **91** 143–57

[91] Stella S 1992 On Hausdorff dimension of recurrent net fractals *Proc. Am. Math.* 389–400 Soc. 116

[92] Swift J 2005 *Gulliver's Travels* (Oxford: Oxford University Press)

[93] Ueda Y 1979 Randomly transitional phenomena in the system governed by Duffing's equation *J. Stat. Phys.* **20** 181–96

[94] Ueda Y and Abraham R 2000 *The Chaos Avant-Garde: Memories of the Early Days of Chaos Theory* (Singapore: World Scientific)

[95] Valery P 1975 *Collected Works of Paul Valéry* (Princeton, NJ: Princeton University Press)

[96] Van der Waerden B L 1930 *Moderne Algebra, 2 Volumes* (Berlin: Springer)

[97] Wiggins S 1988 Global bifurcation and chaos *Analytical Methods* (New York: Springer)

[98] Winther R G 2016 *The Structure of Scientific Theories* (Stanford, CA: Stanford University) (The Stanford Encyclopedia of Philosophy, Stanford University)

[99] Zaslavsky G M, Edelman M and Niyazov B A 1997 Self-similarity, renormalization, and phase space nonuniformity of Hamiltonian chaotic dynamics *Chaos* **7** 159–81

[100] Zmeskal O, Dzik P and Vesely M 2013 Entropy of fractal systems *Comput. Math. Appl.* **66** 135–46

Part II

Alpha labeling and unpredictability

Chapter 2

Alpha labels are a new mathematical structure

The chapter serves as one of the central components of the book. It begins by explaining the concept of alpha sets and alpha maps. The labeling process is defined within a metric space, resulting in what is referred to as alpha space. Alpha labeling is then applied to yield ultra Poincaré chaos, which derives from the idea of alpha unpredictability. This method is extending to hyperbolic and unbounded dynamics. Lastly, the chapter introduces a definition of abstract fractals using measures in metric spaces. The concepts discussed in this chapter are relevant and applicable to many other sections of the book.

This chapter is one of the book's two cores. It starts by describing what we understand as alpha sets (abstract similarity sets) and alpha maps. The labeling procedure is for a metric space; the resulting image is named alpha space. The explanation provided clarifies why it may not be classified as a metrical system. The alpha labeling is applied to dynamical systems to generate their ultra Poincaré chaos, which is rooted in the idea of alpha unpredictability. This chaos is universal and can be readily generalized to hyperbolic and unbounded dynamics, as demonstrated in this chapter. Additionally, we introduce a new definition of abstract fractals—termed alpha fractals—by applying measures within the metric spaces under investigation. This concept may prove advantageous in addressing engineering challenges related to porous materials [85], turbulence [74], and studies of stochastic processes. The material covered in this chapter will be relevant throughout various sections of the book.

2.1 Sets of alpha labels and alpha map

In this section of the chapter, we will establish the foundations for the applications of alpha dynamics, specifically focusing on abstract similarity sets. These sets act as *images* for the labeling operator. When applied to a dynamical system, we refer to the state space of the dynamics being studied as the *pre-image* of the alpha set.

Consequently, the state space is *structured* through the application of alpha labeling. This approach allows for the assignment of multiple, and even infinite, images to a single element, providing flexibility in labeling for various applications. This assumption enables the pre-image set to contain an uncountable, countable, or even a finite number of elements. For instance, it is particularly useful to investigate random processes with finite state spaces.

2.1.1 Alpha sets and rectangles

Introduce the following *alpha set*,

$$\mathscr{F} = \{\mathscr{F}_{i_1 i_2 \ldots i_n \ldots} : i_k = 1, 2, \ldots, m, \ k = 1, 2, \ldots, \}, \tag{2.1}$$

where m is a natural number. The members of $\mathscr{F}$ are indexed symbols and they are called *alpha labels* in what follows.

To get continuity on the initial value in the research and alpha-unpredictability with sensitivity, we shall need the following sets of alpha labels, *n-dimensional rectangles*,

$$\mathscr{F}_{i_1 i_2 \ldots i_n} = \bigcup_{j_k = 1, 2, \ldots, m} \mathscr{F}_{i_1 i_2 \ldots i_n j_1 j_2 \ldots}, \tag{2.2}$$

where indices $i_1, i_2, \ldots, i_n$, are fixed.

The rectangles satisfy the following relations,

$$\mathscr{F} \supseteq \mathscr{F}_{i_1} \supseteq \mathscr{F}_{i_1 i_2} \supseteq \cdots \supseteq \mathscr{F}_{i_1 i_2 \ldots i_n} \supseteq \mathscr{F}_{i_1 i_2 \ldots i_n i_{n+1}} \cdots, \ i_k = 1, 2, \ldots, m, \ k = 1, 2, \ldots,$$

for arbitrary choice of the indices $i_1 i_2 \ldots i_n i_{n+1} \ldots$.

2.1.2 Alpha maps and self-similarity

Next, we shall determine the map $\varphi: \mathscr{F} \to \mathscr{F}$ such that

$$\varphi(\mathscr{F}_{i_1 i_2 \ldots i_n \ldots}) = \mathscr{F}_{i_2 i_3 \ldots i_n \ldots}. \tag{2.3}$$

Iterating the map on an n-dimensional rectangle, obtain that

$$\varphi^n(\mathscr{F}_{i_1 i_2 \ldots i_n}) = \mathscr{F} \tag{2.4}$$

for arbitrary natural number n and $i_k = 1, 2, \ldots, m, \ k = 1, 2, \ldots, n$a. The definition (2.3) and the equality (2.4) make it reasonable to call φ *abstract similarity map* or *similarity map*. We shall also call it *alpha map*, and the number n is *order of similarity*. Correspondingly, the *alpha set* is called *abstract similarity set*.

One needs a metric for the couple $(\mathscr{F}, \varphi)$ to define alpha dynamics. In the following subsection, we shall show how a distance function can be inherited from the dynamics under investigation. We assume that the metric δ is assigned for the alpha set such that the triple $(\mathscr{F}, \varphi, \delta)$ is a dynamical system. Also, be aware that the distance function δ cannot necessarily be a metric since alpha labeling may not be one-to-one. We shall discuss the possibility further, later in the book.

2.2 Alpha dynamics

We continue this chapter with the book's most abstract construction: alpha dynamics. Formally, the concept reminds us of symbolic dynamics but is much more general in application. The approach will also be developed to strengthen the results obtained by the Markov partition technique.

2.2.1 Algorithm of alpha labeling for dynamical systems

In what follows, we shall show the mechanism of alpha labeling and, correspondingly, see how the alpha distance δ for dynamics $(\mathscr{F}, \varphi, \delta)$ can be designed. the abstract couple $(\mathscr{F}, \varphi)$ can be accompanied with an element of a dynamical model, namely the distance function, such that the abstract object became ready for the research of processes. In the present paper, our investigation is focused on chaos. The construction of abstract similarity dynamics (alpha dynamics) is strongly joined with the procedure of alpha labeling itself.

Consider a metric space (F, d) where d is the distance. It is assumed that set F can be finite, countable or uncountable. We shall say that the abstract similarity set $\mathscr{F}$ is an *alpha label set* for F, if each element of F is labeled by at least one member of $\mathscr{F}$, such that every element in $\mathscr{F}$ serves as a label for F. It is important that if $\mathscr{F}_{i_1 i_2 \ldots i_n \ldots}$ and $\mathscr{F}_{j_1 j_2 \ldots j_n \ldots}$ are alpha labels for an element f, the relations $i_n = j_n$ may not be correct for all $n = 1, 2, \ldots$. In what follows, $\mathscr{L}: F \to \mathscr{F}$ denotes an *alpha operator* such that $\mathscr{F}_{i_1 i_2 \ldots i_n \ldots} = \mathscr{L}(f)$ if $\mathscr{F}_{i_1 i_2 \ldots i_n \ldots}$ is a label for f. We shall call it the *label operator* also.

Definition 2.2.1. *A non-negative function* $\delta: \mathscr{F} \times \mathscr{F} \to \mathbb{R}$, *is said to be alpha distance in the alpha set* $\mathscr{F}$ *of the set F, if* $\delta(\mathscr{F}_{i_1 i_2 \ldots i_n \ldots}, \mathscr{F}_{j_1 j_2 \ldots j_n \ldots}) = d(f_1, f_2)$, *where* $\mathscr{F}_{i_1 i_2 \ldots i_n \ldots}$ *and* $\mathscr{F}_{j_1 j_2 \ldots j_n \ldots}$ *are alpha labels of elements* f_1 *and* f_2, *such that* $\delta(\mathscr{F}_{i_1 i_2 \ldots i_n \ldots}, \mathscr{F}_{j_1 j_2 \ldots j_n \ldots}) = 0$ *for distinct alpha labels of the same point in F.*

One can remark that the alpha distance is a metric if the label operator is one-to-one, and then the couple $(\mathscr{F}, \delta)$ is a metric space. Now, we introduce *alpha dynamics* or *abstract similarity dynamics* as a triple $(\mathscr{F}, \varphi, \delta)$, where $\mathscr{F}$, δ, and φ are an alpha set, distance, and map, respectively.

Let the metric space (F, d) be a domain for a map ψ such that (F, ψ, d) is a dynamical system [66, 75].

Definition 2.2.2. *The maps* ψ *and* φ *agree, if for each element f of F, it is true that alpha labels of f and* $\psi(f)$ *are* $\mathscr{F}_{i_1 i_2 \ldots i_n \ldots}$ *and* $\mathscr{F}_{i_2 \ldots i_n \ldots}$, *respectively. That is,* $\mathscr{L}(\psi(f)) = \varphi(\mathscr{L}(f))$.

The last equality will be called *image condition* in what follows. The alpha labeling is not an equivalence for the two dynamics, since there is no one-to-one

$$\mathcal{F}_{i_1 i_2 \ldots} \xrightarrow{\quad \varphi \quad} \mathcal{F}_{i_2 i_3 \ldots}$$

$$\mathcal{L} \uparrow \qquad\qquad\qquad \uparrow \mathcal{L}$$

$$f \xrightarrow[\psi]{\qquad\qquad} g$$

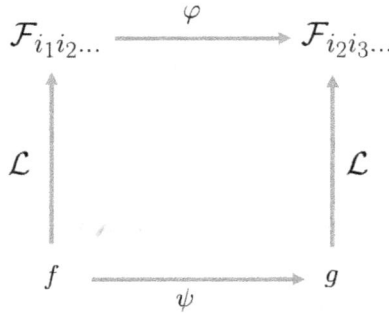

Figure 2.1. Diagram illustrates the image condition for alpha labeling.

correspondence, in general. It is isometric if the uniqueness is correct. The image condition is illustrated by the diagram in figure 2.1.

Definition 2.2.3. *The triple* $(\mathcal{F}, \varphi, \delta)$ *is an alpha dynamics for the dynamical system* (F, ψ, d) *if the maps* ψ *and* φ *agree.*

The triple $(\mathcal{F}, \varphi, \delta)$ is referred to as the *dynamics on alpha labels* or *abstract similarity dynamics* induced by the dynamical system (F, ψ, d). It is also known as *domain structured dynamics* or simply *alpha dynamics*. While we may refer to the triple $(\mathcal{F}, \varphi, \delta)$ as a dynamics, it should not be classified as a dynamical system because the operator $\mathcal{L}$ may not be a bijection, and the alpha distance $\delta: \mathcal{F} \times \mathcal{F} \to \mathbb{R}$ may not satisfy the conditions of a metric. Specifically, the requirement for indiscernible identity, a fundamental aspect of metric definitions, may not hold true. Consequently, $(\mathcal{F}, \delta)$ may not qualify as a metric space.

When discussing this concept in isolation, we refer to it as *alpha space*, and when examining it in the context of a dynamical system, we designate it as *alpha label space*. The function δ that is applied to elements of $\mathcal{F}$ is termed the *alpha distance*. This framework allows for the development of new topological concepts, such as alpha convergence, as well as alpha complete and compact spaces, all of which are grounded in the alpha distance.

Moving forward, we will utilize the fact that an alpha compact space is also alpha complete. If the alpha labeling operator $\mathcal{L}$ is both one-to-one and an isometric map, then the alpha space qualifies as a metric space where all alpha-related properties can be considered as metrical properties.

Finally, we describe how to apply alpha labeling in research of complexity for problems, which are non-dynamical originally. For example, it can be the task of chaos arrangement in fractals. So, consider a metric space (F, d), which is not assigned with a mapping. Suppose that one can find an alpha label set $\mathcal{F}$ for set F and, correspondingly alpha space $(\mathcal{F}, \delta)$ with the similarity map φ such that the triple $(\mathcal{F}, \varphi, \delta)$ admits certain dynamical properties. For instance, the separation and diameter conditions, which are determined below. That is, the abstract dynamics are chaotic. Then, by introducing the map ψ in space (F, d), which agrees with φ, one obtains that the dynamics (F, ψ, d) is chaotic, too.

The following feature for alpha spaces, combined with the action of the similarity map, is equivalent to what is said to be a continuous dependence on the initial value for dynamics determined by differential or difference equations.

Definition 2.2.4. *It will be said that an alpha space $(\mathscr{F}, \delta)$ satisfies diameter condition if for all rectangles $\mathscr{F}_{i_1 i_2 \dots i_n}$ it is valid that*

$$\sup_{i_k = 1, 2, \dots, n} \operatorname{diam} \mathscr{F}_{i_1 i_2 \dots i_n} \to 0 \ \text{ as } \ n \to \infty, \tag{2.5}$$

where $\operatorname{diam}(A) = \sup\{\delta(\mathbf{x}, \mathbf{y})\colon \mathbf{x}, \mathbf{y} \in A\}$ *for a set A in $\mathscr{F}$.*

From the diameter condition, it implies that for arbitrary positive ε, there exists a finite ε-net, and consequently, an alpha complete label space is alpha compact. Finally, let us remark that the diameter condition implies the existence of finite ε-nets with arbitrary positive ε, and consequently the space is alpha bounded. The topological properties of the alpha space will be utilized for chaotic dynamics in what follows.

The following example strongly suggests that the alpha dynamics technique is at least as powerful as symbolic dynamics. Consequently, all the results obtained by the last one can be repeated by using alpha labeling, but not vice versa. That is, through our suggestions, we can obtain new results since their simplicity is valid for alpha dynamics applications.

Example 2.2.1. [21] *Remember the space of infinite symbolic strings of several symbols* [40, 87]. *We consider the simplest case of two numbers 0 and 1 strings, which are defined as*

$$\Sigma = \{s_1 s_2 s_3 \dots \colon s_k = 0 \text{ or } 1].$$

The distance in Σ is defined by

$$d(s, t) = \sum_{k=1}^{\infty} \frac{|s_k - t_k|}{2^{k-1}}, \tag{2.6}$$

if $s = s_1 s_2 \dots$ and $t = t_1 t_2 \dots$ are two elements of Σ.

Let us denote members of the space as $\mathscr{F}_{s_1 s_2 \dots} = \mathscr{L}(s_1 s_2 \dots)$, and describe the nth order rectangles in $\mathscr{S}$ by

$$\mathscr{F}_{s_1 s_2 \dots s_n} = \{\mathscr{L}(s_1 s_2 \dots s_n s_{n+1} s_{n+2} \dots)\colon s_k = 0 \text{ or } 1]),$$

where $s_1, s_2, \dots, s_n$ are fixed symbols. One can show that distance between any two alpha labels of $\mathscr{F}_{s_1 s_2 \dots s_n}$ is less than $\frac{1}{2^{n-1}}$. Moreover,

$$\delta(\mathscr{F}_{s_1 s_2 \dots s_n 000\dots}, \mathscr{F}_{s_1 s_2 \dots s_n 111\dots}) = d(s_1 s_2 \dots s_n 000 \dots , s_1 s_2 \dots s_n 111 \dots) = \frac{1}{2^{n-1}}.$$

That is, $\mathrm{diam}(\mathscr{F}_{s_1 s_2 \dots s_n}) = \frac{1}{2^{n-1}}$. *Consequently,*

$$\lim_{n\to\infty} \mathrm{diam}\, \mathscr{F}_{s_1 s_2 \dots s_n} = 0.$$

One can easily see that the similarity map for the set of alpha labels agrees with the Bernoulli shift, $\sigma(s_1 s_2 s_3 \dots) = s_2 s_3 s_4 \dots$. *That is,*

$$\varphi(\mathscr{F}_{s_1 s_2 s_3 \dots}) = \mathscr{L}(\sigma(s_1 s_2 s_3 \dots)).$$

Thus, $(\mathscr{F}, \varphi, \delta)$ *is an abstract similarity dynamics for the symbolic dynamical system* (Σ, σ, d).

Remark 2.2.1. *Operator $\mathscr{L}$ in the last example is one-to-one and an isometric transformation. Moreover, the symbolic dynamics is compact [87]. This is why the couple $(\mathscr{F}, \delta)$ is a compact metric space. Moreover, it is perfect [66, 87]. This example is not purely illustrative, since research of dynamics on the basis of alpha labeling is much easier, and consequently more productive, than that with symbolic dynamics, and the existence of alpha dynamics demonstrates that former results can be reconsidered on the new fundamentals and fresh opportunities will appear. In the last example, we discuss the space of sequences with two symbols since it is the most essential instrument in the analysis. However, it is easier to consider alpha labeling with any symbolic dynamics and their applications.*

The present approach to chaos investigation emphasizes the foundational role of the domain over the chaotic map [22]. Traditionally, chaos description starts with a map that possesses certain properties, like unimodality, hyperbolicity, period three, and topological conjugacy to a standard chaotic map such as the Bernoulli shift. For instance, in Li–Yorke chaos, one first defines a map that exhibits period three, and then shapes the chaos domain (scrambled set) accordingly. Similarly, for unimodal maps, the chaos domain is typically a Cantor set. In the case of the Smale horseshoe, both the map and the domain are determined together, with the domain being constructed step by step starting from an initial set influenced by the map. Thus, the structure of the domain and the nature of the map are interdependent. In symbolic dynamics, the domain is primarily described as infinite sequences of symbols, while the map is introduced as a shift within this space. Nonetheless, the map retains priority since the properties of the sequences are defined relative to the map. Differing from this, our approach begins with constructing an alpha set in a metric space, designed to be conducive to chaotic behaviors. Subsequently, an alpha map is developed based on the invariance and self-similarity properties of the domain to depict an abstract motion. This ensures the map is suitable for the abstract alpha set and other geometries with self-similarity, facilitating the proof of chaos for these sets. Moreover, the continuity of the motion is not a requisite since a chaotic map does not necessarily need to be continuous [1, 57]. For example, paper [57] discusses chaos conditions on the product of semi-flows without assuming the continuity of certain chaotic maps. While continuity of a chaotic map is beneficial analytically for proving chaos presence [87], it must be rigorously correct to treat it as inherent to chaos.

2.2.2 How to apply alpha dynamics

The abstract similarity dynamics method promises to become a general tool of theoretical analysis and applications. It is another, higher level of abstraction for sophisticated dynamics research. For the moment we can see the following three ways such that alpha dynamics can be applied. They depend on how the label operator relates to sets, where dynamics happen.

- A metric space (F, d) is provided and there is no a map given for the space. Then one can insert abstract similarity dynamics in the space by alpha labeling. Thus, an *alpha induced dynamical system* can be constructed.
- For a dynamical system (F, ψ, d), there exists an alpha space $(\mathscr{F}, \delta)$ such that the alpha map φ agrees with the map ψ.
- For a random dynamics with a state space F one can find an alpha dynamics which trajectories mirrors all or some realizations of the stochastic process. This is also a sort of alpha induced dynamics, but is realized on individual level.

Consider the three modes in more a detailed way. The first, when a metric space (F, d) is not given a map. Then, by considering the dynamics $(\mathscr{F}, \varphi, \delta)$, chosen properly for the metric space, we establish a desired behavior. For example, chaos. In our papers, this has been done for fractals. We believe that the results are of significant interest since they demonstrate that fractals can be domains of chaos. Moreover, it is shown in our research that chaos can be arranged in finite-dimensional cubes and, consequently, for any topologically equivalent set. Another surprise is that chaos can be formulated for finite sets, and even two-point sets. This creates nice opportunities to look for chaotic ingredients in stochastic processes. Generally speaking, the approach is a fundamental contribution to recognizing chaos as the world's most generic type of motion. It is very important to search chaotic dynamics in neural networks since it is needed for the effectiveness of brain activity and artificial intelligence.

One more way of applying alpha dynamics is for spaces that are already equipped with mappings or vector fields. That is, when a dynamics (F, ψ, d) is under analysis and can be formalized with an abstract similarity dynamics $(\mathscr{F}, \varphi, \delta)$ such that the last one reflects certain features of the original motion, and one can make a decision on their existence. In this case, the presentation means that each element of the set F admits an alpha label in $\mathscr{F}$. Generally, we do not assume that uniqueness is required for the relation. Moreover, we keep the distance d for the set $\mathscr{F}$ considering $\delta(\mathscr{F}_{i_1 i_2 \ldots i_n \ldots}, \mathscr{F}_{j_1 j_2 \ldots j_n \ldots}) = d(f_1, f_2)$, if $\mathscr{F}_{i_1 i_2 \ldots i_n \ldots}$ and $\mathscr{F}_{j_1 j_2 \ldots j_n \ldots}$ are alpha labels for elements f_1 and f_2 of the set F, respectively. This alpha chaos applies to the models in three modes. The first one relates to those which have not been proven with any chaos, and the alpha dynamics method is to be tried. Others which are shown one of the conventional types of chaos, but not all of them. Finally, we can consider models such that the alpha dynamics just confirms what is known already by conservative methods, but does this more efficiently using additional ultra Poincaré chaos presence.

2.2.3 Alpha dynamics versus symbolic dynamics

The symbolic dynamics was initiated in the famous paper [64] by Marston Morse and Gustav Hedlund. In many senses, the deep analysis of the obtained structures is a theoretical investigation that repeats those for continuous dynamics in [29]. The recurrence and transitivity are proven and specified very carefully since the particular properties of the symbolic spaces allow this. The theory of alpha dynamics can be developed similarly to the research of Morse and Hedlund, but on a new, much higher level of generalization. We hope that the significant research results in [64] will help to structure the chaos and support deeper understanding of the alpha unpredictable motion in future papers.

Next, we shall discuss what are principal novelties of alpha similarity dynamics.

The method of conventional abstract symbolic dynamics is beneficial for discrete-time systems with continuous state spaces. The research starts with *finite* partitioning of a state space and each part is assigned with its own label. Then, each point of the state space generates an infinite sequence of symbols: the label of the original part to which the start point belongs, the symbol of the part to which the first iteration belongs, and so forth. The original dynamics cannot be followed entirely in this way by the symbolic dynamics. One reason is that the dynamics can be drastically simplified. Let us introduce the algorithm of the simplification. If x is an element of the state space, and $f(x)$ the iteration, then we need a function of states to symbols, g, to create the symbol sequence $s = g(x)g(f(x))g(f^2(x))g(f^3(x))\ldots$. That is, $s_0 s_1 s_2 s_3 \ldots$. So, the dynamics is now reduced to shifting in sequences to the right. If function g is a homeomorphism or diffeomorphism, then a qualitative result for the symbolic dynamics can be easily reformulated for the original iterations of the map f. Thus, the method's main difficulty is finding the map and proving it is a homeomorphism. In alpha labeling, we do not consider the symbolic sequences at all. We replace the mapping with alpha labeling for *all* elements of the state space. One can say that partitioning has been extended ultimately, from the finite to infinite, and even *uncountable* partitioning. Accordingly, an abstract similarity map (alpha mapping) is introduced. Next, we request that the map of the original dynamical system under investigation agrees with the alpha map. At first glance, the task of such partitioning (alpha labeling) and choice of an abstract similarity map seems difficult, even impossible, but surprisingly, good results have been achieved in this direction. Moreover, we believe that the class of dynamics, which domain can be alpha labeled, is significantly large. One more important remark on alpha labeling isthat, since it concerns each point of the domain, one can say that the totality of alpha labeling reduces the study of dynamical systems to the analysis of alpha dynamics. At the same time, conservative symbolic methods consider only specific properties. To make the alpha method valuable for chaos investigation, we have formulated for the set of alpha labels the separation and diameter conditions, which guarantee Poincaré, Devaney, and Li–Yorke chaos for the dynamics. That is, alpha chaos means that the three types are present, and we believe that the method will provide in future much more interesting results. Ultimately speaking, all three sorts can be equivalent. The totality of alpha labeling means it is not necessary to map state spaces to the space of sequences, and, thus, the theoretical discussion becomes much more transparent and fruitful.

It is essential to clarify the principal differences between alpha and symbolic dynamics and the advantages of the former; these are emphasized below.

- The state space of symbolic dynamics consists of infinite sequences of symbols. A state space of alpha dynamics is a set of any nature, if its elements can be alpha labeled as we suggested above. We consider also multiple labeling such that even finite sets can be researched as a domain of models. Thus, finite, countable and uncountable sets can be considered for alpha dynamics, while their symbolic counterparts request uncountable domains, since homeomorphism/diffeomorphism is the equivalence condition.

- The results of symbolic dynamics, when applied to models with topological equivalence, offer unique insights. In contrast, alpha dynamics, with its direct labeling, provides a different perspective that doesn't require any additional conditions. This distinction can enlighten researchers in their understanding of dynamical systems.

- There is a perception that finding a generating partition and one-to-one correspondence between states and the symbol sequences makes research in the symbolic dynamics entirely equivalent to the study of the original dynamical system. This is not correct in the case of a finite partition. The coincidence of all dynamical properties is valid for the alpha dynamics since the total alpha labeling assumes that the image condition is valid. Alpha labeling is for each point of the domain, not only for a finite number of its regions. Another benefit of the present approach is that alpha labeling is not necessarily one-to-one. We have applied one-to-several and one-to-infinity alpha labels in the research. This supports essential developments of chaotic ingredients for random processes.

- Bernoulli shift is directed motion along a sequence of symbols. The abstract similarity map is the motion among elements of a space. There is no a special direction. That is, the alpha similarity map is a universal transformation, restricted only with possibility and choice of alpha labeling. Moreover, we can apply various and distinct alpha sets for the same space set, choosing them when convenient to solve a specific problem, or separately if needed.

- By considering abstract hyperbolic sets for alpha dynamics with bi-infinite indexes, we prove time two-directional chaos for Smale horseshoes and two-dimensional Baker transformations in an easier way than conservative methods do. It is not difficult to guess that alpha hyperbolic dynamics can be recognized in other cases such as Plykin attractor, Anosov diffeomorphism, etc.

- Alpha dynamics proves to be an effective tool for uncovering random processes' chaotic features. Our papers have introduced laws of large unpredictable strings for Bernoulli schemes, alpha-unpredictability for Markov chains, and processes with continuous time through modular chaos. This effectiveness should instill confidence in the utility of alpha dynamics for researchers in the field.

2.3 Alpha chaos

2.3.1 Basic conditions for alpha chaos

Next, we shall describe a feature for alpha dynamics, which is basically important for complexity in mathematics. We shall call it *separation condition*. It causes the alpha unpredictability of Poincaré chaos and sensitivity of Lorentz chaos, if they are results of alpha labeling.

Consider the distance between two nonempty bounded sets A and B in $\mathscr{F}$ as the function $\delta(A, B) = \inf\{\delta(\mathbf{x}, \mathbf{y})\colon \mathbf{x} \in A, \mathbf{y} \in B\}$.

Definition 2.3.1. *The alpha set $\mathscr{F}$ satisfies the separation condition of degree n, if there exist a positive number ε_0 and a natural n such that for arbitrary $i_1 i_2 \ldots i_n$ one can find indices $j_1 j_2 \ldots j_n$ such that*

$$\delta(\mathscr{F}_{i_1 i_2 \ldots i_n}, \mathscr{F}_{j_1 j_2 \ldots j_n}) > \varepsilon_0. \qquad (2.7)$$

The positive number ε_0 is called separation constant.

Remark 2.3.1. *The separation condition combined with the action of the similarity map is an abstraction of the Lorenz sensitivity. Thus, the diameter and separation conditions create chaos, which, for dynamical systems, is caused by continuity in the initial value and sensitivity. The conditions are only what we need to guarantee chaos on alpha dynamics. Nevertheless, it is of intense interest to search for other sufficient criteria. The conditions can be elaborated further, as well as new structures similar to alpha spaces introduced for complexity research.*

It is well accepted that the first appearance of chaos in dynamical systems was observed by H. Poincaré [72]. He introduced the concept of Poisson stability as a theoretical criterion for complexity in celestial mechanics. At that time, invention of new fundamental concepts to describe irregularities in the classical deterministic dynamics had been abandoned till the moment when the leading role was given to Lorenz sensitivity, which emerged in the numerical analysis of fluid dynamics. The concept became a rescue for those who want to be busy with sophisticated motions and not involved in complex theoretical analysis of Poincaré–Birkhoff theory [29]. Nevertheless, the direction initiated by E. Lorenz was significantly widened, as is usual in mathematics, since, based on sensitivity, new concepts of Li–Yorke and Devaney chaos were introduced and realized in numerous investigations. The concepts are now in the flow of exciting developments, making up the main body of chaos research. Moreover, numerical methods based on positive Lyapunov exponents and bifurcation diagrams have been involved effectively in the analysis. Our recent studies contain suggestions to come back to the tops of Poincaré–Birkhoff theory [29, 72], and continue the line of recurrent motions such that the Poisson stability coexists with the alpha-unpredictability. Thus, new recurrent motions have been created, keeping all Poisson features' stable dynamics. That is the quasi-minimal set with uncountable Poisson stable motions dense in the set [66, 75]. Alpha-unpredictability causes Lorenz sensitivity [5, 14], and one can claim that *alpha-unpredictability implies chaos.*

Definition 2.3.2. *A point $\mathscr{F}_{i_1 i_2 \ldots}$ from alpha set $\mathscr{F}$ is alpha-unpredictable, if there exist a positive number ε_0 and sequences t_m, s_m, of natural numbers, both of which diverge to infinity such that $\varphi^{t_m}(\mathscr{F}_{i_1 i_2 \ldots i_k \ldots}) \to \mathscr{F}_{i_1 i_2 \ldots i_k \ldots}$ as $m \to \infty$ and $\delta(\varphi^{t_m + s_m}(\mathscr{F}_{i_1 i_2 \ldots i_k \ldots}), \varphi^{s_m}(\mathscr{F}_{i_1 i_2 \ldots i_k \ldots})) \geqslant \varepsilon_0$ for each natural number m.*

We call the dynamics, which exists because of the alpha unpredictability point, *ultra Poincaré chaos.* The existence of infinitely many alpha unpredictable Poisson stable trajectories that lie in a compact set meets all requirements of chaos. Based on this, chaos can appear in the dynamics on a quasi-minimal set, which is the closure of a Poisson stable trajectory. Therefore, the ultra Poincaré chaos is referred to as the dynamics on the quasi-minimal set of trajectory initiated from the alpha unpredictable point.

2.3.2 Theorems by alpha dynamics

To validate the discussions above we suggest the following several assertions.

Theorem 2.3.1. [21] *If the diameter and separation conditions are valid, then alpha dynamics $(\mathscr{F}, \varphi, \delta)$ possesses alpha-unpredictable points. Moreover, if $(\mathscr{F}, \delta)$ is a alpha-complete space, then the alpha dynamics admits ultra Poincaré chaos.*

Proof. We start with shaping an alpha-unpredictable point, an easy task similar to the technique of symbolic dynamics, when transitivity is verified. Fix an element $\mathscr{F}_{i_1 i_2 \ldots i_k \ldots}$ of the alpha dynamics such that a sequence of indices $i_1 i_2 \ldots i_k \ldots$ is collected such that the first m symbols are ordered numbers 1 to m. Next, write subsequently all couples of the m numbers. Then place in the sequence all triples of the chosen numbers. Thus, the infinite sequence consists of subsequently placed in row all possible strings of length one, two, etc, to infinity. The sample of this construction is seen in example 2.2.5 for the two symbolic alphabet cases. It is easy to verify by the diameter condition that the point is Poisson stable. That is, there exists a sequence of integers, t_m, divergent to infinity such that $\varphi^{t_m}(\mathcal{F}_{i_1 i_2 \ldots i_k \ldots}) \to \mathcal{F}_{i_1 i_2 \ldots i_k \ldots}$ as $n \to \infty$. Let us look for the separation property now. We will show that there is a sequence s_m divergent to infinity such that $\delta(\varphi^{t_m + s_m}(\mathcal{F}_{i_1 i_2 \ldots i_k \ldots}), \varphi^{t_m}(\mathcal{F}_{i_1 i_2 \ldots i_k \ldots})) > \varepsilon_0$, where ε_0 is the separation constant for the space $(\mathcal{F}, \delta)$. Assume, on the contrary, that there is no such sequence. Then, $\delta(\varphi^{t_m + i}(\mathcal{F}_{i_1 i_2 \ldots i_k \ldots}), \varphi^{t_m}(\mathcal{F}_{i_1 i_2 \ldots i_k \ldots})) < \varepsilon_0$, for all natural i. This contradicts the diameter condition and the choice of the initial point, because it means absence of the initial string of length n in the sequence of indices for the point $\varphi^{t_m}(\mathcal{F}_{i_1 i_2 \ldots i_k \ldots})$. One can remark that the diameter property implies that the alpha-complete space is alpha-compact. Next, we will refer to the theorems in [14], which confirm that there is the alpha-quasi-minimal set for the dynamics as well as an uncountable set of dense alpha-unpredictable trajectories, and each of the orbits is dense in the quasi-minimal set. That is, there exists ultra Poincaré chaos. $\square$

Now, the following assertion is valid.

Theorem 2.3.2. [21] *If the diameter, separation and image conditions for the dynamics $(\mathcal{F}, \varphi, \delta)$ hold, then the dynamical system (F, ψ, d) possesses alpha-unpredictable points. Moreover, if the space (F, d) is complete, then the dynamical system admits ultra Poincaré chaos.*

In the theorem below, we prove that the similarity map φ possesses the ingredients of Devaney chaos, density of periodic points, transitivity, and sensitivity. A point $\mathcal{F}_{i_1 i_2 i_3 \ldots}$ of $\mathcal{F}$ is periodic with period n if its index consists of endless repetitions of a block of n terms

Theorem 2.3.3. [21] *If the diameter and separation conditions hold, then the abstract similarity dynamics $(\mathcal{F}, \varphi, \delta)$ is chaotic in the sense of Devaney.*

Proof. Fix a member $\mathcal{F}_{i_1 i_2 \ldots i_n \ldots}$ of $\mathcal{F}$ and a positive number ε. Find a natural number k such that $\operatorname{diam}(\mathcal{F}_{i_1 i_2 \ldots i_k}) < \varepsilon$ and choose a k-periodic element $\mathcal{F}_{i_1 i_2 \ldots i_k i_1 i_2 \ldots i_k \ldots}$ of the alpha rectangle $\mathcal{F}_{i_1 i_2 \ldots i_k}$. It is clear that the periodic point is an ε-approximation for the considered element of $\mathcal{F}$. The density of periodic points is thus proved.

Next, utilizing the diameter condition, the transitivity will be proved if we show the existence of an element $\mathcal{F}_{i_1 i_2 \ldots i_n \ldots}$ of $\mathcal{F}$ such that for any subset $\mathcal{F}_{i_1 i_2 \ldots i_k}$ there exists a sufficiently large integer p so that $\varphi^p(\mathcal{F}_{i_1 i_2 \ldots i_n \ldots}) \in \mathcal{F}_{i_1 i_2 \ldots i_k}$. This is true since we can construct the sequence $i_1 i_2 \ldots i_n \ldots$ such that it contains all sequences of the type $i_1 i_2 \ldots i_k$ as blocks. For sensitivity, fix a point $\mathcal{F}_{i_1 i_2 \ldots}$ in $\mathcal{F}$ and an arbitrary positive number ε. Due to the diameter condition, there exist an integer k and element

$\mathscr{F}_{i_1 i_2 \ldots i_k j_{k+1} j_{k+2} \ldots} \neq \mathscr{F}_{i_1 i_2 \ldots i_k i_{k+1} i_{k+2} \ldots}$ such that $\delta(\mathscr{F}_{i_1 i_2 \ldots i_k i_{k+1} \ldots}, \mathscr{F}_{i_1 i_2 \ldots i_k j_{k+1} j_{k+2} \ldots}) < \varepsilon$. We choose $j_{k+1}, j_{k+2}, \ldots$ such that $\delta(\mathscr{F}_{i_{k+1} i_{k+2} \ldots i_{k+n}}, \mathscr{F}_{j_{k+1} j_{k+2} \ldots j_{k+n}}) > \varepsilon_0$, by the separation condition. This proves sensitivity. $\square$

From the last assertion the following theorem easily is implied.

Theorem 2.3.4. [21] *If the diameter and separation conditions for the dynamics* $(\mathscr{F}, \varphi, \delta)$ *hold, then the dynamical system* (F, ψ, d) *is Devaney chaotic.*

In addition to Devaney and ultra Poincaré chaos, it can be shown that Li–Yorke chaos also takes place in the dynamics of the map φ. The proof of the following theorem is similar to that of theorem 6.35 in [38] for the shift map defined on the space of symbolic sequences. One can also see the arguments in paper [77]. Firstly, we shall introduce needed definitions.

Definition 2.3.3. *A couple* $(\mathscr{F}_{i_1 i_2 \ldots}, \mathscr{F}_{j_1 j_2 \ldots})$ *of elements from* $\mathscr{F}$ *is called proximal, if*

$$\liminf_{n \to \infty} \delta(\varphi^n(\mathscr{F}_{i_1 i_2 \ldots}), \varphi^n(\mathscr{F}_{j_1 j_2 \ldots})) = 0.$$

Definition 2.3.4. *A couple* $(\mathscr{F}_{i_1 i_2 \ldots}, \mathscr{F}_{j_1 j_2 \ldots})$ *of elements from* $\mathscr{F}$ *is frequently separated if*

$$\limsup_{n \to \infty} \delta(\varphi^n(\mathscr{F}_{i_1 i_2 \ldots}), \varphi^n(\mathscr{F}_{j_1 j_2 \ldots})) > 0.$$

Definition 2.3.5. *A subset of* $\mathscr{F}$ *is called scrambled if each of its couples is proximal and frequently separated.*

Definition 2.3.6. *The dynamics* $(\mathscr{F}, \varphi, \delta)$ *is Li–Yorke chaotic, if*
- *there exists an n-periodic orbit for φ with arbitrary natural period;*
- *there is an uncountable scrambled subset;*
- *each element of the scrambled set makes a frequently separated couple with any periodic point.*

Theorem 2.3.5. *The alpha dynamics* $(\mathscr{F}, \varphi, \delta)$ *is Li–Yorke chaotic if the diameter and separation conditions hold.*

The last assertion easily implies the next result.

Theorem 2.3.6. *If the diameter and separation conditions hold for* $(\mathscr{F}, \varphi, \delta)$ *and the image condition is valid, then* (F, ψ, d) *possesses Li–Yorke chaos.*

We shall call the chaos for dynamics $(\mathscr{F}, \phi, \delta)$ *abstract similarity chaos* or *alpha chaos*. Logically, the chaos discovered for (F, ψ, d) by the alpha labeling is said to be *domain structured chaos* or *chaos by alpha labeling*. In what follows, we shall call it *alpha chaos* as it originated from abstract similarity chaos. Thus, alpha chaos is suitable for both abstract similarity dynamics and dynamics by alpha labeling. It is a significant result for dynamical systems theory since it confirms that all three types, ultra Poincaré, Li–Yorke, and Devaney chaos, are present if a model is alpha chaotic.

Remark 2.3.2. *The relative simplicity of the abstract similarity dynamics $(\mathscr{F}, \varphi, \delta)$ implies that for the existence of Li–Yorke or Devaney chaos, there is no need for alpha-compactness of the alpha space. In contrast, ultra Poincaré chaos discussion demands the condition for the quasi minimal set, and consequently, chaos existence [29, 66, 75]. So, this can be seen in the formulation of corresponding theorems. One can develop chaos theory for alpha dynamics without distance from a dynamic system. In this case, we are obliged to introduce a function of alpha-distance with all metric properties except that the distance can be equal to zero, even for distinct alpha labels. This way of describing the dynamics can possibly be done in the future as a new organization of the theory.*

Let us provide an initial definition of alpha dynamics, which is not induced by any dynamical system. Assume that an alpha set $\mathscr{F}$ and similarity map φ on the set are given.

Definition 2.3.7. *A function $\delta\colon \mathscr{F} \times \mathscr{F} \to \mathbf{R}$ is alpha-distance for the set $\mathscr{F}$ provided that*

- $\delta(\mathscr{F}_{i_1 i_2 i_3 \ldots}, \mathscr{F}_{s_1 s_2 s_3 \ldots}) > 0$, *for any couple of elements of the set;*
- $\delta(\mathscr{F}_{i_1 i_2 i_3 \ldots}, \mathscr{F}_{s_1 s_2 s_3 \ldots}) = \delta(\mathscr{F}_{s_1 s_2 s_3 \ldots}, \mathscr{F}_{i_1 i_2 i_3 \ldots})$, *for any couple of elements of the set;*
- $\delta(\mathscr{F}_{i_1 i_2 s_i \ldots}, \mathscr{F}_{s_1 s_2 s_3 \ldots}) \leqslant \delta(\mathscr{F}_{i_1 i_2 i_3 \ldots}, \mathscr{F}_{j_1 j_2 j_3 \ldots}) + \delta(\mathscr{F}_{j_1 j_2 j_3 \ldots}, \mathscr{F}_{s_1 s_2 s_3 \ldots})$, *for any three members of the set.*

It is easily seen that the property $\delta(\mathscr{F}_{i_1 i_2 i_3 \ldots}, \mathscr{F}_{1 s_2 s_3 \ldots}) = 0$, if and only if $i_k = s_k$ for all $k = 1, 2, 3, \ldots$ of metric spaces is ignored in the definition. It is assumed for alpha labeling that two labels, which indices do not coincide, are different ones. Now, the couple $(\mathscr{F}, \delta)$ is an alpha space. Theorems 2.3.4, 2.3.6 and 2.3.12 are valid for the dynamics $(\mathscr{F}, \varphi, \delta)$, where the alpha distance is not necessarily inherited from a dynamical system.

2.4 Examples of alpha chaotic models

Here we provide several examples to emphasis alpha chaos universality and effectiveness.

2.4.1 Symbolic dynamics

Example 2.4.1. [21] *In example 2.2.5 we have shown that for the symbolic dynamics alpha labeling can be easily constructed. Next, we will continue with this dynamics. One can see that $\mathscr{F} = \mathscr{F}_0 \cup \mathscr{F}_1$, for the set of alpha labels, $\mathscr{F}_0 = \{\mathscr{F}_{0s_2s_3...}\}$, $\mathscr{F}_1 = \{\mathscr{F}_{1s_2s_3...}\}$. Consequently,*

$$
\begin{aligned}
d(\mathscr{F}_0, \mathscr{F}_1) &= \inf\{d(s, t): s \in \mathscr{F}_0, t \in \mathscr{F}_1] \\
&= d(000 \ldots , 1000 \ldots) \\
&= d(0111 \ldots , 111 \ldots) = 1.
\end{aligned}
$$

Thus, the separation condition of degree 1 holds with the constant $\varepsilon_0 = 1$. The diameter condition has been proven in example 2.2.5. Thus, the Bernoulli shift is ultra Poincaré, Li–Yorke, and Devaney chaotic. One can accept the discussion as another confirmation of chaos for the dynamics [4, 5, 40, 87].

The last example clearly demonstrates how alpha dynamics and alpha chaos can be applied in a particular case. Let us now discuss the method more generally. Let $\mathscr{F}$ be a set of alpha labels for a set F. Assume that the separation and diameter conditions are valid for the dynamics on alpha labels ($\mathscr{F}$, φ, δ) and by theorems 2.3.4, 2.3.6 and 2.3.12 the dynamics admits alpha chaos, i.e., it is ultra Poincaré, Li–Yorke, and Devaney chaotic. Consider the dynamical system (F, ψ, d), which is alpha labeled by ($\mathscr{F}$, φ, δ). Assume that the mappings ψ and φ agree such that $\mathscr{F}_{i_1i_2...} = \varphi(\mathscr{F}_{i_1i_2...})$ is an alpha label for $\psi(f)$, if $\mathscr{F}_{i_1i_2...}$ is that for an element f of F, i.e., the image condition is valid. One can easily see that if the agreement is valid then the dynamics (F, ψ, d) is also ultra Poincaré, Li–Yorke, and Devaney chaotic. That is, theorems 2.3.5, 2.3.7 and 2.3.13 are valid, and the dynamics is alpha chaotic.

We have proved in the last example that the symbolic dynamics is alpha chaotic or chaotic by alpha labeling. Assume that a metric space (F, d) is not provided a dynamics. Then, we say that the metric space is assigned with alpha chaos or chaos by alpha labeling, if there exists a set of alpha labels $\mathscr{F}$ for F such that ($\mathscr{F}$, φ, δ) admits abstract similarly chaos. In the present research, this means, that the diameter and separation conditions are valid for the alpha dynamics. The alpha chaos is assumed to be that for the dynamical system (F, ψ, d), where the image condition determines the mapping ψ such that $\mathscr{L}(\psi(f)) = \varphi(\mathscr{L}(f))$. For simplicity, assume that the labeling is a bijection, then the mapping is determined very clearly, $\psi(f) = \mathscr{L}^{-1}(\varphi(\mathscr{L}(f)))$.

2.4.2 Alpha chaos in finite sets

It is of intense interest to consider the problem of domain for chaos. The original comprehension of sets where chaos may happen is that they should be fractal-like. This is seen most clearly and has been researched for domains of Lorenz, Chua, and Rössler models. One can also consider trajectories of the celestial three-body problem, among others. The dynamics on alpha labels technique provides exceptional opportunities for building domains of chaos with very sophisticated as well as simple geometrical structures. The next model confirms the idea.

Example 2.4.2. *Assume that the finite metric space* (F, d), *of* m *elements* f_l, $l = 1, \ldots, m$, *is given. Alpha label the members infinitely many times with* $\mathscr{F}_{li_2\ldots}$, $l = 1, \ldots, m$, *such that* $i_k = 1, \ldots, m$, *for all* $k = 2, 3, \ldots$ *It is clear that the separation and diameter conditions are valid for the dynamics* $(\mathscr{F}, \varphi, \delta)$. *Now, introduce in the space* (F, d) *the mapping* ψ, *which agrees with* φ. *This means, that* $\psi(f_l) = f_{i_2}$, *for an element* f_l *of the set* F, *if* $\mathscr{F}_{li_2\ldots}$ *is the alpha label for* f_l. *Accordingly to theorems of the section, alpha chaos is valid for the dynamical system* (F, ψ, d). *For example, if* $\mathscr{F}_{i_1 i_2 \ldots i_k \ldots}$ *is an alpha-unpredictable point of the alpha dynamics, then* $f_{i_1}, f_{i_2}, f_{i_3}, \ldots$ *is an alpha-unpredictable trajectory. The result is of significant use for chaos observation in random processes.*

2.4.3 Cantor middle set

Next, we shall realize the opportunity of chaos arrangement in geometrical sets by exploiting the famous Cantor set. Traditionally, the logistic map is helpful to prove chaos, but it is tiresome as it requests topological equivalence to the symbolic dynamics [40]. We apply the alpha labeling procedure, which is straightforward and flexible. The reader is invited to estimate the advantage of the alpha labeling approach in the following example.

Example 2.4.3. *Consider the Cantor middle set,* $\mathscr{C}$, *in the section* $[0, 1]$. *By the known algorithm* [40, 87], *assign an alpha label* $\tau_{i_1 i_2\ldots}$, $i_k = 0, 1$, $k = 1, 2, \ldots$ *to each point of the set, and denote* $\mathscr{T}$ *the union of alpha labels. Precisely, rectangles* $\mathscr{T}_1$ *and* $\mathscr{T}_2$ *are for points of the Cantor set in sections* $[0, 1/3]$ *and* $[2/3, 1]$, *respectively, rectangles* $\mathscr{T}_{11}$ *and* $\mathscr{T}_{12}$ *are for points of the Cantor set in sections* $[0, 1/9]$ *and* $[2/9, 1/3]$, *respectively, etc. Introduce the distance*

$$d(\tau_{i_1 i_2\ldots}, \tau_{j_1 j_2\ldots}) = |c_1 - c_2|,$$

where $|\cdot|$ *is the absolute value, and* $\tau_{i_1 i_2\ldots}$ *with* $\tau_{j_1 j_2\ldots}$ *are alpha labels for points* c_1 *and* c_2 *in Cantor set. The alpha labeling procedure confirms the abstract similarity chaos. Correspondingly, the chaos by alpha labels is confirmed for the Cantor set. Thus, we have obtained the chaos for the geometrical object without topological equivalence of the logistic map and the symbolic dynamics* [40]. *That is possibly the shortest way to arrange the chaos for the set.*

Let us consider another version of the discussion. One can apply the distance

$$\delta(\tau_{i_1 i_2\ldots}, \tau_{j_1 j_2\ldots}) = \sum_{k=1}^{\infty} \frac{|i_k - j_k|}{2^{k-1}}$$

for the set of alpha labels. Make the analysis similar to the above with distance $d(c_1, c_2) = \delta(\tau_{i_1 i_2\ldots}, \tau_{j_1 j_2\ldots})$ *in* $\mathscr{C}$. *Now, the consider equivalence of the two metrics,*

$$1/3|c_1 - c_2| \leqslant \sum_{k=1}^{\infty} \frac{|i_k - j_k|}{2^{k-1}} \leqslant 9/2|c_1 - c_2|,$$

and conclude alpha chaos for $\mathscr{C}$.

2.4.4 Multi-folded Baker map

The following example considers a multi-folded Baker map. It is scalar, but similarly one can discuss vector maps. The case is convenient for analyzing chaos in the decimal system of real numbers, if one uses ten-folded Baker transformation.

Example 2.4.4. *(multi-folded scalar Baker transformation) Consider the Baker map, which is equal to $B(x) = nx - i$ on $(i/n, (i + 1)/n]$, $i = 0, 1, \ldots, n - 1$, where n is a fixed natural. In figure 2.2, a graph of the map is seen.*

Denote $\mathscr{F}_i$, $i = 0, 1, \ldots, n - 1$, the first order rectangles of the alpha set. They are for points in intervals $(i/n, (i + 1)/n]$, $i = 0, 1, \ldots, n - 1$, respectively. Next, determine that $\mathscr{F}_j = \cup_{k=0}^{n-1} \mathscr{F}_{jk}$, $j = 0, 1 \ldots, n - 1$, where $\varphi(\mathscr{F}_{jk}) = \mathscr{F}_j$, $k = 0, 1, \ldots, n - 1$, and $\mathscr{F}_{jk}$ are rectangles such that each of them for fixed j and k is points in the interval $(j/n + k/n^2, j/n + (k + 1)/n^2]$, $k = 0, 1, \ldots, n - 1$. Continue the process, considering inductively, $\mathscr{F}_{i_1 i_2 \ldots i_{n-1} i_n} = \cup_{k=0}^{n-1} \mathscr{F}_{i_1 i_2 \ldots i_n k}$, such that $\varphi(\mathscr{F}_{i_1 i_2 \ldots i_n k}) = \mathscr{F}_{i_1 i_2 \ldots i_{n-1} k}$, $k = 0, 1, \ldots, n - 1$, if all rectangles $\mathscr{F}_{i_1 i_2 \ldots i_{n-1} i_n}$ for a fixed n are already determined. The alpha set $\mathscr{F}$ is collection of elements $\mathscr{F}_{i_1 i_2 \ldots}$ such that a label with indices $i_1 i_2 \ldots$ belongs to $\mathscr{F}_{i_1 i_2 \ldots i_n}$ for all natural n. The reader can easily verify that the set is an alpha set for the interval. The separation condition is valid for rectangles of second order $\mathscr{F}_{ij}$, $i, j = 0, 1$, with $\varepsilon_0 = 1/(2n)$. Moreover, if $n > 2$, then the separation condition is of the first degree with $\varepsilon_0 = 1/n$. The final and essential remark is that the Baker map B and alpha map φ agree at each point of the domain such that the alpha label for $B(x)$ is $\varphi(\mathscr{F}_{i_1 i_2 \ldots})$, if the number x in $[0, 1]$ is alpha labeled with $\mathscr{F}_{i_1 i_2 \ldots}$. Thus, alpha chaos is proven for the Baker transformation. It is valuable that domain of chaos is the unit section, and it is not a fractal set. More precisely, it is true since the interval is a closure for an alpha-unpredictable trajectory. That is, it has been proved that the unit section is a quasi-minimal set for the Baker map. This

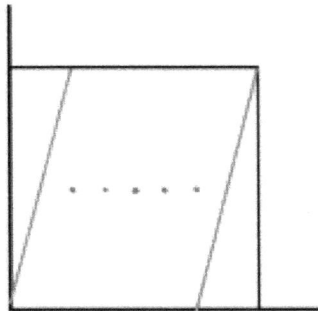

Figure 2.2. The graph of the multi-folded Baker map.

example provides another vision of quasi-minimal sets for Poisson stable dynamics. Our results shed more light on the important property of recurrence, since this time it is combined with the separation property. Thus, we should recognize that the classical concept of ergodicity [68] is not sufficient for a description of the process and should be developed with new features.

2.4.5 The logistic map

Finally, we consider the map, which is very traditional to illustrate achievements of chaos theory.

Example 2.4.5. *(the logistic map) Next, we will prove alpha chaos for the logistic map $L(x) = mx(1 - x)$ with the coefficient m larger than four. One can see a graph of the map in figure 2.3. We shall alpha label by $\mathcal{F}$ the set of all positively invariant points of the unit section. That is, those which are initial points for infinite positive semi-trajectories in the unit. For convenience, we do not define the points of the Cantor set, and their alpha labels, in what follows. The part of the set in the left and right halves of the unit section denote $\mathcal{F}_0$ and $\mathcal{F}_1$, respectively. Then, find that $\mathcal{F}_j = \mathcal{F}_{j0} \cup \mathcal{F}_{j1}, j = 0, 1$, where $L(\mathcal{F}_{jk}) = \mathcal{F}_j, k = 0, 1$. Continue the process, considering inductively, $\mathcal{F}_{i_1 i_2 \ldots i_{n-1} i_n} = \mathcal{F}_{i_1 i_2 \ldots i_n 0} \cup \mathcal{F}_{i_1 i_2 \ldots i_n 1}$, such that $L(\mathcal{F}_{i_1 i_2 \ldots i_n k}) = \mathcal{F}_{i_1 i_2 \ldots i_{n-1} k}, k = 0, 1$, if all sets $\mathcal{F}_{i_1 i_2 \ldots i_{n-1} i_n}$ for a fixed n are determined. The alpha label-set $\mathcal{F}$ is the collection of points $\mathcal{F}_{i_1 i_2 \ldots}$. To prove the diameter property, consider the points $x_n \to 0$ as $n \to \infty$ such that $L^n(x_n) = m/4$. The existence of the sequence proves the property. The separation condition is valid for $\mathcal{F}_j, j = 0, 1$. Thus, alpha chaos is proven for the dynamics.*

Remark 2.4.1. *The last three examples illustrate that proof of chaos presence for conventional models eases significantly, if we use alpha dynamics. We suggest, also, the following two open problems: period three implies alpha chaos; prove the theorem in [58] by alpha labeling.*

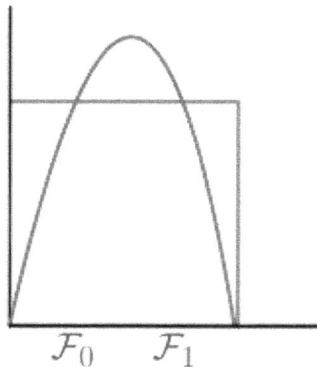

Figure 2.3. The graph of the logistic map.

2.5 Hyperbolic alpha labeling

In this section a version for abstraction of hyperbolic sets [32, 81] is suggested. We strongly believe that by the *dynamics on alpha labels*, one can make the analysis of hyperbolic dynamics much more effective and productive since it is true for many new concepts, which were invented to envelop the experience and knowledge accumulated in science. The suggestions are more general, and request less information for their description, than former definitions, and consequently, they are more productive in the theoretical sense and more effective in applications. We provide this in the present research for hyperbolic sets through dynamics on alpha labels. The results are, first of all, about the consistency of negative and positive directions of time. Moreover, we show that the research is not only about consistency but about alpha-unpredictability and alpha chaos being present for both directions of time, while previously only positive time changes have been considered. Another novelty is the construction of distinctly new hyperbolic sets, such as finite sets of points. We hope these proposals can provide new material for applying results accumulated in the theory of hyperbolic dynamics.

2.5.1 Hyperbolic alpha dynamics

Let us consider a metric space (F, d) with distance d. In our research, set F can be finite, countable, or uncountable. Beside it, consider the following hyperbolic set of alpha labels

$$\mathscr{F} = \{\mathscr{F}_{...i_{-2}i_{-1}i_0 \cdot i_1 i_2...} : i_k = 1, 2, ..., m, \ k = ...-2, -1, 0, 1, 2, ...\}, \qquad (2.8)$$

where m is a fixed natural number and the symbol $\cdot$ is used to determine the dynamics of the present section.

We assume that each element of F is alpha labeled at least by one member of set $\mathscr{F}$, and every element of $\mathscr{F}$ necessarily presents a member of F. In what follows, we shall denote $\mathscr{L}$ the alpha labeling operator. That is $\mathscr{L}: F \to \mathscr{F}$, and $\mathscr{L}(f) = \mathscr{F}_{...i_{-2}i_{-1}i_0 \cdot i_1 i_2...}$ if $\mathscr{F}_{...i_{-2}i_{-1}i_0 \cdot i_1 i_2...}$ is an alpha label for f.

Definition 2.5.1. *The alpha-distance is the function* $\delta(\mathscr{F}_{...i_{-2}i_{-1}i_0 \cdot i_1 i_2...}, \mathscr{F}_{...j_{-2}j_{-1}j_0 \cdot j_1 j_2...}) = d(f_1, f_2)$ *of elements* $\mathscr{F}_{...i_{-2}i_{-1}i_0 \cdot i_1 i_2...}$ *and* $\mathscr{F}_{...j_{-2}j_{-1}j_0 \cdot j_1 j_2...}$ *of* $\mathscr{F}$, *which are alpha labels of members* f_1 *and* f_2 *of F, respectively.*

It is easily seen that the alpha-distance admits the following properties.

- $\delta(\mathscr{F}_{...i_{-2}i_{-1}i_0 \cdot i_1 i_2...}, \mathscr{F}_{...j_{-2}j_{-1}j_0 \cdot j_1 j_2...}) > 0$, for any couple of elements of the hyperbolic set;
- $\delta(\mathscr{F}_{...i_{-2}i_{-1}i_0 \cdot i_1 i_2...}, \mathscr{F}_{...j_{-2}j_{-1}j_0 \cdot j_1 j_2...}) = \delta(\mathscr{F}_{...j_{-2}j_{-1}j_0 \cdot j_1 j_2...}, \mathscr{F}_{...i_{-2}i_{-1}i_0 \cdot i_1 i_2...})$ for any couple of elements of the hyperbolic set;
- $\delta(\mathscr{F}_{...i_{-2}i_{-1}i_0 \cdot i_1 i_2...}, \mathscr{F}_{...j_{-2}j_{-1}j_0 \cdot j_1 j_2...}) \leqslant \delta(\mathscr{F}_{...i_{-2}i_{-1}i_0 \cdot i_1 i_2...}, \mathscr{F}_{...s_{-2}s_{-1}s_0 \cdot s_1 s_2...}) + \delta(\mathscr{F}_{...s_{-2}s_{-1}s_0 \cdot s_1 s_2...}, \mathscr{F}_{...j_{-2}j_{-1}j_0 \cdot j_1 j_2...})$, for any three members of the hyperbolic set.

The last three conditions can be applied as laws for the distance, if hyperbolic alpha space is determined axiomatically.

One can see that

$$\delta(\mathscr{F}_{...i_{-2}i_{-1}i_0 \cdot i_1 i_2...}, \; \mathscr{F}_{...j_{-2}j_{-1}j_0 \cdot j_1 j_2...}) = 0$$

for two different alpha labels of the same member of F. If the alpha labeling is one-to-one then the operator $\mathscr{L}$ is isometric and the couple $(\mathscr{F}, \delta)$ is a metric space. It is clear that for the alpha space $(\mathscr{F}, \delta)$ all the topological properties of the section 2.2 are valid.

The set $\mathscr{F}$ is called the *abstract hyperbolic set* or *alpha hyperbolic set*, and its elements are called *abstract hyperbolic points* or *alpha hyperbolic points*. The collection

$$\mathscr{F}_{i_1 i_2...} = \{\mathscr{F}_{...j_{-2}j_{-1}j_0 \cdot i_1 i_2 ... i_n...} : j_k = 1, 2, ..., m, \; k = ...-2, -1, 0\} \qquad (2.9)$$

is the *stable set* for a fixed point $\mathscr{F}_{...i_{-k} ... i_{-2}i_{-1}i_0 \cdot i_1 i_2 ... i_n...}$, and the union

$$\mathscr{F}_{...i_{-k} ... i_{-2}i_{-1}i_0} = \{\mathscr{F}_{...i_{-2}i_{-1}i_0 \cdot j_1 j_2...} : j_k = 1, 2, ..., m, \; k = 1, 2, ...\} \qquad (2.10)$$

is said to be the *unstable set* of the point.

The following sets, we call them rectangles, are needed in what follows:

$$\mathscr{F}_{i_{-k}i_{-k+1} ... i_0 \cdot i_1 i_2 ... i_n} = \bigcup_{i_s = 1,2,...,m} \mathscr{F}_{...i_{-k-1}i_{-k}i_{-k+1} ... i_0 \cdot i_1 i_2 ... i_n i_{n+1}...}, \qquad (2.11)$$

where indices i_{-k}, i_{-k+1}, ... , i_0, i_1, i_2, ... , i_n, are fixed,

$$\mathscr{F}_{i_1 i_2 ... i_n} = \bigcup_{i_s = 1,2,...,m} \mathscr{F}_{...i_{-k-1}i_{-k}i_{-k+1} ... i_0 \cdot i_1 i_2 ... i_n i_{n+1}...}, \qquad (2.12)$$

where indices i_1, i_2, ... , i_n, are fixed, and

$$\mathscr{F}_{i_{-k}i_{-k+1} ... i_0} = \bigcup_{i_s = 1,2,...,m} \mathscr{F}_{...i_{-k-1}i_{-k}i_{-k+1} ... i_0 \cdot i_1 i_2 ... i_n i_{n+1}...}, \qquad (2.13)$$

where indices i_{-k}, i_{-k+1}, ... , i_0 as well as the number k are fixed.

Definition 2.5.2. *We will say that the diameter condition is valid for the sets, if it is true that*

$$\sup_{i_{-n}i_{-n+1} ... i_0 \cdot i_1 i_2 ... i_n} \mathrm{diam} \, \mathscr{F}_{i_{-n}i_{-n+1} ... i_0 \cdot i_1 i_2 ... i_n} \to 0 \;\; as \;\; n \to \infty, \qquad (2.14)$$

where $\mathrm{diam}(A) = \sup\{\delta(\mathbf{x}, \mathbf{y}) : \mathbf{x}, \mathbf{y} \in A\}$ *for a subset A of $\mathscr{F}$.*

Determine the following one-to-one and onto map $\varphi: \mathscr{F} \to \mathscr{F}$ such that

$$\varphi(\mathscr{F}_{...i_{-k} ... i_{-2}i_{-1}i_0 \cdot i_1 i_2 ... i_n...}) = \mathscr{F}_{...i_{-k} ... i_{-2}i_{-1}i_0 i_1 \cdot i_2 ... i_n...}.$$

Consider a point $\mathscr{F}_{\dots i_{-k} \dots i_{-2} i_{-1} i_0 \cdot i_1 i_2 \dots i_n \dots}$ and an arbitrary element, $\mathscr{F}_{\dots j_{-2} j_{-1} j_0 \cdot i_1 i_2 \dots i_n \dots}$, from the stable set of the point. From the definition of the stable set and the diameter property it follows that

$$d(\varphi^n(\mathscr{F}_{\dots j_{-2} j_{-1} j_0 \cdot i_1 i_2 \dots i_n \dots}), \; \varphi^n(\mathscr{F}_{\dots i_{-k} \dots i_{-2} i_{-1} i_0 \cdot i_1 i_2 \dots i_n \dots})) \to 0$$

as $n \to \infty$.

Similarly, it is true that

$$d(\varphi^n(\mathscr{F}_{\dots i_{-2} i_{-1} i_0 \cdot j_1 j_2 \dots j_n \dots}), \; \varphi^n(\mathscr{F}_{\dots i_{-k} \dots i_{-2} i_{-1} i_0 \cdot i_1 i_2 \dots i_n \dots})) \to 0$$

as $n \to -\infty$, for points of the unstable set. The last two limits are why the sets are called the stable and unstable sets of the point.

It is clear that

$$\mathscr{F} \supseteq \mathscr{F}_{i_1} \supseteq \mathscr{F}_{i_1 i_2} \supseteq \cdots \supseteq \mathscr{F}_{i_1 i_2 \dots i_n} \supseteq \mathscr{F}_{i_1 i_2 \dots i_n i_{n+1}} \cdots$$

and

$$\mathscr{F} \supseteq \mathscr{F}_{i_0} \supseteq \mathscr{F}_{i_{-1} i_0} \supseteq \cdots \supseteq \mathscr{F}_{i_{-n} i_{-n+1} \dots i_0} \supseteq \mathscr{F}_{i_{-n-1} i_{-n} \dots i_0} \cdots .$$

Considering iterations of the map, one can verify that

$$\varphi^n(\mathscr{F}_{i_1 i_2 \dots i_n}) = \mathscr{F}, \tag{2.15}$$

and

$$\varphi^{-n}(\mathscr{F}_{i_{-n} i_{-n+1} \dots i_0}) = \mathscr{F}, \tag{2.16}$$

for arbitrary natural number n. The relations (2.15) and (2.16) give us a reason to call φ a *abstract similarity map* or *alpha map* and the number n the *degree of similarity*. The triple $(\mathscr{F}, \varphi, \delta)$ is said to be *abstract hyperbolic dynamics* or *alpha hyperbolic dynamics*.

2.5.2 Abstract hyperbolic chaos

Denote by $\delta(A, B) = \inf\{\delta(\mathbf{x}, \mathbf{y}): \mathbf{x} \in A, \mathbf{y} \in B\}$ the function of two bounded subsets A and B in an abstract hyperbolic set $\mathscr{F}$.

Definition 2.5.3. *An alpha hyperbolic space $(\mathscr{F}, \delta)$ satisfies the positive separation condition of degree n if there exist a number $\varepsilon_0 > 0$ and a natural n such that for arbitrary $i_1 i_2 \dots i_n$ one can find indices $j_1 j_2 \dots j_n$, which satisfy the inequality,*

$$\delta(\mathscr{F}_{i_1 i_2 \dots i_n}, \mathscr{F}_{j_1 j_2 \dots j_n}) > \varepsilon_0. \tag{2.17}$$

Definition 2.5.4. *Negative separation condition of degree n is valid for space $(\mathscr{F}, \delta)$, if there exist a positive number ε_0 and a natural n such that for arbitrary $i_{-n+1} i_{-n+2} \dots i_0$ one can find indices $j_{-n+1} j_{-n+2} \dots j_0$ for which it is true that*

$$\delta(\mathscr{F}_{i_{-n+1} i_{-n+2} \dots i_0}, \mathscr{F}_{j_{-n+1} j_{-n+2} \dots j_0}) > \varepsilon_0. \tag{2.18}$$

We say that separation condition is valid if both positive and negative instances are fulfilled.

A point $\mathscr{F}_{\ldots i_{-k} \ldots i_{-2}i_{-1}i_0 \cdot i_1 i_2 \ldots i_n \ldots}$ from $\mathscr{F}$ is periodic with period n if its index set consists of endless repetitions of a block of n terms. We shall say that a dynamics exhibits *positive Devaney chaos* if the three ingredients of chaos are proper for the discrete-time increasing to infinity. That is, the chaos is valid for the abstract similarity map φ. A more precise description of the chaos will be seen in the proof of the next theorem. Similarly, a dynamics exhibits *negative Devaney chaos* if it happens for the time decreasing to negative infinity. That chaos is valid for the inverse map φ^{-1}. If a dynamics has negative and positive Devaney chaos, the hyperbolic dynamics is said to be with Devaney chaos. That is, the chaos is valid for both the abstract similarity map and its inverse. One can remark that the positive chaos is the traditional one, but to keep the symmetry we decide that, terminologically, the positive and negative chaos must be considered. We do not define Devaney chaos [40], since it seen in the proof below.

Theorem 2.5.1. *If diameter and positive separation conditions are valid, then the alpha hyperbolic dynamics is chaotic in the sense of Devaney.*

Proof. The transitivity will be proved if there exists an element $\mathscr{F}_{\ldots i_{-k} \ldots i_{-2}i_{-1}i_0 \cdot i_1 i_2 \ldots i_n \ldots}$ of $\mathscr{F}$ such that for any finite sequence $j_{-k} \ldots j_{-2}j_{-1}j_0 \cdot j_1 j_2 \ldots j_n$ there exists sufficiently large p, for which $\varphi^p(\mathscr{F}_{\ldots i_{-k} \ldots i_{-2}i_{-1}i_0 \cdot i_1 i_2 \ldots i_n \ldots})$ belongs to the rectangle $\mathscr{F}_{j_{-k} \ldots j_{-2}j_{-1}j_0 \cdot j_1 j_2 \ldots j_n}$. This is true, since one can construct a sequence $\ldots i_1 i_2 \ldots i_n \ldots$, which contains all the sequences of the type $j_{-k} \ldots j_{-2}j_{-1}j_0 \cdot j_1 j_2 \ldots j_n$ as blocks [14, 87].

Fix a member $\mathscr{F}_{\ldots i_{-k} \ldots i_{-2}i_{-1}i_0 \cdot i_1 i_2 \ldots i_n \ldots}$ of the alpha set and a positive number κ. Find a sufficiently large number k such that $\mathrm{diam}(\mathscr{F}_{i_{-k}i_{-k+1} \ldots i_0 \cdot i_1 i_2 \ldots i_k}) < \kappa$ and choose $2k$-periodic element $\mathscr{F}_{\ldots i_{-k}i_{-k+1} \ldots i_0 \cdot i_1 i_2 \ldots i_k \ldots}$ of $\mathscr{F}_{i_{-k}i_{-k+1} \ldots i_0 \cdot i_1 i_2 \ldots i_k}$. It is clear that the periodic point is a κ-approximation for the considered member. Thus, the density of periodic points is proved.

Now, fix a point $\mathscr{F}_{\ldots i_{-k} \ldots i_{-2}i_{-1}i_0 \cdot i_1 i_2 \ldots i_n \ldots}$ of the alpha set and an arbitrary positive number ε. By the diameter condition there exist an integer k and point

$$\mathscr{F}_{\ldots i_{-k} \ldots i_{-2}i_{-1}i_0 \cdot i_1 i_2 \ldots i_k j_{k+1} j_{k+2} \ldots} \neq \mathscr{F}_{\ldots i_{-k} \ldots i_{-2}i_{-1}i_0 \cdot i_1 i_2 \ldots i_n \ldots}$$

such that $\delta(\mathscr{F}_{\ldots i_{-k} \ldots i_{-2}i_{-1}i_0 \cdot i_1 i_2 \ldots i_n \ldots}, \mathscr{F}_{\ldots i_{-k} \ldots i_{-2}i_{-1}i_0 \cdot i_1 i_2 \ldots i_k j_{k+1} j_{k+2} \ldots}) < \varepsilon$. The indices $j_{k+1}, j_{k+2}, \ldots$ are chosen such that the separation condition implies $\delta(\mathscr{F}_{i_{k+1}i_{k+2} \ldots i_{k+n}}, \mathscr{F}_{j_{k+1}j_{k+2} \ldots j_{k+n}}) > \varepsilon_0$. This proves the sensitivity. $\square$

The following theorems can be proved very similarly to the last assertion.

Theorem 2.5.2. *If the diameter and negative separation conditions are valid, then the abstract hyperbolic dynamics is negative chaotic in the sense of Devaney.*

Theorem 2.5.3. *If the diameter and separation conditions are valid, then the abstract hyperbolic dynamics is Devaney chaotic.*

Remark 2.5.1. *It is obvious that instead of positive and negative Devaney chaos in the last three theorems, one can say that Devaney chaos for maps is φ and φ^{-1}, respectively.*

Definition 2.5.5. *A point $\mathscr{F}_{\ldots i_{-k} \ldots i_{-2} i_{-1} i_0 \cdot i_1 i_2 \ldots i_n \ldots}$ from the hyperbolic set $\mathscr{F}$ is positive alpha-unpredictable if there exist a positive number ε_0 and sequences t_m, s_m, of natural numbers both of which diverge to infinity such that $\varphi^{t_m}(\mathscr{F}_{\ldots i_{-k} \ldots i_{-2} i_{-1} i_0 \cdot i_1 i_2 \ldots i_n \ldots}) \rightarrow \mathscr{F}_{\ldots i_{-k} \ldots i_{-2} i_{-1} i_0 \cdot i_1 i_2 \ldots i_n \ldots}$ as $m \to \infty$ and*

$$\delta(\varphi^{t_m + s_m}(\mathscr{F}_{\ldots i_{-k} \ldots i_{-2} i_{-1} i_0 \cdot i_1 i_2 \ldots i_n \ldots}), \varphi^{s_m}(\mathscr{F}_{\ldots i_{-k} \ldots i_{-2} i_{-1} i_0 \cdot i_1 i_2 \ldots i_n \ldots})) > \varepsilon_0$$

for each natural number m.

Definition 2.5.6. *A point $\mathscr{F}_{\ldots i_{-k} \ldots i_{-2} i_{-1} i_0 \cdot i_1 i_2 \ldots i_n \ldots}$ from the set $\mathscr{F}$ is negative alpha-unpredictable, if there exist a positive number ε_0 and sequences t_m, s_m, of natural numbers both of which diverge to infinity such that $\varphi^{-t_m}(\mathscr{F}_{\ldots i_{-k} \ldots i_{-2} i_{-1} i_0 \cdot i_1 i_2 \ldots i_n \ldots}) \rightarrow \mathscr{F}_{\ldots i_{-k} \ldots i_{-2} i_{-1} i_0 \cdot i_1 i_2 \ldots i_n \ldots}$ as $m \to \infty$ and*

$$\delta(\varphi^{-t_m - s_m}(\mathscr{F}_{\ldots i_{-k} \ldots i_{-2} i_{-1} i_0 \cdot i_1 i_2 \ldots i_n \ldots}), \varphi^{-s_m}(\mathscr{F}_{\ldots i_{-k} \ldots i_{-2} i_{-1} i_0 \cdot i_1 i_2 \ldots i_n \ldots})) > \varepsilon_0$$

for each natural number m.

Definition 2.5.7. *A point $\mathscr{F}_{\ldots i_{-k} \ldots i_{-2} i_{-1} i_0 \cdot i_1 i_2 \ldots i_n \ldots}$ is alpha-unpredictable if it is positive and negative alpha-unpredictable.*

For future research one can apply the following definitions of Poisson stability.

Definition 2.5.8. *A point $\mathscr{F}_{\ldots i_{-k} \ldots i_{-2} i_{-1} i_0 \cdot i_1 i_2 \ldots i_n \ldots}$ is positive Poisson stable if there exists a sequence t_m of natural numbers, which diverges to infinity such that $\varphi^{t_m}(\mathscr{F}_{\ldots i_{-k} \ldots i_{-2} i_{-1} i_0 \cdot i_1 i_2 \ldots i_n \ldots}) \rightarrow \mathscr{F}_{\ldots i_{-k} \ldots i_{-2} i_{-1} i_0 \cdot i_1 i_2 \ldots i_n \ldots}$ as $m \to \infty$.*

Definition 2.5.9. *A point $\mathscr{F}_{\ldots i_{-k} \ldots i_{-2} i_{-1} i_0 \cdot i_1 i_2 \ldots i_n \ldots}$ is negative Poisson stable if there exists a sequence t_m of natural numbers, which diverges to infinity such that $\varphi^{-t_m}(\mathscr{F}_{\ldots i_{-k} \ldots i_{-2} i_{-1} i_0 \cdot i_1 i_2 \ldots i_n \ldots}) \rightarrow \mathscr{F}_{\ldots i_{-k} \ldots i_{-2} i_{-1} i_0 \cdot i_1 i_2 \ldots i_n \ldots}$ as $m \to \infty$.*

Definition 2.5.10. *A point $\mathscr{F}_{\ldots i_{-k} \ldots i_{-2} i_{-1} i_0 \cdot i_1 i_2 \ldots i_n \ldots}$ is Poisson stable if it is positive and negative Poisson stable.*

Theorem 2.5.4. *There exists a positive alpha-unpredictable member of the abstract hyperbolic dynamics $(\mathscr{F}, \varphi, \delta)$, if diameter and positive separation conditions are valid.*

Theorem 2.5.5. *There exists a negative alpha-unpredictable member of the abstract hyperbolic dynamics $(\mathscr{F}, \varphi, \delta)$, if diameter and negative separation conditions are valid.*

Theorem 2.5.6. *There exists an alpha-unpredictable member of the abstract hyperbolic dynamics $(\mathscr{F}, \varphi, \delta)$, if diameter and separation conditions are valid.*

Correspondingly, theorems on ultra Poincaré chaos can be formulated.

Theorem 2.5.7. *If diameter and positive separation conditions are valid and $(\mathscr{F}, \delta)$ is an alpha-complete space, then dynamics $(\mathscr{F}, \psi, \delta)$ admits negative ultra Poincaré chaos.*

Theorem 2.5.8. *The dynamics $(\mathscr{F}, \psi, \delta)$ is negative ultra Poincaré chaotic if diameter and negative positive separation properties are valid and $(\mathscr{F}, \delta)$ is an alpha-complete space.*

Theorem 2.5.9. *If diameter and separation conditions are valid and $(\mathscr{F}, \delta)$ is an alpha-complete space, then dynamics $(\mathscr{F}, \psi, \delta)$ admits hyperbolic ultra Poincaré chaos.*

Beside the last three theorems, the following theorems can be useful for the dynamics research.

Theorem 2.5.10. *There exists a positive Poisson stable member of the abstract hyperbolic dynamics $(\mathscr{F}, \varphi, \delta)$ if diameter and positive separation conditions are valid.*

Theorem 2.5.11. *There exists a negative Poisson stable member of the abstract hyperbolic dynamics $(\mathscr{F}, \varphi, \delta)$ if diameter and negative separation conditions are valid.*

Theorem 2.5.12. *There exists a Poisson stable member of the abstract hyperbolic dynamics $(\mathscr{F}, \varphi, \delta)$ if diameter and separation conditions are valid.*

Remark 2.5.2. *The definitions of positive and negative alpha-unpredictable and Poisson stable points for abstract hyperbolic dynamics are given not only to exploit the symmetry of time in the dynamics but also to satisfy the classical definition of*

Poisson stability [29, 66, 75], which is given for both directions, while chaos in science is only what we call positive chaos, that is, with increasing time. One can also show the existence of quasi-minimal sets as the closure of the alpha-unpredictable orbits in abstract hyperbolic dynamics, and we decided to omit the discussion since it is routine, and request additional explanations related to the non-uniqueness of the alpha labeling operator. One can provide the definition easily and consider the quasi-minimal sets for dynamics subdued to the hyperbolic alpha labeling.

Similarly to that which has been done for abstract similarity dynamics, one can consider two-directional Li–Yorke chaos for hyperbolic abstract dynamics. Let us introduce the following definitions.

Definition 2.5.11. *A couple $\mathscr{F}_{...i_{-2}i_{-1}i_0 \cdot i_1 i_2...}$, $\mathscr{F}_{...j_{-2}j_{-1}j_0 \cdot j_1 j_2...}$ of elements from $\mathscr{F}$ is called positive proximal if*

$$\liminf_{n\to\infty} \delta(\varphi^n(\mathscr{F}_{...i_{-2}i_{-1}i_0 \cdot i_1 i_2...}), \varphi^n(\mathscr{F}_{...j_{-2}j_{-1}j_0 \cdot j_1 j_2...})) = 0.$$

Definition 2.5.12. *A couple $\mathscr{F}_{...i_{-2}i_{-1}i_0 \cdot i_1 i_2...}$, $\mathscr{F}_{...j_{-2}j_{-1}j_0 \cdot j_1 j_2...}$ of elements from $\mathscr{F}$ is positive frequently separated if*

$$\limsup_{n\to\infty} \delta(\varphi^n(\mathscr{F}_{...i_{-2}i_{-1}i_0 \cdot i_1 i_2...}), \varphi^n(\mathscr{F}_{...j_{-2}j_{-1}j_0 \cdot j_1 j_2...})) > 0.$$

Definition 2.5.13. *A subset of $\mathscr{F}$ is called positive scrambled if each couple of the set is positive proximal and positive frequently separated.*

Definition 2.5.14. *The dynamics $(\mathscr{F}, \varphi, \delta)$ is positive Li–Yorke chaotic if*
- *there exist orbits with an arbitrary natural period;*
- *there exists an uncountable positive scrambled subset;*
- *each element of the positive scrambled set makes a positive frequently separated couple with any periodic one.*

Theorem 2.5.13. *The abstract hyperbolic dynamics $(\mathscr{F}, \varphi, \delta)$ is positive Li–Yorke chaotic if diameter and positive separation conditions hold.*

Definition 2.5.15. *A couple $\mathscr{F}_{...i_{-2}i_{-1}i_0 \cdot i_1 i_2...}$, $\mathscr{F}_{...j_{-2}j_{-1}j_0 \cdot j_1 j_2...}$ of elements from $\mathscr{F}$ is called negative proximal if*

$$\liminf_{n\to\infty} \delta(\varphi^{-n}(\mathscr{F}_{...i_{-2}i_{-1}i_0 \cdot i_1 i_2...}), \varphi^{-n}(\mathscr{F}_{...j_{-2}j_{-1}j_0 \cdot j_1 j_2...})) = 0.$$

Definition 2.5.16. *A couple* $\mathcal{F}_{\ldots i_{-2} i_{-1} i_0 \cdot i_1 i_2 \ldots}$, $\mathcal{F}_{\ldots j_{-2} j_{-1} j_0 \cdot j_1 j_2 \ldots}$ *of elements from* $\mathcal{F}$ *is negative frequently separated if*

$$\lim \sup_{n \to \infty} \delta(\varphi^{-n}(\mathcal{F}_{\ldots i_{-2} i_{-1} i_0 \cdot i_1 i_2 \ldots}), \varphi^{-n}(\mathcal{F}_{\ldots j_{-2} j_{-1} j_0 \cdot j_1 j_2 \ldots})) > 0.$$

Definition 2.5.17. *A subset of* $\mathcal{F}$ *is called negative scrambled if each couple of the set is negative proximal and positive frequently separated.*

Definition 2.5.18. *A dynamics* $(\mathcal{F}, \varphi, \delta)$ *is negative Li–Yorke chaotic if*
- *there exist n-periodic orbits for map* φ^{-1} *with an arbitrary natural period;*
- *there is an uncountable negative scrambled subset;*
- *each element of the negative scrambled subset makes a negative frequently separated couple with any periodic point.*

Definition 2.5.19. *A couple* $\mathcal{F}_{\ldots i_{-2} i_{-1} i_0 \cdot i_1 i_2 \ldots}$, $\mathcal{F}_{\ldots j_{-2} j_{-1} j_0 \cdot j_1 j_2 \ldots}$ *of elements from* $\mathcal{F}$ *is called proximal if it is positive and negative proximal.*

Definition 2.5.20. *A couple* $\mathcal{F}_{\ldots i_{-2} i_{-1} i_0 \cdot i_1 i_2 \ldots}$, $\mathcal{F}_{\ldots j_{-2} j_{-1} j_0 \cdot j_1 j_2 \ldots}$ *of elements from* $\mathcal{F}$ *is frequently separated if it is positive and negative frequently separated.*

Definition 2.5.21. *A subset of* $\mathcal{F}$ *is called scrambled if each couple of the set is proximal and frequently separated.*

Let us finalize the discussion with the following assertion.

Theorem 2.5.14. [21] *An alpha dynamics* $(\mathcal{F}, \varphi, \delta)$ *is negative Li–Yorke chaotic if diameter and negative separation conditions hold.*

The proof of theorems 2.5.29 and 2.5.37 is similar to that of theorem 6.35 in [38] for a Bernoulli map defined on a space of symbolic sequences. See also paper [76].

If the separation and diameter condition are valid for the abstract dynamics, we shall say that it is *alpha hyperbolic chaotic*.

2.5.3 Alpha hyperbolic chaos

Consider a dynamical system (F, ψ, d) with the state space F, and mapping $\psi: F \to F$, such that (F, d) is a complete metric space. The map ψ is not required to be invertable, and this is beneficial in analysis of random processes. Assume that an abstract hyperbolic set $\mathcal{F}$ is an alpha set for F. Keeping that $\delta(\mathcal{F}_{\ldots i_{-k} \ldots i_{-2} i_{-1} i_0 \cdot i_1 i_2 \ldots i_n \ldots}, \mathcal{F}_{\ldots j_{-2} j_{-1} j_0 \cdot j_1 j_2 \ldots j_n \ldots}) = d(f, g)$, for alpha labels of elements

f and g, we introduce the hyperbolic abstract similarity dynamics $(\mathscr{F}, \varphi, \delta)$. We say that maps ψ and ϕ agree if the alpha label of an image is the value of a similarity map on the alpha label for preimage. That is, $\mathscr{L}(\psi(f)) = \varphi(\mathscr{L}(f))$, for a member f of F.

In what follows, the equality will be called *image condition*. That is, we say that *image condition* is valid if the mappings agree. Space $(\mathscr{F}, \delta)$ is compact if the diameter condition is assumed to be valid space and (F, d) is complete. If abstract hyperbolic dynamics are chaotic and the maps agree, we shall say that (F, ψ, d) is hyperbolic chaotic by alpha labels. It is reasonable to say both dynamics are *alpha hyperbolic chaotic* if the chaos is valid. In what follows, arguments for that will be given.

Theorem 2.5.15. *The dynamics (F, ψ, d) is alpha (positive) chaotic in the ultra Poincaré sense if diameter, positive separation, and image conditions are valid and (F, d) is a complete space.*

Theorem 2.5.16. *If diameter, positive separation, and image conditions hold for $(\mathscr{F}, \varphi, \delta)$, then the dynamics (F, ψ, d) possesses alpha (positive) Li–Yorke chaos.*

Theorem 2.5.17. *If diameter, positive separation, and image conditions are valid, then the dynamics (F, ψ, d) is alpha (positive) Devaney chaotic.*

Theorem 2.5.18. *The dynamics (F, ψ, d) is alpha negative chaotic in the ultra Poincaré sense if diameter, negative separation, and image conditions are valid and (F, d) is a complete space.*

Theorem 2.5.19. *If diameter, negative separation, and image conditions hold, then ψ^{-1} is alpha Li–Yorke chaotic.*

Theorem 2.5.20. *If diameter, negative separation, and image conditions hold, then dynamics (F, ψ, d) is alpha negative chaotic in the sense of Devaney.*

Theorem 2.5.21. *The dynamics (F, ψ, d) is alpha hyperbolic chaotic in ultra Poincaré sense, if diameter, separation, and image conditions are valid and (F, d) is a complete space.*

Theorem 2.5.22. *If diameter, separation, and image conditions hold for $(\mathscr{F}, \varphi, \delta)$, then the dynamics (F, ψ, d) possesses alpha hyperbolic Li–Yorke chaos.*

Theorem 2.5.23. *If diameter, separation, and image conditions are valid, then the dynamics (F, ψ, d) admits alpha hyperbolic Devaney chaos.*

2.5.4 Examples of alpha hyperbolic chaos

The following analysis of popular models illustrates the might of our suggestions.

Example 2.5.1. *(symbolic dynamics)*[14] *Let us take into account the following space of bi-infinite sequences* [87],

$$\Sigma^2 = \{s = (\dots s_{-2}s_{-1} \cdot s_0 s_1 s_2 \dots) : s_j = 0 \text{ or } 1 \text{ for each } j]$$

with metric

$$d(s, \bar{s}) = \sum_{k=-\infty}^{\infty} \frac{|s_k - \bar{s}_k|}{2^{|k|}},$$

where $s = (\dots s_{-2}s_{-1} \cdot s_0 s_1 s_2 \dots)$, $\bar{s} = (\dots \bar{s}_{-2}\bar{s}_{-1} \cdot \bar{s}_0 \bar{s}_1 \bar{s}_2 \dots) \in \Sigma^2$. *The shift map* $\sigma : \Sigma^2 \to \Sigma^2$ *is defined as*

$$\sigma(\dots s_{-2}s_{-1} \cdot s_0 s_1 s_2 \dots) = (\dots s_{-2}s_{-1}s_0 \cdot s_1 s_2 \dots).$$

Let us alpha label the sequences as $\mathscr{F}_{\dots s_{-2}s_{-1} \cdot s_0 s_1 s_2 \dots} = \mathscr{L}(\dots s_{-2}s_{-1} \cdot s_0 s_1 s_2 \dots)$, *and describe nth order rectangles in* $\mathscr{S}$ *by*

$$\mathscr{F}_{s_{-n}s_{-n+1} \dots s_0 \cdot s_1 s_2 \dots s_n} = \{\mathscr{F}(\dots s_{-n}s_{-n+1} \dots s_0 \cdot s_1 s_2 \dots s_n s_{n+1}s_{n+2} \dots) : s_k = 0 \text{ or } 1],$$

where $s_{-n}s_{-n+1} \dots s_0, s_1, s_2, \dots , s_n$ *are fixed symbols. One can show that distance between any two alpha labels of* $\mathscr{F}_{s_{-n}s_{-n+1} \dots s_0 \cdot s_1 s_2 \dots s_n}$ *is less than* $\frac{1}{2^{n-2}}$. *Moreover,*

$$\delta(\mathscr{F}_{\dots 00 s_{-n}s_{-n+1} \dots s_0 \cdot s_1 s_2 \dots s_n 00 \dots}, \mathscr{F}_{\dots 11 s_{-n}s_{-n+1} \dots s_0 \cdot s_1 s_2 \dots s_n 11 \dots}) =$$

$$d(\dots 00 s_{-n}s_{-n+1} \dots s_0 \cdot s_1 s_2 \dots s_n 00 \dots , \dots 11 s_{-n}s_{-n+1} \dots s_0 \cdot s_1 s_2 \dots s_n 11 \dots) = \frac{1}{2^{n-2}}.$$

That is, $\text{diam}(\mathscr{F}_{s_{-n}s_{-n+1} \dots s_0 \cdot s_1 s_2 \dots s_n} = \frac{1}{2^{n-2}}$. *Consequently,*

$$\lim_{n \to \infty} \text{diam } \mathscr{F}_{s_{-n}s_{-n+1} \dots s_0 \cdot s_1 s_2 \dots s_n} = 0.$$

One can easily see that the similarity map for the set of alpha labels agrees with the Bernoulli shift. That is,

$$\varphi(\mathscr{F}_{\dots s_{-2}s_{-1} \cdot s_0 s_1 s_2 \dots}) = \mathscr{L}(\sigma(\dots s_{-2}s_{-1}s_0 s_1 \cdot s_2 \dots)).$$

The discussion above implies that the symbolic dynamics (Σ^2, σ, d) *is alpha hyperbolic chaotic, and consequently all the three basic types of chaos are valid for the dynamics in both, negative and positive, directions.*

Example 2.5.2. *(finite metric space) Consider the space* (F, d), *with elements* f_l, $l = 1, \dots , m$, *and distance d, and perform alpha labeling with infinite multiplicity* $\mathscr{F}_{\dots i_{-k} \dots i_{-2}i_{-1}i_0 \cdot l i_2 \dots i_n \dots}$, $l = 1, \dots , m$, *such that* $i_n = 1, \dots , m$, *for all* $n = 2, 3, \dots$. *One can easily verify that separation and diameter conditions are valid for alpha dynamics* $(\mathscr{F}, \varphi, \delta)$. *Introduce in F mapping* ψ, *which agrees with* φ. *That is,*

$\psi(f_l) = f_{i_2}$, for the element f_l, $l = 1, \ldots, m$, of the set F, if $\mathscr{F}_{\ldots i_{-k} \ldots i_{-2}i_{-1}i_0 \cdot li_2 \ldots i_n \ldots}$ is the alpha label for f_l. Conditions for all types of chaos are valid for the newly constructed dynamics (F, ψ, d). Thus, it is demonstrated that hyperbolic dynamics can be realized for simple finite systems. Possibly, the simplest one is of two members. This example is helpful for research of hyperbolic dynamics in random processes [11–13].

Example 2.5.3. (*Smale horseshoe*) As an example for hyperbolic chaos, consider the model by Smale [80]. To prove the dynamics, a topologically equivalent bi-infinite symbolic counterpart has been used traditionally [87]. We will demonstrate the existence of chaos by hyperbolic alpha labeling. Take the closed unit square in figure 2.4(a). Contract the set vertically with a coefficient k less than $1/2$, and stretch it horizontally with a factor $1/k$. Then bend it into a horseshoe-shaped figure and place it on the unit square as it is seen in figure 2.4(b). We denote S the map and consider further Smale hyperbolic sets, that is, all bi-infinite trajectories united in the square.

Let us start with vertical rectangles $V_{.0}$ and $V_{.1}$, which are seen in figure 2.5(a). Their images are horizontal rectangles $H_{0.}$ and $H_{1.}$, such that $S(V_{.j}) = H_{j.}$, $j = 0, 1$ in figure 2.5(b). The rectangles are compact as are all figures in what follows. We use indexes $i_j = 0$ or 1 with integer j and $\cdot$ as a dot-index. One can determine recurrently the set of horizontal rectangles $H_{i_{-n}i_{-n+1} \ldots i_{-1}i_0.}$ such that $S^{-1}(H_{i_{-n}i_{-n} \ldots i_{-1}i_0.}) = H_{i_{-n}i_{-n+1} \ldots i_{-1}.}$, and $H_{i_{-n} \ldots i_{-1}i_0.} \subset H_{i_{-n}i_{-n+1} \ldots i_{-1}.}$. Similarly, obtain vertical rectangles $V_{.i_1 i_2 \ldots i_n i_n}$ such that $S(V_{.i_1 i_2 \ldots i_n i_n}) = V_{.i_2 \ldots i_{n-1}i_n}$, $V_{.i_1 \ldots i_n} \subset V_{.i_2 \ldots i_n}$. The first members of the series are seen in figures 2.5.

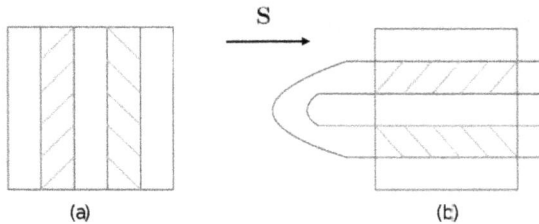

Figure 2.4. Design of the Smale horseshoe map.

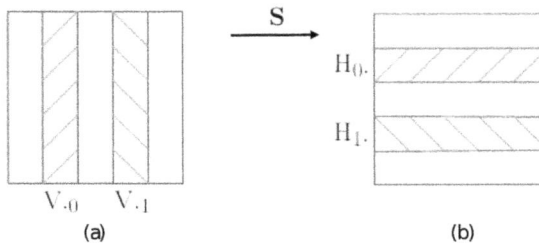

Figure 2.5. First rectangles of the horseshoe construction.

There exists the unique horizontal section $h_{...i_{-n}i_{-n+1}...i_{-1}i_0}$, an intersection of nested rectangles, H_{i_0}, $H_{i_{-1}i_0}$, $H_{i_{-2}i_{-1}i_0}$, ... Analogously, construct vertical line $v_{i_1i_2...i_ni_{n+1}...}$ as intersection of infinite rectangles V_{i_1}, $V_{i_1i_2}$, $V_{i_1i_2i_3}$.... Let $\mathscr{F}_{...i_{-n}i_{-n+1}...i_{-1}i_0 \cdot i_1i_2i_3...}$ be an alpha label for intersection point of lines $v_{i_1i_2...i_ni_{n+1}...}$ and $h_{...i_{-n}i_{-n+1}...i_{-1}i_0}$. The diameter condition is correct, since rectangle $\mathscr{F}_{i_{-n}i_{-n+1}...i_{-1}i_0 \cdot i_1i_2i_3...i_n}$ consists of points included in the intersection of the rectangles $H_{i_{-n}i_{-n+1}...i_{-1}i_0}$ and $V_{i_1i_2...i_{n-1}i_n}$. The separation condition is fulfilled for sets with alpha labels in $\mathscr{F}_{i\cdot j}$, $i, j = 0, 1$. Thus, hyperbolic chaos is valid for the alpha label-set with abstract similarity map $\varphi\colon \mathscr{F} \to \mathscr{F}$. Finally, by using this method of constructing rectangles, one can observe that $S(x) = y$ if and only if x and y are labeled by $\mathscr{F}_{...i_{-2}i_{-1}i_0 \cdot i_1i_2i_3...}$ and $\mathscr{F}_{...i_{-1}i_0i_1 \cdot i_2i_3...}$, respectively. Indeed, $\varphi(\mathscr{F}_{...i_{-2}i_{-1}i_0 \cdot i_1i_2i_3...}) = \mathscr{F}_{...i_{-2}i_{-1}i_0i_1 \cdot i_2i_3...}$. But, $\mathscr{F}_{...i_{-2}i_{-1}i_0 \cdot i_1i_2i_3...}$ is the label for intersection of lines $v_{i_1i_2...}$ and $h_{...i_{-1}i_0}$, and $\mathscr{F}_{...i_{-n}i_{-n+1}...i_{-1}i_0i_1 \cdot i_2i_3...}$ is the label for intersection of lines $v_{i_2...}$ and $h_{...i_{-1}i_0i_1}$. Moreover, $S^{-1}(h_{...i_{-1}i_0}) = h_{...i_{-2}i_{-1}}$, and $S(v_{i_1i_2...}) = v_{i_2i_3...}$. That is, $\varphi(\mathscr{F}_{...i_{-n}i_{-n+1}...i_{-1}i_0 \cdot i_1i_2i_3...}) = \mathscr{F}_{...i_{-n}i_{-n+1}...i_{-1}i_0i_1 \cdot i_2i_3...}$. Hence, the Smale model now exhibits hyperbolic alpha labeling, and there are ultra Poincaré, Li–Yorke, and Devaney chaos for the map. In other words, alpha hyperbolic chaos. Each point of the set is an intersection of horizontal and vertical lines, stable and unstable sets.

Example 2.5.4. *(two-dimensional Baker transformation) We will apply the alpha labeling algorithm to the popular two-dimensional Baker map [68, 79]. Consider it for the unit square in figure 2.6(a).*

The rectangle is stretched to twice as long as it is in figure 2.6(b). Then it is cut vertically in the middle, and the pieces are joined to form the square in figure 2.6(c).

Considering the notations of horizontal rectangles, one can see that H_0 is split in H_{00} and H_{10}, rectangle H_1 is split into H_{01} and H_{11}, such that H_{00} and H_{01} are in H_0 and H_{10} and H_{11}, are in H_1.

Proceeding in this way, we obtain an infinite nested set of horizontal rectangles such that $H_{i_0i_1...i_{n-1}i_n} \subset H_{i_0i_1...i_{n-1}}$, $n = 1, 2, ... , i_j = 0, 1$. Next, denote $h_{i_0i_1...i_{n-1}i_n...}$, the horizontal section, as the intersection of all rectangles. The union of such sections is an uncountable set, the square.

The Baker map transforms the pre-image in figure 2.7(a) to the original square in figure 2.7(c). The pre-image contains vertical rectangles V_0 and V_1 such that $B(V_0) = H_0$ and $B(V_1) = H_1$, where B is for the map. One can see that it is true for

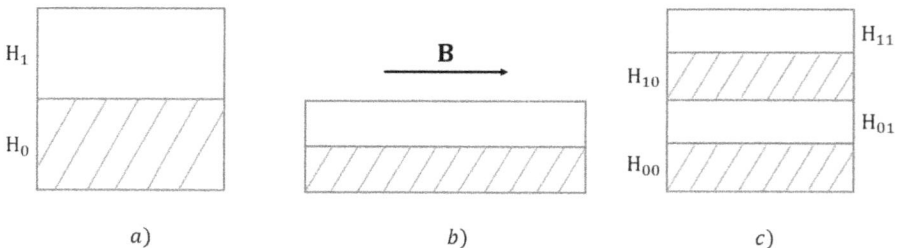

$a)$ $b)$ $c)$

Figure 2.6. Design of the baker transformation.

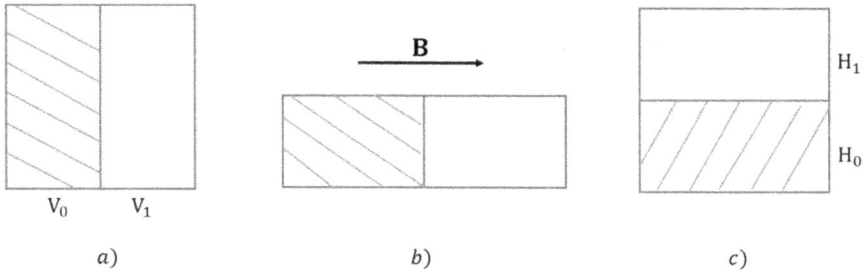

Figure 2.7. The pre-image of the original square.

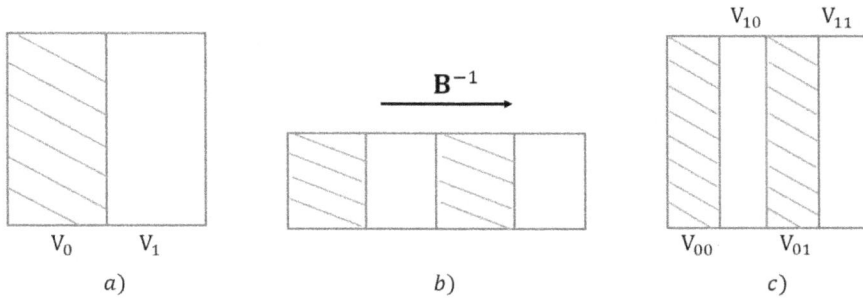

Figure 2.8. The second inverse iteration.

the rectangles in figure 2.8 that the inverse map B^{-1} splits V_0 to rectangles V_{00} and V_{01}, while V_1 to rectangles V_{10} and V_{11}. Repeating application of the inverse map infinitely many times, one constructs a nested sequence of rectangles such that $V_{i_n i_{n-1} \ldots i_1} \subset V_{i_{n-1} \ldots i_1}$, $n = 2, \ldots$, $i_j = 0, 1$. Denote their intersection as $v_{\ldots i_{-n} i_{-n+1} \ldots i_{-1}}$. It is a vertical section. The set of sections makes the square. Then label the intersection of $h_{i_0 i_1 \ldots i_{n-1} i_n \ldots}$ and $v_{\ldots i_{-n} i_{-n+1} \ldots -1}$ by $\mathcal{F}_{\ldots i_{-n} i_{-n+1} \ldots i_{-1} i_0 \cdot i_1 i_2 \ldots i_n \ldots}$ and unite all the alpha labels as a hyperbolic alpha set $\mathcal{F}$.

One can verify that the relation $B(u) = v$ is correct if and only if points u and v are labeled with $\mathcal{F}_{\ldots i_{-n} i_{-n+1} \ldots i_{-1} \cdot i_0 i_1 i_2 \ldots i_n \ldots}$ and $\mathcal{F}_{\ldots i_{-n} i_{-n+1} \ldots i_{-1} i_0 \cdot i_1 \cdot i_2 \ldots i_n \ldots}$, respectively, for some sequences $\ldots i_{-n} i_{-n+1} \ldots i_{-1} i_0 \cdot i_0 i_1 \ldots i_n \ldots$. That is, alpha map φ and Baker B map agree with each other, and the image condition is valid. Each rectangle $\mathcal{F}_{i_{-k} i_{-n+1} \ldots i_{-1} i_0 \cdot i_1 \ldots i_n}$ is a label set for points from the intersection of rectangles $H_{i_{-k} i_{-n+1} \ldots i_{-1} i_0}$ and $V_{i_1 \ldots i_n}$. This immediately implies that the diameter condition is valid. The separation is also correct with second degree and $\varepsilon_0 = 1/4$. That is hyperbolic abstract chaos is proven for the Baker transformation.

Remark 2.5.3. *The last result is complimentary to what is known in literature* [68, 79]. *We have discovered ultra Poincaré chaos by applying alpha-unpredictability. What is significant is that chaos in the negative direction of discrete time is proven in a very detailed way. That is, recurrence as Poisson stability is shown, and divergence is* *described not only by the Lyapunov exponent but also through sequences of convergence and separation for alpha-unpredictability. We do not consider the analytical presentation of the Baker map as it is done in [68, 79], and use the geometrical discussion, which has been demonstrated to be effective. It is correct for any coefficient of stretching.*

We have suggested an abstraction for two-sided dynamical processes with discrete time. A method of chaos generation is provided as dynamics of an abstract similarity map. Thus, all three types of ultra Poincaré, Li–Yorke, and Devaney chaos are proven in domains. It is easily seen that homoclinic and heteroclinic points can be discussed further in the present research framework. The approach is also practical and solid. Thus, comprehension of dynamical complexities has been deepening. If one assumes the backward separation condition, chaotic behavior must be extended for decreasing time. This time, we introduce the concept of hyperbolic chaos by alpha labels, and it has been verified for the Smale horseshoe map and two-dimensional Baker transformation. We are confident this approach can highlight other hyperbolic dynamics such as Anosov diffeomorphisms, Plykin attractors, etc. More benefits exist for multidimensional generalizations, which are possible based on our proposals. Hyperbolic dynamics considers processes with bi-directional infinite chaotic history. This approach does not exclude the possibility of random dynamics analysis [23].

2.6 Modular chaos

In the present section, we apply the method of alpha labeling for dynamics with orbits whirling among infinitely many modules, which are alpha label sets in their turn. Thus, the conservative conditions of chaos existence are extended for modeling of newly constructed irregular dynamics. The research proves that ultra Poincaré chaos is of exceptional use for analysis of stochastic processes. Another significant achievement of the approach is research of *unbounded* motions, though it is suitable for bounded dynamics, too. It is possibly the first time in literature since studies of chaos that already exist are always about compact domains. Examples illustrating the results of modular dynamics are provided.

2.6.1 Introduction

The paradigm of chaos relies on two distance antagonists, proximity and separation. They utilize metrical or topological closeness and divergence considering either couples of orbits [40, 58] or a single trajectory [5]. In the present research, we have renounced the idea that periodic motions' transitivity and density have to be mainly described metrically and topologically. Now, it is suggested for the first time in literature that the closeness can be considered not in metrical sense, but through *indexing*, although separation still entirely relies on distance. That is, the antagonists are still present, but not in the way of conventional theories. The process of indexing is adjacent with the similarity map. More precisely, it is an instrument for mappings to be chaotic. Trajectories move through various alpha- and even-metric spaces when time increases, and each of these spaces (modules) eventually admits a structure suitable for chaos. This is why we say that *modular chaos*—or more precisely, *modular alpha chaos*—is the focus of the present section. Accordingly, if chaos is recognized in dynamical systems by alpha labeling and modules, we shall call it *modular chaos by alpha labels* or *modular alpha chaos*. Undoubtedly, one can consider more complex structures, such as hyperbolic modular alpha chaos and other complicated motions, in the future, but this depends on the needs for applications and the perspective development of the theory.

The method of chaos by alpha labels is initiated in [21, 23]. Results were obtained, which rely on dynamics in a single metrical or topological space, fractals, and bounded random processes such as Bernoulli schemes [15, 21, 23]. Precise and complete information on alpha labeling and its applications is provided in previous sections of the chapter. This time, the approach is enriching research of complexity in two senses. First, we consider infinite multiples of chaotically structured modules. Second, conservative chaos conditions are weakened by considering indexing for closeness. This helps us to determine unbounded deterministic chaotic dynamics and find structures for continuous-time random processes, bounded and unbounded. We have obtained chaos presence in discrete time stochastic dynamics as realizations which coincide with orbits of modular similarity maps, and explain this for continuous time random processes. One must say that ultra Poincaré chaos and alpha-unpredictable sequences [5, 14] are principal for the exploration.

We hope that the suggestions in the chapter will be extended further by considering multidimensional indexing and infinitely dimensional dynamics. Additionally, the definitions can be considered for processes, which are only partially stochastic for hybrid dynamics, where intervals of deterministic and/or stochastic behavior are intermingled.

Let us highlight the probabilistic essence of chaos relations with random processes. Considering modular ultra Poincaré chaos for a stochastic process, one can see that there exists a modular alpha-unpredictable orbit such that any finite realization is an arc of that one. Consequently, we have to conclude that in any experiment, one can observe modular alpha-unpredictable orbits as certain events. This is the main argument for why we compare deterministic and stochastic dynamics.

The study of similarity in science goes back to G. Leibniz, who introduced the notion of recursive self-similarity [88]. The idea of self-similar sets was first considered by Moran [63]. He gave a mathematical definition of geometric construction as a collection of sets satisfying specific conditions. Further definitions of self-similarity and related problems were discussed in papers [30, 43, 48, 50, 51, 54, 55, 65, 71, 82, 83]. Our research is based on a point-set structure in metric spaces and this is why we call it *abstract self-similarity* [21] or *alpha-similarity*. It does not rely on special functions, and the *similarity map* or *alpha map* is naturally aligned with the concept. Bernoulli shift [87] is an example of the abstract similarity map. We suppose that the next extension of results in the book can be obtained on the basis of papers [30, 43, 48, 50, 51, 54, 55, 65, 82].

The potential applications of our research in modular chaos extend beyond theoretical implications. They naturally relate to problems of control and synchronization [47], enriching the theory of stochastic differential equations [69] and Markov chains, [46]. By involving modular chaos results in these studies, we can replicate chaos and alpha dynamics [4, 5, 12, 15, 24–26], opening up new avenues for practical applications. This aspect of our research is particularly inspiring, as it demonstrates the real-world relevance of our findings.

It is of intense interest to study how close and related to each other deterministic and random dynamics are. More precisely, to find features of deterministic chaos in stochastic processes and to recognize probabilistic characteristics in deterministic motions. If the first kind of investigations can be found, for instance, in paper [70], the research [44] argues for deterministic features in random processes. Our analysis is in the second stream, when random processes can be assigned with properties of deterministic chaos. Our results rely not on the statistical parameters, but on ultra Poincaré chaos, alpha-unpredictability and -dynamics [5, 14, 15], which have been introduced and developed in the last four years.

2.6.2 Modular chaos in alpha dynamics

Let a countable collection of compact metric spaces $(F_j, d_j), j = 1, 2, \ldots$ be given with distances d_j. Assume that for every set F_j the following presentation exists

$$\mathscr{F}^j = \left\{ \mathscr{F}^j_{i_1 i_2 \ldots i_n \ldots} : i_k = 1, 2, \ldots, m, \ k = 1, 2, \ldots \right\}, j = 1, 2, \ldots, \qquad (2.19)$$

where m is a natural number not smaller than two and common for all these sets. It means that each element of set F_j is alpha labeled by at least one member of $\mathscr{F}^j$, and every element of $\mathscr{F}^j$ presents a member from F_j. The uniqueness is not necessarily required for the relation. That is, if $\mathscr{F}^j_{i_1 i_2 \ldots i_n \ldots}$ and $\mathscr{F}^j_{j_1 j_2 \ldots j_n, \ldots}$ present the same element f in F_j, it is not necessary that $i_n = j_n$ for all $n = 1, 2, \ldots$. The set $\mathscr{F}^j$ is said to be the *abstract similarity set* or *alpha label set* for F_j. Moreover, we determine that functions $\delta_j(\mathscr{F}^j_{i_1 i_2 \ldots i_n \ldots}, \mathscr{F}^j_{j_1 j_2 \ldots j_n, \ldots}) = d_j(f_1, f_2), j = 1, 2, \ldots,$ if $\mathscr{F}^j_{i_1 i_2 \ldots i_n \ldots}$ and $\mathscr{F}^j_{j_1 j_2 \ldots j_n, \ldots}$ correspond to elements f_1 and f_2 of the set F_j, respectively, such that $\delta_j(\mathscr{F}^j_{i_1 i_2 \ldots i_n \ldots}, \mathscr{F}^j_{j_1 j_2 \ldots j_n, \ldots}) = 0$ for different presentations of a point in F_j.

The following sets (we call them *rectangles* are needed:

$$\mathscr{F}^j_{i_1 i_2 \ldots i_n} = \bigcup_{j_k = 1, 2, \ldots, m} \mathscr{F}^j_{i_1 i_2 \ldots i_n j_1 j_2 \ldots}, \qquad (2.20)$$

where indices $i_1, i_2, \ldots, i_n$, are fixed. It is clear that

$$\mathscr{F}^j \supseteq \mathscr{F}^j_{i_1} \supseteq \mathscr{F}^j_{i_1 i_2} \supseteq \cdots \supseteq \mathscr{F}^j_{i_1 i_2 \ldots i_n} \supseteq \mathscr{F}^j_{i_1 i_2 \ldots i_n i_{n+1}} \cdots, \ i_k = 1, 2, \ldots, m, \ k = 1, 2, \ldots.$$

That is, the rectangles form a nested sequence.

We will say that *diameter condition* is valid for sets $\mathscr{F}^j_{i_1 i_2 \ldots i_n}, j = 1, \ldots$ if

$$\sup_{i_1 i_2 \ldots i_n} \operatorname{diam} \mathscr{F}^j_{i_1 i_2 \ldots i_n} \to 0 \text{ as } n \to \infty, \qquad (2.21)$$

where diam $A = \sup\{\delta_j(\mathbf{x}, \mathbf{y}): \mathbf{x}, \mathbf{y} \in A\}$, for a set A in $\mathscr{F}_j$.

Determine function $\delta_j(A, B) = \inf\{\delta_j(\mathbf{x}, \mathbf{y}): \mathbf{x} \in A, \mathbf{y} \in B\}$, for two nonempty bounded sets A and B in $\mathscr{F}_j$. The set $\mathscr{F}^j, j = 1, 2, \ldots$, satisfies the *separation condition* of degree n if there exist a positive number ε_0^j and natural $n = n(j)$ such that for arbitrary indices $i_1 i_2 \ldots i_n$ one can find indices $j_1 j_2 \ldots j_n$ for which

$$\delta_j\left(\mathscr{F}^j_{i_1 i_2 \ldots i_n}, \mathscr{F}^j_{j_1 j_2 \ldots j_n}\right) > \varepsilon_0^j. \tag{2.22}$$

If the diameter and separation conditions are valid, then $\mathscr{F}^j$ is said to be *abstract chaotic set* for $F_j, j = 1, 2, \ldots$, and *abstract chaotic module* for F.

In what follows, we shall denote δ the alpha-distance which is specified as alpha-distance δ_j for each fixed $j = 1, 2, \ldots$. correspondingly the *modular abstract label space* or *modular alpha space* is the couple $(\mathscr{F}, \delta)$.

Let us introduce maps $\varphi_j: \mathscr{F}^j \to \mathscr{F}^{j+1}, j = 1, 2, \ldots$, such that

$$\varphi_j(\mathscr{F}^j_{i_1 i_2 \ldots i_n \ldots}) = \mathscr{F}^{j+1}_{i_2 i_3 \ldots i_n \ldots}. \tag{2.23}$$

That is, for a fixed integer j, the map φ_j transfers elements of the module $\mathscr{F}^j$ to elements of the neighbor module $\mathscr{F}^{j+1}$ with shifting all the sub-indices by one place to the right. Moreover, we refer to map φ on set $\mathscr{F}$, which is specified by equation (2.23). That is, $\varphi(\mathscr{F}^j) = \varphi_j(\mathscr{F}^j), j = 1, 2, \ldots$. Similarly, accept that $\delta(\mathscr{F}^j_{i_1 i_2 \ldots i_n \ldots}, \mathscr{F}^j_{j_1 j_2 \ldots j_n \ldots}) = \delta_j(\mathscr{F}^j_{i_1 i_2 \ldots i_n \ldots}, \mathscr{F}^j_{j_1 j_2 \ldots j_n \ldots}), j = 1, 2, \ldots$.

We shall say that $(\mathscr{F}, \delta)$ is *alpha modular chaotic space* if the diameter condition is valid for all $j = 1, 2, \ldots$, and there exists a positive separation constant $\varepsilon_0 = \inf_{j=1, 2, \ldots} \varepsilon_0^j$.

Considering iterations of the map, one can verify that

$$(\varphi_{j+n}(\ldots(\varphi_j(\mathscr{F}^j_{i_1 i_2 \ldots i_n})) \ldots) = \varphi^n(\mathscr{F}^j_{i_1 i_2 \ldots i_n}) = \mathscr{F}^{j+n}, \tag{2.24}$$

for arbitrary natural number n and $i_k = 1, 2, \ldots, m, k = 1, 2, \ldots$. That is, any rectangle of order n by iterations of the map is transformed to the whole module. Formulas (2.23) and (2.24) make it reasonable to call φ *affine similarity map*, since the shifts for sub- and sup-indexes are common. One can call it also *modular similarity map* or *modular alpha map*. Moreover, the number n is named *order of similarity* for rectangles. It is clear that the map cannot be considered as a modification of a Bernoulli map, since it is not a shift map. There is no direction in the phase space for the mapping. Shifts happen in indexes, but not in elements of dynamics. Thus, we can say about *modular alpha label dynamics* $(\mathscr{F}, \varphi, \delta)$ or *modular alpha dynamics* that if $\mathscr{F}^j$ does not depend on j, then the last dynamics is just alpha dynamics, and the modular similarity map φ is the *alpha map* of sections 2.1 and 2.2.

Now, let us consider a principal novelty, when closeness is determined through indexing. We shall say that alpha label $\mathscr{F}^j_{i_1 i_2 \ldots i_n \ldots}$ is in (k, m)-neighborhood of $\mathscr{F}^p_{s_1 s_2 \ldots s_n \ldots}, j < p$, if $\mathscr{F}^{j+k}_{i_{k+1} i_{k+2} \ldots i_{k+m}} = F^p_{s_1 s_2 \ldots s_m}$. In other words, $j + k = p, i_{k+1} = s_1, \ldots, i_{k+m} = s_m$.

It is easily seen that the closeness is supported by the diameter condition.

Since the concepts are based on Bernoulli shift [87], known for the symbolic dynamics, it is easily to see that the following chaotic properties are valid.

A point $\mathscr{F}^j_{i_1 i_2 i_3 \dots}, j = 1, 2, \dots$ is periodic with period p if its lower indices are endless repetitions of a block with p terms. The map φ is said to be *modular-periodic*, since $\varphi^{kp}(\mathscr{F}^j_{i_1 i_2 \dots i_n \dots}) = \mathscr{F}^{j+kp}_{i_1 i_2 \dots i_n \dots}$ for any natural k, if point $\mathscr{F}^j_{i_1 i_2 \dots i_n \dots}$ is p-periodic.

Analogously, the map is *modular-dense* or, in other words, *modular-transitive*, since there exists lower sequence-index $i_1 i_2 i_3 \dots$, such that for any sequence $j_1 j_2 j_3 \dots$ and naturals k and l one can find positive integer $j, j < k$, such that $\varphi^{k-j}(\mathscr{F}^j_{i_1 i_2 \dots i_n \dots}) = \mathscr{F}^k_{j_1 j_2 \dots j_n \dots}$, and $i_s = j_s, s = 1, \dots, l$. That is, $\mathscr{F}^j_{i_1 i_2 \dots i_n \dots}$ is in $(j - k, l)$-neighborhood of $\mathscr{F}^k_{j_1 j_2 \dots j_n \dots}$. In other words, we accept that a dense trajectory of the modular dynamics is the union of orbits, which initial points are $\mathscr{F}^j_{i_1 i_2 \dots i_n \dots}$ with $j = 1, 2, 3, \dots$ where the sequence $i_1 i_2 \dots i_n \dots$ is that one specially constructed for transitivity in symbolic dynamics [87] and in our research.

It is easily seen that we can also prove modular transitivity in the following way. Because of the diameter condition for any point $\mathscr{F}^k_{j_1 j_2 \dots j_n \dots}$ and positive number ε one can find a natural j such that $\delta_k(\varphi^{k-j}(\mathscr{F}^j_{i_1 i_2 \dots i_n \dots}), \mathscr{F}^k_{j_1 j_2 \dots j_n \dots}) < \varepsilon$, where the sequence $i_1 i_2 \dots i_n \dots$ is that specific one.

Finally, one can say that periodic points are *modular-dense* in $\mathscr{F}$, since for any point $\mathscr{F}^k_{j_1 j_2 \dots j_n \dots}$ of the set and natural number l there exists a periodic point $\mathscr{F}^m_{i_1 i_2 \dots i_n \dots}, m \leqslant k$ such that it is in $(k - m, l)$-neighborhood of $\mathscr{F}^k_{j_1 j_2 \dots j_n \dots}$. Again, for density of periodic points, we consider those of all modules in the construction, and it can be formulated in terms of alpha distance δ_k.

It is clear that couples $(\mathscr{F}^j, \delta_j), j = 1, 2, \dots$ are not, in general, metric spaces. Nevertheless, the map φ is convenient to extend all attributes of ultra Poincaré, Li–Yorke, and Devaney chaos [15, 40, 58] for the dynamics, since (F_j, d_j) are metric spaces. We shall say that the map φ is *modular chaotic in the Devaney sense* if it is dense with respect to modular-periodic trajectories, modular-transitive, and *modular-sensitive*. The modular-sensitivity means that there exists a positive number ε_0 such that for each element $\mathscr{F}^j_{i_1 i_2 \dots i_n \dots}$ and arbitrary positive κ one can find an element $\mathscr{F}^j_{j_1 j_2 \dots j_n \dots}$ and a natural number k, which satisfy $\delta(\varphi^k(\mathscr{F}^j_{i_1 i_2 \dots i_n \dots}), \varphi^k(\mathscr{F}^j_{j_1 j_2 \dots j_n \dots})) > \varepsilon_0$, despite $\delta(\mathscr{F}^j_{i_1 i_2 \dots i_n \dots}, \mathscr{F}^j_{j_1 j_2 \dots j_n \dots}) < \kappa$.

Theorem 2.6.1. *If $(\mathscr{F}, \delta)$ is a strong alpha modular chaotic space, then dynamics of φ is modular chaotic in sense of Devaney.*

Proof. Modular-transitivity and density for modular-periodicity are valid because of the definitions. For modular-sensitivity, fix a point $\mathscr{F}^j_{i_1 i_2 \dots} \in \mathscr{F}$ and an arbitrary positive number κ. Due to the diameter and separation conditions, there exist an integer k and element $\mathscr{F}^j_{i_1 \dots i_k j_{k+1} \dots}$ such that $0 < \delta(\mathscr{F}^j_{i_1 \dots i_k i_{k+1} \dots}, \mathscr{F}^j_{i_1 \dots i_k j_{k+1} \dots}) < \kappa$ and $\delta(\mathscr{F}^j_{i_{k+1} \dots i_{k+n}}, \mathscr{F}^j_{j_{k+1} \dots j_{k+n}}) > \varepsilon_0$. This proves the sensitivity and the theorem. $\square$

In [5, 14], Poisson stable motion is utilized to differentiate chaotic behavior from periodic motions in Devaney and Li–Yorke types. The dynamics is named ultra Poincaré chaos in the book. Let us provide the definition adapted for the dynamics of present research.

Definition 2.6.1. *A point $\mathscr{F}^j_{i_1 i_2 \ldots} \in \mathscr{F}$, $j = 1, 2, \ldots$, is said to be a modular alpha unpredictable point for an affine similarity map φ, if there exist a positive number ε_0 and two unbounded sequences of natural numbers, κ_n and ζ_n, $n = 1, 2, \ldots$, such that the following two conditions are valid:*

$$\lim_{n\to\infty}\delta_{j+\kappa_n}(\varphi^{\kappa_n}(\mathscr{F}^j_{i_1 i_2\ldots}), \mathscr{F}^{j+\kappa_n}_{i_1 i_2\ldots}) = 0, \quad (\alpha)$$

$$\delta_{j+\kappa_n+\zeta_n}(\varphi^{\kappa_n+\zeta_n}(\mathscr{F}^j_{i_1 i_2\ldots}), \mathscr{F}^{j+\kappa_n+\zeta_n}_{i_1 i_2\ldots}) > \varepsilon_0. \quad (\beta)$$

The relations (α) and (β) above are called *convergence* and *divergence conditions*, respectively.

Following the definition of ultra Poincaré chaos, we shall say that the dynamics of an affine similarity map φ is *modular chaotic in the sense of ultra Poincaré*, if there is a modular alpha unpredictable point for each $j = 1, 2, \ldots$.

Theorem 2.6.2. *If $(\mathscr{F}, \delta)$ is a alpha modular chaotic space, then map φ possesses an alpha modular point for each natural j.*

Theorem 2.6.3. *If $(\mathscr{F}, \delta)$ is an alpha modular chaotic space and each alpha space $(\mathscr{F}^j, \delta_j)$ is alpha complete for each natural j, then map φ possesses modular ultra Poincaré chaos.*

Proof of the last assertion is based on the verification of theorem 2.3.4 applied to the modular similarity map.

Beside modular Devaney and ultra Poincaré chaos, Li–Yorke chaos [58] can be defined in the following way. Additionally to modular-periodic points of dynamics, assume that there exists an uncountable *scrambled set* of non-periodic points in $\mathscr{F}^j$, $j = 1, 2, \ldots$, such that

$$\limsup_{k\to\infty}\delta(\varphi^k(\mathscr{F}^j_{i_1 i_2\ldots}), \varphi^k(\mathscr{F}^j_{j_1 j_2\ldots})) > 0 \quad (\gamma),$$

and

$$\liminf_{k\to\infty}\delta(\varphi^k(\mathscr{F}^j_{i_1 i_2\ldots}), \varphi^k(\mathscr{F}^j_{j_1 j_2\ldots})) = 0 \quad (\delta),$$

for each couple in the set, if $\delta(\mathscr{F}^j_{i_1 i_2\ldots}, \mathscr{F}^j_{j_1 j_2\ldots}) \neq 0$. Finally, assume that for each point $\mathscr{F}^j_{i_1 i_2\ldots}$ of the set and a periodic point $\mathscr{F}^j_{j_1 j_2\ldots}$ it is true that $\limsup_{k\to\infty}\delta(\varphi^k(\mathscr{F}^j_{i_1 i_2\ldots}), \varphi^k(\mathscr{F}^j_{j_1 j_2\ldots})) > 0$. The conditions (γ) and (δ) are called *modular-proximality* and *modular-frequent separation* properties for *modular Li–Yorke chaos*, respectively.

We will say that for sets $\mathscr{F}_{i_1 i_2 \dots i_n}^j, j = 1, \dots$, *strong diameter condition* is valid, if

$$\sup_j \max_{i_1 i_2 \dots i_n} \operatorname{diam} \mathscr{F}_{i_1 i_2 \dots i_n}^j \to 0 \quad \text{as} \quad n \to \infty. \tag{2.25}$$

We shall say that an alpha modular space $(\mathscr{F}, \delta)$ is a *strong alpha modular chaotic space* for the union F of all sets $F_j, j = 1, 2, \dots$, if the strong diameter condition is valid for all $j = 1, 2, \dots$, and there exists a positive separation constant $\varepsilon_0 = \inf_{j=1, 2, \dots} \varepsilon_0^j$. We shall call the map Φ *modular chaotic in the sense of Li–Yorke*, if the corresponding map φ is modular chaotic in the sense of Li–Yorke.

Theorem 2.6.4. *If $(\mathscr{F}, \delta)$ is a strong alpha modular chaotic space, then map φ is modular chaotic in the Li–Yorke sense.*

The proof of the assertion is similar to that of theorem 6.35 in [38] for the shift map defined in the space of symbolic sequences.

2.6.3 Modular chaos through alpha labeling

Let us determine map $\Phi\colon F_j \to F_{j+1}, j = 1, 2, \dots$, a geometrical presentation of the similarity map. Fix a number j and a point f from F_j. Then $\Phi(f)$ is the collection of all elements in F_{j+1}, which alpha labels are values of the function φ at alpha labels of f. It is clear that Φ is a set-valued mapping, but we shall call it a map for simplicity. In what follows, we shall describe chaotic properties of map Φ in terms of dynamics of the map φ. For example, map Φ is *modular chaotic in the sense of ultra Poincaré* if corresponding map φ is modular chaotic in that sense.

The following theorems can be of strong interest:

Theorem 2.6.5. *If $(\mathscr{F}, \delta)$ is an alpha modular chaotic space and each space $(F_j, d_j), j = 1, 2, \dots$ is complete, then map Φ possesses modular ultra Poincaré chaos.*

Theorem 2.6.6. *If $(\mathscr{F}, \delta)$ is a strong alpha modular chaotic space, then the map Φ is modular chaotic in the Li–Yorke sense.*

Theorem 2.6.7. *If $(\mathscr{F}, \delta)$ is an alpha modular chaotic space, then dynamics of Φ is modular chaotic in the Devaney sense.*

One can easily see that, in the case when (F_j, d_j) does not depend on $j = 1, 2, \dots$, we obtain alpha chaos, which was introduced in previous sections. That is, alpha modular chaos can be considered a generalization of alpha chaos.

2.6.4 Examples of modular chaos

Let us provide examples of deterministic modular chaos.

Example 2.6.1. *Take into account metric spaces (F^j, d_j), $j = 1, 2, \ldots$ of sequences $f = (i_1^j, i_2^j, \ldots)$, which coordinates are elements of finite sets $(a_1^j, a_2^j, \ldots, a_m^j)$, m is a natural number, and metrics*

$$d_j(f_1, f_2) = \sum_{l=1}^{\infty} \frac{\rho_j(i_l^j, k_l^j)}{2^{l-1}},$$

where $f_1 = (i_1^j, i_2^j, \ldots)$ and $f_2 = (k_1^j, k_2^j, \ldots)$ are members of F^j, $j = 1, 2, \ldots$ and ρ_j are distances for elements of the alphabets. Considering modules $\mathscr{F}^j$, $j = 1, 2, \ldots$ as sets of alpha labels $\mathscr{F}_{i_1 i_2 \ldots}^j = (i_1^j, i_2^j, \ldots)$ and the abstract similarity map we can deal with alpha modular chaos. It is clear that the diameter condition is valid, but for a strong version of the property some additional conditions on the distances are required. Analogously, the metrics are requested to satisfy extra conditions for the separation property.

Example 2.6.2. *It is known that there exists a Cantor set $\mathscr{C}$ in section $[0, 1]$, such that to each point of the set alpha label $\tau_{i_1 i_2 \ldots}$, $i_k = 0, 1$, $k = 1, 2, \ldots$ is assigned [40, 87]. Consider the alpha modular set $\mathscr{T}$, which consists of the alpha labels, and introduce distance $\delta(\tau_{i_1 i_2 \ldots}, \tau_{j_1 j_2 \ldots}) = |c_1 - c_2|$, where $|\cdot|$ is the absolute value, and $\tau_{i_1 i_2 \ldots}$ with $\tau_{j_1 j_2 \ldots}$ are alpha labels for points c_1 and c_2 in the set. Consider shifts $\mathscr{C}^j$, $j = 1, 2, \ldots$ of the Cantor set $\mathscr{C}$ from the last example along the real axis by natural numbers. That is, $\mathscr{C}^j = \{c + (j - 1) | c \in \mathscr{C}\}$, $j = 1, 2, \ldots$. Accordingly, we alpha label points of the Cantor sets as $\tau_{i_1 i_2 \ldots}^j$, $i_k = 0, 1$, $k, j = 1, 2, \ldots$. Determine the distance*

$$d_j(c_1, c_2) = \delta_j(\tau_{i_1 i_2 \ldots}^j, \tau_{j_1 j_2 \ldots}^j) = \sum_{k=1}^{\infty} \frac{|i_k - j_k|}{2^{k-1}}, j = 1, 2, \ldots,$$

if $\tau_{i_1 i_2 \ldots}^j$ and $\tau_{j_1 j_2 \ldots}^j$ are alpha labels of the points c_1 and c_2 of the set $\mathscr{C}^j$, $j = 1, 2, \ldots$. By considering similarity map

$$\varphi(\tau_{i_1 i_2 \ldots}^j) = \tau_{i_2 i_3 \ldots}^{j+1}$$

we arrange alpha modular chaos on the union of Cantor sets $\mathscr{C}^j$, corresponding to modules $\mathscr{T}^j$, $j = 1, 2, \ldots$. It is clear that the dynamics satisfies strong diameter and separation conditions, and all the three types, ultra Poincaré, Li–Yorke, and Devaney modular chaos are valid.

To find a concrete meaning for the last construction, let us consider the non-autonomous scalar equation $x_{i+1} = (i + 1) + m(x_i - i)(1 - x_i + i)$, $m > 4$. Fix $\mathscr{C}$ as the invariant set for the map $mx(1 - x)$ in the unit section. It is clear that the union of sets $\mathscr{C}^j$, $j = 1, 2, \ldots$ is the domain for the modular chaos of the equations.

The first example above is completely illustrative and similar to that example with symbolic sequences for alpha label dynamics, but the second one can be considered as a way to study unbounded models strictly relating to real world problems. One should remark that besides Example 2.6.9, it is easily possible to construct modular chaos with finite elements spaces. They can be effective for unbounded and bounded stochastic processes.

2.7 Abstract fractals: alpha spaces with measures

Fractals are a class of sophisticated geometric sets with certain properties. One of the main features of the objects is self-similarity, which can be defined as a property whereby parts hold similarity to the whole at any level of magnification [2]. The idea of self-similarity was formulated in the 17th century when Leibniz introduced the notions of recursive self-similarity [88]. The first mathematical definition of self-similar shape was introduced in 1872 by Karl Weierstrass during his study of functions that were continuous but not differentiable. The most famous examples of fractals that display exact self-similarity are the Cantor set, Koch curve, and Sierpinski gasket and carpet, which were discovered by G. Cantor in 1883, Helge von Koch in 1904 and W. Sierpinski in 1916, respectively. Julia sets, discovered by G. Julia and P. Fatou in 1917–19, gained significance in being generated using dynamics of iterative functions. In 1979, B. Mandelbrot visualized the sets and determined fractals as sets with Hausdorff dimension strictly larger than topological one [60].

This time, we are introducing a new mathematical concept called *abstract fractal*. It attempts to establish solid foundation by abstracting the idea of self-similarity. We define abstract fractals as collections of points in metric measure spaces. They are represented through iterative construction algorithms with specific conditions and are introduced to govern the relationship between sets at each iteration. Our approach for construction is based on the concept of porosity rather than the roughness notion introduced by Mandelbrot. The concept is an intrinsic property of materials, and it is usually defined as the ratio of void to total volume [27]. The concept is essential in several research fields such as geology, soil mechanics, material science, civil engineering, etc. [27, 45] Fractal geometry has been widely used to study the properties of porous materials. However, the concept of porosity was not utilized as a criterion for fractal structures, and the relevant researchers have investigated the relationship between porosity and property to be fractal [39]. For instance, several researchers [85] determined the non-integer dimension of some pore structures using their properties. The simplicity and importance of the porosity concept insistently invite us to develop new definitions of fractals. In other words, the property should be involved in fractal theory as a feature, which can be compared with fractional dimension. It needs to specify the concept of porosity for surfaces and lines. In the present study, we ignore equivalence for definitions of fractals in terms of porosity and those through self-similarity and dimension. Instead, we introduce abstract definitions which, we hope, will be helpful to in application areas.

2.7.1 Abstract fractals as alpha induced sets

We shall consider the metric measure space defined by the triple (X, d, μ), where (X, d) is a compact metric space, d is a metric on X and μ is a measure on X.

To construct an abstract fractal, let us consider the initial set $F \subset X$ and fix two natural numbers m and M such that $1 < m < M$. We assume that there exist M nonempty disjoint subsets, F_i, $i = 1, 2, \ldots, M$, such that $F = \cup_{i=1}^{M} F_i$. For each $i = 1, 2, \ldots, m$, again, there exist M nonempty disjoint subsets F_{ij}, $j = 1, 2, \ldots, M$ such that $F_i = \cup_{j=1}^{M} F_{ij}$. Generally, for each $i_1, i_2, \ldots, i_n$, $i_k = 1, 2, \ldots, m$, there exist M nonempty disjoint sets $F_{i_1 i_2 \ldots i_n j}$, $j = 1, 2, \ldots, M$, such that $F_{i_1 i_2 \ldots i_n} = \cup_{j=1}^{M} F_{i_1 i_2 \ldots i_n j}$, for each natural number n. The following conditions are needed:

There exist two positive numbers, r and R, such that for each natural number n we have

$$ r \leqslant \frac{\sum_{j=1}^{m} \mu(F_{i_1 i_2 \ldots i_{n-1} j})}{\sum_{j=m+1}^{M} \mu(F_{i_1 i_2 \ldots i_{n-1} j})} \leqslant R. \tag{2.26} $$

where $i_k = 1, 2, \ldots, m$, $k = 1, 2, \ldots, n-1$. We call the relation (2.26) the *ratio condition*. The numbers r and R in (2.26) are characteristics for porosity. Another condition is the *adjacent condition* and it is formulated as follows: For each $i_1 i_2 \ldots i_n$, $i_k = 1, 2, \ldots, m$ there exists j, $j = m + 1, m + 2, \ldots, M$, such that

$$ \delta(F_{i_1 i_2 \ldots i_n}, F_{i_1 i_2 \ldots i_{n-1} j}) = 0. \tag{2.27} $$

We call $F_{i_1 i_2 \ldots i_{n-1} i_n}$ a complement set of order n if $i_k = 1, 2, \ldots, m$, $k = 1, 2, \ldots, n-1$ and $i_n = m + 1, m + 2, \ldots, M$.

An accumulation point of any couple of complement sets does not belong to any of them. We dub this stipulation the *accumulation condition*.

Let us define the diameter of a bounded subset A in X by $\mathrm{diam}(A) = \sup\{\delta(\mathbf{x}, \mathbf{y}): \mathbf{x}, \mathbf{y} \in A\}$. Considering the above construction, we assume that the *diameter condition* holds for the sets $F_{i_1 i_2 \ldots i_n}$, i.e.,

$$ \sup_{i_k = 1, 2, \ldots, M} \mathrm{diam}\, F_{i_1 i_2 \ldots i_n} \to 0 \quad \text{as} \quad n \to \infty. \tag{2.28} $$

Fix an infinite sequence $i_1 i_2 \ldots i_n \ldots$. The diameter conditions as well as the compactness of X imply that there exists a sequence p_n, such that $p_0 \in F$, $p_1 \in F_{i_1}$, $p_2 \in F_{i_1 i_2}$, $\ldots, p_n \in F_{i_1 i_2 \ldots i_n}$, $n = 1, 2, \ldots$, which converges to a point in X. The point is labeled by $\mathscr{F}_{i_1 i_2 \ldots i_n \ldots}$.

We define *abstract fractal* $\mathscr{F}$ as collection of points $F_{i_1 i_2 \ldots i_n \ldots}$ such that $i_k = 1, 2, \ldots, m$, that is

$$ \mathscr{F} = \{\mathscr{F}_{i_1 i_2 \ldots i_n \ldots} \mid i_k = 1, 2, \ldots, m\}, \tag{2.29} $$

provided that the above four conditions hold. The subsets of $\mathscr{F}$ can be represented by

$$ \mathscr{F}_{i_1 i_2 \ldots i_n} = \{\mathscr{F}_{i_1 i_2 \ldots i_n i_{n+1} i_{n+2} \ldots} \mid i_k = 1, 2, \ldots, m\}, \tag{2.30} $$

where $i_1 i_2 \ldots i_n$ are fixed numbers. We call such subsets *subfractals* of order n. Because of the bijection, the metric δ determined through the metric d and the measure ζ of subsets in $\mathscr{F}$ on the basis of μ can be easily introduced. Thus, we have obtained the triple $(\mathscr{F}, \delta, \zeta)$ as the *abstract fractal* or *alpha-fractal*. If it is necessary to research the geometry with complement sets, one can consider the abstract counterpart for the whole space (X, d, μ).

It is clear that an abstract fractal is with an alpha set, and there is alpha chaos if the separation condition is fulfilled.

2.7.2 Alpha fractals in 2D geometry

In this subsection we find the pattern of abstract fractals in some geometrical well-known sets Sierpinski carpet, Pascal triangles and Koch curve.

2.7.2.1 Assigning an alpha fractal to the Sierpinski Carpet
To construct an abstract fractal corresponding to the Sierpinski carpet, let us consider a square as an initial set F. Firstly, we divide F into nine ($M = 9$) equal squares and denote them by F_i, $i_1 = 1, 2, \ldots, 9$ (see figure 2.9(a)). In the second step, each square F_i, $i = 1, 2, \ldots, 8$ is again divided into nine equal squares denoted as F_{ij}, $j = 1, 2, \ldots, 9$. Figure 2.9(b) illustrates the sub-squares of F_1. We continue in this way such that at the nth step, each set $F_{i_1 i_2 \ldots i_{n-1}}$, $i_k = 1, 2, \ldots, 8$, is divided into nine subsets $F_{i_1 i_2 \ldots i_{n-1} j}$, $j = 1, 2, \ldots, 9$. For the Sierpinski carpet the number m is 8, and the measure ratio (2.26) can be evaluated as follows. If we consider the first order sets F_{i_1}, $i_1 = 1, 2, \ldots, 9$, then

$$\frac{\sum_{j=1}^{8} \mu(F_j)}{\mu(F_9)} = 8.$$

Thus, the ratio condition holds. From the construction, we can see that each $F_{i_1 i_2 \ldots i_n}$, $i_k = 1, 2, \ldots, 8$ has common boundary with $F_{i_1 i_2 \ldots i_{n-1} j}$, $j = 9$. Therefore,

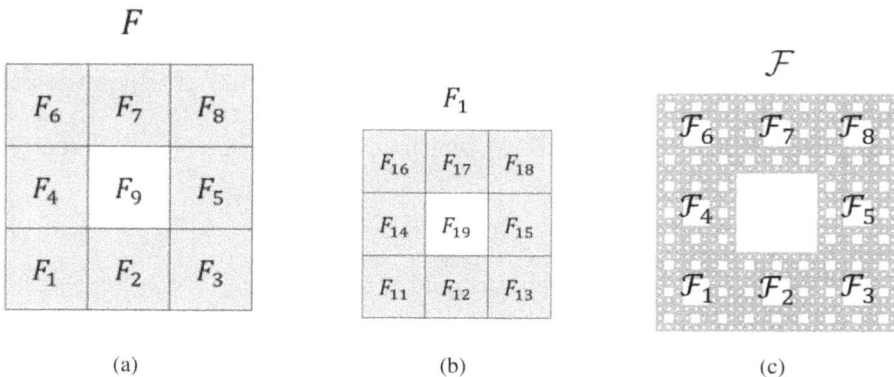

Figure 2.9. On the way to the abstract fractal for the Sierpinski carpet. Reproduced with permission from [89]. CC BY NC ND 4.0.

the adjacent condition holds. Since the construction consists of division into smaller parts, the diameter condition is also valid. Moreover, it is clear that the accumulation condition holds as well.

As a result, the points of the desired abstract fractal $\mathscr{F}$ is a collection of labels for points $F_{i_1 i_2 \ldots i_n \ldots}$,

$$\mathscr{F} = \{\mathscr{F}_{i_1 i_2 \ldots i_n \ldots} | i_k = 1, 2, \ldots, 8\}.$$

Figure 2.9(c) shows the set $\mathscr{F}$ and illustrates its 1st order subfractals.

2.7.2.2 The Pascal triangle admits an abstract fractal structure

Pascal triangle is a mathematical structure that consists of a triangular array of numbers. Triangular fractals can be obtained if these numbers are plotted using specific moduli. The Sierpinski gasket, for instance, is the Pascal's triangle modulo 2. Let us build an abstract fractal on the basis of a fractal associated with Pascal triangle modulo 3. Consider an equilateral triangle as an initial set F. In the first step, we divide F into nine smaller equilateral triangles and denote them as F_i, $i = 1, 2, \ldots, 9$ as shown in figure 2.10(a). Next, each triangle F_i, $i = 1, 2, \ldots, 6$ is again divided into nine equilateral triangles named as F_{ij}, $j = 1, 2, \ldots, 9$. Figure 2.10(a) illustrates the second step for the set F_1. Similarly, the subsequent steps are performed such that at the nth step, each set $F_{i_1 i_2 \ldots i_{n-1}}$, $i_k = 1, 2, \ldots, 6$, is divided into nine subsets $F_{i_1 i_2 \ldots i_{n-1} j}$, $j = 1, 2, \ldots, 9$. In this case we have $m = 6$ and $M = 9$. Therefore,

$$\frac{\sum_{j=1}^{6} \mu(F_j)}{\sum_{j=7}^{9} \mu(F_j)} = 2,$$

and the ratio condition holds. One can also verify that the adjacent, the accumulation, and the diameter conditions are also valid. Based on this, the points of the fractal can be defined by $F_{i_1 i_2 \ldots i_n \ldots}$, and labeled as $\mathscr{F}_{i_1 i_2 \ldots i_n \ldots}$, and thus, corresponding to the Pascal triangle abstract fractal is defined by

$$\mathscr{F} = \{\mathscr{F}_{i_1 i_2 \ldots i_n \ldots} | i_k = 1, 2, \ldots, 6\},$$

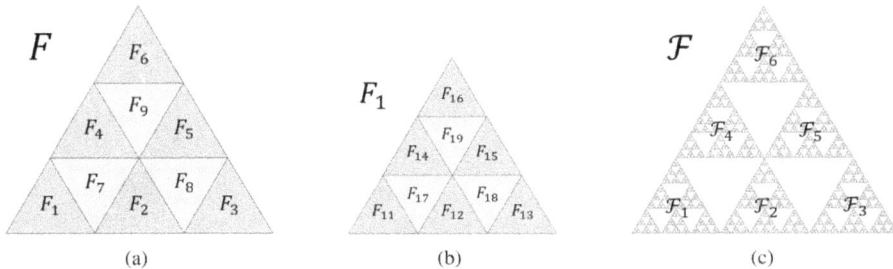

Figure 2.10. Stages for the abstract fractal of the Pascal triangle modulo 3. Reproduced with permission from [89]. CC BY NC ND 4.0.

and the *nth* order subfractals can be written as

$$\mathscr{F}_{i_1 i_2 \ldots i_n} = \{\mathscr{F}_{i_1 i_2 \ldots i_n i_{n+1} \ldots} \mid i_k = 1, 2, \ldots, 6\},$$

where $i_1 i_2 \ldots i_n$ are fixed numbers.

2.7.2.3 An alpha fractal for the Koch curve

In this subsection, we shall show how to build an abstract fractal $\mathscr{F}$ conformable to the Koch curve. For this purpose, we consider the following construction of the Koch curve. Start with with an isosceles triangle F with base angles of $30°$. The first step of the construction consists in dividing F into three equal-area triangles F_1, F_2 and F_3 (see figure 2.11(b)). The triangles F_1 and F_2 are isosceles with base angles of $30°$, whereas the central triangle F_3 is an equilateral one. In the second step, each F_i, $i = 1, 2$ is similarly divided into three triangles, two isosceles, F_{i1} and F_{i2}, and one equilateral, F_{i3}. Figure 2.11(c) illustrates the step. In each subsequent step, the same procedure is repeated for each isosceles triangles resulting from the preceding step. That is, in the *nth* step, each $F_{i_1 i_2 \ldots i_{n-1}}$, $i_k = 1, 2$ is divided into three parts, two isosceles triangles $F_{i_1 i_2 \ldots i_{n-1} j}$, $j = 1, 2$, with base angles of $30°$, and one equilateral triangle $F_{i_1 i_2 \ldots i_{n-1} 3}$. In this construction, we have $m = 2$ and $M = 3$, thus, the measure ratio is

$$\frac{\mu(F_1) + \mu(F_2)}{\mu(F_3)} = 2,$$

and the ratio condition holds. From the construction, it is clear that the adjacent, the accumulation, and the diameter conditions are also valid. Based on this, the points in F can be labeled $\mathscr{F}_{i_1 i_2 \ldots i_n \ldots}$, and thus, a corresponding abstract fractal for the Koch curve is defined by

$$\mathscr{F} = \{\mathscr{F}_{i_1 i_2 \ldots i_n \ldots} \mid i_k = 1, 2\}.$$

The *nth* order subfractals of $\mathscr{F}$ are represented by

$$\mathscr{F}_{i_1 i_2 \ldots i_n} = \{\mathscr{F}_{i_1 i_2 \ldots i_n i_{n+1} i_{n+2} \ldots} \mid i_k = 1, 2\}, \tag{2.31}$$

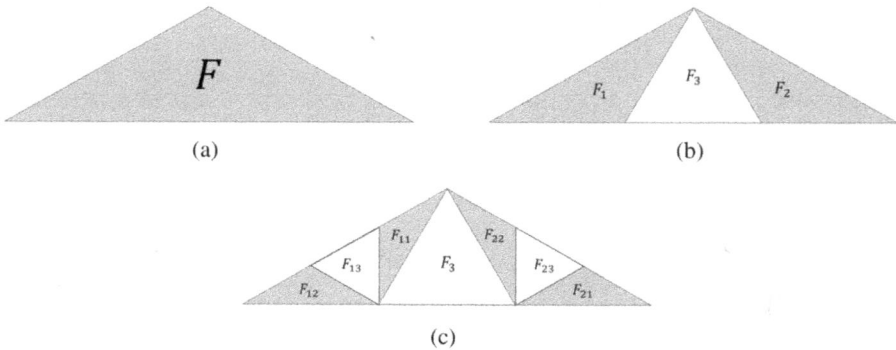

(a)

(b)

(c)

Figure 2.11. The first three steps for the abstract fractal of the Koch curve. Reproduced with permission from [89]. CC BY NC ND 4.0.

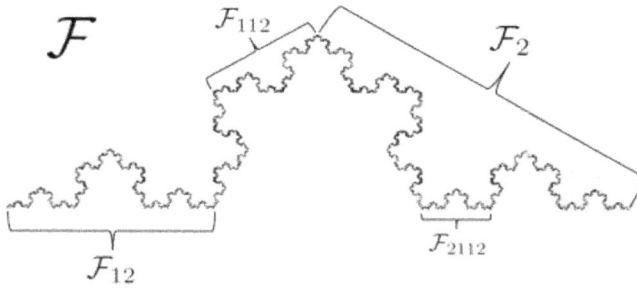

Figure 2.12. Subfractals of the Koch curve. Reproduced with permission from [89]. CC BY NC ND 4.0.

where $i_1 i_2 \ldots i_n$ are fixed numbers. Figure 2.11 illustrates examples of $1st$, $2nd$, $3rd$ and $4th$ subfractals of the abstract Koch curve (figure 2.12).

2.8 Notes

We have announced alpha labeling results in several papers and a book. During this activity, delicate features of the concepts and conditions for constricting the fundamentals have been fine-tuned and improved. More details are involved for an explanation of the theoretical processes. The terminology has been adapted to make the basics more transparent and less confusing. Thus, the present chapter completely describes what we are doing in this direction. Despite the alpha dynamics and chaos, and that abstract fractals are at the highest point of the research, we are still determining if other concepts of the investigation, which here play auxiliary roles, will not be a foundation for other directions of mathematics and industry. This is what we can say about alpha sets, alpha distance, and alpha spaces. The suggestions will be of practical use in other general mathematics areas and various sorts of complexity, such as the theory of algorithms, turbulence, stochastic dynamics, etc. Another remark must be made for alpha modular and alpha hyperbolic dynamics novelties: they are only in their first stage of development. We strongly believe that the necessity to "find order" in unbounded irregular processes will attract attention to what is called modular dynamics and chaos in the present research. Alpha hyperbolic dynamics and chaos will simplify what is known already through many insights. Then, they will give a push not only for new theoretical breakthroughs but also their applications in the solution of essential real-world problems, since the phenomenon of hyperbolic concepts in mathematics is a dialectical reflection of what is happening in the world and is also strictly connected to the chaos, an essential condition of the world existence.

References

[1] Addabbo T, Fort A, Rocchi S and Vignoli V 2011 Digitized chaos for pseudo-random number generation in cryptography *Chaos-Based Cryptography: Theory, algorithms and Applications* (Berlin: Springer) pp 67–97

[2] Addison P S 1997 *Fractals and Chaos: An Illustrated Course* (Bristol: Institute of Physics Publishing)

[3] Adler R, Konheim A and McAndrew M 1965 Topological entropy *Trans. Am. Math. Soc.* **114** 309–19

[4] Akhmet M and Fen M O 2016 *Replication of Chaos in Neural Networks, Economics and Physics* (Berlin: Springer)

[5] Akhmet M and Fen M O 2016 Poincaré chaos and unpredictable functions *Commun. Nonlinear Sci. Numer. Simul.* **48** 85–94

[6] Akhmet M and Fen M O 2016 Unpredictable points and chaos *Commun. Nonlinear Sci. Numer. Simul.* **40** 1–5

[7] Akhmet M and Fen M O 2017 Existence of unpredictable solutions and chaos *Turk. J. Math.* **41** 254–66

[8] Akhmet M and Fen M O 2018 Non-autonomous equations with unpredictable solutions *Commun. Nonlinear Sci. Numer. Simul.* **59** 657–70

[9] Akhmet M, Fen M O and Alejaily E M 2020 *Dynamics with Chaos and Fractals* (Berlin: Springer)

[10] Akhmet M, Fen M O, Tleubergenova M and Zhamanshin A 2019 Unpredictable solutions of linear differential and discrete equations *Turk. J. Math.* **43** 2377–89

[11] Akhmet M 2021 *Domain Structured Dynamics* (Bristol: IOP Publishing)

[12] Akhmet M 2022 Unpredictablity in Markov chains *Carpathian J. Math.* **38** 13–9

[13] Akhmet M and Tola A 2020 Unpredictable strings *Kazakh Math. J.* **20** 16–22

[14] Akhmet M and Fen M O 2016 Unpredictable points and chaos *Commun. Nonlinear Sci. Numer. Simul.* **40** 1–5

[15] Akhmet M, Fen M O and Alejaily E M 2020 *Dynamics with Chaos and Fractals* (Berlin: Springer)

[16] Akhmet M and Fen M O 2014 Entrainment by chaos *J. Nonlinear Sci.* **24** 411–39

[17] Akhmet M and Fen M O 2015 Attraction of Li-Yorke chaos by retarded SICNNs *Neurocomputing* **147** 330–42

[18] Akhmet M U 2009 Creating a chaos in a system with relay *Int. J. Qualit. Th. Diff Equ. Appl.* **3** 3–7

[19] Akhmet M U 2009 Dynamical synthesis of quasi-minimal sets *Int. J. Bifurcation Chaos* **19** 2423–7

[20] Akhmet M U 2009 Shadowing and dynamical synthesis *Int. J. Bifurcation Chaos* **19** 3339–46

[21] Akhmet M and Alejaily E M 2021 Abstract similarity, chaos and fractals *Discrete Continuous Dyn. Syst. Ser.* B **26** 2479–97

[22] Akhmet M and Alejaily E M 2019 Domain-structured chaos in a Hopfield neural network *Int. J. Bifurcation Chaos* **29** 195–205

[23] Akhmet M 2019 Domain-structured chaos in discrete random processes ArXiv e-prints (1912.10478)

[24] Akhmet M, Baskan K and Yesil C 2022 Delta synchronization of Poincaré chaos in gas discharge-semiconductor systems *Chaos* **32** 083137

[25] Akhmet M, Baskan K and Yesil C 2023 Revealing chaos synchronization below the threshold in coupled Mackey-glass systems *Math. Open Access* **11** 3197

[26] Akhmet M, Yesil C and Baskan K 2023 Synchronization of chaos in semiconductor gas discharge model with local mean energy approximation *Chaos Solitons Fractals* **167** 113035

[27] Anovitz L M and Cole D R 2015 Characterization and analysis of porosity and pore structures *Rev. Mineral. Geochem.* **80** 61–164

[28] Brooks J, Cairns G, Davis G, Banks J and Stacey P 1992 On Devaney's definition of chaos *Amer. Math. Monthly* **99** 332–4

[29] Birkhoff G D 1991 *Dynamical Systems* (Providence, RI: Colloquium Publications)

[30] Bandt C and Graf S 1992 Self-similar sets 7. A characterization of self-similar fractals with positive Hausdorff measure *Proc. Am. Math. Soc.* **114** 995–1001

[31] Bowen R 1970 Markov partitions and minimal sets for axiom A diffeomorphisms *Am. J. Math.* **92** 907–18

[32] Bowen R 1970 Markov partitions for axiom a diffeomorphisms *Am. J. Math.* **92** 725–47

[33] Bowen R 1973 Symbolic dynamics for hyperbolic flows *Am. J. Math.* **95** 429–40

[34] Bowen R 1975 A horseshoe with positive measure *Invent. Math.* **29** 203–4

[35] Bowen R and Ruelle D 1975 The ergodic theory of axiom A flows *Invent. Math.* **29** 181–202

[36] Bunimovich L and Ya S 1980 Markov partitions for dispersed billiards *Commun. Math. Phys.* **78** 247–80

[37] Cartwright M and Littlewood J 1945 On nonlinear differential equations of the second order i: the equation $y'' - k(1 - y^2)y' + y = bk \cos(\lambda t + a)$, k large *J. London Math. Soc.* **20** 180–9

[38] Chen G and Huang Y 2011 *Chaotic Maps: Dynamics, Fractals and Rapid Fluctuations, Synthesis Lectures on Mathematics and Statistics* (San Rafael, CA: Morgan & Claypool Publishers)

[39] Davis H T 1989 On the fractal character of the porosity of natural sandstone *Europhys. Lett.* **8** 629–32

[40] Devaney R L 1987 *An Introduction to Chaotic Dynamical Systems* (Menlo Park, CA: Addison-Wesley)

[41] Devaney R L and Nitecki Z 1979 Shift automorphisms in the Henon mapping *Math. Phys.* **67** 137–46

[42] Ditto W L, Murali K and Sinha S 2008 Chaos computing: ideas and implementations *Phil. Trans. R. Soc.* A **366** 653–64

[43] Edgar G A 1990 *Measure, Topology, and Fractal Geometry* (New York: Springer)

[44] Eisencraft M, Monteiro L H A and Soriano D C 2017 White Gaussian chaos *IEEE Commun. Lett.* **21** 17–9

[45] Ganji D D and Kachapi S H H 2015 *Application of Nonlinear Systems in Nanomechanics and Nanofluids: Analytical Methods and Applications* (New York: Elsevier)

[46] Gagniuc P A 2017 *Markov Chains: From Theory to Implementation and Experimentation* (Hoboken, NJ: Wiley)

[47] Gonzáles-Miranda J M 2004 *Synchronization and Control of Chaos* (London: Imperial College Press)

[48] Falconer K J 1995 Sub-self-similar sets *Trans. Amer. Math. Soc.* **347** 3121–9

[49] Hartman P 2002 Ordinary *Differential equations* (New York: SIAM)

[50] Hata M 1985 On the structure of self-similar sets *Japan J. Appl. Math.* **2** 381–414

[51] Hutchinson J 1981 Fractals and self-similarity *Indiana Univ. Math. J.* **30** 713–47

[52] Hilmy H 1936 Sur les ensembles quasi-minimaux dans les systèmes dynamiques *IAnn. Math.* **37** 899–907

[53] Guckenheimer J and Holmes P J 1997 *Nonlinear Oscillations, Dynamical Systems and Bifurcations of Vector Fields* (New York: Springer)

[54] Jorgensen P E T 2006 *Analysis and Probability: Wavelets, Signals, Fractals* Graduate Texts in Mathematics vol 234 (New York: Springer)

[55] Lau K S, Ngai S M and Rao H 2001 Iterated function systems with overlaps and the self-similar measures *J. London Math. Soc.* **63** 99–115

[56] Levinson N 1949 A second order differential equation with singular solutions *Ann. Math.* **50** 127–53

[57] Li R and Zhou X 2013 A note on chaos in product maps *Turk. J. Math.* **37** 665–75

[58] Li T Y and Yorke J A 1975 Period three implies chaos *Am. Math. Mon.* **82** 985–92

[59] Lorenz E N 1984 The local structure of chaotic attractor in four dimensions *Physica* **13D** 90–104

[60] Mandelbrot B B 1983 *The Fractal Geometry of Nature* (New York: Freeman)

[61] Miller A 2019 Unpredictable points and stronger versions of Ruelle-Takens and Auslander-Yorke chaos *Topology Appl.* **253** 7–16

[62] Mischaikow K and Mrozek M 1995 Chaos in the Lorenz equations: a computer assisted proof *Bull. AMS* **32** 66–72

[63] Moran P A P 1946 Additive functions of intervals and Hausdorff measure *Math. Proc. Camb. Phil. Soc.* **42** 15–23

[64] Masoller C, Schifino A C and Sicardi R L 1995 Characterization of strange attractors of Lorenz model of general circulation of the atmosphere *Chaos, Solit. Fract.* **6** 357–66

[65] Ngai S M and Wang Y 2001 Hausdorff dimension of overlapping self-similar sets *J. London Math. Soc.* **63** 665–72

[66] Nemytskii V V and Stepanov V V 1960 *Qualitative Theory of Differential equations* (Princeton, NJ: Princeton University Press)

[67] Newhouse S 1980 *Lectures on Dynamical Systems Progress in Mathematics* (New York: Birkhauser)

[68] Nicolis G and Prigojin I 1989 *Exploring Complexity* (New York: Freeman)

[69] Oksendal B K 2003 *Stochastic Differential equations: An Introduction with Applications* (Berlin: Springer)

[70] Ornstein D 1970 Bernoulli shifts with the same entropy are isomorphic *Adv. Math.* **4** 337–52

[71] Pesin Y and Weiss H 1996 On the dimension of deterministic and random cantor-like sets, symbolic dynamics, and the Eckmann-Ruelle conjecture *Commun. Math. Phys.* **182** 105–53

[72] Poincaré H 1957 *New Methods of Celestial Mechanics Volumes I-III* (New York: Dover)

[73] Robinson C 1995 *Dynamical Systems: Stability, Symbolic Dynamics, and Chaos* (Boca Raton, FL: CRC Press)

[74] Ruelle D 1978 An inequality of the entropy of differentiable maps *Bol. Soc. Bras. Mat—Bull. Braz. Math. Soc.* **9** 83–7

[75] Sell G R 1971 *Topological Dynamics and Ordinary Differential Equations* (London: Van Nostrand Reinhold Company)

[76] Shi Y and Yu P 2007 On chaos of the logistic maps *Dynam. Contin. Discrete Impuls. Syst. Ser. B* **14** 175–95

[77] Shi Y and Yu P 2007 On chaos of the logistic maps *Dynam. Contin. Discrete Impuls. Syst. Ser. B* **14** 175–95

[78] Shilnikov L 1997 Homoclinic chaos *Nonlinear Dynamics, Chaotic and Complex Systems* (Cambridge: Cambridge University Press) pp 39–63

[79] Schuster H G and Just W 2005 *Deterministic Chaos: An Introduction* (Weinheim: Wiley-VCH)

[80] Smale S 1966 Diffeomorphisms with many periodic points *Differential and Combinatorial Topology: A Symposium in Honor of Marston Morse* (Princeton, NJ: Princeton University Press) pp 63–70

[81] Smale S 1967 Differentiable dynamical systems *Bull. Am. Math. Soc.* **73** 747–817

[82] Spear D W 1992 Measure and self-similarity *Adv. Math.* **91** 143–57

[83] Stella S 1992 On Hausdorff dimension of recurrent net fractals *Proc. Amer. Math. Soc.* 116 389–400

[84] Staingrube S, Timme M, Worgotter F and Mannonpong P 2010 Self-organized adaptation of simple neural circuits enables complex robot behavior *Nat. Phys.* **6** 224–30

[85] Tang H P, Wang J Z, Zhu J L, Ao Q B, Wang J Y, Yang B J and Li Y N 2012 Fractal dimension of pore-structure of porous metal materials made by stainless steel powder *Powder Technol.* **217** 383–7

[86] Thakur R and Das R 2020 Strongly Ruelle–Takens, strongly Auslander–Yorke and Poincaré chaos on semiflows *Commun. Nonlinear Sci. Numer. Simulat.* **81** 105018

[87] Wiggins S 1988 Global bifurcation and chaos *Analytical Methods* (New York: Springer)

[88] Zmeskal O, Dzik P and Vesely M 2013 Entropy of fractal systems *Comput. Math. Appl.* **66** 135–46

[89] Akhmet M and Alejaily E M 2021 Abstract Fractals *Discontinuity Nonlinearity Complex* **10** 135–42

IOP Publishing

Ultra Poincaré Chaos and Alpha Labeling
A new approach to chaotic dynamics
Marat Akhmet

Chapter 3

Alpha unpredictability implies a universal mathematical chaos

This chapter provides an in-depth description of a central concept in the book, serving as a foundational guide that introduces essential definitions, rigorously proven theorems, and lemmas. The manuscript extends the topic of recurrence to encompass alpha unpredictable motions, seamlessly integrating the Poincaré–Birkhoff theory to include chaotic characteristics. Here, alpha unpredictability takes the place of Lorenz sensitivity, allowing the analysis of complex motions to focus on the behavior of a single trajectory. This trajectory, when viewed in time-space coordinates, is an alpha unpredictable function.

Let us start this chapter with a vital terminology remark. Accordingly, to the existence of many senses given to the word *unpredictability* in science and industry, we decided that the phenomenon invented and developed in our former research [2, 6, 7, 14] is to be called *alpha unpredictability*. Moreover, we shall call *Poincaré chaos* [6, 7], which is determined on the fundamentals of alpha unpredictability, *ultra Poincaré chaos*. So this makes it different from other types of chaos that are strongly related to the name of the French genius. In our case, the word *ultra* has been chosen considering elaboration of the Poisson stability by alpha unpredictability.

3.1 Poincaré and Lorenz lines in a nutshell

A large part of this book is concerned with the concept of alpha unpredictability, which was introduced in our paper [7]. The concept is a research result of two chaos investigation lines. The first one is issued from and lies in the Poincaré–Birkhoff theory [28, 45]: the *Poincaré line* or *recurrence line*. One can call it the oscillations line, also. Another one starts at the Lorenz sensitivity [38], and accordingly, we call it the *Lorenz line* or *sensitivity line*. Correspondingly, one can say about *ultra Poincaré chaos* and *Lorenz chaos* that if the second one unites all results with sensitivity, Li–Yorke chaos is included, but the first relies only on alpha unpredictability. Let us explain further why the phenomenon is an intersection of the two research lines. Knowledge of the recurrence theory and Lorenz chaos motivated us to prove that sensitivity has to be proven for dynamics in the quasi-minimal set, the closure of a Poisson non-periodic bounded trajectory. We could not achieve the desired result. This is why the decision on alpha unpredictability, new dynamical property of a single trajectory, was made. The intersection of the two lines is the dynamics of Poisson stable points, which admit both sensitivity and alpha unpredictability. Of course, in dialectical thinking, the two properties—convergence in the Poisson sense, and divergence as alpha unpredictability—have to be considered opposites to get a platform for the science of complexity. Nevertheless, alpha unpredictability can be combined with other properties for motions and even researched in isolation to get more interesting dynamical and geometrical consequences.

We start with the definition of the alpha unpredictable point. Let (X, d) be a metric space and suppose that f is a flow on X with continuous or discrete time.

Definition 3.1.1. [7] *A point $p \in X$ and the trajectory through it are alpha unpredictable if there exist a positive number ε_0 (the alpha unpredictability constant) and sequences $\{t_n\}$ and $\{\tau_n\}$, both of which diverge to infinity, such that*
- $\lim_{n \to \infty} f(t_n, p) = p$;
- $d[f(t_n + \tau_n, p), f(\tau_n, p)] > \varepsilon_0$ *for each $n \in \mathbb{N}$.*

The behavior of the sequence $f(t_n, p)$ in the last definition is known as the *Poisson stability*, and it illustrates very clearly what the researchers call recurrent motion. The divergence estimated by ε_0 is said to be *alpha unpredictability property*. The simplest recurrent motion is a fixed point when the point returns to its origin state at each moment. The Poisson stability is the most sophisticated type of recurrence. Simpler kinds of recurrence include periodic, quasi-periodic, almost periodic, and that which is called recurrent motions [44, 47]. One can say that recurrence is weakened for oscillations starting from the fixed point to the Poisson stable point. The last one is the ultimate in the G. Birkhoff classification [28]. Correspondingly, the stability was considered sufficient to properly express the complexity of the celestial mechanics [45]. Some authors accept the dynamics on quasi-minimal sets with Poisson stable orbits as chaotic. We do not agree that this is true, since there is no sensitivity yet. In our research [6, 7] we have prolonged the line of recurrent motions to alpha unpredictable ones by Definition 3.1.1, such that the weakest version of recurrent dynamics has been formulated, and we are not confident that the line cannot be prolonged further. We expressly specified the Poisson stable point to the alpha unpredictable point. The property of alpha unpredictability guarantees the Lorenz sensitivity [7], and consequently, one can say that there is chaos. We named it the Poincaré chaos in [6, 7] for the first time in literature, and ultra Poincaré chaos in this book. In figure 3.1 one can see how the different types of recurrent

Motions

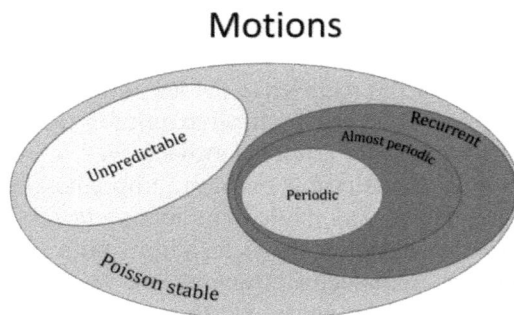

Figure 3.1. This diagram illustrates the most important dynamics of the recurrence theory extended with alpha unpredictable motions, which are introduced in papers [6, 7].

motions relate to each other by Venn diagram. We will proceed with the discussion after the next presentation of the *Lorenz line*, the second in chaos research.

It was in this context that American meteorologist E. Lorenz [38] made a groundbreaking observation. He found that even the slightest alterations in initial conditions could lead to substantial changes in the long-term outcome. This phenomenon, he explained, was the result of a special one-dimensional map derived from the model, marking the discovery of deterministic chaos. This pivotal moment can be considered the birth of the chaotic epoch, or what we now refer to as 'Lorenz chaos'.

The discovery of alpha unpredictability, a result of the intersection of two lines in our research, is a significant and novel breakthrough. This finding, while not ruling out the existence of other intersection points, paves the way for further exploration and discovery in the field of chaos theory. One can remember the dense periodic motions in the dynamics of both lines and existence of a dense trajectory in the domain of chaos [32] and in the quasi-minimal set [28]. The last set is the closure of the Poisson stable trajectory [44, 47] and the alpha unpredictable one [7]. Sensitivity is the basic property for Lorenz chaos, and it was not articulated for the quasi-minimal set [44, 47]. Therefore, it would be interesting to find a property for the dynamics of quasi-minimal sets that has a similar role and generates the same features as similarity. In fact, we discovered this property in our research. The concept of alpha unpredictability is a counterpart for sensitivity in chaos theory. Remarkably, the latter cannot be a consequence of the Poisson stability without alpha unpredictability, which is that additional condition [7], and makes new epoch for dynamics. We proved in the paper that alpha unpredictability of a single trajectory implies sensitivity in quasi-minimal sets. Thus, new ultra Poincaré chaos was shaped, and the meeting of Poincaré and Lorenz lines was completed. This brings, in some sense, the research of Lorenz chaos back to Poincaré–Birkhoff theory and creates new opportunities for both lines. Moreover, one can say that chaos investigation can now be performed in the framework of single motions research as it is true for stable oscillations in dynamical systems and differential equations [27, 29–31, 41, 43]. We have started to work in that direction [12, 13, 15–18, 21, 22].

3.2 An individual chaos: alpha unpredictability

The theory of dynamical systems is closely related to the study of oscillations and recurrence, with Poisson stable motion being one of the most sophisticated types within this theory [47]. In our papers [6, 7], inspired by chaos investigation, we introduced a new type of Poisson stable motion, identifying the initial point as an alpha unpredictable point (or alpha point) and the corresponding trajectory as an alpha unpredictable orbit (or alpha orbit). These innovations enable a connection between homoclinic chaos and recent types of chaos via the concept of alpha unpredictability. For instance, Lorenz sensitivity can be derived from the existence of a single orbit exhibiting this property. Consequently, we refined the concept of Poincaré chaos in our works [6, 7], which has since been further developed and applied in numerous articles and books [1, 2, 14]. To distinguish chaos based on alpha unpredictability from other established definitions linked to Poincaré, we have adopted the term 'ultra Poincaré chaos'. In our approach, by considering the points from functional spaces, we introduced alpha unpredictable functions for analysis. These functions are particularly useful for studying differential equations, serving as both perturbations and solutions. Thus, the theory now includes exciting new opportunities alongside the traditional pursuit of almost periodic and other forms of oscillations. Notably, our results extend beyond the exploration of bounded solutions and oscillations in models to proving the existence of chaos within these models. Discussing alpha unpredictability essentially refers to the existence of chaos and stable recurrent solutions. By chaos, we specifically mean an alpha unpredictable solution that represents chaos. Similarly, in the context of chaos synchronization, we refer to unison alpha unpredictability [8–11]. Our contributions have significantly bridged the fields of differential equations and chaos research, marking a return to the origins of chaos studies within the realm of dynamical systems and differential equations theory.

There are different types of chaos. The first is homoclinic chaos [48], originating from Poincaré's renowned work [45]. Another is Devaney chaos [32], characterized by transitivity, Lorenz sensitivity [38], and the presence of infinitely many unstable periodic motions that are dense within the chaotic attractor. The third type is Li–Yorke chaos [37], defined by a scrambled set where any two distinct motions are proximal and frequently separate. A key difference between homoclinic chaos and others lies in the features of sensitivity and frequent separation. Homoclinic chaos is associated with a homoclinic structure and instability. Lorenz [38] first articulated the concept of neighbor motion divergence as a form of sensitivity, a specific type of instability. Following this, Li and Yorke [37] employed frequent separation and proximity to describe similar phenomena. The lack of a quantitative description of instability sets homoclinic chaos apart from later definitions and poses certain challenges, while modern interpretations of chaos regard sensitivity as a fundamental component. While H. Poincaré recognized the divergence of nearby trajectories, he did not prescribe specific mechanisms for this process. Therefore, one could argue that the framework initiated by the French genius remains incomplete. In our research, we aim to contribute to this ongoing discourse. To this end, we have utilized open Poisson stable motions, which accompany homoclinic chaos [48]. In paper [7], we extended the concept of Poisson stable points to alpha unpredictable points. Alpha unpredictability can be viewed as a form of individual sensitivity. Starting from a single point, we use it as a guiding principle to uncover the phenomenon we term ultra Poincaré chaos [7]. This concept brings all forms of chaos closer together, offering an alternative description of motions in dynamics with homoclinic structures while incorporating elements like transitivity, sensitivity, frequent separation, and proximity. The continuum of Poisson stable orbits serves as a substitute for the presence of infinitely many periodic motions.

Let (X, d) be a metric space and $\pi: \mathbb{T}_+ \times X \to X$, where $\mathbb{T}_+$ is either the set of non-negative real numbers or the set of non-negative integers, be a semi-flow on X, i.e., $\pi(0, x) = x$ for all $x \in X$, $\pi(t, x)$ is continuous in the pair of variables t and x, and $\pi(t_1, \pi(t_2, x)) = \pi(t_1 + t_2, x)$ for all $t_1, t_2 \in \mathbb{T}_+, x \in X$.

A point $x \in X$ is called positively Poisson stable if there exists a sequence $\{t_n\}$ satisfying $t_n \to \infty$ as $n \to \infty$ such that $\lim_{n\to\infty} \pi(t_n, x) = x$ [44]. For a given point $x \in X$, let us denote by Θ_x the closure of the trajectory $T(x) = \{\pi(t, x): t \in \mathbb{T}_+\}$, i.e., $\Theta_x = \overline{T(x)}$. The set Θ_x is a quasi-minimal set if the point x is stable P^+ and $T(x)$ is contained in a compact subset of X [44, 47].

Hilmy established in [36] that if the trajectory associated with a Poisson stable point x lies within a compact subset of X and is neither a rest point nor a cycle, then the quasi-minimal set includes an uncountable set of motions that are everywhere dense and Poisson stable. The subsequent theorem can be demonstrated by modifying the approach outlined in [36, 44].

Theorem 3.2.1. [44, 47] *Suppose that $x \in X$ is positively Poisson stable and $T(x)$ is contained in a compact subset of X. If Θ_x is neither a rest point nor a cycle, then it contains an uncountable set of motions everywhere dense and stable P^+.*

In what follows, we call positively Poisson stability just 'Poisson stability'.

The definitions of an alpha unpredictable point and trajectory are as follows.

Definition 3.2.1. [7] *A point $x \in X$ and the trajectory through it are alpha unpredictable if there exist a positive number ε_0 (the alpha unpredictability constant) and sequences $\{t_n\}$ and $\{\tau_n\}$, both of which diverge to infinity, such that $\lim_{n\to\infty} \pi(t_n, x) = x$ and $d(\pi(t_n + \tau_n, x), \pi(\tau_n, x)) > \varepsilon_0$ for each $n \in \mathbb{N}$.*

Markov [44] proved that a trajectory stable in both Poisson and Lyapunov senses must be an almost periodic one. Since definition 3.2.2 implies instability, an alpha unpredictable motion cannot be almost periodic. In particular, it is neither an equilibrium, nor a cycle.

Lemma 3.2.1. *If $p \in X$ is an alpha unpredictable point, then $T(p)$ is neither a rest point nor a cycle.*

Proof. Let the number ε_0 and the sequences $\{t_n\}$ and $\{\tau_n\}$, be as in definition 3.2.2. Assume that there exists a positive number ω such that $f(t + \omega, p) = f(t, p)$ for all $t \in \mathbb{T}_+$. According to the continuity of $f(t, p)$, there exists a positive number δ such that if $d[p, q] < \delta$ and $0 \leqslant t \leqslant \omega$, then $d[f(t, p), f(t, q)] < \varepsilon_0$. Fix a natural number n such that $d[p_n, p] < \delta$, where $p_n = f(t_n, p)$. One can find an integer m and a number ω_0 satisfying $0 \leqslant \omega_0 < \omega$ such that $\tau_n = m\omega + \omega_0$. In this case, we have that $d[f(\tau_n, pn), f(\tau_n, p)] = d[f(\omega_0, pn), f(\omega_0, p)] < \varepsilon_0$. But, this is a contradiction since $d[f(\tau_n, p_n), f(\tau_n, p)] = d[f(tn + \tau_n, p), f(\tau_r, p)] \geqslant \varepsilon_0$. Consequently, $T(p)$ is neither a rest point nor a cycle.

It is seen from the next lemma that the alpha unpredictability can be transmitted by the flow.

Lemma 3.2.2. *If $p \in X$ is an alpha unpredictable point, then every point of the trajectory $T(p)$ is also alpha unpredictable.*

Proof. Suppose that the number ε_0 and the sequences $\{t_n\}$ and $\{\tau_n\}$, are as in definition 3.2.2. Fix an arbitrary point $q \in T(p)$ such that $q = f(t, p)$ for some $t \in \mathbb{T}_+$. One can verify that $lim_{n\to\infty} f(t_n, q) = lim_{n\to\infty} f(t_n + t, p) = lim_{n\to\infty} f(t, f(t_n, p)) = f(t, p) = q$.

Now, take a natural number n_0 such that $\tau_n > t$ for each $n \geqslant n_0$. If we denote $\zeta_n = \tau_n - t$, then we have for $n \geqslant n_0$ that $d[f(t_n + \zeta_n, q), f(\zeta_n, q)] = d[f(t_n + \zeta_n, f(t, p)), f(\zeta_n, f(t, p))] = d[f(t_n + \tau_n, p), f(\tau_n, p)] \geqslant \varepsilon_0$. Clearly, $\zeta_n \to \infty$ as $n \to \infty$. Consequently, the point q is alpha unpredictable.

Theorem 3.2.2. [7, 14] *The dynamics on Θ_p is sensitive if $p \in X$ is an alpha unpredictable point.*

Proof. Let $\varepsilon_0 > 0$ be the alpha unpredictability constant corresponding to the point p. Fix an arbitrary positive number δ, and take a point $r \in \Theta_p$. First of all, consider the case $r \in T(p)$. By lemma 2.2, there exist sequences $\{t_n\}$ and $\{\tau_n\}$, both of which diverge to infinity, such that $lim_{n\to\infty} f(t_n, r) = r$ and $d[f(t_n + \tau_n, r), f(\tau_n, r)] \geqslant \varepsilon_0$ for each n. Fix a natural number n_0 such that $d[\bar{r}, r] < \delta$, where $\bar{r} = f(t_{n_0}, r)$. In this case, the inequality $d[f(\tau_{n_0}, \bar{r}), f(\tau_{n_0}, r)] \geqslant \varepsilon_0$ is valid. On the other hand, suppose that $r \in \Theta_p \setminus T(p)$. One can find a sequence η_m, $\eta_m \to \infty$ as $m \to \infty$, such that $lim_{m\to\infty} r_m = r$, where $r_m = f(\eta_m, p)$. According to lemma 2.2, for each $m \in N$, there exist sequences s_n^m and ξ_n^m, both of which diverge to infinity, such that $lim_{n\to\infty} r_n^m = r_m$ and $d[f(\xi_n^m, r_n^m), f(\xi_n^m, r_m)] \geqslant \varepsilon_0$, $n \in \mathbb{N}$, where $r_n^m = f(s_n^m, r_m)$. Now, let m_0 be a natural number such that $d[r_{m_0}, r] < \delta/2$. Suppose that there exists a natural number n_1 with

$$d[f(\xi_{n_1}^{m_0}, r_{m_0}), f(\xi_{n_1}^{m_0}, r] \geqslant \varepsilon_0/2.$$

If this is the case, then sensitivity is proved. Otherwise, fix $n_2 \in \mathbb{N}$ satisfying $d[r_{n_2}^{m_0}, r_{m_0}] < \delta/2$ so that $d[r_{n_2}^{m_0}, r] \leqslant d[r_{n_2}^{m_0}, r_{m_0}] + d[r_{m_0}, r] < \delta$. One can confirm that

$$d[f(\xi_{n_2}^{m_0}, r_{n_2}^{m_0}), f(\xi_{n_2}^{m_0}, r)] \geqslant d[f(\xi_{n_2}^{m_0}, r_{m_0})] - d[f(\xi_{n_2}^{m_0}, r_{m_0})], f(\xi_{n_2}^{m_0}, r)] > \varepsilon_0/2.$$

The theorem is proved.

If the positive semi-trajectory of an alpha unpredictable point $p \in X$ is contained in a compact subset of X, then Θ_p contains an uncountable set of everywhere dense Poisson stable motions. Additionally, since $T(p)$ is dense in Θ_p the transitivity is also valid in the dynamics.

We have thus identified a significant dynamical system that incorporates elements traditionally associated with Devaney and Li–Yorke chaos, including sensitivity and transitivity. Although the density of periodic solutions is absent, the presence of alpha unpredictable motions, instead of periodic ones, suggests that this aspect is also present. Consequently, we can now recognize the existence of a new type of chaos in the field.

Definition 3.2.2. [7] *The dynamics on the quasi-minimal set Θ_p is called ultra Poincaré chaotic if p is an alpha unpredictable point.*

Within the context of chaos, there may be infinitely many unstable periodic motions. For instance, chaos in the sense of Devaney [32] and Li–Yorke [37] admits a basis consisting of periodic motions. However, in our case, alpha unpredictable dynamics occur in chaos instead of periodic motions. We do not accept this change as a deficiency since the complexity of the dynamics has increased immensely. The reader must not be confused with Poisson's stability of alpha unpredictable motions. H. Poincaré and G. Birkhoff considered the stability not as a collective characteristic, but one for an individual motion such that to say instability (stability) of Poisson stable motion is better described as Lyapunov instability (stability) of Poisson stable motion.

Example 3.2.1. [7] *Let us take into account the following space of bi-infinite sequences* [51],

$$\Sigma^2 = \{s = (\ldots s_{-2} s_{-1}. s_0 s_1 s_2 \ldots): s_j = 0 \text{ or } 1 \text{ for each } j\}$$

with the metric

$$d[s, \bar{s}] = \sum_{k=-\infty}^{\infty} \frac{|s_k - \bar{s}_k|}{2^{|k|}},$$

where $s = (\ldots s_{-2} s_{-1}. s_0 s_1 s_2 \ldots)$, $\bar{s} = (\ldots \bar{s}_{-2} \bar{s}_{-1}. \bar{s}_0 \bar{s}_1 \bar{s}_2 \ldots) \in \Sigma^2$. The shift map $\sigma: \Sigma^2 \to \Sigma^2$ is defined as

$$\sigma(\ldots s_{-2} s_{-1}. s_0 s_1 s_2 \ldots) = (\ldots s_{-2} s_{-1} s_0. s_1 s_2 \ldots).$$

The map σ is continuous and the metric space Σ^2 is compact [51].

In order to show that the map σ possesses an alpha unpredictable point in Σ^2, we need a collection of finite sequences s_i^m, $m \in \mathbb{N}$, $i = 1, 2, \ldots, 2^m$, consisting of 0's and 1's. Let us denote $s_1^1 = (0)$ and $s_2^1 = (1)$. For each $m \in \mathbb{N}$, we recursively define $s_{2i-1}^{m+1} = (s_i^m 0)$, $s_{2i}^{m+1} = (s_i^m 1)$, $i = 1, 2, \ldots, 2^m$. Here, for each m and i, the finite sequences s_{2i-1}^{m+1} and s_{2i}^{m+1} are obtained by respectively inserting 0 and 1 to the end of the sequence s_i^m of length m. For instance, the sequences with length 2 can be written as $s_1^2 = (s_1^1 0) = (00)$, $s_2^2 = (s_1^1 1) = (01)$, $s_3^2 = (s_2^1 0) = (10)$, $s_4^2 = (s_2^1 1) = (11)$.

Now, consider the following sequence:

$$s* = (\ldots s_8^3 s_6^3 s_4^3 s_2^3 s_4^2 s_2^2. s_1^1 s_1^2 s_3^2 s_1^3 s_3^3 s_5^3 s_7^3 \ldots).$$

It was demonstrated in [51] that the trajectory of $s*$ is dense in Σ^2. We will show that $s*$ is an alpha unpredictable point of the dynamics (Σ^2, σ). For each $n \in \mathbb{N}$, one can find $j \in \mathbb{N}$ such that

$$s_{2j-1}^{2n+2} = (s_{-n}*\ldots s_0*\ldots s_n*0)$$

and

$$s_{2j}^{2n+2} = (s_{-n}^* \ldots s_0*\ldots s_n*1).$$

Therefore, there exists a sequence $\{t_n\}$ with $t_n \geqslant n + \sum_{k=1}^{2n-1} k 2^{k-1}$, $n \in \mathbb{N}$, such that $\sigma^{t_n}(s*) = s_i^*$ for $|i| \leqslant n$. Accordingly, the inequality $d[\sigma^{t_n}(s*), s*] \leqslant 1/2^{n-1}$ is valid so that $\sigma^{t_n}(s*) \to s*$ as $n \to \infty$. Hence, $s*$ is stable P^+. In a similar way, one can confirm that $s*$ is stable P^-. Note that Σ^2 is a quasi-minimal set since $s*$ is Poisson stable. On the other hand, suppose that there exists a natural number n such that

$$\sigma^{t_n + n + 1}(s*)_i = \sigma^{n+1}(s*)_i$$

for each $i \geqslant 0$. Under this assumption we have that $\sigma^{t_n}(s*)_i = s_i^*$ for $i \geqslant -n$. This is a contradiction since the sequence $s*$ is not eventually periodic. For this reason, for each $n \in \mathbb{N}$, there exists an integer $\tau_n \geqslant n + 1$ such that $\sigma^{t_n + \tau_n}(s*)_0 \neq \sigma^{\tau_n}(s*)_0$. Hence, $d[\sigma^{t_n + \tau_n}(s*), \sigma^{\tau_n}(s*)] \geqslant 1$ for each $n \in \mathbb{N}$, and $s*$ is an alpha unpredictable point in Σ^2.

Remark 3.2.1. The reader is invited to compare the result of analysis in the last example 3.2.10 with that performed in example 2.5.1. In both, the same dynamics is under investigation, but in different ways. So, one can see how beneficial the method of alpha labeling is.

Topological conjugacy is an important concept in the theory of dynamical systems that allows us to interpret complicated dynamics using simpler ones. Let

X and Y be metric spaces. A flow (semi-flow) f on X is topologically conjugate to a flow (semi-flow) g on Y if there exists a homeomorphism $h: X \to Y$ such that $h \circ f = g \circ h$ [32, 51]. The following theorem can be verified by using the arguments presented in [26].

Theorem 3.2.3. *Suppose that X and Y are metric spaces and a flow (semi-flow) f on X is topologically conjugate to a flow (semi-flow) g on Y. If there exists an alpha unpredictable point whose trajectory is contained in a compact subset of X, then there also exists an alpha unpredictable point whose trajectory is contained in a compact subset of Y.*

Example 3.2.2. [7] *(Smale horseshoe) Since the shift map σ on Σ^2 is topologically conjugate to the Smale horseshoe [32, 51], one can conclude by using theorem 3.2.12 that the horseshoe map possesses ultra Poincaré chaos.*

Example 3.2.3. [7] *On the other hand, let us consider the Hénon map*

$$x_{n+1} = \alpha - \beta y_n - x_n^2$$
$$y_{n+1} = x_n, \tag{3.1}$$

where $\beta \neq 0$ and $\alpha \geqslant (5 + 2\sqrt{5})(1 + |\beta|)^2/4$. It was proved by Devaney and Nitecki [33] that the map (3.1) possesses a Cantor set in which the map is topologically conjugate to the shift map σ on Σ^2. Therefore, theorem 3.2.12 also implies the presence of an alpha unpredictable point and a trajectory in the dynamics of (3.1).

Example 3.2.4. [7] *Next, as an example of a semi-flow, consider the following space of infinite sequences [32],*

$$\Sigma_2 = \{s = (s_0 s_1 s_2 \ldots): s_j = 0 \text{ or } 1 \text{ for each } j\}$$

with the metric

$$d[s, \bar{s}] = \sum_{k=0}^{\infty} \frac{|s_k - \bar{s}_k|}{2^k},$$

where $s = (s_0 s_1 s_2 \ldots)$, $\bar{s} = (\bar{s}_0 \bar{s}_1 \bar{s}_2 \ldots) \in \Sigma_2$. The shift map $\sigma: \Sigma_2 \to \Sigma_2$ is defined as $\sigma(s_0 s_1 s_2 \ldots) = (s_1 s_2 s_3 \ldots)$. As in the case of the space of bi-infinite sequences, the metric space Σ_2 is compact and the map σ is continuous [32, 51].

Let us take into account the sequence

$$s^* = (\underbrace{0\ 1}_{1\ blocks} | \underbrace{00\ 01\ 10\ 11}_{2\ blocks} | \underbrace{000\ 001\ 010\ 011\ \ldots}_{3\ blocks} | \ldots),$$

which is constructed by successively listing all blocks of 0's and 1's of length n, then length $n + 1$, etc. This sequence is non-periodic and its trajectory $\mathcal{T}(s^*) = \{\sigma^n(s^*): n = 0, 1, 2, \ldots\}$ is dense in Σ_2 [32]. Note that the number of all blocks of length n in s^* is 2^n. Based upon the construction of s^*, there exists a sequence $\{t_n\}$ satisfying $t_n \geqslant \sum_{j=1}^{n} j 2^j$, $n \in \mathbb{N}$, such that $s_i^* = \sigma^{t_n}(s^*)_i$ for each $i = 0, 1, 2, \ldots, n$. Clearly, $t_n \to \infty$ as $n \to \infty$ and $d[\sigma^{t_n}(s^*), s^*] \leqslant 1/2^n$ so that $\sigma^{t_n}(s^*) \to s^*$ as $n \to \infty$. Hence, s^* is stable P^+. In a very similar way to the bi-infinite sequences in example 3.2.10, one can show the existence of a sequence $\{\tau_n\}$, $\tau_n \to \infty$ as $n \to \infty$, such that $d[\sigma^{t_n + \tau_n}(s^*), \sigma^{\tau_n}(s^*)] \geqslant 1$ for each $n \in \mathbb{N}$. Thus, s^* is an alpha unpredictable point in Σ_2.

Example 3.2.5. [7] *It was shown in [46] that the logistic map $x_{n+1} = \mu x_n(1 - x_n)$ possesses an invariant Cantor set $\Lambda \subset [0, 1]$, and the map on Λ is topologically conjugate to σ on Σ_2 for $\mu > 4$. Therefore, the map with $\mu > 4$ possesses an alpha unpredictable point and a trajectory in accordance with theorem 3.2.12.*

3.3 Alpha unpredictable functions: appearance of chaos

In our research, we have expanded the existing catalog of oscillations in dynamical systems ranging from equilibrium to Poisson stable orbits to include alpha unpredictable motions. This addition is expected to stimulate further development in the theory of dynamical systems. Certain properties and laws of dynamical systems can be overlooked or disregarded in applications, particularly when dealing with non-autonomous or non-smooth systems. In such cases, we can utilize alpha unpredictable functions [6]. These functions can be examined as solutions to various equations, as our proposed methods allow for their study within the qualitative framework of differential equations [35].

We utilize the topology of uniform convergence on any compact subset of the real axis to introduce the alpha unpredictable functions. This enables chaos incorporation in the theory of differential equations, initiated in papers [6, 13, 15] and books [14].

The main goal of this chapter is to construct a concrete alpha unpredictable function. To determine the procedure, we start with alpha unpredictable sequences as motions of symbolic dynamics and the logistic map. Then, an alpha unpredictable function is determined as an improper convolution integral with a relay function. Finally, we demonstrated alpha unpredictable functions as solutions of differential equations in examples.

Let us denote by $C(\mathbb{R})$ the set of continuous functions defined on $\mathbb{R}$ with values in $\mathbb{R}^m$, and assume that $C(\mathbb{R})$ has the topology of uniform convergence on compact sets, i.e., a sequence $\{h_k\}$ in $C(\mathbb{R})$ is said to converge to a limit h if for every compact set $\mathcal{U} \subset \mathbb{R}$, the sequence of restrictions $\{h_k |_{\mathcal{U}}\}$ converges to $\{h |_{\mathcal{U}}\}$ uniformly. The topology of the convergence on compact sub-intervals can be metricized [47]. One can define a metric ρ on $C(\mathbb{R})$ as

$$\rho(h_1, h_2) = \sum_{k=1}^{\infty} 2^{-k} \rho_k(h_1, h_2), \tag{3.2}$$

where h_1, h_2 belong to $C(\mathbb{R})$ and

$$\rho_k(h_1, h_2) = \min\{1, \sup_{s\in[-k,k]}\|h_1(s) - h_2(s)\|\}, k \in \mathbb{N}.$$

Let us define the mapping $\pi\colon \mathbb{R}_+ \times C(\mathbb{R}) \to C(\mathbb{R})$ by $\pi(t, h) = h_t$, where $h_t(s) = h(t + s)$. The mapping π is a semi-flow on $C(\mathbb{R})$, and it is called the Bebutov dynamical system [47].

Using the Bebutov dynamical system, we give the descriptions of a Poisson function and an alpha unpredictable function in the next definitions.

Definition 3.3.1. *A Poisson stable point of the Bebutov dynamical system is said to be a Poisson function.*

Definition 3.3.2. [6] *An alpha unpredictable point of the Bebutov dynamical system is said to be an alpha unpredictable function.*

An alpha unpredictable function is a Poisson stable function. According to theorem *III*. 3 [47], a motion $\pi(t, h)$ lies in a compact set if h is a bounded and uniformly continuous function. Therefore, an alpha unpredictable function h determines ultra Poincaré chaos in the functional space if it is bounded and uniformly continuous. Moreover, any system of differential equations that admits a uniformly continuous and bounded alpha unpredictable solution has ultra Poincaré chaos, considering the dynamics through shifts along the time axis.

Based on the above discussions, we shall introduce the following definition, which will be useful in the next chapters of the book.

Definition 3.3.3. [13] *A uniformly continuous and bounded function $v\colon \mathbb{R} \to \mathbb{R}^p$ is alpha unpredictable if there exist positive numbers ε_0, δ and sequences t_n, u_n, both of which diverge to infinity such that $v(t + t_n) \to v(t)$ as $n \to \infty$ uniformly on compact subsets of $\mathbb{R}$ and $\|v(t + t_n) - v(t)\| \geq \varepsilon_0$ for each $t \in [u_n - \delta, u_n + \delta]$ and $n \in \mathbb{N}$.*

To make research of chaos more friendly with discrete equations, we introduce the concept of an alpha unpredictable sequence as a specific alpha unpredictable function on the set of integers. It is convenient to verified this as a solution of a discrete equation.

The following definition of an alpha unpredictable sequence was first mentioned in paper [12] as an analogue for definition 3.3.3 for discrete time.

Definition 3.3.4. *A bounded sequence $\{\varphi_n\}$, $n \in \mathbb{Z}$, in $\mathbb{R}^p$ is called alpha unpredictable if there exist a positive number ε_0 and sequences $\{\zeta_k\}, \{\eta_k\}$, $k \in \mathbb{N}$, of positive integers, both of which diverge to infinity, such that $\|\varphi_{n+\zeta_k} - \varphi_n\| \to 0$ as $k \to \infty$ for each n in bounded intervals of integers and $\|\varphi_{\zeta_k + \eta_k} - \varphi_{\eta_k}\| \geq \varepsilon_0$ for each $k \in \mathbb{N}$.*

Definition 3.3.4 is of main use in the present paper. It is requested by the method of the proof. Nevertheless, in future analyses, another definition may be needed, which can be considered as a direct specification of definition 3.3.4 as well as an analogue of definition 3.3.3.

Definition 3.3.5. *A bounded sequence $\{\varphi_n\}$, $n \in \mathbb{Z}$, in $\mathbb{R}^p$ is called alpha unpredictable if there exist a positive number ε_0 and sequences $\{\zeta_k\}$, $\{\eta_k\}$, $k \in \mathbb{N}$, of positive integers, both of which diverge to infinity, such that $\left\| \varphi_{\zeta_k} - \varphi_0 \right\| \to 0$, as $k \to \infty$, and $\left\| \varphi_{\zeta_k + \eta_k} - \varphi_{\eta_k} \right\| \geqslant \varepsilon_0$ for each $k \in \mathbb{N}$.*

The topology in definitions 3.3.3 and 3.3.4 are metrizable [47]. Consequently, the existence of an alpha unpredictable sequence in the sense of definition 3.3.4 indicates the presence of ultra Poincaré chaos [6, 7]. In what follows, an alpha unpredictable sequence and solution are understood as mentioned in definition 3.3.4.

Next, we will show the presence of an alpha unpredictable point in the symbolic dynamics (Σ_2, σ, d) of example 3.2.15 with a distinguishing feature. In the example, the sequence s^* is proven to be alpha unpredictable. In the proof of the next lemma, an element $s^{**} = (s_0^{**} s_1^{**} s_2^{**} \dots)$ of Σ_2 will be constructed in a similar way to s^*n with the only difference that the order of the blocks will be chosen specially. An extension of the constructed sequence to the left-hand side will also be provided.

Lemma 3.3.1. [6] *For each increasing sequence $\{m_n\}$ of positive integers, there exist a sequence $s^{**} \in \Sigma_2$ and sequences $\{\alpha_n\}$, $\{\beta_n\}$ of positive integers, both of which diverge to infinity, such that*
  (i) $d(\sigma^{\alpha_n + r}(s^{**}), \sigma^r(s^{**})) \leqslant 2^{-m_n}$, $r = -n, -n + 1, \dots, n$,
  (ii) $d(\sigma^{\alpha_n + \beta_n}(s^{**}), \sigma^{\beta_n}(s^{**})) \geqslant 1$ *for each $n \in \mathbb{N}$.*

Proof. Fix an arbitrary increasing sequence $\{m_n\}$ of positive integers. For each $n \in \mathbb{N}$, define $\alpha_n = \sum_{k=1}^{n+m_n} k 2^k$ and $\beta_n = n + m_n + 1$. Clearly, both of the sequences $\{\alpha_n\}$ and $\{\beta_n\}$ diverge to infinity. We will construct a sequence $s^{**} \in \Sigma_2$ such that the inequalities (i) and (ii) are valid.

First of all, we choose the terms s_0^{**}, s_1^{**}, $\dots$, $s_{\alpha_1 - 1}^{**}$ by successively placing the blocks of 0's and 1's in an increasing length, starting from the blocks of length 1 till the end of the ones with length $m_1 + 1$. The order of the blocks with the same length can be arbitrary without any repetitions. Let us take $s_{\alpha_1 + k}^{**} = s_k^{**}$ for $k = 0, 1, \dots, m_1 + 1$, i.e., the terms of the first block $(s_{\alpha_1}^{**} s_{\alpha_1 + 1}^{**} \cdots s_{\alpha_1 + m_1 + 1}^{**})$ of length $m_1 + 2$ is chosen the same as the first $m_1 + 2$ terms of the sequence s^{**}. Moreover, we take the second block of length $m_1 + 2$ in a such a way that its first term $s_{\alpha_1 + \beta_1}^{**}$ is different from $s_{\beta_1}^{**}$. After that we continue placing the remaining blocks of length $m_1 + 2$ and the ones with length greater than $m_1 + 2$ till the last block of length $m_2 + 2$, and again, the blocks of the same length can be in any order without repetitions.

For each $n \geqslant 2$, we set the last block of length $m_n + n$ such that $s^{**}_{\alpha_n - k} = s^{**}_{\alpha_k - k}$ for each $k = 1, 2, \ldots, n - 1$. Then, the terms of the first block of length $m_n + n + 1$ are constituted by taking $s^{**}_{\alpha_n + k} = s^{**}_k$, $k = 0, 1, \ldots, m_n + n$. Moreover, the second block of length $m_n + n + 1$ is chosen such that $s^{**}_{\alpha_n + \beta_n} \neq s^{**}_{\beta_n}$. Lastly, the remaining blocks of 0's and 1's are successively placed similar to the case mentioned above so that the lengths of the blocks in s^{**} are in an increasing order and there are no repetitions within the blocks of the same length. In this way, the construction of the sequence $s^{**} \in \Sigma_2$ is completed. We fix the extension of the sequence s^{**} to the left by choosing $\sigma^{-k}(s^{**})_0 = s^{**}_{\alpha_k - k}$, $k \in \mathbb{N}$. For each $n \in \mathbb{N}$, we have $\sigma^{\alpha_n - n}(s^{**})_k = \sigma^{-n}(s^{**})_k$, $k = 0, 1, \ldots, m_n + 2n$, and $\sigma^{\alpha_n + \beta_n}(s^{**})_0 \neq \sigma^{\beta_n}(s^{**})_0$ so that the inequalities (i) and (ii) are valid.

Using the technique in [32], one can show that the trajectory $T(s^{**}) = \{\sigma^i(s^{**}): i \in \mathbb{Z}\}$ is dense in Σ_2, i.e., $\Theta_{s^{**}} = \Sigma_2$. By lemma 2.2 in [7] any sequence $\sigma^i(s^{**})$, $i \in \mathbb{Z}$, is an alpha unpredictable point of the Bernoulli dynamics on Σ_2, and Σ_2 is a quasi-minimal set. It implies from the last theorem that $T(s^{**})$ is an alpha unpredictable function on $\mathbb{Z}$, i.e., an alpha unpredictable sequence. According to theorem 3.1 presented in [7], the dynamics on Σ_2 is ultra-ultra Poincaré chaotic. Moreover, there are infinitely many alpha unpredictable sequences in the set.

Example 3.3.1. [6] *We will demonstrate the presence of an alpha unpredictable solution of the equation*

$$\eta_{n+1} = F_\mu(\eta_n), \tag{3.3}$$

where $F_\mu(s) = \mu s(1 - s)$ is the logistic map. More precisely, it is proved that for each $\mu \in [3 + (2/3)^{1/2}, 4]$ and sequence of positive numbers $\delta_n \to 0$, there exists a solution $\{\eta_n\}$, $n \in \mathbb{Z}$, of equation (3.3) such that

(i) $|\eta_{i_n + r} - \eta_r| < \delta_n$, $r = -h_0 n, -h_0 n + 1, \ldots, h_0 n$,

(ii) $|\eta_{i_n + j_n} - \eta_{j_n}| \geqslant \varepsilon_0$ *for each $n \in \mathbb{N}$,*

where ε_0 is a positive number, $h_0 > 4$ is a natural number, and $\{i_n\}$, $\{j_n\}$ are integer valued sequences both of which diverge to infinity.

Fix $\mu \in [3 + (2/3)^{1/2}, 4]$ and a sequence $\{\delta_n\}$ of positive real numbers with $\delta_n \to 0$ as $n \to \infty$. Take a neighborhood $U \subset [0, 1]$ of the point $1 - 1/\mu$. There exist a natural number $h_0 > 4$ and a Cantor set $\Lambda \subset U$ such that the map $F_\mu^{h_0}$ on Λ is topologically conjugate to the Bernoulli shift σ on Σ_2. Therefore, there exists a homeomorphism $S: \Sigma_2 \to \Lambda$ such that $S \circ \sigma = F_\mu^{h_0} \circ S$. Since S is uniformly continuous on Σ_2, for each $n \in \mathbb{N}$, there exists a number $\bar{\delta}_n > 0$ such that for each $s^1, s^2 \in \Sigma_2$ with $d(s^1, s^2) < \bar{\delta}_n$, we have $|\eta^1 - \eta^2| < \delta_n/\mu^{h_0 - 1}$, where $\eta^1 = S(s^1)$, $\eta^2 = S(s^2)$.

Let $\{m_n\}$ be an increasing sequence of natural numbers such that $2^{-m_n} < \bar{\delta}_n$ for each $n \in \mathbb{N}$. According to lemma 3.3.6, there exists a sequence $s^{**} \in \Sigma_2$ and sequences $\{\alpha_n\}$,

$\{\beta_n\}$, both of which diverge to infinity such that $d(\sigma^{c_n + r}(s^{**}), \sigma^r(s^{**})) \leqslant 2^{-m_n}$, $r = -n, -n + 1, \ldots, n$, and $d(\sigma^{\alpha_n + \beta_n}(s^{**}), \sigma^{\beta_n}(s^{**})) \geqslant 1$ for each $n \in \mathbb{N}$.

Now, let $\{\eta_n\}$, $n \in \mathbb{Z}$, be the solution of (3.3) with $\eta_{h_0 k} = S(\sigma^k(s^{**}))$, $k \in \mathbb{Z}$. Since the inequality $|F_\mu(u_1) - F_\mu(u_2)| \leqslant \mu|u_1 - u_2|$ is valid for every $u_1, u_2 \in [0, 1]$, we have for each $n \in \mathbb{N}$ that $|\eta_{i_n + r} - \eta_r| < \delta_n$, $r = -h_0 n, -h_0 n + 1, \ldots, h_0 n$, where $i_n = h_0 \alpha_n$. Besides, using the arguments presented in [26], one can verify the existence of a positive number ε_0 such that $|\eta_{i_n + j_n} - \eta_{j_n}| \geqslant \varepsilon_0$, $n \in \mathbb{N}$, where $j_n = h_0 \beta_n$.

Example 3.3.2. *According to the last example, there exists the alpha unpredictable solution ψ_i, $i \in \mathbb{Z}$, of the logistic discrete equation*

$$\lambda_{i+1} = \mu \lambda_i (1 - \lambda_i) \tag{3.4}$$

with $\mu = 3.92$. The sequence belongs to the unit interval $[0, 1]$. There exist a positive number ε_0 and sequences ζ_n, η_n, both of which diverge to infinity such that $|\psi_{i+\zeta_n} - \psi_i| \to 0$ as $n \to \infty$ for each i in bounded intervals of integers and $|\psi_{\zeta_n + \eta_n} - \psi_{\eta_n}| \geqslant \varepsilon_0$ for each $n \in \mathbb{N}$.

Consider the following integral

$$\Theta(t) = \int_{-\infty}^t e^{-2(t-s)}\Omega(s)ds, \; t \in \mathbb{R}, \tag{3.5}$$

where $\Omega(t)$ is a piece-wise constant function defined on the real axis through the equation $\Omega(t) = \psi_i$ for $t \in [i, i + 1)$, $i \in \mathbb{Z}$.

It is worth noting that $\Theta(t)$ is bounded on the whole real axis such that $\sup_{t \in \mathbb{R}} |\Theta(t)| \leqslant 1/2$.

Next, we will show that $\Theta(t)$ is an alpha unpredictable scalar function.

Consider a fixed bounded and closed interval $[\alpha, \beta]$, of the axis and a positive number ε. Without loss of generality one can assume that α and β are integers. Let us fix a positive number ξ and an integer $\gamma < \alpha$, which satisfy the following inequalities $e^{-2(\gamma)} < \frac{\varepsilon}{2}$ and $\xi[1 - e^{-2(\beta-\gamma)}] < \varepsilon$. Let n be a large natural number such that $|\Omega(t + \zeta_n) - \Omega(t)| < \xi$ on $[\gamma, \beta]$. Then for all $t \in [\alpha, \beta]$ we obtain that

$$|\Theta(t + \zeta_n) - \Theta(t)| = |\int_{-\infty}^t e^{-2(t-s)}(\Omega(s + \zeta_n) - \Omega(s))ds| =$$

$$|\int_{-\infty}^\gamma e^{-2(t-s)}(\Omega(s + \zeta_n) - \Omega(s))ds + \int_\gamma^t e^{-2(t-s)}(\Omega(s + \zeta_n) - \Omega(s))ds| \leqslant$$

$$\int_{-\infty}^\gamma e^{-2(t-s)}2ds + \int_\gamma^\beta e^{-2(t-s)}\xi ds \leqslant e^{-2(\gamma)} + \frac{\xi}{2}[1 - e^{-2(\beta-\gamma)}] < \frac{\varepsilon}{2} + \frac{\varepsilon}{2} = \varepsilon.$$

Thus, $|\Theta(t + \zeta_n) - \Theta(t)| \to 0$ as $n \to \infty$ uniformly on the interval $[\alpha, \beta]$. Moreover, the following inequalities are valid, $|\Omega(t + \zeta_n) - \Omega(t)| = |\psi_{\zeta_n + \eta_n} - \psi_{\eta_n}| \geqslant \varepsilon_0$, $t \in [\eta_p, \eta_p + 1)$, $p \in \mathbb{N}$.

Let us fix the numbers κ and n and consider two alternative cases: (i) $|\Theta(\eta_n + \zeta_n) - \Theta(\eta_n)| < \frac{\varepsilon_0}{8}$ and (ii) $|\Theta(\eta_n + \zeta_n) - \Theta(\eta_n)| \geqslant \frac{\varepsilon_0}{8}$.

(i) There exists a positive number $\kappa < 1$ such that $e^{-2\kappa} = \frac{2}{3}$. Using the relation

$$\Theta(t + \zeta_n) - \Theta(t) = \Theta(\eta_n + \zeta_n) - \Theta(\eta_n) +$$
$$\int_{\eta_n}^{t} e^{-2(t-s)}(\Omega(s + \zeta_n) - \Omega(s))ds \qquad (3.6)$$

we obtain that

$$|\Theta(t + \zeta_n) - \Theta(t)| \geqslant |\int_{\eta_n}^{t} e^{-2(t-s)}|\Omega(s + \zeta_n) - \Omega(s))|ds - |\Theta(\eta_n + \zeta_n) - \Theta(\eta_n)| \geqslant$$
$$\int_{\eta_n}^{t} e^{-2(t-s)}\varepsilon_0 ds - \frac{\varepsilon_0}{8} \geqslant \frac{\varepsilon_0}{2}(1 - e^{-2\kappa}) - \frac{\varepsilon_0}{8} = \frac{\varepsilon_0}{24}$$

for $t \in [\eta_n + \kappa, \eta_n + 1)$.

(ii) There exists a positive number $\kappa < 1$ such that $1 - e^{-2\kappa} = \frac{\varepsilon_0}{12}$. From the relation (3.6) we get

$$|\Theta(t + \zeta_n) - \Theta(t)| \geqslant |\Theta(\eta_n + \zeta_n) - \Theta(\eta_n)| - |\int_{\eta_n}^{t} e^{-2(t-s)}(\Omega(s + \zeta_n) - \Omega(s))ds| \geqslant$$
$$\frac{\varepsilon_0}{8} - \int_{\eta_n}^{t} e^{-2(t-s)}2ds \geqslant \frac{\varepsilon_0}{8} - [1 - e^{-2\kappa}] = \frac{\varepsilon_0}{24}$$

for $t \in [\eta_n, \eta_n + \kappa)$.
Thus, $\Theta(t)$ is an alpha unpredictable function.

The reader is invited to visit chapters 4, 7, and 8 for developed examples of alpha unpredictable functions with corresponding numerical simulations, which emphasize the universality and potentials of the functions.

3.4 Notes

This book chapter describes one central concept invented during the last decade. It is an immediate consequence of what was done by H. Poincaré in the dynamics of recurrent motions. The research is considered as an origin of chaos in science. Our invention has brought the idea of ultra Poincaré to its ultimate logical shape in the medium of modern chaos. We believe it can create a universal platform, where everything known as chaos, and even complexity, can be unified. This concerns also the difference between deterministic and random dynamics. Now let us explain why the chaos determined by alpha unpredictability can be considered to be of utmost importance. Of course, this should be discussed under certain formal mathematical circumstances. We use the line of recurrence for this reason. Equilibriums, different periodicity, and Poisson stable motions dot the line. Recurrence is a critical ingredient of any chaos. Unstable periodic orbits, dense in chaotic domains and united, are considered a skeleton of chaos in the definitions of Li–Yorke and

Devaney. They are unstable, observable if one constructs chaos as the cascade of period doubling.

A natural question appears: Is it possible that a skeleton for chaos is of almost periodic and even Poisson-stable motions? Moreover, one may expect a chaotic model to admit periodic and all other sorts of recurrent dynamics [1, 7]. We have obtained the most advanced chaos by modifying the Poisson stable motion to achieve sensitivity in the quasi-minimal set. This is a significant step in the precision of dynamics in the circumstance of the recurrence theorem. By applying alpha labeling, we have proven that chaotic domains of such traditional models as the logistic, Smale horseshoes, and Baker maps admit all three types of chaos. Thus, a couple of novelties, alpha unpredictability, and labeling provide us with an ultimate level of mathematical chaos research.

This research, showing that the periodicity for the shaping of chaos can be replaced by quasi and almost periodicity [1] is of vital importance. The final step in the road was achieved when we proved chaos in quasi-minimal sets constructed by H. Hilmy [36] and G. Birkhoff [28] as the closure of a Poisson stable trajectory. For this, we created a concept of unpredictable point [7]. In this book, we call it the *alpha unpredictable point* to differentiate it from similar terms. An alpha unpredictable point admits the Poisson stability with the additional property of alpha unpredictability (or alpha unpredictability). It was verified that the property immediately makes the dynamics on the quasi-minimal sets with the Lorentz sensitivity such that one can accept that chaos is present. The quasi-minimal set must be densely filled with alpha unpredictable trajectories, which are recurrent as Poisson stable. Thus, we are confident that the recurrence is a fundamental feature for the modern comprehension of mathematical chaos, and the most developed type of alpha unpredictability guarantees that it is ultimate chaos. It is ultra Poincaré chaos. As another argument one can consider the recurrence theorem. Indeed, the fundamental result is most complicated and insightful in dynamical systems theory and is strictly adjoined with problems of ergodicity, comprehension of disorder in the mathematical sense, if discussed deterministically, assuming that trajectories can be prolonged infinitely long. It says that all the trajectories are Poisson stable. We have equipped them with alpha unpredictability, thus closing the question on chaos presence in the circumstances. Further steps in the deterministic comprehension of complexity can be made only if one violates the conditions for classical dynamical systems as continuation of orbits, for example, and consider the problem for more recent models with discontinuities and hybrid systems. We predict that alpha labeling will play a significant, instrumental role in extending the most fundamental insights of collective dynamics because the procedure is applicable for well-arranged classical models and any equations used for change description. We expect that alpha labeling can also be applied for analysis of fractals combining dynamical properties with geometrical complexities more effectively. Combining alpha labeling with measurement theory and metric and topological spaces will provide a powerful instrument for analysis of complexity and even classical regular motions.

One more important remark concerns the relations of deterministic and non-deterministic comprehension of the real world. We have found that alpha unpredictability presents not only in differential, discrete, and hybrid equations, but also in random processes such as Bernoulli schemes and Markov chains. This time it is confirmed for processes with finite space. However, it is clear, if one proceeds from the similarity of the ten-digital theory of numbers and alpha labeling, we can find the phenomenon, and consequently ultra Poincaré chaos, in stochastic dynamics generally. Thus, one can conclude that alpha unpredictability implies ultra Poincaré chaos, which is the ultimate tool to describe the complexity of dynamics mathematically and deterministically at the present state of science development. Another remark on alpha unpredictability is that it makes chaos a proper part of theories of dynamical systems, differential, hybrid, and discrete equations—those mathematical models which have been constructed to study motion and changes in the real world. Those new types of models, which are anticipated to be invented in the future, will also be busy with alpha unpredictable trajectories as inevitable elements of complexity.

The first and most impressive evidence of chaos was discovered by H. Poincare in celestial mechanics [45]. Surprisingly, he showed that there are not stable cycles in the three-body problem. Recurrence, more complex than periodic motions, was observed, and the ultimate recurrent motion was named after Poisson [45]. The closure of a compact Poisson stable orbit is the quasi-minimal set [28, 36, 44] and it is the region where very sophisticated dynamics occur. This had been the ultimate achievement in the recurrence line, before the concept of the alpha unpredictability was introduced in [7].

The next significant discovery was made in meteorology, far from celestial mechanics. American physicist E. Lorenz [38] learned that the dynamics admits what we know as sensitivity. Unfortunately, the sensitivity was not incorporated in quasi-minimal sets for a long time, till in [7] the *alpha unpredictable point* was introduced. From one side it is a specific sort of the Poisson stable point, and from another the alpha unpredictability is a new type of recurrence. We think that it is a new ultimate destination on the line of recurrent motions, which replaced the Poisson stable point in the role, since it is endowed with divergence. We consider the alpha unpredictable point a mild recurrence behind the chaos gate.

This chapter is the introductory part of the book. It introduces alpha unpredictability as a new phenomenon of classical dynamical systems and discusses its appearances in functional spaces.

Emphasizing the ingredients of Devaney [32] and Li–Yorke [37] chaos, one can see that the definitions of chaos have been considered using sets of motions but not through a single motion description. For the first time in the literature, in our papers [6, 7, 14], chaos is initiated from a single motion, which is with sophisticated behavior. Thus, the line of equilibrium, periodic function, quasi-periodic function, almost periodic function, recurrent function, and Poisson stable function is prolonged with the new element: alpha unpredictable motion. This supplement to the line creates the possibility of other functions beyond the known ones.

An essential point to discuss is sensitivity or alpha unpredictability. In the literature, sensitivity has been considered through initially nearby different motions. However, we describe alpha unpredictability as an interior property of a single trajectory. Then chaos appears in a neighborhood of the trajectory. The symbolic dynamics illustrates all the results, and it is an essential tool for the investigation of the complicated dynamics of continuous-time systems such as the Lorenz and Rössler equations [42]. It is worth noting that alpha unpredictable points can be replicated by the techniques summarized in [5].

One can see the proximity of chaos and quasi-minimal sets by comparing their definitions [32, 37, 44]. Transitivity is a common feature, and the closure of a Poisson stable trajectory contains infinitely many Poisson stable orbits. In its turn, we know that a periodic trajectory is also Poisson stable. it is possible that the continuum of Poisson stable trajectories is an ultimate form of infinitely many cycles known for chaos. Is sensitivity proper for any quasi-minimal set? These questions help discover more relationships between quasi-minimal sets and chaos.

We utilize the topology of uniform convergence on any compact subset of the real axis to introduce the alpha unpredictable functions. More precisely, the Bebutov dynamical system [47] has been applied. All these make our duty of incorporating chaos investigation into the theory of differential equations initiated in papers [12, 13, 15] and in the book [14] seem to proceed correctly. More examples and methods to determine alpha unpredictable functions are provided in next chapters of the book [3, 4, 19, 20, 23–25].

In the instrumental sense, discreteness has been the main object of chaos investigation. To check this, it is sufficient to recall definitions of chaos [32, 37, 51], which are based on sequences and maps, as well as Smale Horseshoe and logistic maps, Bernoulli shift [51], which are in the core of the chaos theory. Stroboscopic observation of a motion was the single way to indicate the irregularity in continuous dynamics. Embedding the research in the theory of differential equations requires definitions of chaos for continuous dynamics, which are unrelated to discreteness [5]. The research, as well as the origin of the chaos [45], gave us solid arguments for the development of motions in classical dynamical systems theory [28] by proceeding behind Poisson stable points to alpha unpredictable points [7]. Then, the dynamic has been specified such that a function that is bounded on the real axis is an alpha unpredictable point [13, 15]. In the papers, we have demonstrated that alpha unpredictable functions are easy to analyze as solutions to differential equations. This paradigm is only completed if one considers discrete equations. This is why we deliver discrete analogues for alpha unpredictable functions, calling them alpha unpredictable sequences. The role of the discrete alpha unpredictability cannot be underestimated for applications and theoretical analysis, especially for the modern development of computer technologies, software, and robotics [34, 49].

To protect the reader from misunderstanding, we will comment on the following part in a review of our results in [7], 'We note, however, that these results for a particular trajectory and semi-trajectory are covered by the following rather general result that is easy to prove: Assume X is a compact metric space and consider either a cascade generated by an expansive homeomorphism $X \to X$ or a semi-cascade generated by an expanding map $X \to X$. Then if a point in X is stable P^+ and is not periodic, it is alpha unpredictable'.

Let us remember that a map $f: X \to X$ is expansive if there exists $\varepsilon_0 > 0$ such that, for any $x, y \in X$, $x \neq y$, there exists n such that $d(f^n x, f^n y) > \varepsilon_0$ [32], and the sensitivity does not imply the expansiveness. This is why the suggested assertion can not cover theorem 3.2.1, and the review statement is not correct. Otherwise, the reviewer is invited to clarify their position. At the same time, serious studies already confirm that the research subject has interesting prospects for topological dynamics [39, 40, 50]. This section was reproduced from [14], with permission from Springer Nature.

References

[1] Akhmet M 2020 *Almost Periodicity, Chaos and Asymptotic Equivalence* (Berlin: Springer)

[2] Akhmet M 2021 *Domain Structured Dynamics* (Bristol: IOP Publishing)

[3] Akhmet M 2022 Modular chaos for random processes *Discontinuity Nonlinearity Complexity* **11** 191–201

[4] Akhmet M 2022 Unpredictablity in Markov chains *Carpathian J. Math.* **38** 13–9

[5] Akhmet M and Fen M O 2016 *Replication of Chaos in Neural Networks, Economics and Physics* (Berlin: Springer)

[6] Akhmet M and Fen M O 2016 Poincaré chaos and unpredictable functions *Commun. Nonlinear Sci. Numer. Simul.* **48** 85–94

[7] Akhmet M and Fen M O 2016 Unpredictable points and chaos *Commun. Nonlinear Sci. Numer. Simul.* **40** 1–5

[8] Akhmet M, Baskan K and Yesil C 2022 Delta synchronization of Poincaré chaos in gas discharge-semiconductor systems *Chaos* **32** 083137

[9] Akhmet M, Baskan K and Yesil C 2023 Markovian noise-induced delta synchronization approach for Hindmarsh-Rose model *Chaos Solitons Fractals* **185** 115155

[10] Akhmet M, Baskan K and Yesil C 2023 Revealing chaos synchronization below the threshold in coupled Mackey-glass systems *Math. Open Access* **11** 3197

[11] Akhmet M, Yesil C and Baskan K 2023 Synchronization of chaos in semiconductor gas discharge model with local mean energy approximation *Chaos Solitons Fractals* **167** 113035

[12] Akhmet M and Fen M O 2017 Existence of unpredictable solutions and chaos *Turk. J. Math.* **41** 254–66

[13] Akhmet M and Fen M O 2018 Non-autonomous equations with unpredictable solutions *Commun. Nonlinear Sci. Numer. Simul.* **59** 657–70

[14] Akhmet M, Fen M O and Alejaily E M 2020 *Dynamics with Chaos and Fractals* (Berlin: Springer)

[15] Akhmet M, Fen M O, Tleubergenova M and Zhamanshin A 2019 Unpredictable solutions of linear differential and discrete equations *Turk. J. Math.* **43** 2377–89

[16] Akhmet M, Fen M O, Tleubergenova M and Nugayeva Z 2020 Unpredictable solutions of linear impulsive systems *Math. Open Access* **8** 1–16

[17] Akhmet M, Fen M O, Tleubergenova M and Zhamanshin A 2019 Poincare chaos for a hyperbolic quasilinear system *Miskolc Math. Notes* **20** 33–44

[18] Akhmet M, Fen M O, Tleubergenova M and Zhamanshin A 2019 Unpredictable solutions of linear differential and discrete equations *Turk. J. Math.* **43** 2377–89

[19] Akhmet M, Seilova R D, Tleubergenova M and Zhamanshin A 2020 Shunting inhibitory cellular neural networks with strongly unpredictable oscillations *Commun. Nonlinear Sci. Numer. Simulat.* **89** 105287

[20] Akhmet M, Tleubergenova M and Nugayeva Z 2020 Strongly unpredictable oscillations of Hopfield-type neural networks *Math. Open Access* **8** 1–14

[21] Akhmet M, Tleubergenova M and Zhamanshin A 2020 Quasilinear differential equations with strongly unpredictable solutions *Carpathian J. Math.* **36** 341–9

[22] Akhmet M, Tleubergenova M and Zhamanshin A 2019 Neural networks with Poincare chaos *ACM International Conference Proceeding Series* 34–7 pp

[23] Akhmet M, Tleubergenova M and Zhamanshin A 2021 Modulo periodic poisson stable solutions of quasilinear differential equations *Entropy* **23** 1535

[24] Akhmet M, Tleubergenova M and Zhamanshin A 2023 Unpredictable solutions of duffing type equations with Markov coefficients *arXiv* (2303.17336)

[25] Akhmet M, Tleubergenova M and Zhamanshin A 2023 Compartmental unpredictable functions *Mathematics* **11** 1069

[26] Banks J, Brooks J, Cairns G, Davis G and Stacey P 1992 On Devaney's definition of chaos *Amer. Math. Monthly* **99** 332–4

[27] Besicovitch A S 1954 *Almost Periodic Functions* (Cambridge: Dover)

[28] Birkhoff G D 1991 *Dynamical Systems* (Providence, RI: Colloquium Publications)

[29] Bogolyubov N N 1939 *On Some Arithmetic Properties of Almost Periods* (Kiev: Akademiya Nauk Ukrainian SSR)

[30] Bohl P 1900 Ueber Einige Differentialgleichungen Allgemeinen Charakters, Welche in der Mechanik Anwendbar Sind *PhD Thesis* Dorpat

[31] Bohr H A 1947 *Almost Periodic Functions* (New York: Chelsea Publishing Company)

[32] Devaney R L 1987 *An Introduction to Chaotic Dynamical Systems* (Menlo Park, CA: Addison-Wesley)

[33] Devaney R L and Nitecki Z 1979 Shift automorphisms in the Henon mapping *Math. Phys.* **67** 137–46

[34] Ditto W L, Murali K and Sinha S 2008 Chaos computing: ideas and implementations *Phil. Trans. R. Soc.* A **366** 653–64

[35] Hartman P 2002 *Ordinary Differential Equations* (New York: SIAM)

[36] Hilmy H 1936 Sur les ensembles quasi-minimaux dans les systèmes dynamiques I *Ann. Math.* **37** 899–907

[37] Li T Y and Yorke J A 1975 Period three implies chaos *Am. Math. Mon.* **82** 985–92

[38] Lorenz E N 1963 Deterministic non-periodic flows *J. Atmos. Sci.* **20** 130–41

[39] Mahajan A, Thakur R and Das R 2024 Sensitivity and unpredictability in semiflows on topological spaces *Commun. Nonlinear Sci. Numer. Simulat.* **133** 107949

[40] Miller A 2019 Unpredictable points and stronger versions of Ruelle–Takens and Auslander–Yorke chaos *Topology Appl.* **253** 7–16

[41] Minorsky N 1947 *Introduction to Non-linear Mechanics: Topological Methods, Analytical Methods, Non-linear Resonance, Relaxation Oscillations* (Ann Arbor, MI: J.W. Edwards)

[42] Mischaikow K and Mrozek M 1995 Chaos in the Lorenz equations: a computer assisted proof *Bull. AMS* **32** 66–72

[43] Naife A 1960 *Qualitative Theory of Differential Equations* (Princeton, NJ: Princeton University Press)

[44] Nemytskii V V and Stepanov V V 1960 *Qualitative Theory of Differential Equations* (Princeton, NJ: Princeton University Press)

[45] Poincaré H 1957 *New Methods of Celestial Mechanics Volumes I–III* (New York: Dover)

[46] Robinson C 1995 *Dynamical Systems: Stability, Symbolic Dynamics, and Chaos* (Boca Raton, FL: CRC Press)

[47] Sell G R 1971 *Topological Dynamics and Ordinary Differential Equations* (London: Van Nostrand Reinhold Company)

[48] Shilnikov L 1997 *Nonlinear Dynamics, Chaotic and Complex Systems* (Cambridge: Cambridge University Press) pp 39–63

[49] Staingrube S, Timme M, Worgotter F and Mannonpong P 2010 Self-organized adaptation of simple neural circuits enables complex robot behavior *Nat. Phys.* **6** 224–30

[50] Thakur R and Das R 2020 Strongly Ruelle-Takens, strongly Auslander-Yorke and Poincaré chaos on semiflows *Commun. Nonlinear Sci. Numer. Simulat.* **81** 105018

[51] Wiggins S 1988 *Global Bifurcation and Chaos: Analytical Methods* (New York: Springer)

Part III

Compartmental functions

Chapter 4

Compartmental alpha unpredictable functions

A broad spectrum of recurrent functions ranges from equilibria to Poisson stable functions, playing a crucial role both theoretically and practically, with a storied history. Recently, alpha unpredictable functions have been added to this family. The value of these functions, both in theory and application, is enhanced when supported by rigorously validated research methods, exemplary cases, and numerical simulations. This chapter addresses the complexities of alpha unpredictability by examining functions of two variables on diagonals, where one variable displays regularity and the other exhibits alpha unpredictability. Consequently, the research delves into compartmental unpredictable functions.

A vast family of recurrent functions starts with equilibria and ends with those that are Poisson stable. They are fundamental in theoretical and application senses and have a famous history. Recently, we added alpha unpredictable functions to the family. The research has been performed in several papers and books. The theoretical and application merits of functions increase if one provides rigorously proven efficient research methods, illustrious examples, and numerical simulations. In the present chapter, we meet the challenges for alpha unpredictability by considering the functions of two variables on diagonals. Deterministic and random algorithms have been created. Characteristics are introduced to evaluate the contributions of periodic and unpredictable components to the functions, and they are clearly illustrated in graphs. Definitions of non-periodic compartmental functions are provided as future research suggestions.

Oscillations and recurrence play a special role in studying the dynamics of processes occurring in nature and industry. Numerous results have been obtained in the literature for periodic, quasi-periodic, and almost periodic functions due to established valuable mathematical methods and important applications [41, 46, 61, 62, 73]. On the other hand, recurrent and Poisson stable functions are also crucial for the theory of oscillations [35, 58, 68, 70, 96]. The theory of non-linear dynamics, [35, 41, 61, 79, 91], was focused mainly on periodic motions. Functions which can still be

doi:10.1088/978-0-7503-5162-1ch4
4-1

considered 'periodic' and are sufficiently convenient for strict mathematical analysis are quasi-periodic functions independently introduced and investigated by P. Bohl [37, 38] and E. Esclangon [60]. The fundamental papers of H. Bohr [39] are about the basics of almost periodic functions. Different theories of almost-periodicity were constructed by N. N. Bogolyubov [36], A. S. Besicovitch [34], S. Bochner [40], and V. V. Stepanov [99]. Those functions are essential for developing harmonic analysis of groups, including the Fourier series. The concepts of recurrent motions and Poisson stable points are central in the qualitative theory of dynamical systems. They were considered by H. Poincaré [91] and G. Birkhoff [35] as the main ingredients of complexity in celestial dynamics.

Chaos theory is being widely developed, and more than the classical functions are needed to describe the dynamics of complex systems. It requires not only new models and new solutions of models but also new functions. This is why the *unpredictable* and *compartmental Poisson stable* functions have been introduced in our recent papers [8, 13, 25, 27]. Accordingly, a new *method of included intervals* for the existence of alpha unpredictable and Poisson stable solutions of discrete and differential equations was suggested. Since alpha unpredictability leads to chaos, the role of alpha unpredictable functions is vital in applications. The new functions are essential for studying the dynamics of neural networks. Due to the complexity and non-linearity of neural networks, their behavior is not confined to regular functions. In our papers, alpha unpredictable oscillations in Hopfield-type neural networks [17] as well as shunting inhibitory cellular and Cohen–Grossberg neural networks [16, 18, 19] were investigated.

It is indisputable that when considering functions in applications, one should enlarge the number of methods for construction and numerical presentations, starting with simple algebraic operations and finishing with a Fourier series and the operator theory. The present study suggests a new way of alpha unpredictable functions construction rooted in compartmental functions. This method starts with two variables, one that is alpha unpredictable, and the other being either periodic, quasi-periodic, or almost periodic—or even recurrent or Poisson stable. Then, the domains are narrowed to the diagonals of coordinate spaces, where the arguments are ranged. The method of diagonals is routine, known for quasi-periodic or almost periodic functions [34, 46], but in the study, it is applied for dynamics, which are on an essentially new level, since the dependence on the two variables is not regular. Correspondingly, it is an interesting problem to find such conditions that function on diagonals and admit alpha unpredictability. In the present research, the problem is provided with a particular solution for the case of periodicity, focusing on so-called compartmental periodic functions. Besides the general problem, we discuss elementary algebra for alpha unpredictable functions and alpha unpredictability of compositions.

While working on the new types of recurrence, we have learned that, despite numerous papers on almost periodic and Poisson stable functions, there are no mentions of numerical examples and simulations for the functions nor solutions of differential equations, except quasi-periodic ones. Simultaneously, the needs of industry, and (especially) neuroscience, artificial intelligence, and other modern areas demand numerical presentation of motions, which have already been supported seriously in theories. Our research comprehensively meets the challenges since we have constructed samples of Poisson stable and alpha unpredictable functions utilizing solutions of the logistic equation, as well as determining them randomly through realizations of Bernoulli schemes and Markov chains. That is, deterministic and stochastic routes for the functions have been paved. One should emphasize that, even for Poisson stable functions, which have been in the research for about a century, samples of the concrete functions appeared in our papers [13, 27] for the first time. The numerical experiments are advantageous since they are accompanied by newly developed robust instruments of the functions simulations. For instance, they are suitable for synchronization of chaos, namely, *Delta synchronization*, which works for gas discharge-semiconductor systems [20, 21], where even the generalized synchronization [93] is not practical. A numerical test for alpha unpredictable dynamics has been suggested [11], which discovers strange attractors when conservative methods do not work [12]. Moreover, we have developed algorithms that allow us to observe periodicity contributions and alpha unpredictability for compartmental dynamics. The algorithms are based on the concept of the degree of periodicity. We have learned that very similar time series can be seen in several industrial experiments [76, 88, 103, 105, 106], which is a strong argument for applying our results. One can also believe that the research of the compartmental functions can provide more insights into the problem of the transition from quasi-periodicity to chaos [71, 92].

In section 4.1, one can find complete information on continuous functions, which are the focus of the present research. Section 4.1.1 contains conditions on the novel parameters sufficient to guarantee the alpha unpredictability of compartmental functions with a periodic component. The properties are utilized in the central theoretical part of the article, section 4.1.2, where a new class of alpha unpredictable functions is described. The reader is invited to consider theorem 4.1.14 as the source of weakened conditions for alpha unpredictability. Section 4.1.3 contains examples of concrete compartmental periodic alpha unpredictable functions constructed by applying irregular dynamics features in the logistic equation. The results are rigorously proven by the *method of included intervals*. The degree of periodicity is considered in section 4.1.4. Its role in the analysis is carefully illustrated through several numerical simulations. The route for constructing alpha unpredictable functions by Markov chains is discussed in section 4.1.5. For the first time in a book, discontinuous Poisson stability is discussed; refer to section 4.2, where the complex job of making consistent discrete and continuous components of models to create sophisticated recurrence is accomplished. In the final part, *Miscellanea*, section 4.3, definitions of compartmental quasi-periodic, almost periodic, recurrent, and Poisson stable functions are presented to complete the presentation. Moreover, theorems are proved on the alpha unpredictability of functions being subdued to simple algebraic operations. The section is closed with an example of a compartmental quasi-periodic alpha unpredictable function, which confirms the potential of the research when a diagonalized function is not necessary unpredictable.

This introduction was reproduced with permission from [14]. CC BY 4.0.

4.1 Continuous compartmental unpredictable functions

Throughout the section, $\mathbb{N}$, $\mathbb{Z}$, and $\mathbb{R}$, respectively, stand for the sets of natural numbers, integers, and real numbers. Moreover, we use a Euclidean norm, for vectors.

Definition 4.1.1. *A bounded function $f(t)\colon \mathbb{R} \to \mathbb{R}^n$ is said to be Poisson stable if there exists a sequence t_k, $t_k \to \infty$ as $k \to \infty$, such that the sequence of functions $f(t + t_k)$ uniformly converges to $f(t)$ on each bounded interval of the real axis.*

Definition 4.1.2. [26] *A bounded function $f\colon \mathbb{R} \to \mathbb{R}^n$ is alpha unpredictable if there exist positive numbers ε_0, δ and sequences t_k, s_k, both of which diverge to infinity, such that $\|f(t + t_k) - f(t)\| \to 0$ as $k \to \infty$ uniformly on compact subsets of $\mathbb{R}$ and $\|f(t + t_k) - f(t)\| > \varepsilon_0$ for each $t \in [s_k - \delta, s_k + \delta]$ and $k \in \mathbb{N}$.*

A sequence t_k, $k = 1, 2, \ldots$, in definitions 4.1.1 and 4.16 is said to be the *Poisson* or *convergence sequence* of the function $f(t)$. We call the uniform convergence on compact subsets of $\mathbb{R}$ *the convergence property*, and the existence of the sequence s_k and positive numbers ε_0, δ is called *the separation property*.

Remark 4.1.1. *From the last two definitions, it follows that we consider not only continuous, but also discontinuous alpha unpredictable and Poisson functions. The convergence and separation properties are valid regardless of the continuity. Additionally, we shall use examples with continuous and discontinuous functions. The definition of continuous Poisson stable functions can be found in* [89, 96].

Definition 4.1.3. *A bounded function $f(t, x)\colon \mathbb{R} \times D \to \mathbb{R}^n$, $D \subset \mathbb{R}^n$ is a domain, is alpha unpredictable in t, uniformly, with respect to $x \in D$ if there exist positive numbers ε_0, δ and sequences t_k, s_k, both of which diverge to infinity, such that $\sup_D \|f(t + t_k, x) - f(t, x)\| \to 0$ as $k \to \infty$ uniformly on bounded intervals of t and $\|f(t + t_k, x) - f(t, x)\| > \varepsilon_0$ for $t \in [s_k - \delta, s_k + \delta]$, $x \in D$ and $k \in \mathbb{N}$.*

Definition 4.1.4. [14] *A function $f(t)\colon \mathbb{R} \to \mathbb{R}^n$ is said to be a compartmental periodic unpredictable function if $f(t) = G(t, t)$, where $G(u, v)$ is a continuous bounded function, periodic in u uniformly with respect to v, and alpha unpredictable in v uniformly with respect to u, i.e., there exist positive numbers ω, ε_0, δ and sequences t_k, s_k, both of which diverge to infinity, such that $G(u + \omega, v) = G(u, v)$ for all $u, v \in \mathbb{R}$, $\sup_{u \in \mathbb{R}} \|G(u, v + t_k) - G(u, v)\| \to 0$ as $k \to \infty$ uniformly on bounded intervals of v, and $\|G(u, v + t_k) - G(u, v)\| > \varepsilon_0$ for $v \in [s_k - \delta, s_k + \delta]$, $u \in \mathbb{R}$ and $k \in \mathbb{N}$.*

Definition 4.1.5. [14] *A function $f(t, x): \mathbb{R} \times D \to \mathbb{R}^n$, $D \subset \mathbb{R}^n$ is a domain and is said to be compartmental periodic alpha unpredictable in t uniformly for $x \in D$ function if $f(t, x) = G(t, t, x)$, where $G(u, v, x)$ is a bounded function, periodic in u uniformly with respect to $v \in \mathbb{R}$, $x \in D$, and alpha unpredictable in v uniformly with respect to $u \in R$ and $x \in D$.*

Remark 4.1.2. *What we use as functions on diagonals is the routine technique of multi-periodicity known for almost periodic functions [39]. If the number of periods is finite and incommensurate, then the quasi-periodic functions as a subset of almost periodic functions is shaped. In definitions 4.1.4 and 4.1.5, functions differently depending on two variables since periodicity is not the alpha unpredictability. This time, we consider a more sophisticated case of the technique. The compartmental functions are already assumed to be irregular. However, to clarify that the phenomenon is irregular, we provide conditions for the alpha unpredictability of the functions on diagonals.*

The following subsection is devoted to the kappa property, which plays the role of the periods' relation in quasi-periodicity theory. The property guarantees that the alpha unpredictability in a single variable is inherited by the function on the diagonal of the space of the argument. This is why the new results are the next step in applying the technique.

4.1.1 Kappa property of unbounded sequences

Consider a sequence of positive real numbers t_k, $t_k \to \infty$ as $k \to \infty$. Below, the following two simple lemmas will be of use. They are rather general results adapted to the needs of the present research.

Lemma 4.1.1. *For an arbitrary sequence of positive real numbers t_k, $k = 1, 2, \ldots$, and a positive number ω, there exist a subsequence t_{k_l}, $l = 1, 2, \ldots$, and a number τ_ω, $0 \leqslant \tau_\omega < \omega$, such that $t_{k_l} \to \tau_\omega (mod \ \omega)$ as $l \to \infty$.*

Proof. Consider the sequence τ_k, such that $t_k \equiv \tau_k (mod \ \omega)$ and $0 \leqslant \tau_k < \omega$ for all $k \geqslant 1$. The boundedness of the sequence τ_k implies that there exists a subsequence τ_{k_l}, which converges to a number τ_ω [54].

For fixed $\omega > 0$, by the last lemma, there exist a subsequence t_{k_l} and a number τ_ω, such that $t_{k_l} \to \tau_\omega (mod \ \omega)$ as $l \to \infty$. In what follows, considering applications for alpha unpredictable and Poisson stable functions, we shall call the number τ_ω the *Poisson shift* with respect to the ω. The set of all Poisson shifts $\mathscr{T}_\omega$ is not empty. It can consist of several or even infinite elements. The number $\kappa_\omega = inf \ \mathscr{T}_\omega$, $0 \leqslant \kappa_\omega < \omega$ is said to be the *Poisson number with respect to the number ω*. If $\kappa_\omega = 0$, then we say that the sequence t_k satisfies the *kappa property with respect to the number ω*. The following assertion is useful in the next part of the paper.

Lemma 4.1.2. [27] $\kappa_\omega \in \mathscr{T}_\omega$.

Proof. Assume on the contrary that κ_ω is not in $\mathscr{T}_\omega$. Then, there exists a strictly decreasing sequence τ_m, $m \geqslant 1$, in $\mathscr{T}_\omega$, such that $\tau_m \to \kappa_\omega$. For each natural m, denote by t_i^m a subsequence of t_k, such that $t_i^m \to \tau_m (mod\ \omega)$ as $i \to \infty$.

Fix a sequence of positive numbers ε_n, which converges to the zero. One can find numbers i_n, $n = 1, 2, \ldots$, such that $|t_{i_n}^n - \tau_n| < \varepsilon_n (mod\ \omega)$. It is clear that $t_{i_n}^n \to \kappa_\omega (mod\ \omega)$ as $n \to \infty$.

Next, examples are provided to demonstrate how rich the set of sequences is with respect to the kappa property.

Example 4.1.1. *Let us take the unbounded sequence* $t_k = \frac{(k-1)\omega}{m}$, $k \in \mathbb{N}$, *where* $m \in \mathbb{N}$, $\omega > 0$ *are fixed numbers. If* $m = 1$ *then* $t_k \equiv 0 (mod\ \omega)$, $k \in \mathbb{N}$, *and there exists a unique Poisson shift,* $\tau_\omega = 0$. *If* $m = 2$ *then* $t_k \equiv 0 (mod\ \omega)$ *for even numbers* k, *and* $t_k \equiv \frac{\omega}{2} (mod\ \omega)$, *if* k *is an odd number. There are no other Poisson shifts, therefore,* $\mathscr{T}_\omega = \{0, \frac{\omega}{2}\}$. *Generally, one can find that the set of Poisson shifts is* $\mathscr{T}_\omega = \{0, \frac{\omega}{m}, \frac{2\omega}{m}, \ldots, \frac{(m-1)\omega}{m}\}$. *Thus, we procure, for any* $m = 1, 2, \ldots$, *the sequence* t_k *satisfies the kappa property with respect to* ω.

Example 4.1.2. *Now, consider the sequence* $t_k = k\omega + \mu_k, 0 < \mu_k < \omega, k = 1, 2, \ldots$, μ_k, *where the sequence* μ_k *is determined as follows,*

$$\mu_{2^l - 1} = \frac{\omega}{2^l}, \ \mu_{2^l - l+1} = \frac{2\omega}{2^l}, \ \ldots, \ \mu_{2^{l+1} - l - 2} = \frac{(2^l - 1)\omega}{2^l}, \ l = 1, 2, \ldots.$$

Thus, the sequence μ_k, $k = 1, 2, \ldots$, *is obtained, and each element of the section* $[0, \omega]$ *is a Poisson shift. So, the example when* $\mathscr{T}_\omega$ *is an uncountable set of numbers is considered, and the sequence* t_k *satisfies the kappa property with respect to the number* ω.

4.1.2 Alpha unpredictability of compartmental periodic functions

This part of the chapter concerns the main theoretical achievements of the chapter. Theorem 4.1.14 discusses the weakest sufficient conditions for the alpha unpredictability of compartmental functions with periodicity. It is more theoretical than the results of the following theorems 4.1.16–4.1.20, which are constructive to determine alpha unpredictable functions in examples and experiments.

Theorem 4.1.1. [13, 14] *Assume that* $G(u, v, x)$: $\mathbb{R} \times \mathbb{R} \times D \to \mathbb{R}^n$, $D \subset \mathbb{R}^n$ *is an open and bounded set, which is a continuous function, and* ω-*periodic in* u *uniformly with respect to* v *and* x. *Then, the function* $g(t, x) = G(t, t, x)$ *is alpha unpredictable in* t, *uniformly with respect to* x, *if the following conditions are valid:*

(i) *for each $\varepsilon > 0$ there exists a positive number η such that* $\|G(t + s, t, x) - G(t, t, x)\| < \varepsilon$ *if* $|s| < \eta$, $t \in \mathbb{R}$, $x \in D$;

there exist sequences t_k, s_k, both of which diverge to infinity as $k \to \infty$, and positive numbers ε_0, δ such that
  (ii) *the sequence t_k satisfies the kappa property with respect to the period ω;*
  (iii) $\sup_{I \times D}\|G(t, t + t_k, x) - G(t, t, x)\| \to 0$ *on each bounded interval $I \subset R$;*
  (iv) $\inf_{[s_k - \delta, s_k + \delta] \times D}\|G(t, t + t_k, x) - G(t, t, x)\| > \varepsilon_0$, $k \in \mathbb{N}$.

Proof. Let us fix a positive number ε, and a bounded interval $I \in \mathbb{R}$. Since the sequence t_k satisfies the kappa property, one can write, without loss of generality, that $t_k \to 0 (mod\ \omega)$ as $k \to \infty$. Therefore, by conditions (*i*) and (*iii*), the following inequalities are valid:

$$\sup_{\mathbb{R} \times D}\left\|G(t + t_k, t, x) - G(t, t, x)\right\| < \frac{\varepsilon}{2} \qquad (4.1)$$

and

$$\sup_{I \times D}\left\|G(t, t + t_k, x) - G(t, t, x)\right\| < \frac{\varepsilon}{2}, \qquad (4.2)$$

for sufficiently large k.

Using inequalities (4.1) and (4.2), we obtain:

$\|g(t + t_k, x) - g(t, x)\| = \|G(t + t_k, t + t_k, x) - G(t, t, x)\| \leqslant]$
$\|G(t + t_k, t + t_k, x) - G(t, t + t_k, x)\| + \|G(t, t + t_k, x) - G(t, t, x)\| <$
$\frac{\varepsilon}{2} + \frac{\varepsilon}{2} = \varepsilon,$

for all $t \in I$, $x \in D$. That is, $g(t + t_k, x)$ converges to $g(t, x)$ on each arbitrary bounded time interval uniformly for $x \in D$. Moreover, conditions (*i*) and (*ii*) imply that $\sup_{\mathbb{R} \times D}\|G(t + t_k, t, x) - G(t, t, x)\| < \frac{\varepsilon_0}{2}$ for sufficiently large k. Applying assumption (*iv*), one can obtain:

$\|g(t + t_k, x) - g(t, x)\| = \|G(t + t_k, t + t_k, x) - G(t, t, x)\| \geqslant$
$\|G(t + t_k, t + t_k, x) - G(t + t_k, t, x)\| - \|G(t + t_k, t, x) - G(t, t, x)\| >$
$\varepsilon_0 - \frac{\varepsilon_0}{2} = \frac{\varepsilon_0}{2},$

for all $t \in [s_k - \delta, s_k + \delta]$, $x \in D$, $k \in \mathbb{N}$. The lemma is proved.

The following assertion is a corollary of the theorem 4.1.14.

Theorem 4.1.2. [14] *Assume that a continuous and bounded function* $G(u, v)$: $\mathbb{R} \times \mathbb{R} \to \mathbb{R}^n$, *is ω-periodic in u. The function $f(t) = G(t, t)$ is alpha unpredictable if the following conditions are valid:*

(i) *for each $\varepsilon > 0$ there exists a positive number η such that $\|G(t + s, t) - G(t, t)\| < \varepsilon$ if $|s| < \eta$, $t \in \mathbb{R}$;*

there exist sequences t_k, s_k both of which diverge to infinity as $k \to \infty$, and positive numbers ε_0, δ, such that
    (ii) *the sequence t_k satisfies kappa property with respect to the period ω;*
    (iii) *$\|G(t, t + t_k) - G(t, t)\| \to 0$, uniformly on each bounded interval $I \subset R$ of t;*
    (iv) *$\inf_{[s_k - \delta, s_k + \delta]} \|G(t, t + t_k) - G(t, t)\| > \varepsilon_0$, $k \in \mathbb{N}$.*

Remark 4.1.3. *Conditions (iii) and (iv) in the theorem 4.1.16 are satisfied if $G(t, u)$ is alpha unpredictable in the second argument by definition 4.1.5.*

Theorem 4.1.3. [14] *Assume that a function $G(t, u)$: $\mathbb{R} \times D \to \mathbb{R}^n$, $D \subseteq \mathbb{R}^n$, is ω-periodic in t and satisfies the inequalities $L_1 \|u_1 - u_2\| \leqslant \|G(t, u_1) - G(t, u_2)\| \leqslant L_2 \|u_1 - u_2\|$, where L_1, L_2 are positive constants for all $t \in \mathbb{R}$, $u_1, u_2 \in D$. If $v(t)$: $\mathbb{R} \to D$ is an alpha unpredictable function such that the convergence sequence t_k admits the kappa property with respect to period ω, then $G(t, v(t))$ is an alpha unpredictable function.*

Proof. Consider the function $F(t, u) = G(t, v(u))$. We shall prove that $F(t, u)$ is alpha unpredictable in u. Let us fix positive number ε and a bounded interval I. Because of the kappa property, for a sufficiently large number k we have that $\left\| v(u + t_k) - v(u) \right\| < \frac{\varepsilon}{L_2}$ for $u \in I$. That is why,

$$\|F(t, u + t_k) - F(t, u)\| = \|G(t, v(u + t_k)) - G(t, v(u))\|$$
$$\leqslant L_2 \left\| v(u + t_k) - v(u) \right\| \leqslant L_2 \frac{\varepsilon}{L_2} \leqslant \varepsilon,$$

for all $t \in \mathbb{R}$, and $u \in I$. On the other hand, there exist a sequence s_k and positive numbers ε_0, δ such that $\|v(u + t_k) - v(u)\| > \varepsilon_0$ for $t \in [s_k - \delta, s_k + \delta]$. Therefore, we have:

$$\|F(t, u + t_k) - F(t, u)\| = \|G(t, v(u + t_k)) - G(t, v(u))\|$$
$$\geqslant L_1 \|v(u + t_k) - v(u)\| > L_1 \varepsilon_0,$$

for each $t \in [s_k - \delta, s_k + \delta]$. Thus, one can conclude that the function $G(t, v(t))$ is unpredictable.

Theorem 4.1.4. *Assume that a function $G(t, u)$: $\mathbb{R} \times D \to \mathbb{R}^n$ is alpha unpredictable in t uniformly with respect to u, and satisfies the inequality $\|G(t, u_1) - G(t, u_2)\| \leqslant L \|u_1 - u_2\|$, $t \in \mathbb{R}$, $u_1, u_2 \in D$, where L is a positive* *constant. If $v(t)\colon \mathbb{R} \to D$ is a Poisson stable function with the Poisson sequence t_k common with that for $G(t, u)$, and $2L\sup_{t\in\mathbb{R}}\|v(t)\| < \varepsilon_0$, where ε_0 is a separation constant for $G(t, u)$ then the composition $G(t, v(t))$ is an unpredictable function.*

Proof. Let us fix a positive number ε, and a bounded interval I. Since $G(t, v(t))$ is alpha unpredictable in t, and $v(t)$ is a Poisson stable function, there exists sufficiently large k, such that: $\left\|G(t + t_k, v(t + t_k)) - G(t, v(t + t_k))\right\| < \frac{\varepsilon}{2}$, and $\left\|v(t + t_k) - v(t)\right\| < \frac{\varepsilon}{2L}$ for $t \in I$. That is,

$$\left\|G(t + t_k, v(t + t_k)) - G(t, v(t))\right\| \leqslant \left\|G(t + t_k, v(t + t_k)) - G(t, v(t + t_k))\right\| +$$
$$\left\|G(t, v(t + t_k)) - G(t, v(t))\right\| \leqslant \left\|G(t + t_k, v(t + t_k)) - G(t, v(t + t_k))\right\| +$$
$$L\left\|v(t + t_k) - v(t)\right\| \leqslant \frac{\varepsilon}{2} + L\frac{\varepsilon}{2L} \leqslant \varepsilon,$$

for all $t \in I$. Thus, $G(t + t_k, v(t + t_k)) \to G(t, v(t))$ uniformly on each bounded interval of the real axis. Under assumptions of the theorem, we have that there exists a sequence s_k and positive numbers ε_0, δ such that $\left\|G(t + t_k, v(t + t_k)) - G(t, v(t + t_k))\right\| > \varepsilon_0$ for $t \in [s_k - \delta, s_k + \delta]$. Now, we obtain that:

$$\left\|G(t + t_k, v(t + t_k)) - G(t, v(t))\right\| \geqslant \left\|G(t + t_k, v(t + t_k)) - G(t, v(t + t_k))\right\| -$$
$$\left\|G(t, v(t + t_k)) - G(t, v(t))\right\| \geqslant \left\|G(t + t_k, v(t + t_k)) - G(t, v(t + t_k))\right\| -$$
$$L\left\|v(t + t_k) - v(t)\right\| > \varepsilon_0 - 2L\sup_{t\in\mathbb{R}}\|v(t)\| > 0,$$

for $t \in [s_k - \delta, s_k + \delta]$, and the function $G(t, v(t))$ is unpredictable.

4.1.3 Alpha unpredictable functions related to the logistic equation

This section of the chapter presents the practical applications of two alpha unpredictable functions. These functions, with their alpha unpredictability and additional constructive properties, are key in improving the functionality of various systems. Importantly, lemmas 4.1.22–4.1.28 establish a solid theoretical foundation, inspiring confidence in the potential of future research in industrial and neuroscience problems. The approach encompasses both deterministic and stochastic potentials. Let us consider the logistic map:

$$\lambda_{i+1} = \nu\lambda_i(1 - \lambda_i), \, i \in \mathbb{Z}. \tag{4.3}$$

In [25], it was proved that, for each $\nu \in [3 + (2/3)^{1/2}, 4]$, the equation (4.3) admits an alpha unpredictable solution μ_i, $i \in \mathbb{Z}$. That is, there exist a positive number ε_0, and the sequences ζ_k, η_k, $k \in \mathbb{N}$ of positive integers, both of which diverge to infinity, such that $|\mu_{i+\zeta_k} - \mu_i| \to 0$ as $k \to \infty$ for each i in a bounded interval of integers and $|\mu_{\zeta_k + \eta_k} - \mu_{\eta_k}| > \varepsilon_0$ for each $k \in \mathbb{N}$.

Lemma 4.1.3. *Assume that $\xi(t)$: $(0, h] \to \mathbb{R}^n$, where h is a positive number, is a bounded function. Then, the function $\pi(t) = \mu_i \xi(t - ih)$, $t \in (ih, (i + 1)h]$, $i \in \mathbb{Z}$ is Poisson stable in the sense of definition 4.1.2.*

Proof. Let us fix an interval of real numbers (α, β) and a number $i \in \mathbb{Z}$ such that $(\alpha, \beta) \subset [(i - 1)h, (i + s + 1)h]$, where s is a natural number. Then, for $t_k = \zeta_k h$, $k \in \mathbb{N}$, and $t \in (jh, (j + 1)h]$, $i - 1 \leqslant j \leqslant i + s$, we have $t + \zeta_k h \in ((j + \zeta_k)h, (j + \zeta_k + 1)h]$, and $\xi(t - (j + \zeta_k)h) = \xi(t - jh)$.

Denote $M = \sup_{t \in (0,h]} \|\xi(t)\|$. For a fixed positive number ε, and sufficiently large number k, it is true that $|\mu_{j+\zeta_k} - \mu_j| < \frac{\varepsilon}{M}$, $i - 1 \leqslant j \leqslant i + s$. Therefore, for $t \in (lh, (l + 1)h]$, where l is a fixed integer number from $i - 1$ to $i + s$, one can obtain that

$$\left\|\pi(t + t_k) - \pi(t)\right\| = \left\|\pi(t + \zeta_k h) - \pi(t)\right\| = \left\|\mu_{l+\zeta_k} \xi(t - (l + \zeta_k)h) - \mu_l \xi(t - lh)\right\| = $$
$$|\mu_{l+\zeta_k} - \mu_j| \left\|\xi(t - lh)\right\| \leqslant |\mu_{l+\zeta_k} - \mu_l| M < \varepsilon.$$

The last inequality is valid for all $i - 1 \leqslant l \leqslant i + s$. Consequently, $\left\|\pi(t + t_k) - \pi(t)\right\| < \varepsilon$ if $t \in (\alpha, \beta)$. Thus, the function $\pi(t)$ is Poisson stable. $\blacksquare$

Lemma 4.1.4. *The function $\pi(t) = \mu_i \xi(t - ih)$, $t \in (ih, (i + 1)h]$, $i \in \mathbb{Z}$, is alpha unpredictable if the following condition is satisfied,*
   (A) *there exists a positive number ε_1 such that $\|\xi(t)\| > \varepsilon_1$ for each $t \in (0, h]$.*

Proof. By lemma 4.1.22, the sequence of functions $\pi(t + t_k)$ uniformly converges to $\pi(t)$ on compact subsets of $\mathbb{R}$. It remains to show that the function $\pi(t)$ satisfies the separation property. Due to the alpha unpredictability of the sequence μ_i, there exist a positive number ε_0, and the sequence η_k, $\eta_k \to \infty$ as $k \to \infty$, such that $|\mu_{\zeta_k + \eta_k} - \mu_{\eta_k}| > \varepsilon_0$ for each $k \in \mathbb{N}$.

For $t_k = \zeta_k h$, $k = 1, 2, ...$, and $t \in (\eta_k h, (\eta_k + 1)h]$ we have that $t + t_k = t + \zeta_k h \in ((\zeta_k + \eta_k)h, (\zeta_k + \eta_k + 1)h]$. That is why $\xi(t + t_k) = \xi(t - (\zeta_k + \eta_k)h) = \xi(t - \zeta_k h)$ for all $t \in (\eta_k h, (\eta_k + 1)h]$. So, by using condition (A), we obtain:

$$\left\|\pi(t + t_k) - \pi(t)\right\| = \left\|\mu_{\zeta_k + \eta_k} \xi(t - (\zeta_k + \eta_k)h) - \mu_{\eta_k} \xi(t - \zeta_k h)\right\| = $$
$$|\mu_{\zeta_k + \eta_k} - \mu_{\eta_k}| \left\|\xi(t - \zeta_k h)\right\| > \varepsilon_0 \varepsilon_1 > 0, \tag{4.4}$$

for all $t \in (\eta_k h, (\eta_k + 1)h]$, $k = 1, 2, ...$ Thus, the function $\pi(t)$ is alpha unpredictable with positive numbers $\varepsilon^* = \varepsilon_0 \varepsilon_1$, $\delta = \frac{h}{2}$, and sequences $t_k = \zeta_k h$, $s_k = \eta_k h + \frac{h}{2}$, $k \in \mathbb{N}$. $\blacksquare$

Lemma 4.1.5. *The function $\pi(t) = \mu_i \xi(t - ih)$, $t \in (ih, (i + 1)h]$, $i \in \mathbb{Z}$, is alpha unpredictable if the following condition is valid,*

    (B) *there exist positive numbers δ, s and ε_1 such that $[s - \delta, s + \delta] \subset (0, h]$ and $\|\xi(t)\| > \varepsilon_1$ for each $t \in [s - \delta, s + \delta]$.*

Proof. The convergence property of the function $\pi(t)$ is proved in lemma 4.1.22. Let us show that the function $\pi(t)$ satisfies the separation property. There exist a positive number ε_0, and the sequence η_k, $\eta_k \to \infty$ as $k \to \infty$, such that $|\mu_{\zeta_k + \eta_k} - \mu_{\eta_k}| > \varepsilon_0$ for each $k \in \mathbb{N}$.

From $t_k = \zeta_k h$, $k = 1, 2, \ldots$, and $t \in (\eta_k h + s - \delta, \eta_k h + s + \delta]$ it follows that $t + t_k = t + \zeta_k h \in ((\zeta_k + \eta_k)h + s - \delta, (\zeta_k + \eta_k)h + s + \delta]$. Therefore, $\xi(t + t_k) = \xi(t - (\zeta_k + \eta_k)h) = \xi(t - \eta_k h)$, $k = 1, 2, \ldots$ Applying condition **(B)**, we obtain:

$$\left\|\pi(t + t_k) - \pi(t)\right\| = \left\|\mu_{\zeta_k + \eta_k} \xi(t - (\zeta_k + \eta_k)h) - u_{\eta_k} \xi(t - \zeta_k h)\right\| =$$
$$|\mu_{\zeta_k + \eta_k} - \mu_{\eta_k}|\left\|\xi(t - \eta_k h)\right\| > \varepsilon_0 \varepsilon_1 > 0, \tag{4.5}$$

for all $t \in (\eta_k h + s - \delta, \eta_k h + s + \delta]$, $k = 1, 2, \ldots$. So, one can conclude that the function $\pi(t)$ is alpha unpredictable with positive numbers $\varepsilon^* = \varepsilon_0 \varepsilon_1$, δ, and sequences $t_k = \zeta_k h$, $s_k = \eta_k h + s$, $k = 1, 2, \ldots$.

Now, let us define a continuous function $\Xi(t) \colon \mathbb{R} \to \mathbb{R}^n$, such that:

$$\Xi(t) = \int_{-\infty}^{t} e^{-\alpha(t-s)} \pi(s) ds, \tag{4.6}$$

where α is a positive real number, and $\pi(t)$ is the unpredictable function, which satisfies one of the conditions (A) or (B). The function $\Xi(t)$ is bounded on the whole real axis, such that $\sup_{t \in \mathbb{R}} \|\Xi(t)\| \leqslant \frac{M_\pi}{\alpha}$, where $M_\pi = \sup_{t \in \mathbb{R}} \|\pi(t)\|$. By applying the alpha unpredictability of the function $\pi(t)$ with condition **(B)**, we will prove the following lemma. One can see that the condition (B) implies the condition (A).

Lemma 4.1.6. *The function $\Xi(t)$ is unpredictable.*

Proof. Consider a fixed bounded and closed interval $[a, b]$, of the axis and a positive number ε. Now, applying the *method of included intervals* [27], we will show that the sequence $\Xi(t + t_k)$ uniformly converges to $\Xi(t)$ on $[a, b]$. Let us fix a positive number ξ and a number $c < a$, which satisfy the following inequalities $\frac{2M_\pi}{\alpha} e^{-\alpha(a-c)} < \frac{\varepsilon}{2}$ and $\frac{\xi}{\alpha}[1 - e^{-\alpha(b-c)}] < \frac{\varepsilon}{2}$. Let k be a large enough number, such that $\|\pi(t + t_k) - \pi(t)\| < \xi$ on $[c, b]$. Then, for all $t \in [a, b]$, we obtain:

$$\left\|\Xi(t + t_k) - \Xi(t)\right\| = \left\|\int_{-\infty}^{t} e^{-\alpha(t-s)}(\pi(s + t_k) - \pi(s))ds\right\| =$$

$$\left\|\int_{-\infty}^{c} e^{-\alpha(t-s)}(\pi(s + t_k) - \pi(s))ds + \int_{c}^{t} e^{-\alpha(t-s)}(\pi(s + t_k) - \pi(s))ds\right\|$$

$$\leqslant \int_{-\infty}^{c} e^{-\alpha(t-s)}2M_\pi \, ds + \int_{c}^{t} e^{-\alpha(t-s)}\xi \, ds$$

$$\leqslant \frac{2M_\pi}{\alpha}e^{-\alpha(a-c)} + \frac{\xi}{\alpha}[1 - e^{-\alpha(b-c)}] < \frac{\varepsilon}{2} + \frac{\varepsilon}{2} = \varepsilon.$$

Thus, $\|\Xi(t + t_k) - \Xi(t)\| \to 0$ as $k \to \infty$ uniformly on the interval $[a, b]$.

According to lemma 4.1.26, we have $\|\pi(t + t_k) - \pi(t)\| > \varepsilon^*$ for $t \in [s_k - \delta, s_k + \delta]$. Fix a natural number k and positive $\delta_1 < \delta$, such that $\frac{2M_\pi \delta_1}{\alpha}[1 - e^{-\alpha\delta_1}] < \frac{\varepsilon^*}{3\alpha}$. Consider two alternative cases: (i) $\left\|\Xi(t_k + s_k) - \Xi(s_k)\right\| < \frac{2\delta_1\varepsilon^*}{3\alpha}$, and (ii) $\left\|\Xi(t_k + s_k) - \Xi(s_k)\right\| \geqslant \frac{2\delta_1\varepsilon^*}{3\alpha}$.

It is easily seen that the following relation holds:

$$\Xi(t + t_k) - \Xi(t) = \Xi(t_k + s_k) - \Xi(s_k) + \int_{s_k}^{t} e^{-\alpha(t-s)}(\pi(s + t_k) - \pi(s))ds. \quad (4.7)$$

(i) From the last relation, we obtain:

$$\left\|\Xi(t + t_k) - \Xi(t)\right\| \geqslant \left\|\int_{s_k}^{t} e^{-\alpha(t-s)}(\pi(s + t_k) - \pi(s))ds\right\| - \left\|\Xi(t_k + s_k) - \Xi(s_k)\right\| >$$

$$\int_{s_k}^{t} e^{-\alpha(t-s)}\varepsilon^* ds - \frac{2\delta_1\varepsilon^*}{3\alpha} \geqslant \frac{\delta_1\varepsilon^*}{\alpha} - \frac{2\delta_1\varepsilon^*}{3\alpha} = \frac{\delta_1\varepsilon^*}{3\alpha} \quad (4.8)$$

for $t \in [s_k - \delta_1, s_k + \delta_1]$.

(ii) Using the relation (4.7) we get that

$$\left\|\Xi(t + t_k) - \Xi(t)\right\| \geqslant \left\|\Xi(t_k + s_k) - \Xi(s_k)\right\| - \left\|\int_{s_k}^{t} e^{-\alpha(t-s)}(\pi(s + t_k) - \pi(s))ds\right\| >$$

$$\frac{2\delta_1\varepsilon^*}{3\alpha} - \int_{s_k}^{t} e^{-\alpha(t-s)}2M_\pi \, ds \geqslant \frac{2\delta_1\varepsilon^*}{3\alpha} - \frac{2M_\pi \delta_1}{\alpha}[1 - e^{-\alpha\delta_1}] > \frac{\delta_1\varepsilon^*}{3\alpha} \quad (4.9)$$

for $t \in [s_k - \delta_1, s_k + \delta_1]$. Thus, the inequalities (4.8) and (4.9) prove finally that the function $\Xi(t)$ is alpha unpredictable with positive numbers $\varepsilon_1 = \frac{\delta_1\varepsilon^*}{3\alpha}$, δ_1 and sequences t_k, s_k.

4.1.4 Degree of periodicity and numerical simulations

This section introduces a quantitative characteristic, the degree of periodicity, for functions with the kappa property. It presents examples with graphs of compartmental alpha unpredictable functions related to the logistic equation and discusses the dependence of their trajectories on the degree of periodicity.

To illustrate the dynamics of the compartmental alpha unpredictable function, we will use the function $\Xi(t)$, which is defined by (4.6), with $\alpha = -3$, and the function $\pi(t) = \mu_i \xi(t - ih)$, $t \in (ih, (i+1)h]$, $i \in \mathbb{N}$, where $\xi(t) \equiv 1$. The function $\Xi(t)$ is bounded such that $\sup_{t \in \mathbb{R}} |\Xi(t)| \leqslant \frac{1}{3}$, and is the exponentially stable alpha unpredictable solution of the differential equation $\Xi' = -3\Xi + \pi(t)$. This is why, for the numerical simulations of the function, we will use solutions of the equation.

The number h is said to be *the length of step* of the functions $\pi(t)$ and $\Xi(t)$. For compartmental alpha unpredictable functions, the ratio of the period and the length of step, $\nabla = \omega/h$, is called *the degree of periodicity*.

Next, we shall construct the function, which is alpha unpredictable due to the kappa property. Consider the following function, $G(t, u) = (5\sin^2(0.1t) + 0.1)\arctan(\Xi(u)) + 0.5\Xi(u)^3$, which is 10π-periodic in t, uniformly with respect to u. The function $\arctan(u)$ satisfies Lipschitz conditions with $L_1 = 3/4$ and $L_2 = 1$ if $|u| \leqslant \frac{1}{3}$. This is why, according theorems 4.3.3 and 4.3.6, the component-functions $\arctan(\Xi(u))$ and $0.5\Xi^3(u)$ are unpredictable.

Consider the function

$$f_1(t) = G(t, t) = (5\sin^2(0.1t) + 0.1)\arctan(\Xi(t)) + 0.5\Xi^3(t). \qquad (4.10)$$

where h—the length of step, is a parameter.

We will show that the assumptions of theorem 4.1.16 are valid for the function $f_1(t)$. The uniform continuity of $G(t, u)$ implies the condition (i) Since of $\omega = 10\pi$, one can consider the function $\Xi(t)$ with the convergence sequence and separation sequences $t_k = \zeta_k h$, $s_k = \eta_k h + s$, $k = 1, 2, \ldots$ such that condition (ii) is valid. Let us fix a bounded interval $I \subset \mathbb{R}$. The sequence $\Xi(t + t_k)$ uniformly converges to $\Xi(t)$ on the interval. This is why

$$|G(t, t + t_k) - G(t, t)| \leqslant |5\sin^2(0.1t) + 0.1||\arctan(\Xi(t + t_k)) - \arctan(\Xi(t))| +$$
$$0.5|\Xi(t + t_k) - \Xi(t)||\Xi^2(t + t_k) + \Xi(t + t_k)\Xi(t) + |\Xi^2(t)| \leqslant 5.27|\Xi(t + t_k) - \Xi(t)|,$$

and the sequence of functions $G(t, t + t_k)$ converges to $G(t, t)$ uniformly on I. That is, condition (*iii*) is satisfied.

According lemma 4.1.28, for $t \in [s_k - \delta_1, s_k + \delta_1]$, we have $|\Xi(t + t_k) - \Xi(t)| > \varepsilon_1$, $k = 1, 2, \ldots$. That is,

$$|G(t, t + t_k) - G(t, t)| = |(5\sin^2(0.1t) + 0.1)\arctan(\Xi(t + t_k)) + 0.5\Xi(t + t_k) -$$
$$\sin^2(0.1t)\arctan(\Xi(t)) - 0.5\Xi(t)| \geqslant |5\sin^2(0.1t) + 0.1|L_1|\Xi(t + t_k) - \Xi(t)| -$$
$$0.5|\Xi^2(t + t_k) + \Xi(t + t_k)\Xi(t) + |\Xi^2(t)||\Xi(t + t_k) - \Xi(t)| \geqslant$$
$$(5.1L_1 - \frac{0.5}{3})|\Xi(t + t_k) - \Xi(t)| > 3.65|\Xi(t + t_k) - \Xi(t)| > 3.65\varepsilon_1.$$

The last inequality implies that condition (*iv*) is valid. Thus, all conditions of theorem 4.1.16 are correct, and the function $f_1(t)$ is unpredictable. Moreover, the arguments of theorem 4.1.16 indicate that $f_1(t)$ is a compartmental alpha unpredictable function.

In figure 4.1, the graph of function $f_1(t)$, where the length of step $h = 0.1\pi$, and degree of periodicity $\nabla = 200$, is shown.

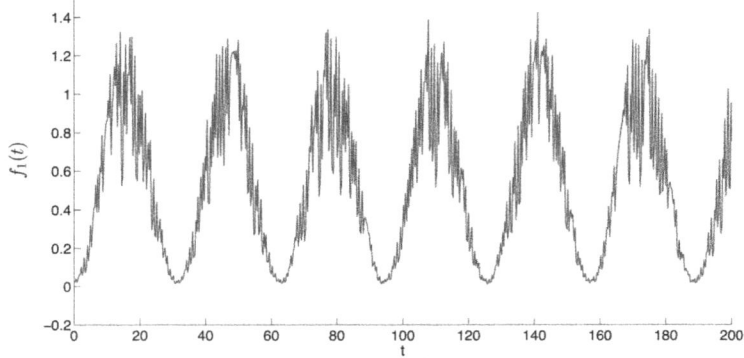

Figure 4.1. The graph of compartmental periodic alpha unpredictable function $f_1(t)$. The length of step $h = 0.1\pi$, and degree of periodicity $\nabla = 200$. Reproduced from [13]. CC BY 4.0.

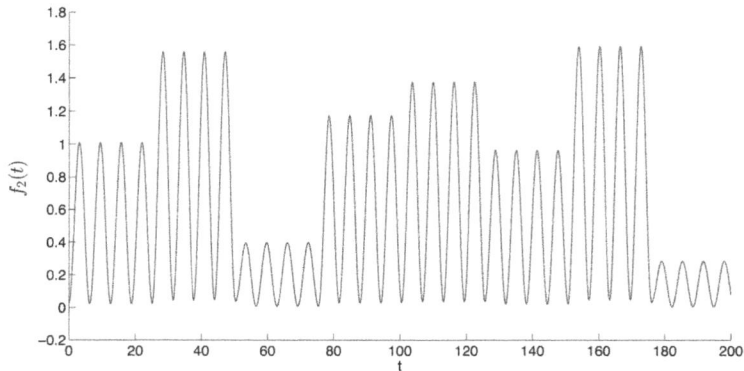

Figure 4.2. The graph of function $f_2(t)$ with the length of step $h = 8\pi$, and the degree of periodicity $\nabla = 0.25$. Reproduced from [13]. CC BY 4.0.

Similarly to the function $f_1(t)$, it can be shown that the compartmental periodic alpha unpredictable function, $f_2(t) = G(t, t)$, where $G(t, u) = (5\sin^2(0.5t) + 0.1)$ $\arctan(u) + 0.5u^3$, and $u = \Xi(t)$, is unpredictable. The function $G(t, u)$ is 2π-periodic in t, uniformly with respect to u. Figure 4.2 depicts the graph of function $f_2(t)$ with length of step $h = 8\pi$, so that the degree of periodicity $\nabla = 0.25$.

Compartmental periodic alpha unpredictable function $f_3(t) = (5\sin^2(t) + 0.1)$ $\arctan(\Xi(t)) + 0.5\Xi^3(t)$, with the period π and the length of step $h = \pi$ is presented in figure 4.3. According theorem 4.1.16, the function $f_3(t)$ is unpredictable.

From the results of the last three simulations, let us observe how the degree of periodicity affects the shape of graphs of compartmental periodic alpha unpredictable functions. By the graphs in figure 4.1, when $\nabla > 1$, the solution admits a periodic shape enveloped by irregularity with small amplitude. Oppositely, if $\nabla \leqslant 1$, one can see in figures 4.2 and 4.3 that the periodicity has lost its dominance and alpha unpredictability prevails. More precisely, periodicity appears only locally on separated intervals if $\nabla < 1$. That is, the periodicity envelopes the alpha unpredictability in this case. The periodicity is not seen for $\nabla = 1$ in figure 4.3.

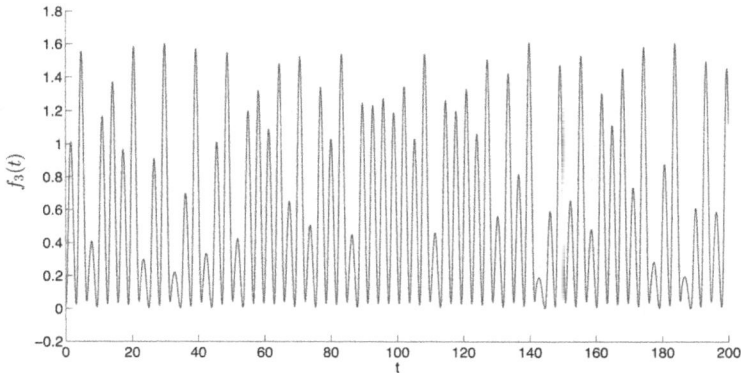

Figure 4.3. The graph of compartmental periodic alpha unpredictable function $f_3(t)$. The degree of periodicity $\nabla = 1$. Reproduced from [13]. CC BY 4.0.

So, the unit is the boundary value between the dominance of regularity and irregularity, which are present with periodicity and alpha unpredictability, respectively. The conclusions can be helpful for analysis of experiments [76, 88, 103, 105, 106].

4.1.5 Randomly determined compartmental unpredictable functions

This section presents practical algorithms for constructing alpha unpredictable functions using Markov chains with finite state spaces. To be applied in the *Miscellanea*, these algorithms are crucial for understanding compartmental quasi-periodic unpredictable functions.

A Markov chain is a stochastic model that describes a sequence of possible events, such that the probability of each event depends only on the state attained in the previous one [56, 65, 86].

Since we expect the alpha unpredictable dynamics realizations to be bounded, the particular Markov chain with boundaries is constructed below. Let the real-valued scalar dynamics be:

$$X_{n+1} = X_n + Y_n, n \geqslant 0, \tag{4.11}$$

such that $Y_n = \{-0.5; 0.5\}$ is a random variable. The probability distribution $P(0.5) = P(-0.5) = 1/2$ if $X_n \neq 1, 4$, and certain events $Y_n = -0.5$ if $X_n = 4$ and $Y_n = 0.5$ if $X_n = 1$. To satisfy the construction of the present research, we will make the following agreements. First of all, denote $s_0 = 1$, $s_1 = 1.5$, $s_2 = 2$, $s_3 = 2.5$, $s_4 = 3$, $s_5 = 3.5$, $s_6 = 4$. Consider the state space of the process $S = \{s_0, s_1, s_2, s_3, s_4, s_5, s_6\}$, and the value $X_n \in S$ is the state of the process at time n. The Markov chain is a random process, which satisfies the property $P\{X_{n+1} = s_j | X_0, \ldots, X_n\} = P\{X_{n+1} = s_j | X_n\}$ for all $s_i, s_j \in S$ and $n \geqslant 0$. Moreover, $P\{X_{n+1} = s_j | X_n = s_i\} = p_{ij}$, where p_{ij} is the transition probability that the chain jumps from state i to state j. It is clear that $\sum_{j=0}^{6} p_{ij} = 1$ for all $i = 0, \ldots, 6$. The alpha unpredictability of infinite realizations of the dynamics is proved by theorem 2.2 [29].

Next, we shall introduce randomly determined alpha unpredictable function $\rho(t) = X_n \xi(t - nh)$, if $t \in [hn, h(n + 1))$, where $\xi(t): (0, h] \to \mathbb{R}$, is a bounded function. In figures 4.4 and 4.5, the graph of the function $\rho(t)$ with $\xi(t) \equiv 1$ and $\xi(t) = \sin(t - 0.5n\pi)$, respectively, for all $t \in [hn, h(n + 1))$, is drawn.

Now, let us show construction of continuous alpha unpredictable functions through the Markov process. Consider the ordinary differential equation:

$$W'(t) = \alpha W(t) + \tanh(\rho(t)), \qquad (4.12)$$

where α is a negative number. The equation (4.12) admits a unique exponentially stable alpha unpredictable solution [29]. It is impossible to specify the initial value of the solution, but, by applying the property of exponential stability, one can consider any solution as arbitrarily close. In figure 4.6, the graph of the solution, $W(t)$, $W(0) = 0.4$ of equation (4.12), where the parameter α is equal to -2.5, and $\rho(t) = X_n$ for $t \in [n, n + 1)$, is shown.

This section was reproduced with permission from [14]. CC BY 4.0.

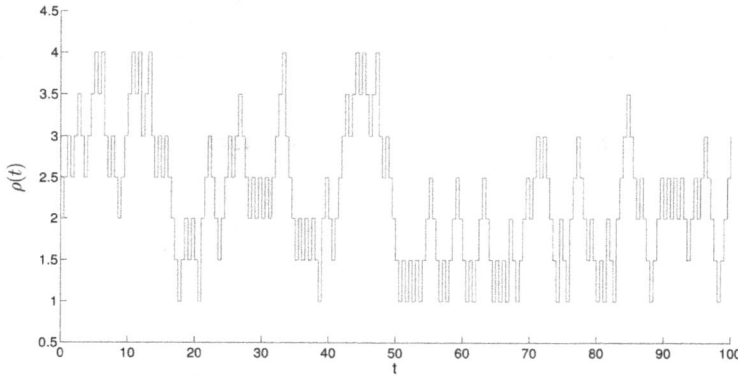

Figure 4.4. The graph of the piecewise constant alpha unpredictable function $\rho(t) = X_n \xi(t - nh)$, $t \in [hn, h(n + 1))$, $n = 0, 1, 2, \ldots$. The vertical lines are drawn for better visibility. Reproduced from [13]. CC BY 4.0.

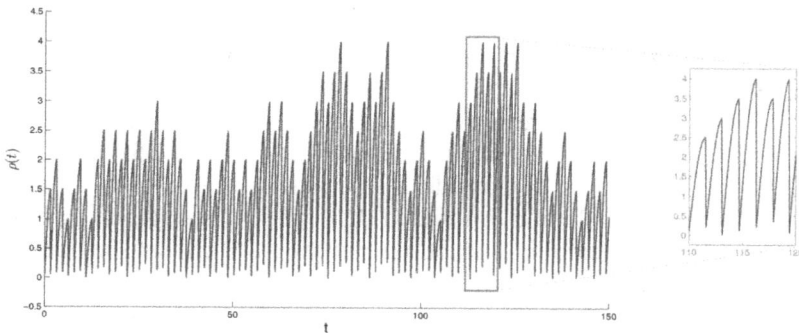

Figure 4.5. The graph of the piecewise continuous alpha unpredictable function $\rho(t) = X_n \sin(t - 0.5n\pi)$, $t \in [hn, h(n + 1))$, $n = 0, 1, 2, \ldots$. The vertical lines are drawn for better visibility. Reproduced from [13]. CC BY 4.0.

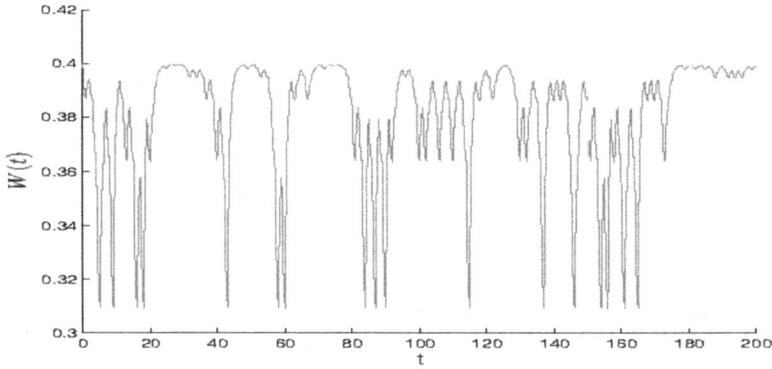

Figure 4.6. The solution $W(t)$ of equation (4.12) with initial value $W(0) = 0.4$ exponentially approaches the alpha unpredictable Markovian function. Reproduced from [13]. CC BY 4.0.

4.2 Discontinuous compartmental periodic Poisson stable functions

Among recurrent functions, the most sophisticated are Poisson stable functions. For discontinuous functions, there are very few results for the stability. Discontinuous compartmental Poisson stable functions are the focus of this section. Particular time sequences, Poisson sequences, are considered the functions' discontinuity points. This is the first time that the discontinuous functions of two compartments, periodic and Poisson stable, have been investigated. To combine periodicity and Poisson stability, in the case of continuous functions, a convergence sequence with a particular kappa property was used [14, 27]. More than this property is needed for discontinuous functions; we also should consider the discontinuity points of the function. For this reason, we need a new concept known as a Poisson couple: a couple of a sequence of discontinuity points and a convergence sequence with the kappa property. Moreover, we meet the challenges for stability by considering functions on diagonals in the space of arguments. Examples of Poisson stable functions are given to illustrate the theoretical results. The method and results can be effectively used in the study of different types of functional differential equations, impulsive differential equations, and differential equations with generalized piecewise constant argument, as well as their application.

Discrete, continuous, or discontinuous functions create a mathematical model for any phenomenon. Therefore, extension methods and numerical representations should follow the theory of functions. Beginning from simple algebraic operations, they can also be Fourier series and even results of operators' theory. The construction and numerical analysis of the discontinuous Poisson stable functions are given in the section. A new way to determine discontinuous Poisson stable functions is proposed. Discontinuous functions with two compartments, periodic and Poisson stable, are the focus of the research. To establish the correspondence between periodicity and Poisson stability, the special *kappa property* is utilized [14, 27]. For functions that depend on the two variables, the method of diagonals of arguments is used [34, 39, 46]. Poisson stability is investigated based on the *B*-topology [2].

Due to the widespread chaos theory, the properties of periodic or Poisson stable functions need to be revised to describe the behavior of nonlinear dynamical systems fully. In this regard, a class of new recurrent functions with the separation property was introduced [5, 26]. Alpha unpredictable functions cause Poincaré chaos. The study of alpha unpredictable solutions of differential equations has led to the possibility of studying chaos based on the laws of the qualitative theory of differential equations. In recent years, the existence and stability of solutions of discrete, linear, and quasilinear differential equations with alpha unpredictable solutions have been studied and significant results have been obtained [5, 6]. Thus, compartmental functions are needed for the study of sophisticated processes. For the first time, a theoretical connection was established between the recurrent functions in [27]. The functions, formed by a combination of Poisson stable, quasi-periodic, and periodic functions, are called compartmental functions, and their properties have been studied in articles [14, 27]. The effectiveness of applying the compartmental method to alpha unpredictability was given by analyzing contributions of alpha unpredictability and periodicity [14]. It was just the first step towards the application of the study since the next ones will be related to the control of chaos that will be used for the parameters of the compartments individually. In this paper, we want to extend the class with discontinuous compartmental functions. That is, we consider discontinuous compartmental periodic Poisson stable functions.

4.2.1 Poisson sequences

This section contains the basic definitions of special time sequences. Correspondingly, we investigate the properties of these sequences that will be utilized to study discontinuous compartmental functions.

Definition 4.2.1. [96] *A bounded sequence μ_i, $i \in \mathbb{Z}$ in $\mathbb{R}$ is said to be Poisson stable, provided that there exists a sequence $l_m \to \infty$, $m \in \mathbb{N}$ of positive integers, which satisfies $\mu_{i+l_m} \to \mu_i$ as $m \to \infty$ on bounded intervals of integers.*

Fix two sequences of real numbers t_m, θ_i, $m \in \mathbb{N}$, $i \in \mathbb{Z}$, strictly increasing with respect to the index and unlimited in both directions, which we call *Poisson sequences* throughout this book. Moreover, it is expected that there exists a number $\underline{\theta} > 0$ such that $\underline{\theta} < \theta_{i+1} - \theta_i$ for all $i \in \mathbb{Z}$.

Definition 4.2.2. [2] *A sequence τ_i, $i \in \mathbb{Z}$, is called with (w, p)-property, provided that there exist an integer p and a real number $w > 0$, which satisfy $\tau_{i+p} - \tau_i = w$ for all $i \in \mathbb{Z}$.*

Definition 4.2.3. *A couple (t_m, θ_i) of the sequences t_m, θ_i, $m \in \mathbb{N}$, $i \in \mathbb{Z}$, is called a Poisson couple, provided that there exists a sequence l_m, $m \in \mathbb{N}$ of integers, which diverges to infinity, such that*

$$\theta_{i+l_m} - t_m - \theta_i \to 0 \text{ as } m \to \infty \tag{4.13}$$

uniformly on each bounded interval of integer i.

To investigate compartmental discontinuous functions, we need sufficient conditions that connect periodicity with Poisson stability.

Lemma 4.2.1. [7] *Suppose that the couple* (t_m, θ_i) *of sequences* t_m, θ_i, $m \in \mathbb{N}$, $i \in \mathbb{Z}$, *satisfies the following conditions*

(i) $t_m = mw$, *where* $m \in \mathbb{N}$, $w \in \mathbb{R}$;

(ii) θ_i *has the* (w, p)-*property.*

Then the couple (t_m, θ_i) *is a Poisson couple.*

Proof. Because of (w, p)-property it is true that $\theta_{i+p} = \theta_i + w$, $i \in \mathbb{Z}$. By taking $l_m = mp$ for $m \in \mathbb{N}$, one can get $\theta_{i+mp} = \theta_i + mw$, $i \in \mathbb{Z}$.

We proceed to prove that

$$\theta_{i+l_m} - t_m - \theta_i = \theta_{i+mp} - mw - \theta_i = \theta_i + mw - mw - \theta_i = 0.$$

Since the sequence consists of zeros, it follows that

$$\lim_{m \to \infty}(\theta_{i+l_m} - t_m - \theta_i) = 0,$$

as $m \to \infty$, on each bounded interval of integers i. So, the condition (4.13) is satisfied. $\square$

According to the lemma 4.1.8, for a number $\omega > 0$ and an arbitrary sequence of positive real numbers t_m, $m = 1, 2, \ldots$, one can find a subsequence t_{m_l}, $l = 1, 2, \ldots$, and a number τ_ω, $0 \leqslant \tau_\omega < \omega$, which satisfies $t_{m_l} \to \tau_\omega (mod \ \omega)$ as $l \to \infty$.

Next, we will consider the application of this assertion in the proof of the Poisson stability. The number τ_ω is called as the *Poisson shift* with respect to the ω. The set of Poisson shifts $\mathscr{T}_\omega$ consists of an infinite number of terms, that is, it is not empty. Denote infimum of the set $\mathscr{T}_\omega$ as κ_ω, and call it the Poisson number with respect to ω or, in short, the Poisson number. We say that a sequence t_m admits the kappa property with respect to the number ω, if the Poisson number is equal to 0.

We will use the following assertion [27].

Lemma 4.2.2. [27] $\kappa_\omega \in \mathscr{T}_\omega$.

Lemma 4.2.3. [7] *If the couple* (t_m, θ_i) *consists of the sequence* t_m, $m \in \mathbb{N}$, *which has the kappa property with respect to w, and the sequence* θ_i. $i \in \mathbb{Z}$ *with the* (ω, p)-*property, then it is a Poisson couple.*

Proof. Taking into account that the sequence t_m has the kappa property, and using the (ω, p)-property of the sequence θ_i, we get that

$$\lim_{m \to \infty} |\theta_{i+l_m} - t_m - \theta_i| = 0$$

as $m \to \infty$, on each bounded interval of integers i. $\square$

4.2.2 Compartmental functions

A discontinuous function $u(t)$: $\mathbb{R} \to \mathbb{R}$, is called *conditionally uniform continuous*, if for any number $\varepsilon > 0$ there exists a number $\sigma > 0$ such that $|u(t_1) - u(t_2)| < \varepsilon$ whenever the points t_1 and t_2 belong to the same continuity interval and $|t_1 - t_2| < \sigma$ [3].

Denote by $\mathscr{G}$ the space of piecewise continuous functions $u(t)$: $\mathbb{R} \to \mathbb{R}$ with a countable set of discontinuity moments of the first kind. The functions are left-continuous and conditionally uniformly continuous. The sets of discontinuity moments are unbounded from both sides and do not have finite accumulation endpoints. Moreover, the set of discontinuity points is strictly ordered and enumerated with integers.

Definition 4.2.4. *An element $f(t)$ of $\mathscr{G}$ is called discontinuous periodic if there exist an integer p and a real number $\omega > 0$, such that the set of its moments of discontinuities θ_i, $i \in \mathbb{Z}$, satisfy (ω, p)-property and $f(t + \omega) = f(t)$ for $t \in \mathbb{R}$.*

The functions $\phi(t)$ and $\psi(t)$ from $\mathscr{G}$, are called *ε-equivalent* on a bounded open interval J if the moments of discontinuities of $\phi(t)$ and $\psi(t)$ in J can be numerated with multiplicity one, θ_i^ϕ and θ_i^ψ, $i = 1, 2, \ldots, l$, such that $|\theta_i^\phi - \theta_i^\psi| < \varepsilon$ for each $i = 1, 2, \ldots, l$, and $|\phi(t) - \psi(t)| < \varepsilon$, for all $t \in J$, except possibly those between θ_i^ϕ and θ_i^ψ, $i = 1, 2, \ldots, l$. If ϕ, ψ are ε-equivalent on J, then we say that the functions are in *ε-neighborhoods* of each other on J. The topology defined on the ε-neighborhoods basis is said to be *B-topology* [2].

Definition 4.2.5. [7] *An element $g(t)$ of $\mathscr{G}$ is called discontinuous Poisson stable if there exist a sequence $t_m \to \infty$ of real numbers such that (t_m, θ_i) is a Poisson couple and $g(t + t_m) \to g(t)$ as $m \to \infty$ on each bounded interval of real numbers in B-topology.*

Definition 4.2.6. [7] *A function $h(t) = f(t) + g(t)$, where $f(t)$ and $g(t)$ are members of $\mathscr{G}$ with a common set of discontinuities, is called a discontinuous modulo periodic Poisson stable function if $f(t)$ is a discontinuous periodic function and $g(t)$ is a discontinuous Poisson stable function.*

Definition 4.2.7. [7] *A product $f(t)g(t)$, where $f(t)$ and $g(t)$ are members of $\mathscr{G}$ with a common set of discontinuities, is called a discontinuous factor periodic Poisson stable function if $f(t)$ is a discontinuous periodic function and $g(t)$ is a discontinuous Poisson stable function.*

Definition 4.2.8. [7] *A function $h(t)$: $\mathbb{R} \to \mathbb{R}$ is called discontinuous compartmental periodic Poisson stable if $h(t) = Q(t, t)$, where $Q(u, v)$ is a discontinuous function with a common set of discontinuities for u and v, discontinuous periodic in u uniformly with respect to v, and discontinuous Poisson stable in v uniformly with respect to u. That is,* *there exists number ω such that $Q(u + \omega, v) = Q(u, v)$ and there exists a sequence $t_m \to \infty$, which satisfies $Q(u, v + t_m) \to Q(u, v)$ as $m \to \infty$ uniformly on each bounded interval of v in B-topology.*

Remark 4.2.1. *Since the compartments in definitions 4.2.9- 4.2.11 have a common set of discontinuities, it satisfies the (ω, p)-property.*

Next, we investigate properties of discontinuous compartmental periodic Poisson stable functions.

Theorem 4.2.1. [7] *A discontinuous periodic function is Poisson stable.*

Proof. Consider discontinuous periodic function $f(t)$ with discontinuity moments θ_i, $i \in \mathbb{Z}$. Together with the sequence θ_i, we fix a sequence t_m, $m \in \mathbb{Z}$, such that (t_m, θ_i), is a Poisson couple in the sense of definition 4.2.3. For a fixed $m \in \mathbb{Z}$, consider the function $f(t + t_m)$, if $\theta'_i \leqslant t < \theta'_{i+1}$, where $\theta'_i = \theta_{i+l_m} - t_m$, $i \in \mathbb{Z}$.

According to lemma 4.2.4, the set of discontinuity moments θ'_{i+1}, $i \in \mathbb{Z}$ of $f(t + t_m)$ coincide with the set θ_i, $i \in \mathbb{Z}$. Taking $t_m = m\omega$ for $n \in \mathbb{N}$, it is easy to prove that the functions $f(t + t_m) - f(t)$ are both zeros.

So, the Poisson stability of $f(t)$ is proved. $\square$

Theorem 4.2.2. [7] *Suppose that a bounded and piecewise continuous function $Q(u, v)$ is ω-periodic in u. Then $h(t) = Q(t, t)$ is discontinuous Poisson stable, with common set of discontinuities θ_i, $i \in \mathbb{Z}$, for u and v, which admits the (ω, p)- property such that*

(a) *for each $\varepsilon > 0$ there exists a number $\sigma > 0$ which satisfies $|Q(t_1, t) - Q(t_2, t)| < \varepsilon$, where the points $t, t + t_r$ are taken from the same continuity interval and $|t_1 - t_2| < \sigma, t \in \mathbb{R}$;*

there exists a sequence t_m, $t_m \to \infty$ as $m \to \infty$ and satisfies following conditions:

(b) *t_m satisfies the kappa property with respect to the period ω;*

(c) *$|Q(t, t + t_m) - Q(t, t)| \to 0$ as $m \to \infty$ on each bounded interval $I \subset \mathbb{R}$ of t in B-topology.*

Proof. Because t_m satisfies the kappa property, there exists a subsequence t_{m_l}, $t_{m_l} \to \tau_\omega (mod \ \omega)$ as $l \to \infty$. Suppose that $t_m \to 0 (mod \ \omega)$ as $m \to \infty$. Let us fix $\varepsilon > 0$ and an interval $I = [b, c] \subset \mathbb{R}$. Consequently, for arbitrarily fixed number $\varepsilon > 0$, the bounded interval I, and by condition (a), one can find sufficiently large m such that

$$|Q(t + t_m, t + t_m) - Q(t, t + t_m)| < \varepsilon/2 \qquad (4.14)$$

for all $t \in \mathbb{R}$, where the points $t, t + t_m$ are taken from the same continuity interval.

Let θ_i and θ'_i, be the discontinuity moments of the functions $Q(t, t), Q(t, t + t_m)$ in I, respectively, where $\theta'_i = \theta_{i+l_m} - t_m$. Assume that $\theta_i \leqslant \theta_{i+l_m} - t_m$ and consider discontinuity moments θ_i, $i = k + 1, k + 2, \ldots, k + r - 1$, of the interval $[b, c]$ such that

$$\theta_k \leqslant b < \theta_{k+1} < \theta_{k+2} < \cdots < \theta_{k+r-1} < c \leqslant \theta_{k+r}.$$

Let us fix i, $i = k, k + 1, \ldots, k + r$, and for fixed i, it follows that $Q(t, t)$, $t \in [\theta_i, \theta_{i+1})$ and $Q(t, t + t_m), t \in [\theta'_i, \theta'_{i+1})$. Thus, for sufficiently large m we have a non-empty interval $(\theta'_i, \theta_{i+1})$. According to (4.13) and by condition (c), the following inequalities are valid:

$$|\theta'_i - \theta_i| < \varepsilon, \tag{4.15}$$

for all $i = k, k + 1, \ldots, k + r$, and

$$|Q(t, t + t_m) - Q(t, t)| < \varepsilon/2 \tag{4.16}$$

for $t \in (\theta'_i, \theta_{i+1})$.

Applying (4.14) and (4.16), we get that

$$|h(t + t_m) - h(t)| = |Q(t + t_m, t + t_m) - Q(t, t)| \leqslant$$
$$|Q(t + t_m, t + t_m) - Q(t, t + t_m)| + |Q(t, t + t_m) - Q(t, t)| \leqslant \varepsilon/2 + \varepsilon/2 = \varepsilon$$

for all $t \in (\theta'_i, \theta_{i+1})$, $i = k, k + 1, \ldots, k + r$. Consequently, $h(t + t_m) \to h(t)$ uniformly on each arbitrary bounded time interval in B-topology. $\square$

Theorem 4.2.3. *Assume that $h(t) = f(t) + g(t)$ is a discontinuous modulo periodic Poisson stable function. The discontinuity moments $\theta_i, i \in \mathbb{Z}$, are common for $f(t)$ and $g(t)$ and satisfy the (ω, p)-property, the sequence t_m satisfies the kappa property with respect to the number ω. Then $h(t)$ is a discontinuous Poisson stable function.*

Theorem 4.2.4. [7] *Assume that $h_1(t) = f(t)g(t)$ is a discontinuous factor periodic Poisson stable function. The discontinuity moments $\theta_i, i \in \mathbb{Z}$, are common for $f(t)$ and $g(t)$ and satisfy the (ω, p)-property, the sequence t_m satisfies the kappa property with respect to the number ω. Then the function $h_1(t)$ is discontinuous Poisson stable.*

The technique of proofs of the theorems 4.2.15, 4.2.16 is similar to that for theorems 4.2.13, 4.2.14, and the B-topology is used.

In the following theorems we consider specific discontinuous compartmental periodic Poisson stable functions.

Fix a Poisson stable sequence $\lambda_i, i \in \mathbb{Z}$, such that there exist a sequence l_m which diverges to infinity, $|\lambda_{i+l_m} - \lambda_i| \to 0$ as $m \to \infty$ for each i in bounded intervals of integers.

Theorem 4.2.5. [14] *Let $\xi(t): (0, d] \to \mathbb{R}$, where $d > 0$, be a bounded function. Then function $\zeta(t) = \lambda_i \xi(t - id), t \in (id, (i + 1)d], i \in \mathbb{Z}$, is discontinuous Poisson stable.*

Proof. Fix a number $i \in \mathbb{Z}$ and an interval (γ, δ) so that $(\gamma, \delta) \subset ((i-1)d, (i+s+1)d]$ for $s \in \mathbb{N}$. For $t_m = l_m d$, $n = 1, 2, \ldots$ from interval $t \in (jd, (j+1)d]$, $i - 1 \leqslant j \leqslant i + s$, we have that $t + l_m d \in ((j + l_m)d, (j + l_m + 1)d]$ and $\xi(t - (j + l_m)d) = \xi(t - jd)$.

Let us denote $N = \sup_{t \in (0, d]} |\xi(t)|$. For an arbitrary $\varepsilon > 0$ and sufficiently large m, it is true that $|\lambda_{j+l_m} - \lambda_j| < \frac{\varepsilon}{N}$, $i - 1 \leqslant j \leqslant i + s$. We fixed integer number l in $i - 1 \leqslant l \leqslant i + s$. If $t \in (ld, (l+1)d]$, then $\zeta(t) = \xi(t - ld) = \lambda_l$ and $\zeta(t + t_m) = \zeta(t + l_m d) = \xi(t - (l + l_m)d) = \lambda_{l+l_m}$. This is why, for $t \in (ld, (l+1)d]$, $l \in [i - 1, i + s]$, we have that

$$|\zeta(t + t_m) - \zeta(t)| = |\zeta(t + l_m d) - \zeta(t)| =$$
$$|\lambda_{l+l_m} \xi(t - (l + l_m)d) - \lambda_l \xi(t - lw)| \leqslant |\lambda_{l+l_m} - \lambda_l| |\xi(t - ld)| \leqslant$$
$$|\lambda_{l+l_m} - \lambda_l| N < \varepsilon.$$

It follows that $l \in [i - 1, i + s]$, and, in consequence, $|\zeta(t + t_m) - \zeta(t)| < \varepsilon$, $t \in (\gamma, \delta)$. Thus, it has been proved that $\zeta(t)$ is a Poisson stable function.

Theorem 4.2.6. [7] *The function* $\eta(t) = \lambda_i$, $\theta_i < t \leqslant \theta_{i+1}$, $i = 0, 1, 2, \ldots$ *is discontinuous Poisson stable with the sequence of convergence* t_m, $m = 1, 2, \ldots$ *if sequences* t_m *and* θ_i *make a Poisson couple* (t_m, θ_i).

Proof. Consider the function $\eta(t + t_m)$ for fixed $m \in \mathbb{N}$. If $\eta(t) = \lambda_i$, $\theta_i < t \leqslant \theta_{i+1}$, then it is possible to show that $\eta(t + t_m) = \lambda_{i+l_m}$ for $\theta'_i \leqslant t < \theta'_{i+1}$, where $\theta'_i = \theta_{i+l_m} - t_m$, $i \in \mathbb{Z}$.

Fix an interval $[b_1, c_1]$ where $c_1 > b_1$, and an arbitrary $\varepsilon > 0$ so that $2\varepsilon < \underline{\theta}$ on this interval. Suppose that $\theta_i \leqslant \theta_{i+l_m} - t_m$ and consider discontinuity moments θ_i, $i = k + 1, k + 2, \ldots, k + r_1 - 1$, in the interval $[b_1, c_1]$ such that

$$\theta_k \leqslant b_1 < \theta_{k+1} < \theta_{k+2} < \cdots < \theta_{k+r_1 - 1} < c_1 \leqslant \theta_{k+r_1}.$$

We will prove that for sufficiently large m the inequalities $|\theta'_i - \theta_i| < \varepsilon$ for all $i = k, k + 1, \ldots, k + r_1$, and $|\eta(t + t_m) - \eta(t)| < \varepsilon$ for each $t \in [b_1, c_1]$ are valid, except those between θ_i and θ'_i for each i.

Fix i, $i = k, k + 1, \ldots, k + r_1$. We have that $\eta(t) = \lambda_i$, for $t \in (\theta_i, \theta_{i+1}]$ and $\eta(t + t_m) = \lambda_{i+l_m}$, $t \in (\theta'_i, \theta'_{i+1}]$. Hence, for sufficiently large m the interval $(\theta'_i, \theta_{i+1})$ is non-empty. From (4.13) it implies that $|\theta'_i - \theta_i| < \varepsilon$ is valid. Moreover, for sufficiently large m,

$$|\eta(t + t_m) - \eta(t)| = |\lambda_{i+l_m} - \lambda_i| < \varepsilon$$

for $t \in (\theta'_i, \theta_{i+1})$. Thus, $\eta(t)$ is a discontinuous Poisson stable function.

4.2.3 Examples with numerical simulations

Next, we will construct examples of discontinuous Poisson stable functions.

In [10, 27], a Poisson stable sequence was constructed as a solution to the logistic map

$$\chi_{j+1} = \nu\chi_j(1 - \chi_j). \tag{4.17}$$

Moreover, it is shown that for each ν from the interval $[3 + (2/3)^{1/2}, 4]$, there exists a Poisson stable solution $z_j, j \in \mathbb{Z}$, of (4.17) so that the sequence belongs to $[0, 1]$. That is, there exists a sequence $l_m, l_m \to \infty$ as $m \to \infty$ such that $|z_{j+l_m} - z_j| \to 0$ as $m \to \infty$ for each j in bounded intervals of integers.

Example 4.2.1. *[7] Consider the function* $\zeta_1(t) = z_i\xi(t - id)$, $t \in (id, (i + 1)d]$, $i = 0, 1, 2, ...$, *with* $\xi(t - id) = 1$, *and* $d = 5$, *such that* z_i *is a Poisson stable solution of (4.17). Prove that the function* $\zeta_1(t)$ *is discontinuous Poisson stable in accordance with theorem 4.2.17.*

Fix an interval (γ_1, δ_1) *and number* i *such that* $(\gamma_1, \delta_1) \subset (5(i - 1), 5(i + s + 1)]$ *for* $s \in \mathbb{N}$. *Assume that* γ_1 *and* δ_1 *are integers. For* $t_m = 5l_m$, $m = 1, 2, ...$ *and* $t \in (5j, 5(j + 1)]$, $i - 1 \leqslant j \leqslant i + s$, *we have that* $t + 5l_m \in (5(j + l_m), 5(j + l_m + 1)]$.

For an arbitrary number $\varepsilon > 0$ *and sufficiently large* m, *the inequality* $|z_{j+l_m} - z_j| < \varepsilon$, $i - 1 \leqslant i \leqslant i + s$ *is fulfilled. For* $t \in (3l, 3(l + 1)]$, $l \in [i - 1, i + s]$, *we have that*

$$|\zeta_1(t + t_m) - \zeta_1(t)| = |\zeta_1(t + 5l_m) - \zeta_1(t)| \leqslant |z_{l+l_m} - z_l| < \varepsilon.$$

From the above it follows that for $l \in [i - 1, i + s]$, *and so* $|\zeta_1(t + t_m) - \zeta_1(t)| \to 0$, *as* $n \to \infty$ *uniformly on the interval* $t \in (\gamma_1, \delta_1)$. *So, the Poisson stability of* $\zeta_1(t)$ *has been proved.*

Example 4.2.2. *[7] Consider a function* $\eta_1(t) = z_i$, $\theta_i < t \leqslant \theta_{i+1}$, *where* $\theta_i = \frac{1}{10}(5i + (-1)^i - 1)$, $i \in \mathbb{Z}$, $i = 0, 1, 2, ...$, *is an unbounded sequence. Let us demonstrate that* $\eta_1(t)$ *is discontinuous Poisson stable.*

Fix a sequence $t_m, m \in \mathbb{Z}$, *such that* (t_m, θ_i) *is a Poisson couple in the sense of definition 4.2.3. One can verify that the sequence* θ_i *satisfies* $(1, 2)$-*property. It is known that, if* $\eta_1(t) = z_i$, $t \in (\theta_i, \theta_{i+1}]$, $i \in \mathbb{Z}$, *then* $\eta_1(t + t_m) = z_{i+l_m}$ *for* $\theta'_i \leqslant t < \theta'_{i+1}$, *where* $\theta'_i = \theta_{i+l_m} - t_m$, $i \in \mathbb{Z}$.

Fix an interval $[b_2, c_2]$, *where* $c_2 > b_2$, *and an arbitrary number* $\varepsilon > 0$ *such that* $2\varepsilon < \underline{\theta}$ *on this interval. Assume that* $\theta_i \leqslant \theta_{i+l_m} - t_m$ *and consider discontinuity points* θ_i, $i = l + 1, l + 2, ..., l + r_2 - 1$, *of the interval* $[b_2, c_2]$.

Let us fix i, $i = l, l + 1, ..., l + r_2$, *and for fixed* i *we have that* $\eta_1(t) = z_i$, *for* $t \in (\theta_i, \theta_{i+1}]$ *and* $\eta_1(t + t_m) = z_{i+l_m}$, $t \in (\theta'_i, \theta'_{i+1}]$. *From (4.13)* $|\theta'_i - \theta_i| < \varepsilon$ *is valid. Moreover it implies that for sufficiently large* n,

$$|\eta_1(t + t_m) - \eta_1(t)| = |z_{i+l_m} - z_i| < \varepsilon$$

for $t \in (\theta'_i, \theta_{i+1})$. *That is* $\eta_1(t)$ *is a Poisson stable function.*

4.3 Algebra and significance for applications

In the beginning of this section, two constructive theorems are provided for alpha unpredictable functions by using simple algebraic operations. The assertion on alpha unpredictability of a composition function is another result of the section. Several new definitions of compartmental alpha unpredictable functions are presented to provide information for the subsequent study development. The reader is invited to use the definitions to find conditions, which imply that the functions on diagonals are alpha unpredictable in the sense of definition 4.16. Numerical simulation demonstrates irregular behavior in a compartmental quasi-periodic alpha unpredictable function example.

Theorem 4.3.1. *If the function* $\psi(t)$: $\mathbb{R} \to \mathbb{R}^n$ *is unpredictable, then the sum* $\psi(t) + C$, *where C is a constant, is also unpredictable.*

Proof. Denote $f(t) = \psi(t) + C$. We have that $\|f(t + t_k) - f(t)\| = \|\psi(t + t_k) - \psi(t)\|$. This is why $f(t + t_k) \to f(t)$ as $k \to \infty$ uniformly on compact subsets of $\mathbb{R}$ and $\|f(t + t_k) - f(t)\| > \varepsilon_0$ for each $t \in [s_k - \delta, s_k + \delta]$ and $k \in \mathbb{N}$.

Theorem 4.3.2. *Assume that* $\psi(t)$: $\mathbb{R} \to \mathbb{R}^n$ *is an alpha unpredictable function. Then, the function* $\psi^{2m+1}(t)$, $m \in \mathbb{N}$, *is unpredictable.*

Proof. There exist numbers $\varepsilon_0, \delta > 0$ and sequences t_k, s_k, both of which diverge to infinity, such that $\psi(t + t_k)$ converges to $\psi(t)$ as $k \to \infty$ uniformly on compact subsets of $\mathbb{R}$ and $\|\psi(t + t_k) - \psi(t)\| > \varepsilon_0$ for each $[s_k - \delta, s_k + \delta]$ and $k \in \mathbb{N}$. The proof of the Poisson stability of $\psi^{2m+1}(t)$ is not difficult, since it follows from uniform continuity of $\psi^{2m+1}(t)$ on a compact set. Now, we will show that $\|\psi^{2m+1}(t + t_k) - \psi^{2m+1}(t)\| > \varepsilon(\varepsilon_0)$ for some positive number $\varepsilon(\varepsilon_0)$ and $t \in [s_k - \delta, s_k + \delta]$. Fix a natural number m. Consider the function $F(x, y) = x^{2m+1} - y^{2m+1}$ for $|x - y| \geqslant \varepsilon_0$. Using Lagrange multipliers, one can find that the minimum of $F(x, y)$ occurs at the points x_0, y_0 with $|x_0| = |y_0| = \frac{\varepsilon_0}{2}$. Therefore, $\left\|\psi^{2m+1}(t + t_k) - \psi^{2m+1}(t)\right\| \geqslant \frac{\varepsilon_0^{2m+1}}{2^{2m}}$ for $m \in \mathbb{N}$, $t \in [s_k - \delta, s_k + \delta]$. Thus, the function $\psi^{2m+1}(t)$ is alpha unpredictable with sequences t_k, s_k and positive numbers δ and $\varepsilon < \frac{\varepsilon_0^{2m+1}}{2^{2m}}$.

Remark 4.3.1. *An alpha unpredictable function to an even degree is not necessary unpredictable. This can be shown by considering the function* $G(x, y) = x^{2m} - y^{2m}$, $m = 1, 2, \ldots$. *Let us write the function* $G(x, y)$ *in the form* $G(x, y) = (x - y)(x + y)g(x, y)$. *Despite* $|x - y| \geqslant \varepsilon_0$, *the sum $x + y$ may be arbitrarily small, and the separation property will not be satisfied.*

Theorem 4.3.3. *Assume that the bounded function* $f(u): \mathbb{R}^n \to \mathbb{R}^n$ *satisfies the inequalities* $L_1 \|u_1 - u_2\| \leqslant \|f(u_1) - f(u_2)\| \leqslant L_2 \|u_1 - u_2\|$, *where* L_1, L_2 *are positive constants, for all* $u_1, u_2 \in \mathbb{R}^n$. *Then, the function* $f(\psi(t))$ *is unpredictable, provided that* $\psi(t): \mathbb{R} \to \mathbb{R}^n$ *is an alpha unpredictable function.*

Proof. Consider the function $g(t) = f(\psi(t))$. Let us fix positive number ε and a bounded interval I. One can find sufficiently large k such that $\left\|\psi(t + t_k) - \psi(t)\right\| < \frac{\varepsilon}{L_2}$ for $t \in I$. This is why

$$\|g(t + t_k) - g(t)\| = \|f(\psi(t + t_k)) - f(\psi(t))\| \leqslant L_2 \|\psi(t + t_k) - \psi(t)\| < \varepsilon,$$

for all $t \in I$. Moreover, there exist a sequence s_k and positive numbers ε_0, δ, such that $\|\psi(t + t_k) - \psi(t)\| > \varepsilon_0$ for $t \in [s_k - \delta, s_k + \delta]$. Then, we obtain:

$$\|g(t + t_k) - g(t)\| = \|f(\psi(t + t_k)) - f(\psi(t))\| \geqslant L_1 \|\psi(t + t_k) - \psi(t)\| > L_1 \varepsilon_0,$$

for all $t \in [s_k - \delta, s_k + \delta]$.

For further research, it is important to consider the most general definitions of Poisson stable and alpha unpredictable functions, excluding their continuity.

Definition 4.3.1. [73] *A continuous function* $f(t): \mathbb{R} \to \mathbb{R}^n$ *is called quasi-periodic with periods* $2\pi/\omega_1, 2\pi/\omega_2, \ldots, 2\pi/\omega_m$ *if for every positive* ε *there exists a positive number* δ *such that a number* ρ *satisfies the inequality* $\sup_{t \in \mathbb{R}} \|f(t + \rho) - f(t)\| < \varepsilon$, *provided that* $|\omega_k \rho| < \delta (\text{mod } 2\pi)$, $k = 1, 2, \ldots, m$.

Definition 4.3.2. *A function* $f(t): \mathbb{R} \to \mathbb{R}^n$ *is said to be compartmental quasi-periodic unpredictable, if* $f(t) = G(t, t)$, *where* $G(u, v)$ *is a continuous bounded function, quasi-periodic in* u *uniformly with respect to* $v \in \mathbb{R}$, *and alpha unpredictable in* v *uniformly with respect to* $u \in \mathbb{R}$.

Definition 4.3.3. [62] *A continuous function* $f(t): \mathbb{R} \to \mathbb{R}^n$ *is said to be almost periodic if, for any positive* ε, *the set* $S(f, \varepsilon) = \{\omega: \|f(t + \omega) - f(t)\| < \varepsilon$ *for all* $t \in \mathbb{R}\}$ *is relatively dense.*

Definition 4.3.4. *A function* $f(t): \mathbb{R} \to \mathbb{R}^n$ *is said to be compartmental almost periodic unpredictable, if* $f(t) = G(t, t)$, *where* $G(u, v)$ *is a continuous bounded function, almost periodic in* u *uniformly with respect to* $v \in \mathbb{R}$, *and alpha unpredictable in* v *uniformly with respect to* $u \in \mathbb{R}$.

Definition 4.3.5. [108] *A continuous function $f(t)$: $\mathbb{R} \to \mathbb{R}^n$ is called recurrent if for any positive ε there can be found a positive number L, such that for each real number t and any interval I of length L there exists a number $\tau \in I$, which satisfies $\|f(t+\tau) - f(t)\| < \varepsilon$.*

Definition 4.3.6. *A function $f(t)$: $\mathbb{R} \to \mathbb{R}^n$ is said to be compartmental recurrent unpredictable if $f(t) = G(t, t)$, where $G(u, v)$ is a function recurrent in u uniformly with respect to $v \in \mathbb{R}$, and alpha unpredictable in v uniformly with respect to $u \in \mathbb{R}$.*

Definition 4.3.7. *A function $f(t)$: $\mathbb{R} \to \mathbb{R}^n$ is said to be compartmental Poisson stable unpredictable if $f(t) = G(t, t)$, where $G(u, v)$ is a Poisson stable in u uniformly with respect to $v \in \mathbb{R}$, and alpha unpredictable in v uniformly with respect to $u \in \mathbb{R}$.*

Next, we formulate definitions of specific compartmental functions.

Definition 4.3.8. *A sum $\phi(t) + \psi(t)$ is said to be a modulo periodic (quasi-periodic, almost periodic, recurrent, Poisson stable) alpha unpredictable function if $\phi(t)$ is a continuous periodic (quasi-periodic, almost periodic, recurrent, Poisson stable) function and $\psi(t)$ is an alpha unpredictable function.*

Definition 4.3.9. *A product $\phi(t)\psi(t)$ is said to be factor periodic (quasi-periodic, almost periodic, recurrent, Poisson stable) alpha unpredictable function, if $\phi(t)$ is continuous periodic (quasi-periodic, almost periodic, recurrent, Poisson stable) and $\psi(t)$ is an alpha unpredictable function.*

It is important to remark that the definitions 4.3.9, 4.3.11, 4.3.13, and 4.3.14 are provided without any theoretical consequences within the present research. We consider them a reason for open problems, such that conditions can be looked for for the alpha unpredictability of the functions, similar to the kappa property. Nevertheless, this time, we suggest the next simulation result to illustrate the irregularity and the possibility of seeing the contribution of quasi-periodic and alpha unpredictable components to the composed dynamics. The graphs of quasi-periodic and compartmental quasi-periodic alpha unpredictable functions are shown in figure 4.7. According to theorem 4.3.6, the component $0.5\tanh(W(t))$ is alpha unpredictable with Lipschitz constants $L_1 = 0.43$ and $L_2 = 0.5$, since $\sup_{t \in \mathbb{R}} |W(t)| < \frac{2}{5}$. Thus, $f(t) = \sin(0.2t) + \cos(0.1\sqrt{2}\,t) + 0.5\tanh(W(t))$ can be accepted as a modulo quasi-periodic alpha unpredictable function. One can see that the irregular graph of the alpha unpredictable function $f(t)$ envelopes the graph of the quasi-periodic function $g(t) = \sin(0.2t) + \cos(0.1\sqrt{2}\,t)$, such that contribution of both quasi-periodicity and irregularity are seen in the dynamics of compartmental function. An analog of the degree of periodicity can be looked at for the quasi-periodicity, and the contributions of the components are discussed more deeply.

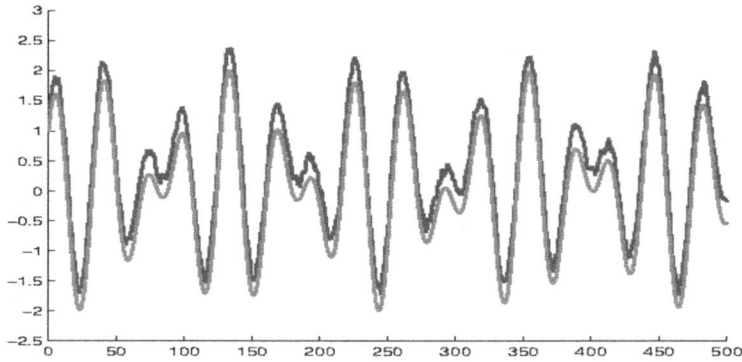

Figure 4.7. The red curve is the graph of quasi-periodic function $g(t) = \sin(0.2t) + \cos(0.1\sqrt{2}\,t)$, and the blue curve is the graph of the randomly determined compartmental quasi-periodic alpha unpredictable function $f(t) = \sin(0.2t) + \cos(0.1\sqrt{2}\,t) + 0.5\tanh(W(t))$. Reproduced from [13]. CC BY 4.0.

In this part of the paper, we explain why the theoretical research of neural network dynamics is helpful for engineering applications. We have found that the simulations of our study are similar to that modeled by neural networks in industrial problems. For example, they are considering tracking periodical or square waves dynamics. This makes us confident that one can learn better what can happen in the dynamics of neural networks with sophisticated environment and structure. Moreover, the information provided here is helpful in that sense. Thus, the instruments for synchronization and control can be improved if one considers our proposals.

To shed more light on the suggestions and provide several issues for the next observations, let us look not only at the simulation results for the networks in the examples above but also at simulations for the Poisson stable functions, especially compared with the industrial simulations in papers [76, 88, 105, 106]. Moreover, it is useful to make a more abstract presentation of the technical parameters of this book. For this reason, let us consider the two-component function $F(t, \kappa) = a\cos(\frac{2\pi t}{\omega}) + b\Theta(t)$, where a and b are real coefficients, $\Theta(t)$ is the Poisson stable function with step h described in the last section. The second component is an ω-periodic function. The parameter κ is the periodicity degree, which has been introduced above, in numerical examples. It can be beneficial for analyzing time series obtained in different areas, including theoretical and engineering. In figures 4.8, 4.9, and 4.10 we have the graphs of functions $F(t, 80)$, $F(t, 1/10)$ and $F(t, 1)$, respectively. The graph of the function $F(t, 80)$ with the periodicity degree $\kappa > 1$ is shown in figure 4.8. We present the function that admits a clear periodic shape, enveloping the irregular Poisson stability.

For the function $F(t, 1/10)$, where $\kappa < 1$, observing its graph in figure 4.9, we have the opposite effect, when the irregular stability dominates and envelopes the periodicity, which appears locally on separated intervals.

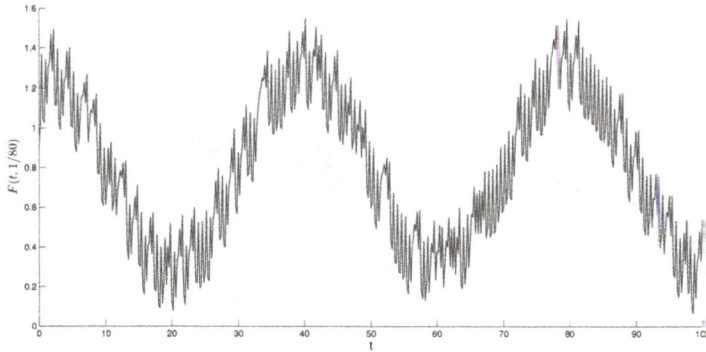

Figure 4.8. The graph of function $F(t, 80)$ with $a = 4$, $b = 0.5$, $\omega = 40$. Reproduced from [3]. CC BY 4.0.

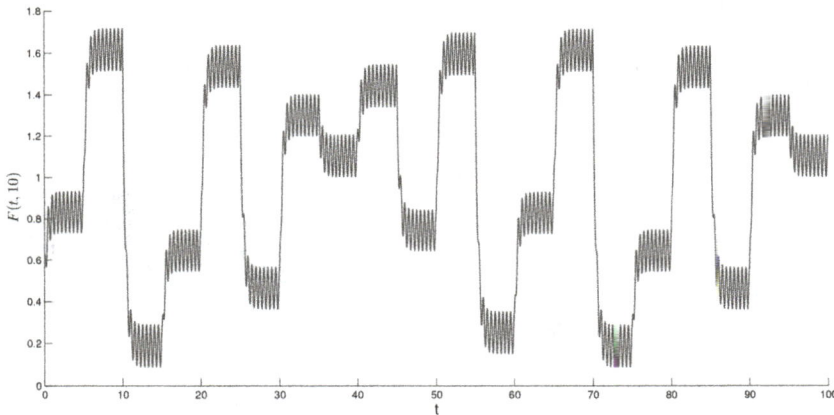

Figure 4.9. The graph of function $F(t, 1/10)$ with $a = 5$, $b = 0.1$, $\omega = 0.5$. Reproduced from [13]. CC BY 4.0.

If $\kappa = 1$, then the graph of function $F(t, 1)$ in figure 4.10 demonstrate that both phenomena, periodicity and Poisson stability, are equally present in the dynamics.

Figure 4.11 (in the paper [106] figure 9(c)) presents an historical hourly weather dataset and is similar to the graph of the function $F(t, 80)$ in figure 4.8 obtained for $\kappa > 1$. That is, one can see dominance of the periodicity over irregularity.

In figure 4.12 (figure 8 in the paper [76]), one can see oscillations similar to Poisson stable ones with $\kappa < 1$ in figure 4.9, which are obtained based on our theoretical issues. If the periodicity degree is smaller than one, $\kappa < 1$, one can say the same about loss of periodicity and dominance of irregularity, which is specified as Poisson stability in our simulations. At the same time, figures 4.12 and 4.9 demonstrate coexistence of the square wave periodicity and Poisson stability. Nevertheless, we believe that one can use the present theoretical results for qualitative characteristics of the square wave tracking with SICNNs [38].

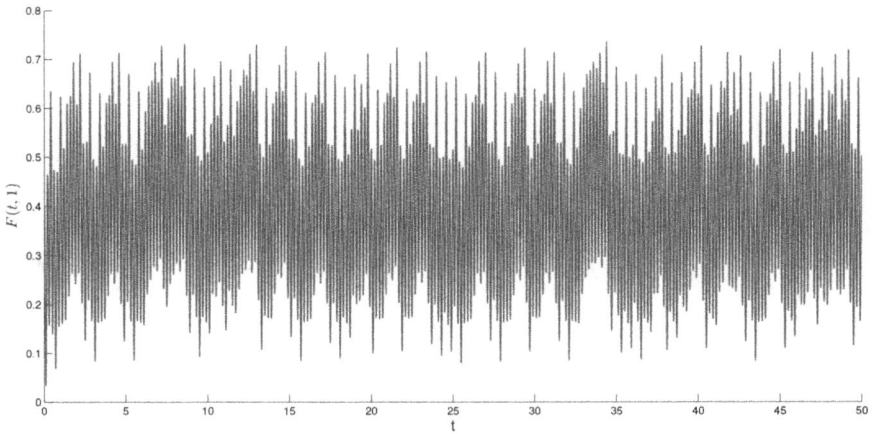

Figure 4.10. The graph of function $F(t, 1)$ with $a = 2$, $b = 0.2$, $\omega = 0.2$. Reproduced from [13]. CC BY 4.0.

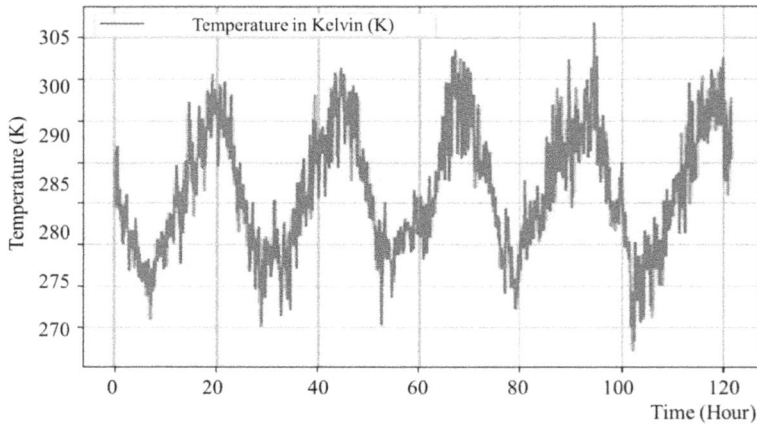

Figure 4.11. Historical hourly weather dataset. Reproduced from [106]. CC BY 4.0.

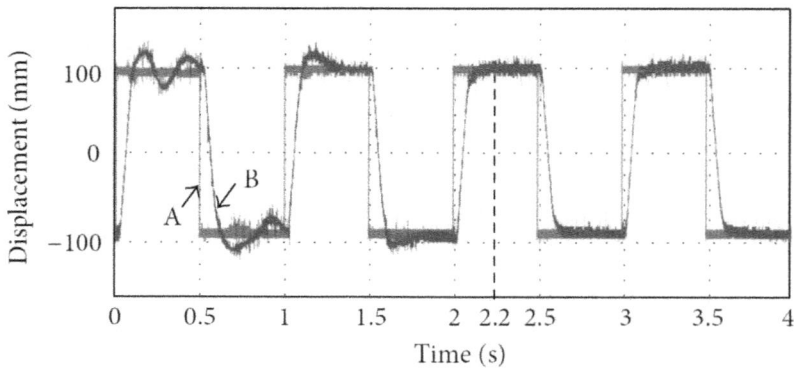

Figure 4.12. Track control of 1 Hz square wave (no load). Reproduced from [76]. CC BY 3.0.

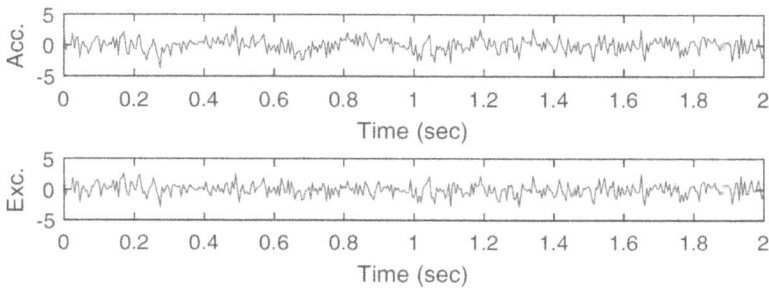

Figure 4.13. Time series of input signals. Reproduced with permission from [105].

When the periodicity degree equals one, $\kappa = 1$, we obtain Poisson stable oscillations as in figure 4.10. Figure 4.13 (in [105] figure 36(a)) presenting the convolutional neural network, is very similar to those which are obtained with the unit periodicity degree. This confirms that specific Poisson stable motions are helpful in representing dynamics of the linear and nonlinear single-degree-of-freedom system and a full-scale three-story multi-degree-freedom steel frame.

This section was reproduced with permission from [14]. CC BY 4.0.

4.4 Notes

The theory of non-linear oscillations is central to research in dynamical systems. Studies of recurrent motions are more abstract and deeper. Researchers engaged in both areas recognize that the dynamics types are often comparable; oscillations can exhibit recurrence, and vice versa. Our research makes a substantial contribution to the theory, akin to the foundational work of H. Poincaré and G. Birkhoff in recurrence dynamics, as well as P. Bohl, E. Esclangon, and H. Bohr in almost periodic functions.

We invite readers to explore the introductory information in this book regarding compartmental alpha unpredictable functions as a means to advance theory similar to the in-depth analyses found in the manuscripts of [34–36, 40, 46, 61, 62, 73, 96, 99]. The potential for our suggested studies is significantly richer than that of almost periodic or even periodic motions, as investigating alpha unpredictability enables researchers to analyze chaotic dynamics. This chapter provides constructive theoretical examples to illustrate how functional investigations can be undertaken using our fundamental proposals.

References

[1] Addabbo T, Fort A, Rocchi S and Vignoli V 2011 Digitzed chaos for pseudo-random number generation in cryptography *Chaos-based Cryptography: Theory, Algorithms and Applications* ed L Kocarev and S Lian (Berlin: Springer) pp 67–97

[2] Akhmet M 2010 *Principles of Discontinuous Dynamical Systems* (New York: Springer)

[3] Akhmet M 2020 *Almost Periodicity, Chaos and Asymptotic Equivalence* (New York: Springer)

[4] Akhmet M and Fen M O 2016 Poincaré chaos and unpredictable functions *Commun. Nonlinear Sci. Numer. Simul.* **48** 85–94

[5] Akhmet M and Fen M O 2018 Non-autonomous equations with unpredictable solutions *Commun. Nonlinear Sci. Numer. Simul.* **59** 657–70

[6] Akhmet M, Fen M O, Tleubergenova M and Zhamanshin A A 2019 Unpredictable solutions of linear differential and discrete equations *Turk. J. Math.* **43** 2377–89

[7] Nugayeva Z, Tleubergenova M and Akhmet M 2020 Discontinous compartmental periodic poisson stable functions *J. Math. Mech. Comput. Sci.* **2** 58–67

[8] Akhmet M, Tleubergenova M and Zhamanshin A 2020 Quasilinear differential equations with strongly unpredictable solutions *Carpathian J. Math.* **36** 341–9

[9] Akhmet M, Tleubergenova M and Zhamanshin A 2021 Modulo periodic Poisson stable solutions of quasilinear differential equations *Entropy* **23** 1535

[10] Akhmet M U 2009 Devaney's chaos of a relay system *Commun. Nonlinear Sci. Numer. Simulat.* **14** 1486–93

[11] Akhmet M, Fen M O and Tola A 2023 A numerical analysis of Poincaré chaos *Discontinuity Nonlinearity Complexity* **12** 183–95

[12] Akhmet M, Fen M O and Tola A 2022 Strange non-chaotic attractors with unpredictable trajectories *J. Vib. Test. Syst. Dyn.* **6** 317–27

[13] Akhmet M, Tleubergenova M and Zhamanshin A 2022 Compartmental Poisson stability in non-autonomous differential equations *Nonlinear Dynamics and Complexity* (Berlin: Springer) pp 1–23

[14] Akhmet M, Tleubergenova M and Zhamanshin A 2023 Compartmental unpredictable functions *Mathematics* **11** 1069

[15] Akhmet M 2021 *Domain Structured Dynamics* (Bristol: IOP Publishing)

[16] Akhmet M, Tleubergenova M and Zhamanshin A 2024 Cohen–Grossberg neural networks with unpredictable and Poisson stable dynamics *Chaos Solitons Fractals* **178** 114307

[17] Akhmet M, Arugaslan Cincin D, Tleubergenova M and Nugayeva Z 2021 Unpredictable oscillations for Hopfield-type neural networks with delayed and advanced arguments *Math. Open Access* **9** 1–19

[18] Akhmet M, Seilova R D, Tleubergenova M and Zhamanshin A 2020 Shunting inhibitory cellular neural networks with strongly unpredictable oscillations *Commun. Nonlinear Sci. Numer. Simulat.* **89** 105287

[19] Akhmet M, Tleubergenova M, Seilova R and Nugayeva Z 2022 Poisson stability in symmetrical impulsive shunting inhibitory cellular neural networks with generalized piecewise constant argument *Entropy* **14** 1754

[20] Akhmet M, Baskan K and Yesil C 2022 Delta synchronization of poincaré chaos in gas discharge-semiconductor systems *Chaos* **32** 083137

[21] Akhmet M, Yesil C and Baskan K 2023 Synchronization of chaos in semiconductor gas discharge model with local mean energy approximation *Chaos Solitons Fractals* **167** 113035

[22] Akhmet M and Alejaily E M 2019 Domain-structured chaos in a Hopfield neural network *Int. J. Bifurcation Chaos* **29** 195–205

[23] Akhmet M and Alejaily E M 2020 Abstract similarity, chaos and fractals *Discrete Continuous Dyn. Syst. Ser.* B **22** 1531–3492

[24] Akhmet M and Alejaily E M 2020 Conservative-progressive growth method for fractal curves and functions (unpublished)

[25] Akhmet M and Fen M O 2016 Poincaré chaos and unpredictable functions *Commun. Nonlinear Sci. Numer. Simul.* **48** 85–94

[26] Akhmet M and Fen M O 2016 Unpredictable points and chaos *Commun. Nonlinear Sci. Numer. Simul.* **40** 1–5

[27] Akhmet M, Tleubergenova M and Zhamanshin A 2021 Modulo periodic Poisson stable solutions of quasilinear differential equations *Entropy* **23** 1535

[28] Akhmet M, Fen M O and Alejaily E M 2018 *Dynamics motivated by Sierpinski fractals* ArXiv e-prints

[29] Akhmet M 2022 Unpredictablity in Markov chains *Carpathian J. Math.* **38** 13–9

[30] Avila A and Moreira C G 2003 Bifurcations of unimodal maps arXiv 1–22

[31] Barnsley M 1988 *Fractals Everywhere* (New York: Academic)

[32] Barnsley M F and Demko S 1985 Iterated function systems and the global construction of fractals *Proc. R. Soc. Ser.* A **399** 243–75

[33] Barnsley M F, Elton J, Hardin D P and Massopust P R 1989 Hidden variable fractal interpolation functions *SIAM J. Math. Anal.* **20** 1218–42

[34] Besicovitch A S 1954 *Almost Periodic Functions* (Cambridge: Dover)

[35] Birkhoff G D 1991 *Dynamical Systems* (Providence, RI: Colloquium Publications)

[36] Bogolyubov N N 1939 *On Some Arithmetic Properties of Almost Periods* (Kiev: Akademiya Nauk Ukrainian SSR)

[37] Bohl P 1900 Ueber Einige Differentialgleichungen Allgemeinen Charakters, Welche in der Mechanik Anwendbar Sind *PhD Thesis* Dorpat

[38] Bohl P 1906 Über eine differentialgleichung der störungstheorie *Grelles J.* **131** 268–321

[39] Bohr H A 1947 *Almost Periodic Functions* (New York: Chelsea Publishing Company)

[40] Bohner S 1962 A new approach to almost periodicity *Proc. Natl Acad. Sci. USA* **45** 195–205

[41] Burton T A 1985 Stability and Periodic Solutions of Ordinary and Functional *Differential equations* (Amsterdam: Elsevier Science)

[42] Bolotin Y, Tur A and Yanovsky V 2013 *Chaos: Concept, Control and Constructive Use* (Berlin: Springer)

[43] Brown R 1928 A brief account of microscopical observations made in the months of June, July and August, 1927, on the particles contained in the pollen of plants; and on the general existence of active molecules in organic and inorganic bodies *Philos. Mag. Ann. Philos. New Ser.* **4** 161–78

[44] Carbone A, Gromov M and Prusinkiewicz P 2000 Pattern Formation in Biology *Vision and Dynamics* (Singapore: World Scientific)

[45] Chen G and Huang Y 2011 *Chaotic Maps: Dynamics, Fractals and Rapid Fluctuations, Synthesis Lectures on Mathematics and Statistics* (San Rafael, CA: Morgan & Claypool Publishers)

[46] Corduneanu C 2009 *Almost Periodic Oscillations and Waves* (New York: Springer)

[47] Devaney R L 1987 *An Introduction to Chaotic Dynamical Systems* (Menlo Park, CA: Addison-Wesley)

[48] Değirmenci N and Koçak Ş 2010 Chaos in product maps *Turk. J. Math.* **34** 593–600

[49] Farkas M 1994 *Periodic Motion* (New York: Springer)

[50] Feigenbaum M J 1980 Universal behavior in nonlinear systems *Los Alamos Sci./Summer* **1** 4–27

[51] Feldman D P 2012 *Chaos and Fractals: An Elementary Introduction* (Oxford: Oxford University Press)

[52] Fink A M 1974 *Almost Periodic Differential Equations* (New York: Springer)

[53] Hahn H K, Georg M and Peitgen H O 2005 Fractal aspects of three-dimensional vascular constructive optimization *Fractals in Biology and Medicine* (Berlin: Birkhäuser) pp 55–66

[54] Haggarty R 1993 *Fundamentals of Mathematical Analysis* (Reading, MA: Addison-Wesley)

[55] Hardin D P and Massopust P R 1986 The capacity for a class of functions *Cormm. Math. Phys.* **105** 455–60

[56] Hajek B 2015 *Random Processes for Engineers* (Cambridge: Cambridge University Press)

[57] Hilborn R C 2000 *Chaos and Nonlinear Dynamics: An Introduction for Scientists and Engineers* (New York: Oxford University Press)

[58] Hino Y 1985 Recurrent solutions for linear almost periodic systems *Funkcialaj Ekvacioj* **28** 117–9

[59] Hunt B R and Kaloshin V Y 2010 *Handbook of Dynamical Systems* ed A Katok and B Hasselblatt (Amsterdam: Elsevier Science) pp 43–87

[60] Esclangon E 1904 *Les Fonctions Quasi-Periodiques* (Paris: Gauthier-Villars)

[61] Farkas M 1994 *Periodic Motion* (New York: Springer)

[62] Fink A M 1974 *Almost Periodic Differential Equations* (New York: Springer)

[63] Ivancevic V G and Ivancevic T T 2007 *High-Dimensional Chaotic and Attractor Systems* (Dordrecht: Springer)

[64] Julia G 1918 Mémoire suv i'itération des fonctions rationelles *J. Math. Pures Appl* **8** 47–245

[65] Karlin S and Taylor H E 2012 *A First Course in Stochastic Processes* (New York: Academic)

[66] Keener J P 1980 Chaotic behavior in piecewise continuous difference equations *Trans. Am. Math. Soc.* **261** 589–604

[67] Kellert S H 1994 *In the Wake of Chaos: Unpredictable Order In Dynamical Systems* (Chicago, IL: University of Chicago Press)

[68] Knight R A 1981 Recurrent and poisson stable flows *Proc. Am. Math. Soc.* **83** 49–53

[69] Korsch H J, Jodl H J and Hartmann T 2008 *Chaos: A Program Collection for the PC* (Berlin: Springer)

[70] Kumar A and Bhagat R P 1987 Poisson stability in product of dynamical systems *Int. J. Math. Math. Sci.* **10** 613–4

[71] Landau L 1944 On the nature of turbulence *Dokl. Acad. Sci. USSR* **44** 311–4

[72] Layek G C 2015 *An Introduction to Dynamical Systems and Chaos* (New Delhi: Springer)

[73] Levitan B and Zhikov V 1983 *Almost Periodic Functions and Differential equations* (Cambridge: Cambridge University Press)

[74] Li R and Zhou X 2013 A note on chaos in product maps *Turk. J. Math.* **37** 665–75

[75] Li T Y and Yorke J A 1975 Period three implies chaos *Am. Math. Mon.* **82** 985–92

[76] Li X, Xu F, Zhang J and Wang S 2013 A multilayer feed forward small-world neural network controller and its application on electrohydraulic actuation system *J. Appl. Math.* **1** 211–44

[77] Lorenz E N 1984 The local structure of chaotic attractor in four dimensions *Physica* **13D** 90–104

[78] Lynch S 2017 *Dynamical Systems with Applications Using Mathematica* 2nd edn (Cham: Springer International Publishing)

[79] Luo A 1966 *Stability of Motion* (New York: Academic)

[80] Mandelbrot B B 1977 *Fractals: Form, Chance, and Dimension* (San Francisco, CA: W. H. Freeman)

[81] Mandelbrot B B 1983 *The Fractal Geometry of Nature* (New York: Freeman)

[82] Masoller C, Schifino A C and Sicardi R L 1995 Characterization of strange attractors of Lorenz model of general circulation of the atmosphere *Chaos Solit. Fract.* **6** 357–66

[83] Massopust P R 1987 Dynamical systems, fractal functions and dimension *Topology Proc.* **12** 93–110

[84] Massopust P R 1997 Fractal functions and their applications *Chaos Solit. Fract.* **8** 171–90

[85] Massopust P R 2004 Fractal functions and wavelets: examples of multiscale theory *Abstract and Applied Analysis, Proceedings of the International Conference, Hanoi, Vietnam* (Singapore: World Scientific) pp 225–47

[86] Meyn S and Tweedie R L 2009 *Markov Chains and Stochastic Stability* (Cambridge: Cambridge University Press)

[87] Mehlhorn K 1984 *Data Structures and Algorithms* **vol 3** (New York: Springer)

[88] Mohammad U, Yasin M, Yousuf R and Anwar I 2019 A novel square wave generator based on the translinear circuit scheme of second generation current controlled current conveyor—cccii *SN Appl. Sci.* **1** 587

[89] Nemytskii V V and Stepanov V V 1960 *Qualitative Theory of Differential Equations* (Princeton, NJ: Princeton University Press)

[90] Osikawa M and Oono Y 1981 Chaos in C^0-endomorphism of interval *Publ. RIMS* **17** 165–77

[91] Poincaré H 1957 *New Methods of Celestial Mechanics Volumes I–III* (New York: Dover)

[92] Ruelle D 1978 An inequality of the entropy of differentiable maps *Bol. Soc. Bras. Mat.— Bull./Braz. Math. Soc.* **9** 83–7

[93] Rulkov N, Sushchik L, Tsimring M and Abarbanel H 1995 Generalized synchronization of chaos in directionally coupled chaotic systems *Phys. Rev.* E **51** 980

[94] Sander E and Yorke J A 2011 Period-doubling cascades galore *Ergod. Theor. Dynam. Syst.* **31** 1249–67

[95] Schöll E and Schuster H G (ed) 2008 *Handbook of Chaos Control* (Weinheim: Wiley-VCH)

[96] Sell G R 1971 *Topological Dynamics and Ordinary Differential Equations* (London: Van Nostrand Reinhold Company)

[97] Singer P and Zajdler P 1999 Self-affine fractal functions and wavelet series *J. Math. Anal. Appl.* **240** 518–51

[98] Skiadas C H and Skiadas C (ed) 2016 *Handbook of Applications of Chaos Theory* (Boca Raton, FL: Chapman and Hall/CRC Press)

[99] Stepanov V V 1925 Impact of network activities on neuronal properties in corticothalamic systems *C. R. l'Acad. Sci.* 90–2

[100] Turing A 1952 The chemical basis of morphogenesis *Phil. Trans. R. Soc. London* B **237** 32–72

[101] Vicsek T 1992 *Fractal Growth Phenomena* (Singapore: World Scientific)

[102] Wang M C and Uhlenbeck G E 1945 On the theory of Brownian motion ii *Rev. Mod. Phys.* **17** 323–42

[103] Weierstrass K 1895 *Uber continuierliche Functionen eines reellen Arguments, die fur keinen Wert des letzteren einen bestimmten Differentialquotienten besitzen, in 'K. Weierstrass, Mathematische Werke'* (Berlin: Mayer & Miiller)

[104] Wiggins S 1988 Global bifurcation and chaos *Analytical Methods* (New York: Springer)

[105] Wu U, Yasin M, Yousuf R and Anwar I 2019 Deep convolutional neural network for structural dynamic response estimation and system identification *J. Eng. Mech. Appl. Sci.* **145** 04018125

[106] Zhang H, Lu H and Nayak A 2020 Periodic time series data analysis by deep learning methodology *IEEE Access* **8** 78–88

[107] Zeraoulia E and Sprott J C 2012 *Robust Chaos and Its Applications* (Singapore: World Scientific)

[108] Zubov V I 1996 *Oscillations and Waves* (Singapore: World Scientific)

Part IV

Differential equations with ultra Poincaré chaos

IOP Publishing

Ultra Poincaré Chaos and Alpha Labeling
A new approach to chaotic dynamics
Marat Akhmet

Chapter 5

Alpha unpredictable differential equations

A new type of recurrence, known as alpha unpredictable functions, has been recently introduced. These functions are significant because they are elements of ultra Poincaré chaos. Therefore, proving the existence of alpha unpredictable solutions in differential equations implies that the dynamics are chaotic. This chapter examines alpha unpredictable quasi-linear differential systems, focusing on various specific models. One such model considers all coordinates of the equations as irregular, which applies to the coordinates of the solutions as well. Another noteworthy innovation is the consideration of perturbations as nonlinear functions that are alpha unpredictable over time.

A new type of oscillations, alpha unpredictable functions, has been introduced in [5, 6]. It was shown, in papers [5, 7, 8, 10], that the existence of an alpha unpredictable solution implies *ultra Poincaré chaos* for a dynamics in a functional space. Consequently, studying alpha unpredictable solutions is significantly beneficial for chaos research. In the present chapter we consider oscillations with all components alpha unpredictable. Since the irregularity is observable in all coordinates, chaos is ultimately intense in this case. Another significant novelty is that perturbations are assumed to be nonlinear functions, strongly alpha unpredictable in the time variable.

Observing the line of implications, one can conclude that by reading the manuscript successively, one can pass from very abstract constructions to analysis of most essential real world problems. The last four chapters of the book are delivered to provide strong evidence that alpha chaos is useful for applied mathematics. We have demonstrated by the present research that randomness potentially is a rich source of alpha unpredictable functions, and alpha dynamics is that tool which is utilized for theoretical support of our constructions. The interaction of alpha dynamics with stochastic processes is two-sided. Not only does the latter serve to obtain alpha unpredictable functions, but the alpha unpredictability clarifies structures of random dynamics. Thus, new methods of analysis for

doi:10.1088/978-0-7503-5162-1ch5

stochastic equations can be designed. This chapter will discuss not stochastic, but deterministic differential equations. It is a standard that oscillations are a main subject in research of differential equations. This is reasoned, first of all, by application demands. Theoretically, mathematicians need to find analytical presentations of oscillations. Unfortunately, this task is not easy for motions with weak recurrence. It is exceptionally difficult for Poisson stable functions because one could not find well-developed samples of the functions presented analytically, through algorithms. or even numerical simulations. Samples of Poisson stable functions are an essential achievement of our research—presented here for the first time in literature. They are constructed deterministically and randomly by applying Bernoulli schemes and Markov chains. We have supported our research with graphical illustration of alpha unpredictable functions as solutions of discrete and differential equations. This is done asymptotically by using a new method of visualization for recurrent solutions and confirmed in examples 5.1.1 and 5.1.2. Thus, alpha unpredictable and Poisson stable functions are exemplified and simulated in our research for the first time in history. The achievements are based on our experience in hybrid systems study and more precisely, in differential equations admitting various types of discontinuity. The issue point in the construction of alpha unpredictable functions is the dynamics of the logistic equation and alpha labeling. We are confident that the supply of alpha unpredictable functions will be enlarged significantly by utilizing randomness based on alpha labeling and verifying chaotic ingredients in stochastic processes. This will make theoretical research of alpha unpredictable solutions for dynamical equations critical for applications, especially in modern research areas such as computer sciences, neural networks, brain activity, and others. Mathematically, we predict that the study of alpha unpredictable solutions will be extended through methods of operator theory, averaging, singular differential equations, and numerical methods. Consequently, the results can be easily exported for partial differential, functional-differential and evolution equations, and various types of hybrid systems.

It is worth noting that in the literature, a large number of results are obtained for periodic, quasi, and almost periodic solutions of differential equations due to established mathematical methods and essential applications [39, 42, 44, 45, 47]. On the other hand, recurrent and Poisson stable solutions are also crucial for the theory of differential equations [41, 52, 53]. Our proposals can revive the interest of specialists in differential equations theory for two reasons. The first is related to verifying alpha unpredictability, which requires a more sophisticated technique than that used for recurrent and Poisson stable functions. Thus, the problem of the existence of alpha unpredictable solutions is a challenging one. They can be investigated for partial differential, evolutions, and hybrid equations. That is, our study suggests a new approach for chaos extension in many types of dynamics. The paper [53] contains a general method of comparability for functions by character of their recurrence. In this chapter, and in papers [7–10], we apply a new approach to prove the existence of Poisson stable solutions, the *method of included intervals*, which is much more straightforward, and consequently more effective, than the method of comparability [41, 53]. It can be applied, in future, for various types of dynamics. Our method makes the discussion of alpha unpredictability and Poisson stability as transparent as it is for periodicity and almost periodicity.

We are exceptionally interested in a single alpha unpredictable motion, since it is an isolated product of chaotic processes with particular and constructive properties. The motions are specified as alpha unpredictable functions of continuous and discrete time. Consequently, one can focus on alpha unpredictable functions as solutions of various dynamical equations and perturbations for them; in our research they are also considered for deterministic equations and models with chance.

This chapter investigates the existence and uniqueness of alpha unpredictable solutions in the dynamics of quasi-linear systems. The research is another stream of our activity, where we work to close chaos and theory of differential equations and make them more friendly to each other. This line is initiated in papers [3, 11–18, 30–36], where the method of replication of chaos is developing. It has been proven to be very powerful, and is based on definitions of chaos with continuous time, while all previous descriptions used discrete time.

The paradigm of input-output nodes in the flow of world processes, where a node is a dynamical model, has been considered and continues to be modern science's most important subject. Thus, we try to be in the stream. Additionally to the replication of chaos, alpha unpredictable functions are used as inputs and outputs in our last papers [2, 5, 6, 9, 24]. This chapter is one of the central parts of the research.

In section 5.1, strongly alpha unpredictable functions are considered as solutions. A new type of irregular equation, quasi-linear systems with all coordinates perturbed alpha unpredictably, is the focus of the investigation. The model is critical for applications in processes when all parameters in observations are wanted to be irregular in both inputs and outputs. Correspondingly, the problem is finding conditions on systems that guarantee that all solution components are alpha unpredictable. Thus, we came to the idea of strongly unpredictable perturbations and solutions for differential equations.

In section 5.2, we realized the constructive merits of chapter 4, where new functions with sophisticated behavior are introduced, which combine properties of periodicity and alpha unpredictability. The novelties are realized in research of alpha unpredictable quasi-linear equations.

The final section, 5.3, is devoted to modulo periodic Poisson stable functions and solutions. Quasi-linear differential equations with modulo periodic Poisson stable coefficients are under investigation. The existence and uniqueness of modulo periodic Poisson stable solutions are proved. Numerical simulations, which illustrate the theoretical results, are provided.

Throughout this chapter, $\mathbb{R}$ and $\mathbb{N}$ will stand for the set of real and natural numbers, respectively, and the norm $\|u\|_1 = \sup_{t \in \mathbb{R}} \|u(t)\|$, where $\|u\| = \max_{1 \leqslant i \leqslant p} |u_i|$, $u = (u_1, \ldots, u_p)$, $u_i \in \mathbb{R}$, $i = 1, 2, \ldots, p$, will be used. Correspondingly, for a square matrix $A = \{a_{ij}\}$, $i, j = 1, 2, \ldots, p$, the norm $\|A\| = \max_{i=1,\ldots,p} \sum_{j=1}^{p} |a_{ij}|$, will be utilized.

5.1 Strongly alpha unpredictable solutions

Two principal features form the basis of this section. The first is that all coordinates of perturbations are alpha unpredictable functions. Secondly, we are looking for solutions with the same property. That is, quasi-linear systems with strongly alpha unpredictable inputs and outputs are the focus of the present research.

5.1.1 Definitions and auxiliaries

The following two definitions form the starting point of this chapter.

Definition 5.1.1. [5] *A uniformly continuous and bounded function $v: \mathbb{R} \to \mathbb{R}^p$ is alpha unpredictable if there exist positive numbers ε_0, σ and sequences t_n, u_n, both of which diverge to infinity such that $v(t + t_n) \to v(t)$ as $n \to \infty$ uniformly on compact subsets of $\mathbb{R}$ and $\|v(t + t_n) - v(t)\| \geqslant \varepsilon_0$ for each $t \in [u_n - \sigma, u_n + \sigma]$ and $n \in \mathbb{N}$.*

Definition 5.1.2. [24] *A continuous and bounded function $f(t, x): \mathbb{R} \times G \to \mathbb{R}^p$, $f = (f_1, f_2, \ldots, f_p)$, $G \subset \mathbb{R}^p$ is a bounded domain, is alpha unpredictable in t if it is uniformly continuous in t and there exist positive numbers ε_0, σ and sequences t_n, u_n, both of which diverge to infinity such that $\sup_G \|f(t + t_n, x) - f(t, x)\| \to 0$ as $n \to \infty$ uniformly on compact sets in $\mathbb{R}$ and $\inf_G \|f(t + t_n, x) - f(t, x)\| \geqslant \varepsilon_0$ for $t \in [u_n - \sigma, u_n + \sigma]$ and $n \in \mathbb{N}$.*

We shall need the following two notions, which are analogues to the last two definitions.

Definition 5.1.3. [24] *A uniformly continuous and bounded function $v: \mathbb{R} \to \mathbb{R}^p$ is strongly alpha unpredictable if there exist positive numbers ε_0, σ and sequences t_n, u_n, both of which diverge to infinity such that $v(t + t_n) \to v(t)$ as $n \to \infty$ uniformly on compact subsets of $\mathbb{R}$ and $|v_i(t + t_n) - v_i(t)| \geqslant \varepsilon_0$, $v = (v_1, \ldots, v_p)$ for each $t \in [u_n - \sigma, u_n + \sigma]$, $i = 1, \ldots, p$ and $n \in \mathbb{N}$.*

Definition 5.1.4. [24] *A continuous and bounded function $f(t, x): \mathbb{R} \times G \to \mathbb{R}^p$, $f = (f_1, f_2, \ldots, f_p)$, $G \subset \mathbb{R}^p$ is a bounded domain, is strongly alpha unpredictable in t if it is uniformly continuous in t and there exist positive numbers ε_0, σ and sequences t_n, u_n both of which diverge to infinity such that $\sup_G \|f(t + t_n, x) - f(t, x)\| \to 0$ as $n \to \infty$ uniformly on compact sets in $\mathbb{R}$ and $\inf_{[u_n - \sigma, u_n + \sigma] \times G} |f_i(t + t_n, x) - f_i(t, x)| > \varepsilon_0$ for all $i = 1, \ldots, p$ and $n \in \mathbb{N}$.*

Comparing the definitions 5.1.1 and 5.1.3 as well as 5.1.2 and 5.1.4, one can find that an alpha unpredictable function may admit some coordinates that are not of the type. Meanwhile, each coordinate of a strongly alpha unpredictable function is a member of the same class.

Next, let us remember the definition of a Poisson stable function [52], p 124, adapted to our case.

Definition 5.1.5. *A continuous and bounded function* $\vartheta \colon \mathbb{R} \to \mathbb{R}^m$ *is Poisson stable if there exists a sequence* $\{t_n\}$, $t_n \to \infty$ *as* $n \to \infty$, *such that* $\vartheta(t + t_n) \to \vartheta(t)$ *as* $n \to \infty$ *uniformly on compact subsets of* $\mathbb{R}$.

By comparing definitions 5.1.1 and 5.1.5, one can see that any alpha unpredictable function is Poisson stable. The last type of function specifies the eponymous point considered for dynamical systems in [52], p 85, [50], pp 344 and 345.

The following lemmas can be useful for applications of our results.

Lemma 5.1.1. [19] *Suppose that* $\phi(t) \colon \mathbb{R} \to \mathbb{R}$ *is an alpha unpredictable function. Then the function* $\phi^3(t)$ *is alpha unpredictable.*

Proof. One can find numbers $\varepsilon_0 > 0$, $\delta > 0$ and sequences $\{t_n\}$, $\{u_n\}$ both of which diverge to infinity such that $\phi(t + t_n) \to \phi(t)$ as $n \to \infty$ uniformly on compact subsets of $\mathbb{R}$ and $\|\phi(t + t_n) - \phi(t)\| \geqslant \varepsilon_0$ for each $t \in [u_n - \delta, u_n + \delta]$ and $n \in \mathbb{N}$. It is easy to check that $\phi^3(t + t_n) \to \phi^3(t)$ as $n \to \infty$ uniformly on compact subsets of $\mathbb{R}$, since it follows from the uniform continuity of the cubic function on a compact set.

Fix a natural number n. Let us show that for $t \in [u_n - \delta, u_n + \delta]$ the inequality $\|\phi(t + t_n) - \phi(t)\| \geqslant \varepsilon_0$ implies $\|\phi^3(t + t_n) - \phi^3(t)\| \geqslant \varepsilon_0^3/4$.

We have that

$$|\phi^3(t + t_n) - \phi^3(t)| = \frac{1}{2}|\phi(t + t_n) - \phi(t)|[\phi^2(t + t_n) + \phi^2(t) + (\phi(t + t_n) + \phi(t))^2]$$

$$\geqslant \frac{1}{2}(\phi^2(t + t_n) + \phi^2(t))\varepsilon_0.$$

Consider the function $F(a, b) = a^2 + b^2$ for $|a - b| \geqslant \varepsilon_0$. The minimum of F occurs at the points (a, b) with $|a| = |b| = \varepsilon_0/2$. Therefore, $|\phi^3(t + t_n) - \phi^3(t)| \geqslant \varepsilon_0^3/4$ for $t \in [u_n - \delta, u_n + \delta]$.

Lemma 5.1.2. [19] *If the function* $\phi(t) \colon \mathbb{R} \to \mathbb{R}$ *is alpha unpredictable, then the function* $\phi(t) + c$, *where* c *is a constant, is also alpha unpredictable.*

Proof. There exist positive numbers ε_0, δ and sequences $\{t_n\}$, $\{u_n\}$ both of which diverge to infinity such that $\|\phi(t + t_n) - \phi(t)\| \to 0$ as $n \to \infty$ uniformly on compact subsets of $\mathbb{R}$ and $\|\phi(t + t_n) - \phi(t)\| \geqslant \varepsilon_0$ for each $t \in [u_n - \delta, u_n + \delta]$ and $n \in \mathbb{N}$. Let us denote $\omega(t) = \phi(t) + c$. Then, we have that $\|\omega(t + t_n) - \omega(t)\| = \|\phi(t + t_n) - \phi(t)\| \to 0$ as $n \to \infty$ uniformly on compact subsets of $\mathbb{R}$ and $\|\omega(t + t_n) - \omega(t)\| = \|\phi(t + t_n) - \phi(t)\| \geqslant \varepsilon_0$ for each $t \in [u_n - \delta, u_n + \delta]$ and $n \in \mathbb{N}$. Therefore, the function $\phi(t) + c$ is alpha unpredictable.

5.1.2 The existence theorem

We shall consider the following differential equation,

$$x'(t) = Ax(t) + f(t, x), \tag{5.1}$$

where $t \in \mathbb{R}$, $x \in \mathbb{R}^p$, p is a fixed natural number, all eigenvalues of the constant matrix $A \in \mathbb{R}^{p \times p}$ have negative real parts, $f \colon \mathbb{R} \times G \to \mathbb{R}^p$, $f = (f_1, \ldots f_p)$, $G = \{x \in \mathbb{R}^p, \|x\| < H\}$, where H is a positive number. It is true that there exist two real numbers $K \geqslant 1$ and $\gamma < 0$ such that $\|e^{At}\| \leqslant Ke^{\gamma t}$ for all $t \geqslant 0$.

Definition 5.1.4 implies that there exists a positive number M such that $\sup_{\mathbb{R} \times G} \|f(t, x)\| = M < \infty$.

The following conditions will be needed in the section:
  (C1) the function $f(t, x)$ is strongly alpha unpredictable in the sense of definition 5.1.4;
  (C2) there exists a positive constant L such that the function $f(t, x)$ satisfies the inequality $\|f(t, x_1) - f(t, x_2)\| \leqslant L\|x_1 - x_2\|$ for all $t \in \mathbb{R}$, $x_1, x_2 \in G$;
  (C3) $\gamma < -\frac{KM}{H}$;
  (C4) $\gamma < -KL$.

Our purpose is to prove that system (5.1) possesses a unique strongly alpha unpredictable solution, provided that the function $f(t, x)$ is strongly alpha unpredictable in t. Moreover, we prove that the solution is uniformly globally exponentially stable. Additionally, the existence of an alpha unpredictable solution, which is not strongly alpha unpredictable, is considered for the system.

Let U be the set of all uniformly continuous functions $\psi(t) = (\psi_1, \psi_2, \ldots, \psi_p)$, $\psi_i \in \mathbb{R}$, $i = 1, 2, \ldots, p$, such that $\|\psi\|_1 < H$, and $\psi(t + t_n) \to \psi(t)$ as $n \to \infty$ uniformly on each closed and bounded interval of the real axis, where sequence t_n is the same as for function $f(t, x)$ in system (5.1). In other words, U is a set of Poisson stable functions with the common convergence sequence.

According to the theory of differential equations [43, 47], a bounded on the whole real axis function $\omega(t) = (\omega_1, \omega_2, \ldots, \omega_p)$ is a solution of system (5.1) if and only if the integral equation

$$\omega(t) = \int_{-\infty}^{t} e^{A(t-s)} f(s, \omega(s)) ds, \ t \in \mathbb{R}, \tag{5.2}$$

is satisfied.

Lemma 5.1.3. [24] *Suppose that conditions* $(C1) - (C4)$ *are valid, then the system (5.1) possesses a unique solution* $\omega(t) \in U$.

Proof. Define an operator Π on U such that

$$\Pi\psi(t) = \int_{-\infty}^{t} e^{A(t-s)} f(s, \psi(s)) ds, \, t \in \mathbb{R}. \tag{5.3}$$

Fix an arbitrary function $\psi(t)$ that belongs to U. We have that

$$\|\Pi\psi(t)\| \leqslant \int_{-\infty}^{t} \|e^{A(t-s)}\| \|f(s, \psi(s))\| ds \leqslant \frac{KM}{|\gamma|}$$

for all $t \in \mathbb{R}$. Therefore, by condition $(C3)$ it is true that $\|\Pi\psi\|_1 < H$.

Next, we shall apply the method of included intervals. Fix an arbitrary positive number ε and a closed interval $[a, b]$, $-\infty < a < b < \infty$, of the real axis. We will show that for sufficiently large n it is true that $\|\Pi\psi(t + t_n) - \Pi\psi(t)\| < \varepsilon$ on $[a, b]$. Let us choose two numbers $c < a$, and $\xi > 0$ satisfying

$$\frac{2KM}{|\gamma|} e^{\gamma(a-c)} < \frac{\varepsilon}{2} \tag{5.4}$$

and

$$\frac{K}{|\gamma|} \xi(L + 1)[1 - e^{\gamma(b-c)}] < \frac{\varepsilon}{2}. \tag{5.5}$$

Consider n sufficiently large such that $\|f(t + t_n, x) - f(t, x)\| < \xi$ and $\|\psi(t + t_n) - \psi(t)\| < \xi$ for $t \in [c, b]$ and $x \in G$. Then, the inequality

$$\|\Pi\psi(t + t_n) - \Pi\psi(t)\| \leqslant \int_{-\infty}^{c} \|e^{A(t-s)}\| \|f(s + t_n, \psi(s + t_n)) - f(s, \psi(s))\| ds +$$

$$\int_{c}^{t} \|e^{A(t-s)}\| \|f(s + t_n, \psi(s + t_n)) - f(s, \psi(s))\| ds \leqslant \int_{-\infty}^{c} 2KM e^{\gamma(t-s)} ds + \tag{5.6}$$

$$\int_{c}^{t} \xi(L + 1) K e^{\gamma(t-s)} ds \leqslant \frac{2}{|\gamma|} KM e^{\gamma(a-c)} + \frac{K}{|\gamma|} \xi(L + 1)[1 - e^{\gamma(b-c)}]$$

is correct for all $t \in [a, b]$.

By inequalities (5.4) and (5.5) it is true that $\|\Pi\psi(t + t_n) - \Pi\psi(t)\| < \varepsilon$ for $t \in [a, b]$, and, therefore, $\Pi\psi(t + t_n) \to \Pi\psi(t)$ uniformly as $n \to \infty$ on each closed and bounded interval of $\mathbb{R}$.

It is easy to verify that $\Pi\psi(t)$ is a uniformly continuous function. Thus, the set U is invariant for the operator Π.

We proceed to show that the operator $\Pi: U \to U$ is contractive. Let $u(t)$ and $v(t)$ be members of U. Then, we obtain that

$$\|\Pi u(t) - \Pi v(t)\| \leqslant \int_{-\infty}^{t} \|e^{A(t-s)}\| \|f(s, u(s)) - f(s, v(s))\| ds \leqslant$$

$$\int_{-\infty}^{t} K e^{\gamma(t-s)} L \|u(s) - v(s)\| ds \leqslant \frac{KL}{|\gamma|} \|u(t) - v(t)\|_1$$

for all $t \in \mathbb{R}$. Therefore, the inequality $\|\Pi u - \Pi v\|_1 \leqslant \frac{KL}{|\gamma|} \|u - v\|_1$ holds, and according to condition $(C4)$ the operator $\Pi \colon U \to U$ is a contraction.

The next task is to show that the space U is complete. Consider a Cauchy sequence $\phi_k(t)$ in U, which converges to a limit function $\phi(t)$ on $\mathbb{R}$. Fix a closed and bounded interval $I \subset \mathbb{R}$. We have that

$$\|\phi(t + t_p) - \phi(t)\| < \|\phi(t + t_n) - \phi_k(t + t_n)\| + \\ \|\phi_k(t + t_n) - \phi_k(t)\| + \|\phi_k(t) - \phi(t)\|. \tag{5.7}$$

Now, one can take sufficiently large n and k such that each term on the right-hand-side of (5.7) is smaller than $\frac{\varepsilon}{3}$ for an arbitrary small ε and $t \in I$. The inequality implies that $\|\phi(t + t_n) - \phi(t)\| < \varepsilon$ on I. That is, the sequence $\phi(t + t_n)$ uniformly converges to $\phi(t)$ on I. Likewise, one can check that the limit function is bounded and uniformly continuous [47]. The completeness of U is proved. By contraction mapping theorem there exists the unique fixed point, $\omega(t) \in U$, of the operator Π, which is a unique bounded solution of the system (5.1). The lemma is proved.

Theorem 5.1.1. [24] *If conditions $(C1) - (C4)$ are fulfilled, then system (5.1) admits a unique exponentially stable strongly alpha unpredictable solution.*

Proof. According to lemma 5.1.3, system (5.1) has a unique solution $\omega(t) \in U$. The stability of $\omega(t)$ can be verified as it is done for a bounded solution in [47]. What is left is to verify that the alpha unpredictability property is valid.

One can find a positive number κ, natural numbers l, k and $j = 1,\dots,p$, such that the following inequalities are valid:

$$\kappa < \sigma, \tag{5.8}$$

$$\|\omega(t + s) - \omega(t)\| < \frac{\varepsilon_0}{k}, \quad t \in \mathbb{R}, \, |s| < \kappa. \tag{5.9}$$

$$|f_j(t_n + u_n + s, x) - f_j(u_n + s, x)| \geqslant \varepsilon_0/2, \, \|x\| < H, \, |s| < \kappa, \, n \in \mathbb{N}, \tag{5.10}$$

$$\kappa\left(1/2 - \left(\frac{1}{l} + \frac{2}{k}\right)(L + \|A\|)\right) > \frac{3}{2l}. \tag{5.11}$$

Let the numbers κ, l, k and j as well as a number $n \in \mathbb{N}$, be fixed.

Denote $\Delta = |\omega_i(u_n + t_n) - \omega_i(u_n)|$ and consider two cases (i) $\Delta \geqslant \varepsilon_0/l$, (ii) $\Delta < \varepsilon_0/l$ such that the remaining proof naturally falls into two following (i) and (ii) parts.

(i) For the case $\Delta \geqslant \varepsilon_0/l$, fix an additional positive number κ_1 sufficiently small for

$$\|\omega(t + s) - \omega(t)\| < \frac{\varepsilon_0}{4l}, \quad t \in \mathbb{R}, \, |s| < \kappa_1.$$

Therefore,

$$\|\omega(t_n + t) - \omega(t)\| \geqslant \|\omega(t_n + u_n) - \omega(u_n)\| - \|\omega(u_n) - \omega(t)\| -$$
$$\|\omega(t_n + t) - \omega(t_n + u_n)\| \geqslant \frac{\varepsilon_0}{l} - \frac{\varepsilon_0}{4l} - \frac{\varepsilon_0}{4l} = \frac{\varepsilon}{2l}, \tag{5.12}$$

if $t \in [u_n - \kappa_1, u_n + \kappa_1]$ and $n \in \mathbb{N}$.

(ii) One can find that from (5.9) it follows that

$$\|\omega(t_n + t) - \omega(t)\| < \frac{\varepsilon_0}{l} + \frac{\varepsilon_0}{k} + \frac{\varepsilon_0}{k} = \varepsilon_0\left(\frac{1}{l} + \frac{2}{k}\right) \tag{5.13}$$

if $t \in [u_n, u_n + \kappa]$.

We obtain from (5.8)–(5.11) that

$$|\omega_j(t_n + t) - \omega_j(t)| \geqslant \int_{u_n}^{t} |f_j(t_n + s, \omega(t_n + s)) - f_j(s, \omega(t_n + s))| ds -$$

$$\int_{u_n}^{t} |f_j(s, \omega(t_n + s)) - f_j(s, \omega(s))| ds - \int_{u_n}^{t} |\sum_{j=1}^{p} a_{ji}[\omega_i(t_n + s) - \omega_i(s)]| ds -$$

$$|\omega_i(t_n + u_n) - \omega_i(u_n)| \geqslant \frac{\kappa}{2}\varepsilon_0 - \kappa L \varepsilon_0\left(\frac{1}{l} + \frac{2}{k}\right) - \kappa\|A\|\varepsilon_0\left(\frac{1}{l} + \frac{2}{k}\right) - \frac{\varepsilon_0}{l} \geqslant \frac{\varepsilon_0}{2l}$$

for $t \in [u_n + \kappa/2, u_n + \kappa]$.

Thus, the solution $\omega(t)$ is alpha unpredictable. The theorem is proved.

5.1.3 Alpha unpredictable solutions

We have considered the problem of existence and uniqueness of strongly alpha unpredictable solutions in the last subsection. In what follows, we will search for quasi-linear systems with solutions that are not strongly alpha unpredictable. For this reason, assume that the following condition is valid.

(C5) The function $f(t, x)$ is alpha unpredictable in the sense of definition 5.1.2.

Theorem 5.1.2. [24] *Suppose that the conditions (C2)–(C5) hold. Then the system (5.1) admits a unique exponentially stable alpha unpredictable solution.*

Proof. One can easily see, proceeding in the way of the last theorem, that there exists a unique solution $\omega(t) \in U$ for system (5.1). The solution is asymptotically stable. What is left is to show that the alpha unpredictability is valid.

We have that

$$\omega(t + t_n) - \omega(t) = \omega(u_n + t_n) - \omega(u_n) + \int_{u_n}^{t} A[\omega(s + t_n) - \omega(s)] ds$$
$$+ \int_{u_n}^{t} [f(t_n + s, \omega(s + t_n)) - f(s, \omega(s))] ds, \, t \in \mathbb{R}. \tag{5.14}$$

One can find a positive number κ and natural numbers l, k and $j = 1, \ldots, p$, such that

$$\kappa < \sigma, \tag{5.15}$$

$$\|\omega(t + s) - \omega(t)\| < \frac{\varepsilon_0}{k}, \quad t \in \mathbb{R}, \, |s| < \kappa, \tag{5.16}$$

$$|f_j(t_n + u_n + s, x) - f_j(u_n + s, x)| \geq \varepsilon_0/2, \, \|x\| < H, \, |s| < \kappa, \, n \in \mathbb{N}, \tag{5.17}$$

$$\kappa\left(1/2 - (\frac{1}{l} + \frac{2}{k})(L + \|A\|)\right) > \frac{3}{2l}. \tag{5.18}$$

Denote $\Delta = \|\omega(t_n + u_n) - \omega(u_n)\|$ and consider two alternative cases (i) $\Delta \geq \varepsilon_0/l$ and (ii) $\Delta < \varepsilon_0/l$.

(i) For the case $\Delta \geq \varepsilon_0/l$, fix an additional positive number κ_1 sufficiently small for

$$\|\omega(t + s) - \omega(t)\| < \frac{\varepsilon_0}{4l}, \quad t \in \mathbb{R}, \, |s| < \kappa_1.$$

Therefore,

$$\|\omega(t_n + t) - \omega(t)\| \geq \|\omega(t_n + u_n) - \omega(u_n)\| - \|\omega(u_n) - \omega(t)\| -$$
$$\|\omega(t_n + t) - \omega(t_n + u_n)\| \geq \frac{\varepsilon_0}{l} - \frac{\varepsilon_0}{4l} - \frac{\varepsilon_0}{4l} = \frac{\varepsilon}{2l}, \tag{5.19}$$

if $t \in [u_n - \kappa_1, u_n + \kappa_1]$ and $n \in \mathbb{N}$.

(ii) One can find that from (5.16) it follows that

$$\|\omega(t_n + t) - \omega(t)\| < \frac{\varepsilon_0}{l} + \frac{\varepsilon_0}{k} + \frac{\varepsilon_0}{k} = \varepsilon_0\left(\frac{1}{l} + \frac{2}{k}\right) \tag{5.20}$$

if $t \in [u_n, u_n + \kappa]$.

We obtain from (5.15)–(5.18) that

$$|\omega_j(t_n + t) - \omega_j(t)| \geq \int_{u_n}^{t} |f_j(t_n + s, \omega(t_n + s)) - f_j(s, \omega(t_n + s))| ds -$$

$$\int_{u_n}^{t} |f_j(s, \omega(t_n + s)) - f_j(s, \omega(s))| ds - \int_{u_n}^{t} |\sum_{j=1}^{p} a_{ji}[\omega_i(t_n + s) - \omega_i(s)]| ds -$$

$$|\omega_i(t_n + u_n) - \omega_i(u_n)| \geq \frac{\kappa}{2}\varepsilon_0 - \kappa L\varepsilon_0\left(\frac{1}{l} + \frac{2}{k}\right) - \kappa\|A\|\varepsilon_0\left(\frac{1}{l} + \frac{2}{k}\right) - \frac{\varepsilon_0}{l} \geq \frac{\varepsilon_0}{2l}$$

for $t \in [u_n + \kappa/2, u_n + \kappa]$.

Thus, the solution $\omega(t)$ is alpha unpredictable. The theorem is proved.

5.1.4 Examples with numerical simulations

Example 5.1.1. *First of all, we will construct an alpha unpredictable function using the dynamics of the logistic map. Let us take into account the logistic map*

$$\lambda_{i+1} = F_\mu(\lambda_i), \tag{5.21}$$

where $i \in \mathbb{Z}$ and $F_\mu(s) = \mu s(1 - s)$. The interval $[0, 1]$ is invariant under the iterations of (5.21) for $\mu \in (0, 4]$ [47].

It was shown in [6] that the logistic map (5.21) possesses an alpha unpredictable solution for each $\mu \in [3 + (2/3)^{1/2}, 4]$. Let $\{o_i\}$, $i \in \mathbb{Z}$, be the solution if $\mu = 3.92$, and consider the equation

$$\Theta(t) = \int_{-\infty}^{t} e^{-5(t-s)/2} \Omega(s) ds, \tag{5.22}$$

where $\Omega(t)$ is a piece-wise constant function defined on the real axis through the equation $\Omega(t) = \rho_i$ for $t \in [i, i + 1)$, $i \in \mathbb{Z}$. It is worth noting that $\Theta(t)$ is the unique bounded on the whole real axis globally exponentially stable solution of the differential equation

$$v'(t) = -\frac{5}{2} v(t) + \Omega(t).$$

Additionally, one can confirm that the function $\Theta(t)$ is such that

$$\sup_{t \in \mathbb{R}} |\Theta(t)| \leqslant \frac{2}{5},$$

and it is uniformly continuous since its derivative is bounded. Because the sequence $\{\rho_i\}$ is alpha unpredictable, there exist a positive number ε_0 and sequences $\{\zeta_n\}$, $\{\eta_n\}$, both of which diverge to infinity such that $|\rho_{i+\zeta_n} - \rho_i| \to 0$ as $n \to \infty$ for each i in bounded intervals of integers and $|\rho_{\zeta_n + \eta_n} - \rho_{\eta_n}| \geqslant \varepsilon_0$ for each $n \in \mathbb{N}$. Fix an arbitrary positive number ε and arbitrary real numbers α, β with $\beta > \alpha$. Let N be a sufficiently large natural number satisfying

$$N \geqslant \frac{2}{5} \ln\left(\frac{6}{5\varepsilon}\right).$$

There exists a natural number n_0 such that for each $n \geqslant n_0$ the inequality

$$|\rho_{i+\zeta_n} - \rho_i| < \frac{5\varepsilon}{6}$$

holds for $i = \lfloor \alpha \rfloor - N, \lfloor \alpha \rfloor - N + 1, ..., \lfloor \beta \rfloor$, where $\lfloor \alpha \rfloor$ and $\lfloor \beta \rfloor$ respectively denote the largest integers which are not greater than α and β. Accordingly, if $n \geqslant n_0$, then we have

$$|\Omega(t + \zeta_n) - \Omega(t)| < \frac{5\varepsilon}{6} \tag{5.23}$$

for $t \in [\lfloor \alpha \rfloor - N, \lfloor \beta \rfloor + 1)$. The last inequality implies that $|\Theta(t + \zeta_n) - \Theta(t)| < \varepsilon$ for $t \in [\lfloor \alpha \rfloor, \lfloor \beta \rfloor + 1]$. Hence, $|\Theta(t + \zeta_n) - \Theta(t)| \to 0$ as $n \to \infty$ uniformly on the interval $[\alpha, \beta]$.

On the other hand, one can confirm for each $n \in \mathbb{N}$ that $|\Omega(t + \zeta_n) - \Omega(t)| \geqslant \varepsilon_0$ if $t \in [\eta_n, \eta_n + 1)$. For fixed $n \in \mathbb{N}$, using the equation

$$\Theta(t + \zeta_n) - \Theta(t) = \Theta(\zeta_n + \eta_n) - \Theta(\eta_n) - \frac{5}{2} \int_{\eta_n}^{t} (\Theta(s + \zeta_n) - \Theta(s))ds$$

$$+ \int_{\eta_n}^{t} (\Omega(s + \zeta_n) - \Omega(s))ds$$

we attain that

$$|\Theta(\zeta_n + \eta_n + 1) - \Theta(\eta_n + 1)| \geqslant \left| \int_{\eta_n}^{\eta_n + 1} (\Omega(s + \zeta_n) - \Omega(s))ds \right| - |\Theta(\zeta_n + \eta_n) - \Theta(\eta_n)|$$

$$- \frac{5}{2} \left| \int_{\eta_n}^{\eta_n + 1} (\Theta(s + \zeta_n) - \Theta(s))ds \right|.$$

Therefore, one can verify that

$$\sup_{t \in [\eta_n, \eta_n + 1]} |\Theta(t + \zeta_n) - \Theta(t)| \geqslant \frac{2\varepsilon_0}{9}.$$

Thus, there exists a sequence $\{u_n\}$ with $\eta_n \leqslant u_n \leqslant \eta_n + 1$, $n \in \mathbb{N}$, such that

$$|\Theta(u_n + \zeta_n) - \Theta(u_n)| \geqslant \frac{2\varepsilon_0}{9}.$$

For $t \in [u_n - \delta, u_n + \delta]$, where $\delta = \varepsilon_0/36$, we have

$$|\Theta(t + \zeta_n) - \Theta(t)| \geqslant |\Theta(u_n + \zeta_n) - \Theta(u_n)| - \frac{5}{2} \left| \int_{u_n}^{t} |\Theta(s + \zeta_n) - \Theta(s)|ds \right|$$

$$- \left| \int_{u_n}^{t} |\Omega(s + \zeta_n) - \Omega(s)|ds \right|$$

$$\geqslant \frac{\varepsilon_0}{9} = \bar{\varepsilon}_0.$$

It is clear that $u_n \to \infty$ as $n \to \infty$. Thus, the function $\Theta(t)$ is alpha unpredictable.

Example 5.1.2. [19] *Consider the system*

$$\begin{aligned} x_1' &= -2x_1 + 2x_2 - 50\Theta(t) \\ x_2' &= x_1 - 3x_2 + 5\Theta^3(t), \end{aligned} \tag{5.24}$$

where $\Theta(t)$ is the alpha unpredictable function defined by (5.22). The eigenvalues of the matrix of coefficients are -2 and -0.5. One can confirm that the perturbation function $(-50\Theta(t), 5\Theta^3(t))$ is alpha unpredictable by lemma 5.1.1. By the main result, there exists an asymptotically stable alpha unpredictable solution $(\varphi_1(t), \varphi_2(t))$ of system (5.24). Consequently, any solution of the equation behaves irregularly, ultimately being near $(\varphi_1(t), \varphi_2(t))$. This is seen from simulation of the solution with $x_1(0) = 0, 18, x_2(0) = 0, 01$ in figure 5.1.

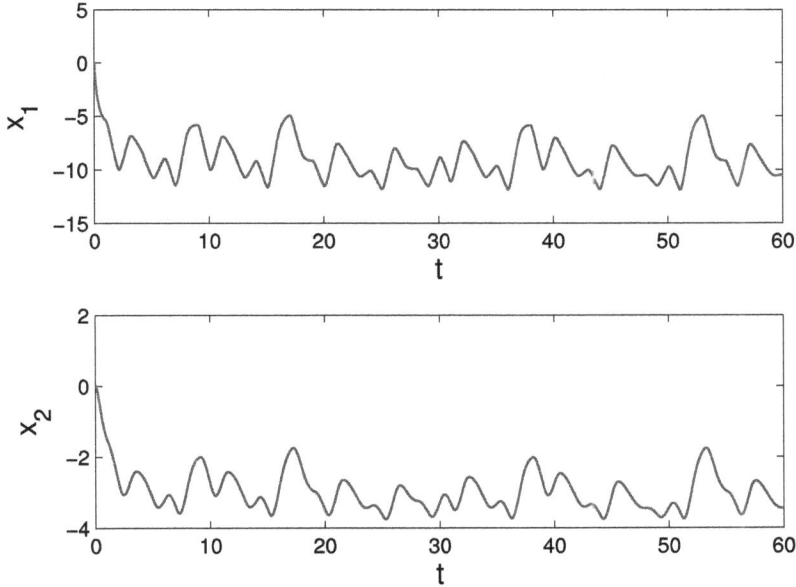

Figure 5.1. The time series of the x_1 and x_2 coordinates of system (5.24) with the initial conditions $x_1(0) = 0, 18$, $x_2(0) = 0, 01$. The figure manifests irregular behavior of the solution. Reproduced with permission from [10].

Example 5.1.3. [24] *Consider the function* $g(t, x) = (arctan(x) + 2)\Theta(t)$ *of two variables* t *and* x*, where* $\Theta(t)$ *is the function from example 5.1.1. It is easy to see that function* $g(t, x)$ *is continuously differentiable if* $t \neq i, i \in \mathbb{Z}$*, and bounded such that* $\sup_{\mathbb{R} \times G} |g(t, x)| = \frac{\pi}{4} + 1$*. Moreover,* $\sup_{\mathbb{R} \times G} |\frac{\partial g(t,x)}{\partial x}| = 1/2, t \neq i, i \in \mathbb{Z}$*.*

Let us fix an arbitrary compact interval $I \subset \mathbb{R}$ *and positive number* ε*. We have that* $|\Theta(t + t_n) - \Theta(t)| < \varepsilon$ *for* $t \in I$ *and sufficiently large* n*. Consequently,*

$$|g(t + t_n, x) - g(t, x)| \leqslant |arctan(x) + 2||\Theta(t + t_n) - \Theta(t)| < (\frac{\pi}{2} + 2)\varepsilon.$$

That is, $g(t + t_n, x) \to g(t, x)$ *as* $n \to \infty$ *uniformly for* $(t, x) \in I \times G$*.*

On the other hand, it is true that $|\Theta(t + t_n) - \Theta(t)| \geqslant \bar{\varepsilon}_0$ *for all* $t \in [u_n - \kappa, u_n + \kappa]$ *and* $n \in \mathbb{N}$*. This is why we obtain that*

$$|g(t + t_n, x) - g(t, x)| = |arctan(x) + 2||\Theta(t + \zeta_n) - \Theta(t)| \geqslant (-\frac{\pi}{2} + 2)\bar{\varepsilon}_0,$$

for $(t, x) \in [u_n - \kappa, u_n + \kappa] \times G, n \in \mathbb{N}$*. Thus,* $g(t, x)$ *is a strongly alpha unpredictable in* t *function.*

Example 5.1.4. [24] *Let us consider the system of differential equations*

$$\begin{aligned} x_1' &= -3x_1 - x_2 - x_3 + 0.51g(t, x_3) \\ x_2' &= -x_1 - 3x_2 - x_3 - 0.62g(t, x_1) \\ x_3' &= x_1 + x_2 - x_3 + 0.51g(t, x_2), \end{aligned} \qquad (5.25)$$

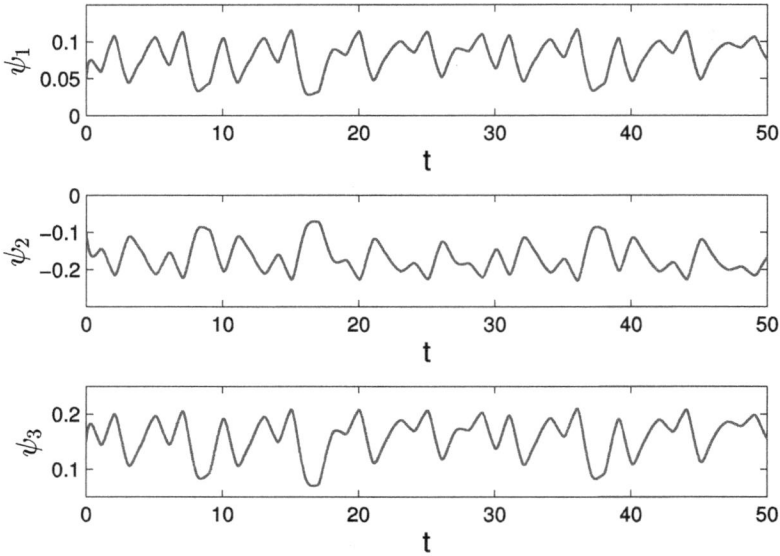

Figure 5.2. The coordinates of the solution, $\psi(t)$, of system (5.25). Reproduced with permission [24].

where $g(t, x)$ is the function from the last example. The eigenvalues of the matrix of coefficients are equal to -2, -2 and -3. One can find that conditions (C3)–(C6) are valid for system (5.25) with $\gamma = -2$, $K = 6$ and $L = 0.31$. According to theorem 5.1.2, there exists the unique asymptotically stable alpha unpredictable solution of system (5.25). The simulation results for the solution of (5.25) with initial data $\psi_1(0) = 0.05$, $\psi_2(0) = -0.1$, $\psi_3(0) = 0.15$ are seen in figure 5.2. They confirm the irregularity of the dynamics.

5.2 Compartmental quasi-linear equations

Many scientists have devoted their research to studying the behavior of recurrent solutions. The most studied are periodic [40, 44] and almost periodic solutions [42, 45, 48]. The Poisson stable motions, which were introduced by H. Poincaré [51], represent the most challenging type of analysis of recurrence [52]. In our papers, *the method of included intervals* has been introduced and developed to prove Poisson stability in differential equations and neural networks [4, 24, 27, 28].

In paper, [6], the Poisson stability has been extended to the alpha unpredictability to make the recurrence a chaotic ingredient. Thus, the ultra Poincaré chaos was determined, and one can say now that *the alpha unpredictability implies chaos* [2, 5]. The concept has been applied for various differential equations problems [24] and neural networks [2, 23, 26]. It is a powerful instrument for indicating chaos.

Unlike paper [24] and section 5.1, where quasi-linear differential equations with constant coefficients and alpha unpredictable perturbations were investigated, in the present section, we consider differential equations with time-varying compartmental coefficients and perturbations. The newly introduced functions combine the properties of periodicity and alpha unpredictability. To prove that the functions are alpha unpredictable, a particular condition of synchrony for characteristics, the kappa property, is introduced, which has not been considered in the literature before.

The rest of the section is organized as follows. In section 5.2.1, the main definitions of the alpha unpredictable and compartmental periodic alpha unpredictable functions (simply compartmental functions, in what follows) are presented, and preliminary results concerning the properties of such functions are provided. The main result of the study is given in section 5.2.2. Under certain conditions, it is rigorously proven that an alpha unpredictable solution, which is asymptotically stable, takes place in the quasi-linear differential equations with compartmental perturbations. In section 5.2.3, examples supporting theoretical results are given. A compartmental function and alpha unpredictable solutions of differential equations are presented. Moreover, in section 5.2.3, the degree of periodicity—a quantitative characteristic that strongly affects periodical and the alpha unpredictable appearance in the dynamics—is considered.

5.2.1 Definitions and auxiliaries

Definition 5.2.1. [2] *A uniformly continuous and bounded function $f: \mathbb{R} \to \mathbb{R}^n$ is alpha unpredictable if there exist positive numbers ε_0, δ and sequences t_k, s_k, both of which diverge to infinity such that $\|f(t + t_k) - f(t)\| \to 0$ as $k \to \infty$ uniformly on compact subsets of $\mathbb{R}$ and $\|f(t + t_k) - f(t)\| > \varepsilon_0$ for each $t \in [s_k - \delta, s_k + \delta]$ and $k \in \mathbb{N}$.*

The sequence t_k, $k = 1, 2, \ldots$ is said to be the *Poisson* or *convergence sequence* of the function $f(t)$. We call the uniform convergence on compact subsets of $\mathbb{R}$, *the convergence property*, and the existence of the sequence s_k and positive numbers ε_0, δ is called *the separation property*.

Definition 5.2.2. [24] *A continuous and bounded function $f(t, x): \mathbb{R} \times D \to \mathbb{R}^n$, $D \subset \mathbb{R}^n$ is an open and bounded set, is alpha unpredictable in t uniformly with respect to $x \in D$ if it is uniformly continuous in t and there exist positive numbers ε_0, δ and sequences t_k, s_k, both of which diverge to infinity such that $\sup_D \|f(t + t_k, x) - f(t, x)\| \to 0$ as $k \to \infty$ uniformly on bounded intervals of t and $x \in D$, and $\|f(t + t_k, x) - f(t, x)\| > \varepsilon_0$ for $t \in [s_k - \delta, s_k + \delta]$, $x \in D$ and $k \in \mathbb{N}$.*

Definition 5.2.3. *A function $f(t): \mathbb{R} \to \mathbb{R}^n$ is said to be a compartmental periodic alpha unpredictable (CPU) function if $f(t) = G(t, t)$, where $G(u, s)$ is a continuous bounded function, periodic in u uniformly with respect to s, and alpha unpredictable in s uniformly with respect to u.*

The next definition is a particular case of the definition 5.2.3.

Definition 5.2.4. *A function* $f(t) = \phi(t) + \psi(t)$ *is said to be the modulo periodic alpha unpredictable (MPU) function if* $\phi(t)$ *is an* ω-*periodic continuous function and* $\psi(t)$ *is an alpha unpredictable function.*

Definition 5.2.5. *A function* $f(t, x): \mathbb{R} \times D \rightarrow \mathbb{R}^n, D \subset \mathbb{R}^n$ *is an open and bounded set, is said to be compartmental periodic alpha unpredictable in t uniformly for x function, if* $f(t, x) = G(t, t, x)$, *where* $G(u, s, x)$ *is a continuous bounded function, periodic in u uniformly with respect to arguments s, x, and alpha unpredictable in s uniformly with respect to arguments u and x.*

Consider an alpha unpredictable function $\psi(t)$ with convergence sequence t_k. For fixed $\omega > 0$ there exist a subsequence t_{k_l} and a number τ_ω such that $t_{k_l} \rightarrow \tau_\omega (mod \ \omega)$ as $l \rightarrow \infty$. In what follows, we shall call the number τ_ω *Poisson shift* for function $\psi(t)$ with respect to ω. The set of Poisson shifts $\mathcal{T}_\omega$ is not empty, it can consist of several or even an infinite number of elements. The number $\kappa_\omega = inf \ \mathcal{T}_\omega, 0 \leqslant \kappa_\omega < \omega$ is said to be the *Poisson number for function* $\psi(t)$ *with respect to* ω. We say that the sequence t_k satisfies the *kappa property with respect to* ω if $\kappa_\omega = 0$.

Remark 5.2.1. A compartmental function is not necessarily alpha unpredictable. It is in the sense of the definition 5.2.1 under a special condition, for example, if its convergence sequence satisfies the kappa property.

Lemma 5.2.1. *Assume that* $G(u, v, x): \mathbb{R} \times \mathbb{R} \times D \rightarrow \mathbb{R}^n, D \subset \mathbb{R}^n$ *is an open and bounded set, is a continuous* ω-*periodic in u uniformly with respect to v and x function. Then the function* $g(t, x) = G(t, t, x)$, *is alpha unpredictable in t, if the following conditions are valid:*

(i) *for each* $\varepsilon > 0$ *there exists a positive number* η *such that* $\|G(t + s, t, x) - G(t, t, x)\| < \varepsilon$ *if* $|s| < \eta, t \in \mathbb{R}, x \in D$; *there exist sequences* t_k, s_k, *both of which diverge to infinity as* $k \rightarrow \infty$, *and positive numbers* ε_0, δ *such that*

(ii) *the sequence* t_k *satisfies the kappa property with respect to the period* ω;

(iii) $\sup_{I \times D} \|G(t, t + t_k, x) - G(t, t, x)\| \rightarrow 0$ *on each bounded interval* $I \subset \mathbb{R}$;

(iv) $\inf_{[s_k - \delta, s_k + \delta] \times D} \|G(t, t + t_k, x) - G(t, t, x)\| > \varepsilon_0, k \in \mathbb{N}$.

Proof. Let us fix a positive number ε, and a bounded interval $I \in \mathbb{R}$. Since the sequence t_k satisfies the kappa property, one can write, without loss of generality, that $t_k \rightarrow 0(mod \ \omega)$ as $k \rightarrow \infty$. Therefore, by conditions (i) and (iii), the following inequalities are valid

$$\sup_{\mathbb{R} \times D} \|G(t + t_k, t, x) - G(t, t, x)\| < \frac{\varepsilon}{2} \tag{5.26}$$

and

$$\sup_{I \times D} \| G(t, t + t_k, x) - G(t, t, x) \| < \frac{\varepsilon}{2}, \tag{5.27}$$

for sufficiently large k.

Using inequalities (5.26) and (5.27), we get that

$$\| g(t + t_k, x) - g(t, x) \| = \| G(t + t_k, t + t_k, x) - G(t, t, x) \| \leq$$
$$\| G(t + t_k, t + t_k, x) - G(t, t + t_k, x) \| + \| G(t, t + t_k, x) - G(t, t, x) \| <$$
$$\frac{\varepsilon}{2} + \frac{\varepsilon}{2} = \varepsilon,$$

for all $t \in I$, $x \in D$. That is, $g(t + t_k, x)$ converges to $g(t, x)$ on each arbitrary bounded time interval uniformly for $x \in D$. Moreover, conditions (*i*) and (*ii*) imply that $\sup_{\mathbb{R} \times D} \| G(t + t_k, t, x) - G(t, t, x) \| < \frac{\varepsilon_0}{2}$ for sufficiently large k. Applying assumption (*iv*), one can obtain that

$$\| g(t + t_k, x) - g(t, x) \| = \| G(t + t_k, t + t_k, x) - G(t, t, x) \| \geq$$
$$\| G(t + t_k, t + t_k, x) - G(t + t_k, t, x) \| - \| G(t + t_k, t, x) - G(t, t, x) \| >$$
$$\varepsilon_0 - \frac{\varepsilon_0}{2} = \frac{\varepsilon_0}{2},$$

for all $t \in [s_k - \delta, s_k + \delta]$, $x \in D$, $k \in \mathbb{N}$. The lemma is proved.

Along with lemma 5.2.1, the next assertion, which can be proved in the similar way, will be needed in what follows.

Lemma 5.2.2. *A compartmental periodic alpha unpredictable in t function $g(t, x)$ is alpha unpredictable in t provided that its convergence sequence t_k admits the kappa property.*

5.2.2 The existence theorem

The main object of the present subsection is the system of quasi-linear differential equations,

$$x'(t) = (A(t) + B(t))x + g(t, x), \tag{5.28}$$

where $t \in \mathbb{R}$, $x \in \mathbb{R}^n$, n is a fixed natural number; $A(t)$ and $B(t)$ are n-dimensional square matrices; $g \colon \mathbb{R} \times D \to \mathbb{R}^n$, $D = \{x \in \mathbb{R}^n, \|x\| < H\}$, where H is a fixed positive number.

We assume that the following conditions are satisfied.

(C1) $A(t)$ is an ω-periodic matrix for a fixed positive ω;

(C2) $B(t)$ is an alpha unpredictable matrix, and $g(t, x)$ is compartmental function;

(C3) the entries of the matrix $B(t)$ and function $g(t, x)$ are with common convergence sequence t_k;

(C4) the convergence sequence t_k satisfies the kappa property with respect to the period ω.

Let us consider the homogeneous system associated with (5.28):

$$x'(t) = A(t)x(t). \tag{5.29}$$

Let $X(t)$, $t \in \mathbb{R}$, be the fundamental matrix of the system (5.29) such that $X(0) = I$, and I is the $n \times n$ identical matrix. Moreover, $X(t, s)$ is a transition matrix of the system (5.29), which is equal to $X(t)X^{-1}(s)$, and $X(t + \omega, s + \omega) = X(t, s)$ for all $t, s \in \mathbb{R}$.

Assume that the following assumption is valid.

(C5) All multipliers of the system (5.29) in modulus are less than one.

It follows from the last condition that there exist positive numbers $K \geqslant 1$ and α such that

$$\|X(t, s)\| \leqslant Ke^{-\alpha(t-s)}, \tag{5.30}$$

for $t \geqslant s$ [43].

The next lemma is necessary for further reasoning.

Lemma 5.2.3. [27] *If the inequality (5.30) is satisfied, then the following estimation is correct:*

$$\|X(t + \tau, s + \tau) - X(t, s)\| \leqslant \max_{t \in \mathbb{R}} \|A(t + \tau) - A(t)\| \frac{2K^2}{\alpha^2 e} e^{-\frac{\alpha}{2}(t-s)}, \tag{5.31}$$

for $t \geqslant s$ and arbitrary real number τ.

The additional conditions on system (5.28) are required:

(C6) there exists a positive constant L such that $\|g(t, x_1) - g(t, x_2)\| \leqslant L\|x_1 - x_2\|$ for all $t \in \mathbb{R}$, $x_1, x_2 \in D$;

(C7) $K(b + \frac{m_g}{H}) < \alpha$;

(C8) $K(b + L) < \alpha$,

where $b = \sup_{t \in \mathbb{R}} \|B(t)\|$ and $m_g = \sup_{\mathbb{R} \times D} \|g(t, x)\|$.

According to [43], a bounded on the real axis function $y(t)$ is a solution of (5.28), if and only if it satisfies the equation

$$y(t) = \int_{-\infty}^{t} X(t, s)[B(s)y(s) + g(s, y(s))]ds, \ t \in \mathbb{R}. \tag{5.32}$$

Theorem 5.2.1. *If conditions (C1)–(C8) are valid, then the system (5.28) possesses a unique asymptotically stable alpha unpredictable solution.*

Proof. Let t_k be the convergence sequence of the function $g(t, x)$ in the system (5.28). We denote by $\mathscr{B}$ the set of functions $\psi(t) = \mathbb{R} \to \mathbb{R}^n$, which satisfy the convergence property with common sequence t_k, and $\|\psi\|_1 < H$.

Let us show that $\mathscr{B}$ is a complete space. Consider a Cauchy sequence $\theta_m(t)$ in $\mathscr{B}$, which converges to a limit function $\theta(t)$ on $\mathbb{R}$. We have that

$$\|\theta(t + t_k) - \theta(t)\| < \|\theta(t + t_k) - \theta_m(t + t_k)\| + \|\theta_m(t + t_k) - \theta_m(t)\| + \|\theta_m(t) - \theta(t)\|. \tag{5.33}$$

for a fixed closed and bounded interval $I \subset \mathbb{R}$. Now, one can take sufficiently large m and k such that each term on the right hand-side of (5.33) is smaller than $\frac{\varepsilon}{3}$ for a fixed positive ε and $t \in I$. That is, the sequence $\theta(t + t_k)$ uniformly converges to $\theta(t)$ on I. Likewise, one can check that the limit function is uniformly continuous [43]. The completeness of $\mathscr{B}$ is shown.

Define the operator Π on $\mathscr{B}$ such that

$$\Pi\nu(t) = \int_{-\infty}^{t} X(t, s)[B(s)\psi(s) + g(s, \psi(s))]ds, \; t \in \mathbb{R}. \tag{5.34}$$

Fix a function $\psi(t)$ that belongs to $\mathscr{B}$. We have that

$$\|\Pi\psi(t)\| \leqslant \int_{-\infty}^{t} \|X(t, s)\|(\|B(s)\|\|\psi(s)\| + \|g(s, \psi(s))\|)ds \leqslant \frac{K(bH + m_g)}{\alpha}$$

for all $t \in \mathbb{R}$. Therefore, by the condition (C7) it is true that $\|\Pi\psi\|_1 < H$.

Now, applying the method of included intervals [24], we will show that $\Pi\psi(t + t_k) \to \Pi\psi(t)$ as $k \to \infty$, uniformly on compact subsets of $\mathbb{R}$. Let us fix a positive number ε and an interval $[a, b]$, $-\infty < a < b < \infty$. There exist numbers c, ξ such that $c < a$ and $\xi > 0$, which satisfy the following inequalities:

$$\frac{4K^2\xi}{\alpha^3 e}(bH + m_g) < \frac{\varepsilon}{3}, \tag{5.35}$$

$$\frac{2K}{\alpha}(2bH + m_g)e^{-\alpha(a-c)} < \frac{\varepsilon}{3}, \tag{5.36}$$

and

$$\frac{K(H + b + L)\xi}{\alpha}[1 - e^{-\alpha(b-c)}] < \frac{\varepsilon}{3}. \tag{5.37}$$

By using the condition (C4), for periodic matrix $A(t)$ one can attain that $\|A(t + t_k) - A(t)\| < \xi, t \in \mathbb{R}$. Moreover, since $\psi(t)$ belongs to the set $\mathscr{B}$, the matrix $B(t)$ is alpha unpredictable and $g(t, x)$ is alpha unpredictable in t then for sufficiently large k we have that $\|\psi(t + t_k) - \psi(t)\| < \xi$, $\|B(t + t_k) - B(t)\| < \xi$, and $\|g(t + t_k, \psi(t)) - g(t, \psi(t))\| < \xi$ for $t \in [c, b]$. Applying lemma 5.2.3, we get that

$$\|\Pi\psi(t + t_k) - \Pi\psi(t)\| =$$

$$\left\| \int_{-\infty}^{t} X(t + t_k, s + t_k)(B(s + t_k)\psi(s + t_k) + g(s + t_k, \psi(s + t_k)))ds - \right.$$

$$\left. \int_{-\infty}^{t} X(t, s)(B(s)\psi(s) + g(s, \psi(s)))ds \right\| \leqslant$$

$$\int_{-\infty}^{t} \|X(t + t_k, s + t_k) - X(t, s)\|(\|B(s + t_k)\|\|\psi(s + t_k)\| + \|g(s + t_k, \psi(s + t_k))\|)ds +$$

$$\int_{-\infty}^{t} \|X(t, s)\|(\|B(s + t_k) - B(s)\|\|\psi(s + t_k)\| + \|B(s)\|\|\psi(s + t_k) - \psi(s)\| +$$

$$\|g(s + t_k, \psi(s + t_k)) - g(s, \psi(s))\|)ds \leqslant$$

$$\int_{-\infty}^{t} \|X(t + t_k, s + t_k) - X(t, s)\|(\|B(s + t_k)\|\|\psi(s + t_k)\| + \|g(s + t_k, \psi(s + t_k))\|)ds +$$

$$\int_{-\infty}^{c} \|X(t, s)\|(\|B(s + t_k) - B(s)\|\|\psi(s + t_k)\| + \|B(s)\|\|\psi(s + t_k) - \psi(s)\| +$$

$$\|g(s + t_k, \psi(s + t_k)) - g(s, \psi(s))\|)ds \leqslant$$

$$\int_{c}^{t} \|X(t, s)\|(\|B(s + t_k) - B(s)\|\|\psi(s + t_k)\| + \|B(s)\|\|\psi(s + t_k) - \psi(s)\| +$$

$$\|g(s + t_k, \psi(s + t_k)) - g(s, \psi(s))\|)ds \leqslant \int_{-\infty}^{t} \frac{2K^2\xi}{\alpha^2 e} e^{-\frac{\alpha}{2}(t-s)}(bH + m_g)ds +$$

$$\int_{-\infty}^{c} Ke^{-\alpha(t-s)}(2bH + 2bH + 2m_g)ds + \int_{c}^{t} Ke^{-\alpha(t-s)}(\xi H + b\xi + L\xi)ds \leqslant$$

$$\frac{4K\xi}{\alpha^3 e}(bH + m_g) + \frac{2K}{\alpha}(2bH + m_g)e^{-\alpha(a-c)} + \frac{K(H + b + L)\xi}{\alpha}[1 - e^{-\alpha(b-c)}],$$

for all $t \in [a, b]$. From inequalities (5.35) to (5.37) it follows that $\|\Pi\psi(t + t_k) - \Pi\psi(t)\| < \varepsilon$ for $t \in [a, b]$. Therefore, $\Pi\psi(t + t_k)$ uniformly converges to $\Pi\psi(t)$ on a bounded interval of $\mathbb{R}$.

It is easy to verify that $\Pi\psi(t)$ is a uniformly continuous function, since its derivative is a uniformly bounded function on the real axis. Summarizing the above discussion, the set $\mathscr{B}$ is invariant for operator Π.

We proceed to show that operator $\Pi: \mathscr{B} \to \mathscr{B}$ is contractive. Let $u(t)$ and $v(t)$ be members of $\mathscr{B}$. Then, we obtain that

$$\|\Pi u(t) - \Pi v(t)\| \leqslant \int_{-\infty}^{t} \|X(t, s)\|(\|B(s)\|\|u(s) - v(s)\| + g(s, u(s)) - g(s, v(s))\|)ds \leqslant$$

$$\int_{-\infty}^{t} Ke^{-\alpha(t-s)}(b + L)\|u(s) - v(s)\|ds \leqslant \frac{K(b + L)}{\alpha}\|u(t) - v(t)\|_1,$$

for all $t \in \mathbb{R}$. Therefore, the inequality $\|\Pi u - \Pi v\|_1 \leqslant \frac{K(b+L)}{\alpha}\|u - v\|_1$ holds, and according to the condition $(C8)$ the operator $\Pi: \mathscr{B} \to \mathscr{B}$ is a contraction.

By the contraction mapping theorem, there exists the unique fixed point $z(t) \in \mathscr{B}$ of the operator Π, which is the unique bounded solution of the system (5.28).

Next, we will show that the solution $z(t)$ is alpha unpredictable. According to condition (C2) and lemma 5.2.2, the function $g(t, z(t))$ is alpha unpredictable in t, and there exists positive numbers δ and ε_0 such that $\|g(t + t_k, z(t)) - g(t, z(t))\| > \varepsilon_0$ for $t \in [s_k - \delta, s_k + \delta]$. Denote

$K_0 = \inf\{\|X(t, s): t > s\}\| > 0$. The number exists because of the periodicity. One can find numbers $l \in \mathbb{N}$ and $\delta_1 > 0$ which satisfy the following inequalities:

$$\delta_1 < \delta, \tag{5.38}$$

$$\|X(t + t_k, s + t_k) - X(t, s)\| \leqslant \frac{\varepsilon_0}{l}, t \in \mathbb{R}, |s| < \delta_1, \tag{5.39}$$

$$\delta_1(K_0 - \frac{bH + m_g}{l}) > \frac{1}{2l}\left(3 + \frac{K(H + b + L)}{2\alpha}\right), \tag{5.40}$$

$$\|z(t + s) - z(t)\| \leqslant \frac{\varepsilon_0}{4l}, t \in \mathbb{R}, |s| < \delta_1, \tag{5.41}$$

$$\|B(t + s) - B(t)\| \leqslant \frac{\varepsilon_0}{4l}, t \in \mathbb{R}, |s| < \delta_1, \tag{5.42}$$

Assume that the numbers l, δ_1 and $k \in \mathbb{N}$ are fixed. Denote by Δ the value of $\|z(s_k + t_k) - z(s_k)\|$, and consider two alternative cases: (1) $\Delta \geqslant \frac{\varepsilon_0}{l}$, and (2) $\Delta < \frac{\varepsilon_0}{l}$.

(1) If $\Delta \geqslant \frac{\varepsilon_0}{l}$, it is easily found from (5.41), that

$$\|z(t + t_k) - z(t)\| \geqslant \|z(s_k + t_k) - z(s_k)\| - \|z(s_k) - z(t)\| -$$

$$\|z(t + t_k) - z(s_k + t_k)\| > \frac{\varepsilon_0}{l} - \frac{\varepsilon_0}{4l} - \frac{\varepsilon_0}{4l} = \frac{\varepsilon_0}{2l},$$

for $t \in [s_k - \delta_1, s_k + \delta_1]$.

(2) Applying the relation

$z(t + t_k) - z(t) = z(s_k + t_k) - z(s_k) +$

$\int_{s_k}^{t} X(t + t_k, s + t_k)(B(s + t_k)z(s + t_k) + g(s + t_k, z(s + t_k)))ds -$

$\int_{s_k}^{t} X(t, s)(B(s)z(s) + g(s, z(s)))ds = z(s_k + t_k) - z(s_k) +$

$\int_{s_k}^{t} X(t + t_k, s + t_k)[(B(s + t_k) - B(s))z(s + t_k)) + B(s)(z(s + t_k) - z(s))]ds +$

$\int_{s_k}^{t} [X(t + t_k, s + t_k) - X(t, s)]B(s)z(s)ds +$

$\int_{s_k}^{t} X(t + t_k, s + t_k)[g(s + t_k, z(s + t_k)) - g(s, z(s + t_k))]ds +$

$\int_{s_k}^{t} X(t + t_k, s + t_k)[g(s, z(s + t_k)) - g(s, z(s))]ds +$

$\int_{s_k}^{t} [X(t + t_k, s + t_k) - X(t, s)]g(s, z(s))ds,$

and inequalities (5.38)–(5.42), one can obtain that

$$\|z(t+t_k) - z(t)\| \geq \int_{s_k}^{t} \|X(t+t_k, s+t_k)\| \|g(s+t_k, z(s+t_k)) - g(s, z(s+t_k))\| ds -$$

$$\int_{s_k}^{t} \|X(t+t_k, s+t_k)\| [\|B(s+t_k) - B(s)\| \|z(s+t_k)\| + \|B(s)\| \|z(s+t_k) - z(s)\|] ds -$$

$$\int_{s_k}^{t} \|X(t+t_k, s+t_k) - X(t, s)\| \|B(s)\| \|z(s)\| ds - \|z(s_k+t_k) - z(s_k)\| -$$

$$\int_{s_k}^{t} \|X(t+t_k, s+t_k)\| \|g(s, z(s+t_k)) - g(s, z(s))\| ds -$$

$$\int_{s_k}^{t} \|X(t+t_k, s+t_k) - X(t, s)\| \|g(s, z(s))\| ds \geq \int_{s_k}^{t} K_0 \varepsilon_0 ds -$$

$$\int_{s_k}^{t} K e^{-\alpha(t-s)}(H+b)\frac{\varepsilon_0}{4l} ds - \int_{s_k}^{t} \frac{\varepsilon_0}{l} bH ds - \frac{\varepsilon_0}{l} - \int_{s_k}^{t} K e^{-\alpha(t-s)} L \frac{\varepsilon_0}{4l} ds - \int_{s_k}^{t} \frac{\varepsilon_0}{l} m_g ds \geq$$

$$\delta_1 K_0 \varepsilon_0 - \frac{K(H+b+L)\varepsilon_0}{4\alpha l} - \delta_1 \frac{\varepsilon_0}{l}(bH + m_g) - \frac{\varepsilon_0}{l} > \frac{\varepsilon_0}{2l},$$

for each $t \in [s_k, s_k + \delta_1]$. Thus, the solution $z(t)$ is alpha unpredictable.

Finally, we will study the asymptotic stability of the solution $z(t)$ of the system (5.28). It is true that

$$z(t) = X(t, t_0)z(t_0) + \int_{t_0}^{t} X(t, s)(B(s)z(s) + g(s, z(s)))ds,$$

for $t \geq t_0$.

Let $x(t)$ be another solution of system (5.28). One can write

$$x(t) = X(t, t_0)x(t_0) + \int_{t_0}^{t} X(t, s)(B(s)x(s) + g(s, x(s)))ds.$$

Making use of the relation

$$z(t) - x(t) = X(t, t_0)(z(t_0) - x(t_0)) +$$
$$\int_{t_0}^{t} X(t, s)(B(s)(z(s) - x(s)) + g(s, z(s)) - g(s, x(s)))ds,$$

we obtain that

$$\|z(t) - x(t)\| \leq \|X(t, t_0)\| \|z(t_0) - x(t_0)\| +$$
$$\int_{t_0}^{t} \|X(t, s)\| (\|B(s)\| \|z(s) - x(s)\| + \|g(s, z(s)) - g(s, x(s))\| ds \leq$$
$$K e^{-\alpha(t-t_0)} \|z(t_0) - x(t_0)\| + \int_{t_0}^{t} K(b+L)e^{-\alpha(t-s)} \|z(s) - x(s)\| ds.$$

Now, applying Gronwall-Bellman Lemma, one can attain that

$$\|z(t) - x(t)\| \leq K e^{-(K(b+L))(t-t_0)} \|z(t_0) - x(t_0)\|, \quad t \geq t_0. \tag{5.43}$$

The last inequality and condition (C8) confirm that the alpha unpredictable solution $z(t)$ is asymptotically stable. The theorem is proved.

5.2.3 Examples with numerical simulations

Let ψ_i, $i \in \mathbb{Z}$, be a solution of the logistic map

$$\lambda_{i+1} = \mu\lambda_i(1 - \lambda_i), \tag{5.44}$$

where $i \in \mathbb{Z}$, and $\mu \in [3 + (2/3)^{1/2}, 4]$ is a fixed parameter. The section $[0, 1]$ is invariant with respect to (5.44) for the considered values of μ.

Now, consider the following integral function

$$\Theta(t) = \int_{-\infty}^{t} e^{-3(t-s)}\Omega(s)ds, \tag{5.45}$$

where $\Omega(t)$ is a piecewise constant function defined on the real axis through the equation $\Omega(t) = \psi_i$ for $t \in [i, i + 1)$, $i \in \mathbb{Z}$. It is worth noting that $\Theta(t)$ is bounded on the whole real axis such that $\sup_{t \in \mathbb{R}}|\Theta(t)| \leqslant \frac{1}{3}$. In [2], it was proved that the function $\Theta(t)$ is alpha unpredictable.

In what follows, the piecewise constant function, $\Omega(t)$, will be defined for $t \in [hi, h(i + 1))$, where $i \in \mathbb{Z}$, and h is a positive real number. The number h is said to be *the length of step* of the functions $\Omega(t)$ and $\Theta(t)$. The ratio of the period and the length of step, $\nabla = \omega/h$, we call *the degree of periodicity*. The dependence of the dynamics of the compartmental functions on the degree of periodicity is shown below.

Example 5.2.1. *Before considering samples of alpha-alpha unpredictable systems, let us show how the degree of periodicity can affect shape of graphs trajectory of alpha-alpha unpredictable functions.*

Consider the following modulo periodic alpha unpredictable function

$$f(t) = a \sin\left(\frac{2\pi t}{\omega}\right) + b\Theta(t), \tag{5.46}$$

where a and b are real coefficients, $\Theta(t)$ is the alpha unpredictable function with step h, and the first component is an ω-periodic function. In figure 5.3 the graph of function $f(t)$ with degree of periodicity $\nabla = 0.25 < 1$ is shown. It is seen that periodicity appears locally on intervals of 10 units' length approximately.

Figure 5.4 depicts the graph of alpha unpredictable function $f(t)$ with $\nabla = 1$. In this case, the alpha unpredictability dominates in the dynamics, and the periodicity is not seen.

Figure 5.5 demonstrates the trajectory of the function $f(t)$ with degree of periodicity $\nabla = 60 > 1$. We obtain the function, which admits a periodic shape and is shadowed by alpha unpredictability.

Analyzing simulations with different degrees of periodicity, we can say that there is a loss of global periodicity and irregularity dominates if ∇ is less than or equal to one. If ∇ is larger than one, we have the opposite effect, when periodicity in the trajectory is observed and envelopes the irregularity.

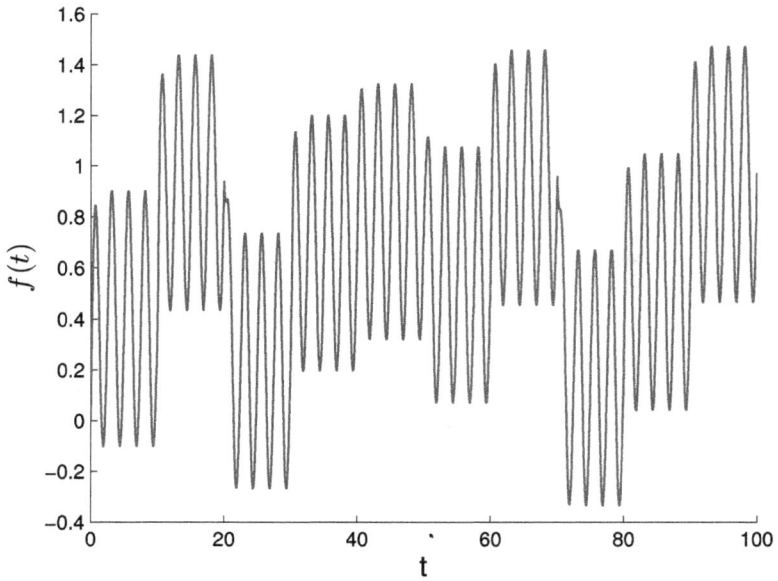

Figure 5.3. The graph of function $f(t)$ with $a = 0.5$, $b = 3$, $\omega = 2.5$, and degree of periodicity $\nabla = 0.25$. Reproduced with permission from [54].

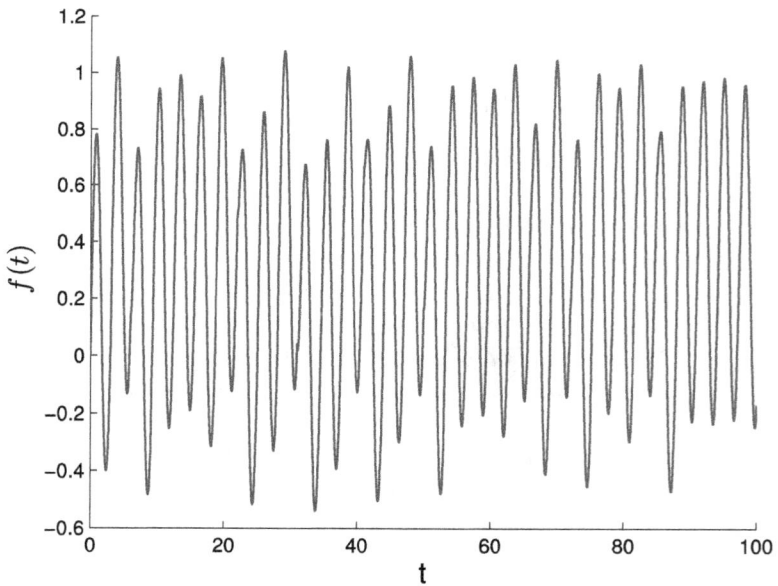

Figure 5.4. The graph of function $f(t)$ with $a = 0.6$, $b = 1.5$, $\omega = \pi$, and degree of periodicity $\nabla = 1$. Reproduced with permission from [54].

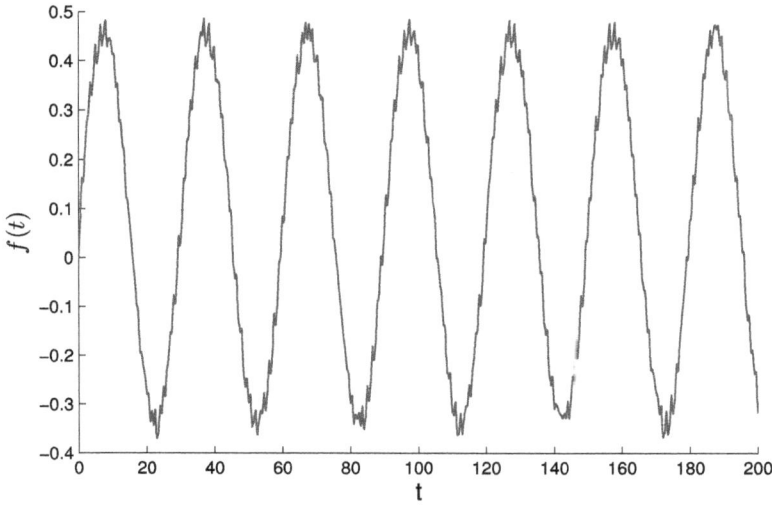

Figure 5.5. The graph of function $f(t)$ with $a = 0.4$, $b = 0.3$, $\omega = 30$, and degree of periodicity $\nabla = 60$. Reproduced with permission from [54].

Example 5.2.2. *Now, let us consider the following alpha unpredictable system,*

$$x_1' = (-3 + \cos(t) + 0.3\Theta(t))x_1 - 0.4\sin(2t) + 0.25arctg(x_2) + 1.2\Theta(t),$$
$$x_2' = (-2 + \sin(2t) - 0.2\Theta(t))x_2 + 0.2\cos(2t) + 0.3arctg(x_1) + 0.9\Theta(t), \quad (5.47)$$

where $\Theta(t)$ is the alpha unpredictable function with the length of step $h = 8\pi$, described above in (5.45),

$$A(t) = \begin{pmatrix} -3 + \cos(t) & 0 \\ 0 & -2 + \sin(2t) \end{pmatrix}, \quad B(t) = \begin{pmatrix} 0.3\Theta(t) & 0 \\ 0 & -0.2\Theta(t) \end{pmatrix},$$

$$g(t, x) = \begin{pmatrix} -0.4\sin(2t) + 0.25arctg(x_2) + 1.2\Theta(t) \\ 0.2\cos(2t) + 0.3arctg(x_1) + 0.9\Theta(t) \end{pmatrix}.$$

The common period of the matrix and the components of $g(t, x)$ is equal to 2π. The degree of periodicity is equal to 0.25. All conditions of the theorem 5.2.1 are satisfied with $L = 0.3$, $K = 1$, $\alpha = 4\pi$, $b = 0.12$, $m_g = 1.54$ and $\tilde{H} = 0.5$. The coordinates of the solution $x(t)$, which asymptotically converges to the alpha unpredictable solution are shown in figure 5.6.

Figure 5.7 demonstrates the time series of the coordinates $x_1(t)$, $x_2(t)$ of the solution $x(t)$ of the system (5.47) with initial values $x_1(0) = 0$, $x_2(0) = 0.4$, when the degree of periodicity $\nabla = 1$.

Now, let us obtain simulation results for the following system,

$$x_1' = (-3 + \cos(t) + 0.3\Theta(t))x_1 - 0.4\sin(0.2t) + 0.25arctg(x_2) + 1.2\Theta(t),$$
$$x_2' = (-2 + \sin(2t) - 0.2\Theta(t))x_2 + 0.2\cos(0.2t) + 0.3arctg(x_1) + 0.9\Theta(t). \quad (5.48)$$

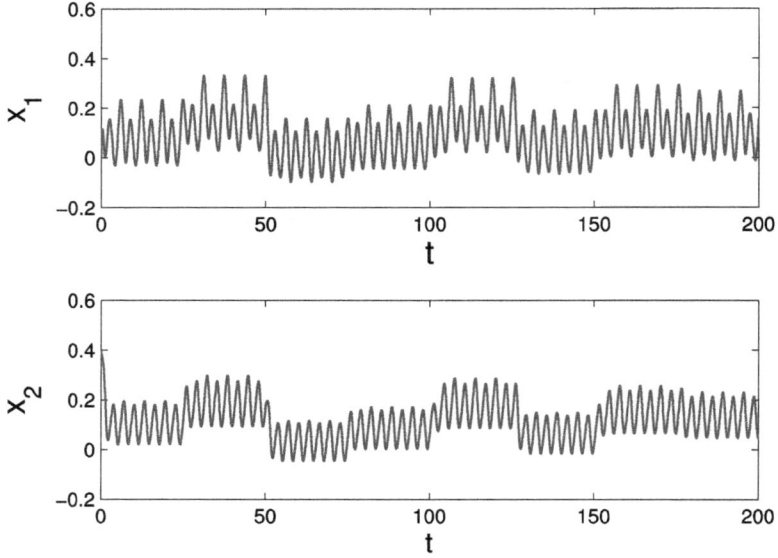

Figure 5.6. The coordinates of the solution $x(t)$ of system (5.47), which asymptotically converge to the coordinates of the alpha unpredictable solution. The degree of periodicity is equal to $\nabla = 0.25$. Reproduced with permission from [54].

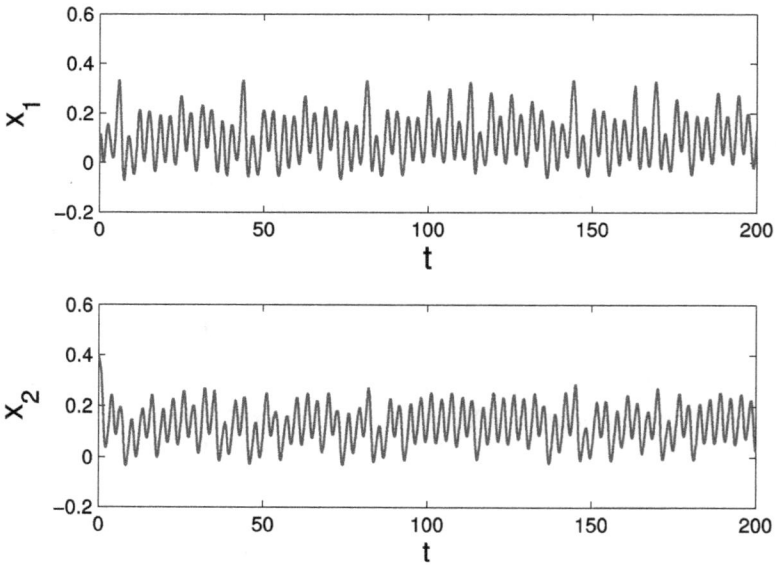

Figure 5.7. The time series of the coordinates $x_1(t)$, $x_2(t)$ of the solution of system (5.47) with $\nabla = 1$, which asymptotically approach the coordinates of the alpha unpredictable solution. Reproduced with permission from [54].

All coefficients are the same as in system (5.47), with the only difference that the common period of the coefficients and perturbations is equal to 10π. The alpha unpredictable function $\Theta(t)$ with the length of step $h = 0.1\pi$. Thereby, the degree of

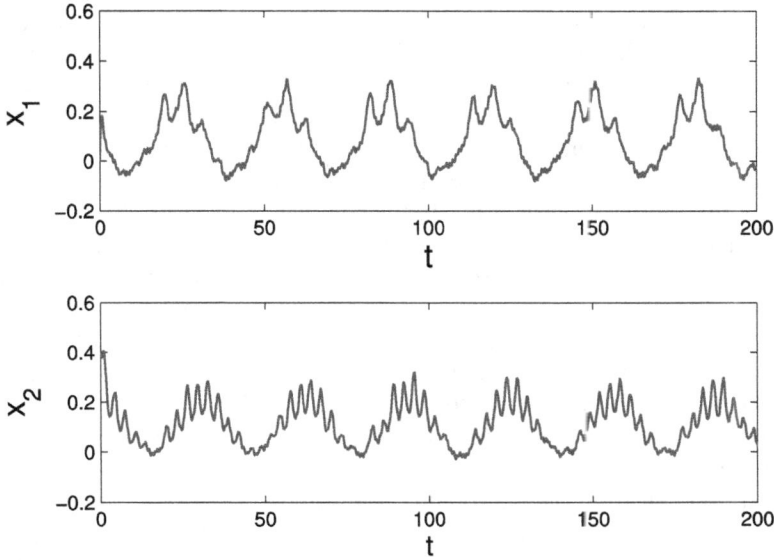

Figure 5.8. The coordinates $x_1(t)$, $x_2(t)$ of the solution of system (5.48) with initial values $x_1(0) = 0$, $x_2(0) = 0.4$. The degree of periodicity is equal to $\nabla = 100$. Reproduced with permission from [54].

periodicity, ∇, is equal to 100. In this case, the solution admits periodic shape, which envelopes the alpha unpredictability, as seen in figure 5.8.

Observation of the figures 5.6, 5.7, and 5.8 helps to conclude that the degree of periodicity is useful to decide how large the presence of periodicity and/or irregularity is in the chaotic processes. That is, if the degree is less than one, then irregularity dominates and envelopes a periodicity, while for the degree larger than one, periodicity is vividly seen and it envelopes an irregularity. One can suppose that the interaction of periodicity and irregularity can provide rich industrial opportunities as well as can be extended to interactions of quasi-periodicity and almost periodicity with irregularity in further investigations.

5.3 Modulo periodic Poisson stable solutions

The theory of differential equations is a doctrine on oscillations and recurrence, which are basic in science and technique. Oscillations are most preferable in engineering [49], while recurrence originates in celestial mechanics [51]. The ultimate recurrence is the Poisson stability [38, 50, 52]. Presently, needs for functions with irregular behavior are exceptionally strong in neuroscience and celestial dynamics, which is still in the developing mode. In the present research, we have decided to combine periodic dynamics with the phenomenon of Poisson stability. That is, one of the simplest forms of oscillations is amalgamated with the most sophisticated type of recurrence. We hope that the choice can give a new push for the nonlinear analysis, which faces challenging problems of the real world and industry. The present products of the design are *modulo periodic Poisson stable functions*.

In paper [6], we extended the Poisson stability to the alpha unpredictability property to strengthen the role of recurrence as a chaotic ingredient. Thus, the ultra Poincaré chaos has been determined, and one can now say that the *alpha unpredictability implies chaos*. The alpha unpredictable point of the Bebutov dynamics is the alpha unpredictable function. In papers [1, 2, 5, 7, 9, 20, 24, 28], we provided a dynamical way to construct Poisson stable functions. Deterministic and stochastic dynamics have been used. Deterministically alpha unpredictable functions have been constructed as solutions of hybrid systems, consisting of discrete and differential equations [9, 24], and are the random results of the Bernoulli process inserted into a linear differential equation [1, 2, 29]. Alpha unpredictable oscillations in neural networks have been researched in [2, 4, 9, 21, 22].

In papers [1, 24, 28] and books [2, 9], discussing the existence of alpha unpredictable solutions, we have developed a new way to approve Poisson stable solutions, since alpha unpredictable functions are a subset of Poisson stable functions, and to verify the alpha unpredictability, one must determine whether the Poisson stability is valid. The method is distinctly different than the *comparability method by the character of recurrence*, which was introduced in [53]. Unlike papers [1, 2, 4, 7, 9, 21, 22, 24, 28, 29], the present research is busy with the new type of Poisson stable functions. Correspondingly, it is the first time in literature when quasilinear equations with Poisson stable coefficients are under investigation. Finally, the systems are proved with modulo-periodic Poisson stable solutions. The newly invented method of verification of the Poisson stability, joined with the presence of periodic components in the recurrence, has made it possible to extend the class of studied differential equations. In papers [41, 53], quasilinear systems have constant matrices of coefficients; in our case, we research systems with periodic and even with Poisson stable coefficients. Another significant novelty is the numerical simulation of the Poisson stable functions and solutions [2, 9, 24]. We believe that all together, the present suggestions can shape a new, interesting direction for science, not only in the theoretical study of differential equations, but also to provide rich opportunities for applications in mechanics, electronics, artificial neural networks, and neuroscience.

5.3.1 Preliminaries

Definition 5.3.1. [52] *A continuous and bounded function $\psi(t)$: $\mathbb{R} \to \mathbb{R}^n$ is called Poisson stable if there exists a sequence t_k, which diverges to infinity such that the sequence $\psi(t + t_k)$ converges to $\psi(t)$ uniformly on bounded intervals of $\mathbb{R}$.*

The sequence t_k in the last definition is said to be a *Poisson sequence* of the function $\psi(t)$.

Definition 5.3.2. *The sum $\phi(t) + \psi(t)$ is said to be a modulo periodic Poisson stable (MPPS) function if $\phi(t)$ is a continuous periodic and $\psi(t)$ is a Poisson stable function.*

We shall call the function $\phi(t)$ the *periodic component* and the function $\psi(t)$ the *Poisson component* of the *MPPS* function in what follows.

Remark 5.3.1. Duo to lemma 5.3.3, an *MPPS* function is Poisson stable if κ_ω equals zero. Otherwise, without loss of generality, the sequence $\phi(t + t_k) + \psi(t + t_k)$ converges on all compact subsets of the real axis to the function $\phi(t + \tau_\omega) + \psi(t)$, where τ_ω is a nonzero Poisson shift for the sequence t_k. Because of the periodicity of the function $\phi(t)$, one can accept the last convergence as a special form of recurrence. In the next section, we shall consider it as a result of theorem 5.3.1.

Lemma 5.3.1. *For arbitrary sequence of positive real numbers t_k, $k = 1, 2, \ldots$, and a positive number ω there exist a subsequence t_{k_l}, $l = 1, 2, \ldots$, and a number τ_ω, $0 \leqslant \tau_\omega < \omega$, such that $t_{k_l} \to \tau_\omega (mod\ \omega)$ as $l \to \infty$.*

Proof. Consider the sequence τ_k such that $t_k \equiv \tau_k(mod\ \omega)$, and $0 \leqslant \tau_k < \omega$ for all $k \geqslant 1$. The boundedness of the sequence τ_k implies that there exists a subsequence τ_{k_l}, which converges to a number τ_ω [46].

Remark 5.3.2. The last assertion implies that if $\kappa_\omega = 0$, then there exists a subsequence t_{k_l} such that $t_{k_l} \to 0(mod\ \omega)$ as $l \to \infty$.

By lemma 5.3.1, for a positive fixed ω there exists a subsequence t_{k_l} of the Poisson sequence t_k and a number τ_ω such that $t_{k_l} \to \tau_\omega(mod\ \omega)$ as $l \to \infty$. We shall call the number τ_ω the *Poisson shift* for the Poisson sequence t_k with respect to the ω. It is not difficult to find that for the fixed ω, the set of all Poisson shifts, $\mathcal{T}_\omega$, is not empty, and it can consist of several and even infinite number elements. The number $\kappa_\omega = inf\ T_\omega$, $0 \leqslant \kappa_\omega < \omega$, is said to be *the Poisson number* for the Poisson sequence t_k with respect to the number ω.

Lemma 5.3.2. $\kappa_\omega \in \mathcal{T}_\omega$.

Proof. Assume, on the contrary, that κ_ω is not in $\mathcal{T}_\omega$. Then there exists a strictly decreasing sequence τ_m, $m \geqslant 1$, in $\mathcal{T}_\omega$, such that $\tau_m \to \kappa_\omega$. For each natural m, denote by t_i^m a subsequence of t_k such that $t_i^m \to \tau_m(mod\ \omega)$ as $i \to \infty$.

Fix a sequence of positive numbers ε_n, which converges to zero. One can find numbers i_n, $n = 1, 2, \ldots$, such that $|t_{i_n}^n - \tau_n| < \varepsilon_n(mod\ \omega)$. It is clear that $t_{i_n}^n \to \kappa_\omega(mod\ \omega)$ as $n \to \infty$.

Lemma 5.3.3. *If* $f(t) = \phi(t) + \psi(t)$ *is an MPPS function, and* $\kappa_\omega = 0$, *then the function* $f(t)$ *is Poisson stable.*

Proof. According to lemma 5.3.2, there exists a subsequence t_{k_l}, which tends to zero in modulus ω as $l \to \infty$. Without loss of generality, assume that $t_k \to 0(mod\ \omega)$ as $k \to \infty$. Fix a positive number ε and bounded interval $I \subset \mathbb{R}$. The periodic function $\phi(t)$ is uniformly continuous on $\mathbb{R}$. Consequently, there exists a number k_1 such that

$$\|\phi(t + t_k) - \phi(t)\| < \frac{\varepsilon}{2},$$

for all $t \in \mathbb{R}$ and $k > k_1$. Moreover, there exists an integer k_2, such that

$$\|\psi(t + t_k) - \psi(t)\| < \frac{\varepsilon}{2},$$

for $t \in I$, $k > k_2$. This is why,

$$\|f(t + t_k) - f(t)\| \leqslant \|\phi(t + t_k) - \phi(t)\| + \|\psi(t + t_k) - \psi(t)\| < \varepsilon,$$

if $t \in I$ and $k > \max(k_1, k_2)$. That is, the function $f(t)$ is Poisson stable.

Lemma 5.3.4. *Assume that* $\psi(t)$ *is a Poisson stable function. If* $\kappa_\omega = 0$, *for some positive number* ω, *then* $\psi(t)$ *is MPPS function.*

Proof. Let us write $\psi(t) = g(t) + (\psi(t) - g(t))$, where $g(t)$ is a continuous ω-periodic function. Since $\kappa_\omega = 0$, then the subtraction $\psi(t) - g(t)$ is Poisson stable by lemma 5.3.3.

Remark 5.3.3. *The last result is a source for the optimization problem of how to choose the function* $g(t)$ *and the period* ω *to minimize the difference* $\psi(t) - g(t)$. *In other words, the problem of approximation of Poisson stable functions with periodic ones. It is of exceptional interest for celestial mechanics [51].*

Lemma 5.3.5. *Assume that a function* $G(t, u)$: $\mathbb{R} \times U \to \mathbb{R}^n$, $U \subseteq \mathbb{R}^n$ *is a Poisson stable function in* t *and satisfies the inequality* $\|G(t, u_1) - G(t, u_2)\| \leqslant L\|u_1 - u_2\|$, *where* L *is a positive constant, for all* $t \in \mathbb{R}$, $u_1, u_2 \in U$. *Moreover,* $v(t)$: $\mathbb{R} \to U$ *is* ω-*periodic in* t. *If the Poisson sequence and period* ω *are such that the Poisson number* κ_ω *equals zero, then the function* $G(t, v(t))$ *is Poisson stable.*

Proof. By the lemma 5.3.2 there exists a subsequence t_{k_l}, such that $t_{k_l} \to 0(mod\ \omega)$ as $l \to \infty$. We assume, without loss of generality, that the sequence t_k itself satisfies the condition $t_k \to 0(mod\ \omega)$ as $k \to \infty$.

Let us fix a positive number ε, and a bounded interval I. Because of the property of the sequence t_k, we have that for sufficiently large k, it is true that $\|G(t + t_k, v(t + t_k)) - G(t, v(t + t_k))\| < \frac{\varepsilon}{2}$ for all $t \in \mathbb{R}$, and $\|v(t + t_k) - v(t)\| < \frac{\varepsilon}{2L}$ for $t \in I$, and

$$\|G(t + t_k, v(t + t_k)) - G(t, v(t))\| \leqslant \|G(t + t_k, v(t + t_k)) - G(t, v(t + t_k))\| +$$
$$\|G(t, v(t + t_k)) - G(t, v(t))\| \leqslant \|G(t + t_k, v(t + t_k)) - G(t, v(t + t_k))\| +$$
$$L\|v(t + t_k) - v(t)\| \leqslant \frac{\varepsilon}{2} + L\frac{\varepsilon}{2L} \leqslant \varepsilon,$$

for all $t \in I$. That is, $G(t, v(t))$ is a Poisson stable function.

Lemma 5.3.6. *Assume that a function $G(t, u)$: $\mathbb{R} \times U \to \mathbb{R}^n$, $U \subseteq \mathbb{R}^n$ is ω-periodic in t and satisfies the inequality $\|G(t, u_1) - G(t, u_2)\| \leqslant L\|u_1 - u_2\|$, where L is a positive constant for all $t \in \mathbb{R}$, $u_1, u_2 \in U$. Moreover, $v(t)$: $\mathbb{R} \to U$ is a Poisson stable function. If the Poisson sequence and period ω are such that the Poisson number κ_ω equals zero, then the function $G(t, v(t))$ is Poisson stable.*

Proof. Since $\kappa_\omega = 0$, the lemma 5.3.2 implies that there exists a subsequence t_{k_l}, such that $t_{k_l} \to 0(mod\ \omega)$ as $l \to \infty$. For simplicity, we assume that the sequence t_k itself satisfies the condition $t_k \to 0(mod\ \omega)$ as $k \to \infty$. Therefore, $G(t + t_k, v)$ uniformly converges to $G(t, v)$ as $k \to \infty$, for all $t \in \mathbb{R}$ and $v \in \mathcal{U}$.

Consequently, for arbitrarily fixed positive number ε and a bounded interval I one can find a sufficiently large number k such that $\|G(t + t_k, v(t + t_k)) - G(t, v(t + t_k))\| < \frac{\varepsilon}{2}$ for all $t \in \mathbb{R}$, and $\|v(t + t_k) - v(t)\| < \frac{\varepsilon}{2L}$ for $t \in I$. Finally, we have that

$$\|G(t + t_k, v(t + t_k)) - G(t, v(t))\| \leqslant \|G(t + t_k, v(t + t_k)) - G(t, v(t + t_k))\| +$$
$$\|G(t, v(t + t_k)) - G(t, v(t))\| \leqslant \|G(t + t_k, v(t + t_k)) - G(t, v(t + t_k))\| +$$
$$L\|v(t + t_k) - v(t)\| \leqslant \frac{\varepsilon}{2} + L\frac{\varepsilon}{2L} \leqslant \varepsilon,$$

for all $t \in I$. That is, $G(t, v(t))$ is a Poisson stable function.

Lemma 5.3.7. *Assume that a function $G(t, u)$: $\mathbb{R} \times U \to \mathbb{R}^n$, $U \subseteq \mathbb{R}^n$, is Poisson stable in t and satisfies the inequality $\|G(t, u_1) - G(t, u_2)\| \leqslant L\|u_1 - u_2\|$, where L is a positive constant for all $t \in \mathbb{R}$, $u_1, u_2 \in U$. Moreover, $v(t)$: $\mathbb{R} \to U$ is a Poisson stable function. If there exists a Poisson sequence common for the functions $G(t, u)$ and $v(t)$, then the function $G(t, v(t))$ is Poisson stable.*

Proof. Let us fix a positive number ε and a bounded interval I. Since $G(t, v(t))$ is Poisson stable in t, and $v(t)$ is a Poisson stable function, there exists sufficiently large

k such that $\|G(t + t_k, v(t + t_k)) - G(t, v(t + t_k))\| < \frac{\varepsilon}{2}$ for all $t \in \mathbb{R}$, and $\|v(t + t_k) - v(t)\| < \frac{\varepsilon}{2L}$ for $t \in I$. That is,

$$\|G(t + t_k, v(t + t_k)) - G(t, v(t))\| \leqslant \|G(t + t_k, v(t + t_k)) - G(t, v(t + t_k))\| +$$
$$\|G(t, v(t + t_k)) - G(t, v(t))\| \leqslant \|G(t + t_k, v(t + t_k)) - G(t, v(t + t_k))\| +$$
$$L\|v(t + t_k) - v(t)\| \leqslant \frac{\varepsilon}{2} + L\frac{\varepsilon}{2L} \leqslant \varepsilon,$$

for all $t \in I$. Thus, $G(t, v(t))$ is a Poisson stable function.

Remark 5.3.4. *The last lemma implies, in particular, that the sum and product of Poisson stable functions with common Poisson sequences are Poisson stable functions.*

5.3.2 A linear non-homogeneous system

Consider the following system

$$x'(t) = A(t)x(t) + \phi(t) + \psi(t), \tag{5.49}$$

where $t \in \mathbb{R}$, $x \in \mathbb{R}^n$, $n \in \mathbb{N}$, $\phi(t): \mathbb{R} \to \mathbb{R}^n$ and $\psi(t): \mathbb{R} \to \mathbb{R}^n$ are continuous functions, $A(t)$ is a continuous $n \times n$ matrix.

We assume that the following conditions are satisfied.

(C1) $A(t)$ is an ω-periodic matrix for a fixed positive ω;
(C2) $\phi(t)$ is an ω-periodic function, and $\psi(t)$ is a Poisson stable function with a Poisson sequence t_k;
(C3) the Poisson number κ_ω for the sequence t_k is equal to zero.

According to definition 5.3.2 and condition (C2), the sum $\phi(t) + \psi(t)$ is an *MPPS* function, i.e., the linear system (5.49) admits *MPPS* perturbation.

Let us consider the homogeneous system associated with (5.49),

$$x'(t) = A(t)x(t). \tag{5.50}$$

Let $X(t)$, $t \in \mathbb{R}$ be the fundamental matrix of the system (5.50) such that $X(0) = I$, and I is the $n \times n$ identical matrix. Moreover, $X(t, s)$ is a transition matrix of the system (5.50), which is equal to $X(t)X^{-1}(s)$, and $X(t + \omega, s + \omega) = X(t, s)$ for all $t, s \in \mathbb{R}$.

We assume that the following additional assumption is valid.

(C4) The multipliers of the system (5.50) in modulus are less than one.

It follows from the last condition that there exist positive numbers $K \geqslant 1$ and α such that

$$\|X(t, s)\| \leqslant K e^{-\alpha(t-s)}, \tag{5.51}$$

for $t \geqslant s$ [47].

Lemma 5.3.8. [27] *If the inequality (5.51) is satisfied, then the following estimation is correct*

$$\|X(t + \tau, s + \tau) - X(t, s)\| \leqslant \max_{t \in \mathbb{R}} \|A(t + \tau) - A(t)\| \frac{2K^2}{\alpha^2 e} e^{-\frac{\alpha}{2}(t-s)}, \quad (5.52)$$

for $t \geqslant s$ and arbitrary real number τ.

Proof. Since

$$\frac{dX(t + \tau, s + \tau)}{dt} = A(t)X(t + \tau, s + \tau) + (A(t + \tau) - A(t))X(t + \tau, s + \tau),$$

we have that

$$X(t + \tau, s + \tau) = X(t, s) + \int_s^t X(t, u)(A(u + \tau) - A(u))X(u + \tau, s + \tau)du.$$

That is why,

$$\|X(t + \tau, s + \tau) - X(t, s)\| \leqslant$$

$$\int_s^t \|X(t, u)\| \|A(u + \tau) - A(u)\| \|X(u + \tau, s + \tau)\| du \leqslant$$

$$\max_{t \in \mathbb{R}} \|A(t + \tau) - A(t)\| \int_s^t K^2 e^{-\alpha(t-s)} du =$$

$$\max_{t \in \mathbb{R}} \|A(t + \tau) - A(t)\| \frac{K^2}{\alpha} e^{-\alpha(t-s)}(t - s) =$$

$$\max_{t \in \mathbb{R}} \|A(t + \tau) - A(t)\| \frac{K^2}{\alpha} e^{-\frac{\alpha}{2}(t-s)} e^{-\frac{\alpha}{2}(t-s)}(t - s).$$

Since $\sup_{u \geqslant 0} e^{-\frac{\alpha}{2}u} u = \frac{2}{\alpha e}$, the lemma is proved.

Theorem 5.3.1. [27] *Assume that conditions (C1), (C2), and (C4) are valid. Then the system (5.49) admits a unique asymptotically stable MPPS solution.*

Proof. The bounded solution of system (5.49) has the form [47]

$$x(t) = \int_{-\infty}^t X(t, s)[\phi(s) + \psi(s)]ds, \ t \in \mathbb{R}. \quad (5.53)$$

One can write that $x(t) = x_\phi(t) + x_\psi(t)$, where $x_\phi(t) = \int_{-\infty}^t X(t, s)\phi(s)ds$ and $x_\psi(t) = \int_{-\infty}^t X(t, s)\psi(s)ds$.

It is not difficult to show that the function $x_\phi(t)$ is ω-periodic [44].

Next, we prove that the function $x_\psi(t)$ is Poisson stable. Fix arbitrary positive number ε and interval $[a, b]$, $-\infty < a < b < \infty$. We will show that for a large k it is true that $\|x_\psi(t + t_k) - x_\psi(t)\| < \varepsilon$ on $[a, b]$. Let us choose two numbers c and ξ such that $c < a$ and ξ is positive, satisfying the following inequalities,

$$\frac{4K^2 m_\psi}{\alpha^3 e}\xi < \frac{\varepsilon}{3}, \tag{5.54}$$

$$\frac{2K m_\psi}{\alpha}e^{-\alpha(a-c)} < \frac{\varepsilon}{3}, \tag{5.55}$$

and

$$\frac{K\xi}{\alpha}[1 - e^{-\alpha(b-c)}] < \frac{\varepsilon}{3}, \tag{5.56}$$

with $m_\psi = \sup_{t \in \mathbb{R}} \|\psi(t)\|$. By applying condition (C4), without loss of generality, for sufficiently large k we obtain that $\|A(t + t_k) - A(t)\| < \xi$ for all $t \in \mathbb{R}$, and $\|\psi(t + t_k) - \psi(t)\| < \xi$ for $t \in [c, b]$. Using lemma 5.2.10 we attain that

$$\|x_\psi(t + t_k) - x_\psi(t)\| = \|\int_{-\infty}^{t} (X(t + t_k, s + t_k)\psi(s + t_k) - X(t, s)\psi(s))ds\| \leq$$

$$\int_{-\infty}^{t} \|X(t + t_k, s + t_k) - X(t, s)\|\|\psi(s + t_k)\|ds+$$

$$\int_{-\infty}^{t} \|X(t, s)\|\|\psi(s + t_k) - \psi(s)\|ds=$$

$$\int_{-\infty}^{t} \|X(t + t_k, s + t_k) - X(t, s)\|\|\psi(s + t_k)\|ds+$$

$$\int_{-\infty}^{c} \|X(t, s)\|\|\psi(s + t_k) - \psi(s)\|ds + \int_{c}^{t} \|X(t, s)\|\|\psi(s + t_k) - \psi(s)\|ds \leq$$

$$\int_{-\infty}^{t} \frac{2K^2\xi}{\alpha^2 e}e^{-\frac{\alpha}{2}(t-s)}m_\psi ds + \int_{-\infty}^{t} 2Ke^{-\alpha(t-s)}m_\psi ds + \int_{-\infty}^{t} Ke^{-\alpha(t-s)}\xi ds \leq$$

$$\frac{4K^2\xi}{\alpha^3 e}m_\psi + \frac{2K m_\psi}{\alpha}e^{-\alpha(a-c)} + \frac{K\xi}{\alpha}[1 - e^{-\alpha(b-c)}].$$

Now, the inequalities (5.54) to (5.56) imply that $\|x_\psi(t + t_k) - x_\psi(t)\| < \varepsilon$ for $t \in [a, b]$. Therefore, the sequence $x_\psi(t + t_k)$ uniformly converges to $x_\psi(t)$ on each bounded interval. Thus, according to the definition 5.3.2 the solution $x(t)$ of the system (5.49) is an *MPPS* function with the periodic component $x_\phi(t)$ and the Poisson component $x_\psi(t)$. The asymptotic stability of the *MPPS* solution can be verified in the same way as for the bounded solution of a linear inhomogeneous system [44].

5.3.3 Examples with numerical simulations

The following examples show the validity of the obtained theoretical result.

Example 5.3.1. *Let us consider the following linear inhomogeneous system,*

$$x_1' = (-1 + 0.5\sin(2t))x_1 + 2.5\cos(t) + 5.5\Theta^2(t),$$
$$x_2' = (-2 + 0.25\cos(t))x_2 + 2\sin(2t) + 1.7\Theta(t), \tag{5.57}$$

where $\Theta(t) = \int_{-\infty}^{t} e^{-3(t-s)}\Omega_{(3.85;6\pi)}(s)ds$ *is the Poisson stable function described in section 5.2.3. The perturbation is an MPPS function with the periodic component* $\phi(t) = (2.5\cos(t), 2\sin(2t))^T$ *and the Poisson component* $\psi(t) = (5.5\Theta^2(t), 1.7\Theta(t))^T$. *The common period of the coefficient* $A(t)$ *and the periodic component* $\phi(t)$ *is* 2π.

Since the function $\Omega_{(3.85,6\pi)}(t)$ *is constructed on the intervals* $[6\pi i, 6\pi(i + 1)), i \in \mathbb{Z}$, *for the Poisson sequence* t_k *of the function* $\Theta(t)$ *there exists a subsequence* t_{k_l} *such that* $t_{k_l} \to 0(mod\ 2\pi)$. *Therefore, the Poisson number* $\kappa_\omega = 0$. *Condition (C4) is valid with the multipliers* $\rho_1 = e^{-2\pi}$ *and* $\rho_2 = e^{-4\pi}$. *According to theorem 5.3.1, the system admits a unique asymptotically stable MPPS solution,* $z(t)$. *Since it is impossible to determine the initial value of the solution, we simulate a solution, which asymptotically approaches* $z(t)$ *as time increases. We depict in figure 5.9 the coordinates of the solution* $x(t)$, *with initial values* $x_1(0) = 2.5$ *and* $x_2(0) = 1.5$, *which visualizes the MPPS solution approximately. In figure 5.10 the trajectory of the solution* $x(t)$ *is shown.*

In the next example, the periodic component $\phi(t)$ of the MPPS perturbation is absent, but the condition (C2) is correct, since a constant function is of an arbitrary period. It is remarkable to say that the absence of a proper non-constant periodic component makes the dynamics more irregular; this is seen in figures 5.11 and 5.12.

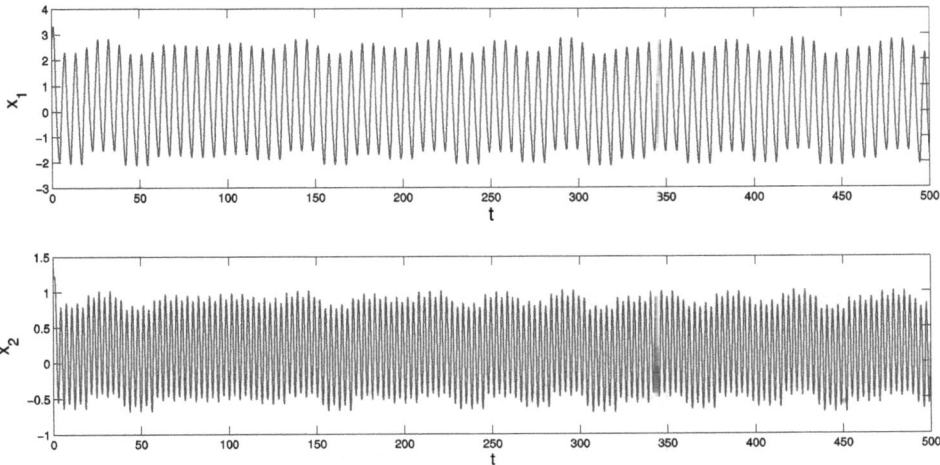

Figure 5.9. Coordinates of the solution $x(t)$ of system (5.57) with initial values $x_1(0) = 2.5$ and $x_2(0) = 1.5$, which asymptotically converge to the coordinates of the *MPPS* solution $z(t)$ of the system. Reproduced from [27]. CC BY 4.0.

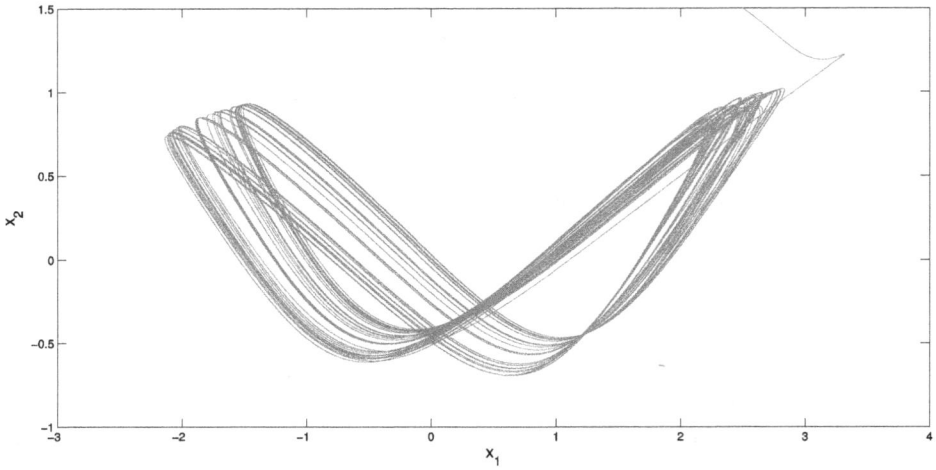

Figure 5.10. The trajectory of the solution $x(t)$ of the equation (5.57), which asymptotically approaches the MPPS solution $z(t)$ of the system. Reproduced from [27]. CC BY 4.0.

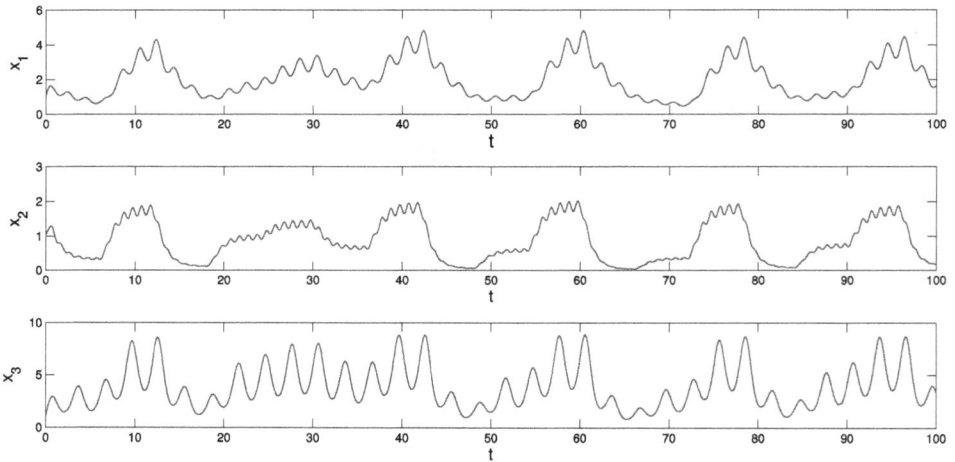

Figure 5.11. Coordinates of the solution $x(t)$, with initial values $x_1(0) = 1$, $x_2(0) = 1$ and $x_3(0) = 1$, which asymptotically converge to the coordinates of the *MPPS* solution of system (5.58). Reproduced from [27]. CC BY 4.0.

Example 5.3.2. *Consider the inhomogeneous linear system*

$$
\begin{aligned}
x_1' &= (-0.25 + 0.5\cos(\pi t))x_1 + 12\Theta^3(t),\\
x_2' &= (-1.5 + \sin^2(\pi t))x_2 + 8\Theta^2(t),\\
x_3' &= (-0.5 + \cos(\tfrac{2\pi}{3}t))x_3 + 6\Theta(t),
\end{aligned}
\tag{5.58}
$$

where $\Theta(t) = \displaystyle\int_{-\infty}^{t} e^{-2(t-s)}\Omega_{(3.9;6)}(s)ds.$

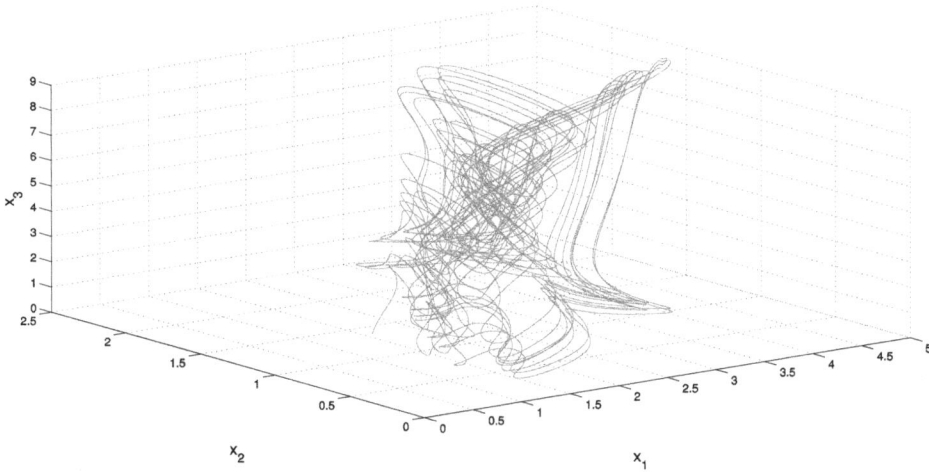

Figure 5.12. The trajectory of the solution, $x(t)$, of equation (5.58), which asymptotically approaches the MPPS solution of the equation. Reproduced from [27]. CC BY 4.0.

The conditions (C1)–(C3) are satisfied, and condition (C4) is valid with multipliers $\rho_1 = e^{-0.75}$, $\rho_2 = e^{-3}$ and $\rho_3 = e^{-1.5}$. Consequently, there exists the unique asymptotically stable MPPS solution of the system (5.58). Figure 5.11 presents the coordinates of the solution $x(t)$ with initial values $x_1(0) = 1$, $x_2(0) = 1$ and $x_3(0) = 1$. The coordinates of solution $x(t)$ approximate the coordinates of the MPPS solution. The trajectory of the solution $x(t)$ is shown in figure 5.12.

5.3.4 A quasi-linear system

The main object of the present section is the system of quasilinear differential equations

$$x'(t) = A(t)x + g(t, x) + \phi(t) + \psi(t), \qquad (5.59)$$

where $t \in \mathbb{R}$, $x \in \mathbb{R}^n$, n is a fixed natural number; $A(t)$ is an n-dimensional square matrix and satisfies to the condition (C1) and inequality (5.51); $g: \mathbb{R} \times U \to \mathbb{R}^n$, $g = (g_1, \ldots, g_n)$, $U = \{x \in \mathbb{R}^n, \|x\| < H\}$, where H is a fixed positive number; the functions $\phi(t)$ and $\psi(t)$ satisfy conditions (C2) and (C3).

The following conditions on system (5.59) are required.

(C5) the function $g(t, x)$ is continuous and ω-periodic in t;

(C6) there exists a positive constant L such that $\|g(t, x_1) - g(t, x_2)\| \leqslant L\|x_1 - x_2\|$ for all $t \in \mathbb{R}$, $x_1, x_2 \in U$.

We denote $\sup_{\mathbb{R} \times U} \|g(t, x)\| = m_g$, $\max_{t \in \mathbb{R}} \|\phi(t)\| = m_\phi$ and $\sup_{t \in \mathbb{R}} \|\psi(t)\| = m_\psi$.

The following additional conditions will be needed:

(C7) $\dfrac{K(m_g + m_\phi + m_\psi)}{H} < \alpha$;

(C8) $KL < \alpha$.

For simplicity, we use the notation $F(t, x) = g(t, x) + \phi(t) + \psi(t)$ in what follows.

According to [47], a bounded on the real axis function $y(t)$ is a solution of (5.59) if and only if it satisfies the equation

$$y(t) = \int_{-\infty}^{t} X(t, s)F(s, y(s))ds, \ t \in \mathbb{R}. \tag{5.60}$$

Theorem 5.3.2. [27] *If conditions (C1)–(C8) are valid, then the system (5.59) possesses a unique asymptotically stable Poisson stable solution.*

Proof. Let t_k be the Poisson sequence of the function $\psi(t)$ in the system (5.59). We denote by B the set of all Poisson stable functions $\nu(t) = (\nu_1, \nu_2, \ldots, \nu_n)$, $\nu_i \in \mathbb{R}$, $i = 1, 2, \ldots, n$, with common Poisson sequence t_k, which satisfy $\|\nu\|_1 < H$.

Let us show that the B is a complete space. Consider a Cauchy sequence $\theta_m(t)$ in B, which converges to a limit function $\theta(t)$ on $\mathbb{R}$. We have that

$$\|\theta(t + t_k) - \theta(t)\| < \|\theta(t + t_k) - \theta_m(t + t_k)\| + \|\theta_m(t + t_k) - \theta_m(t)\| + \\ \|\theta_m(t) - \theta(t)\|. \tag{5.61}$$

for a fixed closed and bounded interval $I \subset \mathbb{R}$. Now, one can take sufficiently large m and k such that each term on the right hand-side of (5.61) is smaller than $\frac{\varepsilon}{3}$ for a fixed positive ε and $t \in I$, i.e., the sequence $\theta(t + t_k)$ uniformly converges to $\theta(t)$ on I. Likewise, one can check that the limit function is uniformly continuous [47]. The completeness of B is shown.

Define the operator Π on B such that

$$\Pi\nu(t) = \int_{-\infty}^{t} X(t, s)F(s, \nu(s))ds, \ t \in \mathbb{R}. \tag{5.62}$$

Fix a function $\nu(t)$ that belongs to B. We have that

$$\|\Pi\nu(t)\| \leqslant \int_{-\infty}^{t} \|X(t, s)\| \|F(s, \nu(s))\| ds \leqslant \frac{K(m_g + m_\phi + m_\psi)}{\alpha}$$

for all $t \in \mathbb{R}$. Therefore, by the condition (C7) it is true that $\|\Pi\nu\|_1 < H$.

Fix a positive number ε and an interval $[a, b]$, $-\infty < a < b < \infty$. Let us choose two numbers $c < a$, and $\xi > 0$ satisfying the inequalities

$$\frac{4K^2\xi}{\alpha^3 e}(m_g + m_\phi + m_\psi) < \frac{\varepsilon}{3}, \tag{5.63}$$

$$\frac{2K}{\alpha}(m_g + m_\phi + m_\psi)e^{-\alpha(a-c)} < \frac{\varepsilon}{3}, \tag{5.64}$$

and

$$\frac{K\xi}{\alpha}[1 - e^{-\alpha(b-c)}] < \frac{\varepsilon}{3}. \tag{5.65}$$

Using the condition (C4) and lemmas 5.2.4 and 5.3.14, without loss of generality, we obtain that $\|A(t + t_k) - A(t)\| < \xi$ for all $t \in \mathbb{R}$, and $\|F(t + t_k, \nu(t + t_k)) - F(t, \nu(t))\| < \xi$ for $t \in [c, b]$ and sufficiently large k. Then, applying the inequality (5.31), we obtain:

$$\|\Pi\nu(t + t_k) - \Pi\nu(t)\| =$$

$$\left\| \int_{-\infty}^{t} X(t + t_k, s + t_k)F(s + t_k, \nu(s + t_k))ds - \int_{-\infty}^{t} X(t, s)F(s, \nu(s))ds \right\| \leqslant$$

$$\int_{-\infty}^{t} \|X(t + t_k, s + t_k) - X(t, s)\|\|F(s + t_k, \nu(s + t_k))\|ds +$$

$$\int_{-\infty}^{c} \|X(t, s)\|\|F(s + t_k, \nu(s + t_k)) - F(t, s)\|ds +$$

$$\int_{c}^{t} \|X(t, s)\|\|F(s + t_k, \nu(s + t_k)) - F(t, s)\|ds \leqslant$$

$$\int_{-\infty}^{t} \frac{2K^2\xi}{\alpha^2 e}e^{-\frac{\alpha}{2}(t-s)}(m_g + m_\phi + m_\psi)ds +$$

$$\int_{-\infty}^{t} 2Ke^{-\alpha(t-s)}(m_g + m_\phi + m_\psi)ds + \int_{-\infty}^{t} Ke^{-\alpha(t-s)}\xi ds \leqslant$$

$$\frac{4K\xi}{\alpha^3 e}(m_g + m_\phi + m_\psi) + \frac{2K}{\alpha}(m_g + m_\phi + m_\psi)e^{-\alpha(a-c)} + \frac{K\xi}{\alpha}[1 - e^{-\alpha(b-c)}],$$

for all $t \in [a, b]$. From inequalities (5.63)–(5.65) it follows that $\|\Pi\nu(t + t_k) - \Pi\nu(t)\| < \varepsilon$ for $t \in [a, b]$. Therefore, $\Pi\nu(t + t_k)$ uniformly converges to $\Pi\nu(t)$ on the bounded interval of $\mathbb{R}$.

It is easy to verify that $\Pi\nu(t)$ is a uniformly continuous function, since its derivative is a uniformly bounded function on the real axis. Summarizing the above discussion, the set B is invariant for the operator Π.

We proceed to show that the operator $\Pi: B \to B$ is contractive. Let $u(t)$ and $v(t)$ be members of B. Then, we obtain that

$$\|\Pi u(t) - \Pi v(t)\| \leqslant \int_{-\infty}^{t} \|X(t, s)\|\|F(s, u(s)) - F(s, v(s))\|ds \leqslant$$

$$\int_{-\infty}^{t} Ke^{-\alpha(t-s)}L\|u(s) - v(s)\|ds \leqslant \frac{KL}{\alpha}\|u(t) - v(t)\|_1,$$

for all $t \in \mathbb{R}$. Therefore, the inequality $\|\Pi u - \Pi v\|_1 \leqslant \frac{KL}{\alpha}\|u - v\|_1$ holds, and according to the condition (C8) the operator $\Pi: B \to B$ is contractive.

By the contraction mapping theorem there exists the unique fixed point, $\bar{x}(t) \in B$, of the operator Π, which is the unique bounded Poisson stable solution of the system (5.59).

Finally, we will study the asymptotic stability of the Poisson stable solution $\bar{x}(t)$ of the system (5.59). It is true that

$$\bar{x}(t) = X(t, t_0)\bar{x}(t_0) + \int_{t_0}^{t} X(t, s)(g(s, \bar{x}(s)) + \phi(s) + \psi(s))ds,$$

for $t \geqslant t_0$.

Let $x(t)$ be another solution of system (5.59). One can write

$$x(t) = X(t, t_0)x(t_0) + \int_{t_0}^t X(t, s)(g(s, x(s)) + \phi(s) + \psi(s))ds.$$

Making use of the relation

$$\bar{x}(t) - x(t) = X(t, t_0)(\bar{x}(t_0) - x(t_0)) + \int_{t_0}^t X(t, s)(g(s, \bar{x}(s)) - g(s, x(s)))ds,$$

we obtain that

$$\|\bar{x}(t) - x(t)\| \leqslant \|X(t, t_0)\|\|\bar{x}(t_0) - x(t_0)\| + \int_{t_0}^t \|X(t, s)\|\|g(s, \bar{x}(s)) - g(s, x(s))\|ds \leqslant$$

$$Ke^{-\alpha(t-t_0)}\|\bar{x}(t_0) - x(t_0)\| + \int_{t_0}^t KLe^{-\alpha(t-s)}\|\bar{x}(s) - x(s)\|ds.$$

Now, applying Gronwall–Bellman Lemma, one can attain that

$$\|\bar{x}(t) - x(t)\| \leqslant Ke^{-(\alpha KL)(t-t_0)}\|\bar{x}(t_0) - x(t_0)\|, \; t \geqslant t_0. \tag{5.66}$$

The last inequality and condition (C8) confirm that the Poisson stable solution $\bar{x}(t)$ is asymptotically stable. The theorem is proved.

Remark 5.3.5. *According to the lemma 5.3.11, the Poisson stable solution $\bar{x}(t)$ of the system (5.59) is an MPPS function.*

Example 5.3.3. *Consider the quasi-linear system.*

$$x_1' = (-1.5 + 2\sin(2t))x_1 + 0.01\cos(2t)\arctan(x_2) + 1.2\sin(8t) - 10.5\Theta^3(t),$$
$$x_2' = (-3.5 + 3\sin^2(2t))x_2 + 0.03\sin(4t)\arctan(x_3) - 1.5\cos(8t) + 2.5\Theta(t), \tag{5.67}$$
$$x_3' = (-1.5 + 2\cos^2(t))x_3 - 0.02\sin(2t)\arctan(x_1) + \sin(4t) + 7.2\Theta^2(t),$$

where $\Theta(t) = \int_{-\infty}^t e^{-3(t-s)}\Omega_{(3.86,3\pi)}(s)ds$ is the Poisson stable function, which is described similarly to that in example 5.3.25. Since the piecewise constant function $\Omega_{(3.86;3\pi)}(t)$ is given on intervals $[3\pi i, 3\pi(i+1))$, for the Poisson sequence t_k of the function $\Theta(t)$ there exists a subsequence t_{k_l} such that $t_{k_l} \to 0(\mathrm{mod}\ \pi)$, that is, the condition (C3) is valid. The common period of the matrix $A(t)$ and functions $g(t, x), \phi(t)$ is equal to π. We have that the function $g(t, x) = (0.01\cos(2t)\arctan(x_2), 0.03\sin(4t)\arctan(x_3), -0.02\sin(2t)\arctan(x_1))^T$ is continuous and π-periodic in t and satisfies condition (C6) with $L = 0.03$. The sum of $\phi(t) = (1.2\sin(8t), -1.5\cos(8t), \sin(4t))^T$ and $\psi(t) = (10.5\Theta^3(t), 2.5\Theta(t), 7.2\Theta^2(t))^T$ is an MPPS function, which meets conditions (C2), (C3). The assumptions (C4)–(C8) are valid with $m_g = 0.048$, $m_\phi = 1.5$, $m_\psi = 0.84$, $\rho_1 = e^{-1.5\pi}$, $\rho_2 = e^{-2\pi}$, $\rho_3 = e^{-0.5\pi}$, $\alpha = 0.5\pi$, $K = 1$, and $H = 4.8$. Thus, all conditions for the last theorem have been verified, and there is the Poisson stable solution of the system, which is asymptotically stable. It is worth noting that the simulation of the Poisson stable solution, $\bar{x}(t)$, is not possible, since the initial value is not

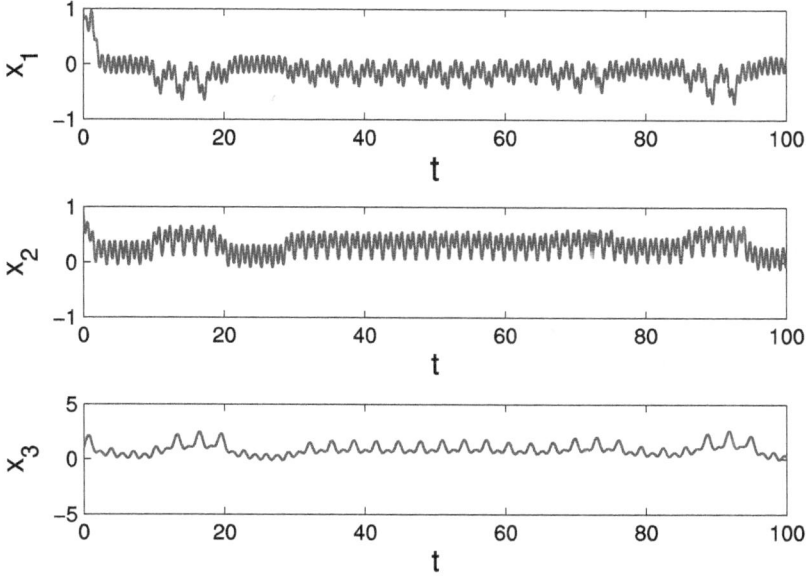

Figure 5.13. The coordinates of the solution $x(t)$, with $x_1(0) = 1$, $x_2(0) = 1$, $x_3(0) = 1$, which is asymptotic for the Poisson stable solution of the system (5.67). Reproduced from [27]. CC BY 4.0.

known precisely. For this reason, we will consider the solution $x(t)$ of the system (5.67), with initial values $x_1(0) = 1, x_2(0) = 1$ and $x_3(0) = 1$. Using the inequality (5.43) one can obtain that $\|\bar{x}(t) - x(t)\| \leqslant e^{-1.54}\|\bar{x}(0) - x(0)\|$ for $t \geqslant 0$. The last inequality shows that $\|\bar{x}(t) - x(t)\|$ decreases exponentially. Consequently, the graph of the solution $x(t)$ asymptotically approaches the Poisson stable solution $\bar{x}(t)$ of the system (5.67) as time increases. Figure 5.13 demonstrates the coordinates of the solution $x(t)$, which illustrate the Poisson stability of the system (5.67). In figure 5.14, the trajectory of the function $x(t)$ is depicted.

5.3.5 Modulo periodic Poisson stable coefficients

Let us consider the quasilinear equation (5.59) with $A(t) = B(t) + D(t)$, where $B(t)$ is a continuous ω-periodic matrix, and $D(t)$ is a Poisson stable matrix with the convergence sequence t_k. That is, the coefficient is an *MPPS* matrix and the system (5.59) is of the form

$$x'(t) = (B(t) + D(t))x + g(t, x) + \phi(t) + \psi(t), \tag{5.68}$$

where the functions $\phi(t)$ and $\psi(t)$ satisfy conditions (C2) and (C3) and their sum is an *MPPS* function. The function $g(t, x)$ satisfies conditions (C5), (C6).

Denote $G(t, x) = D(t)x + g(t, x) + \phi(t) + \psi(t)$ and rewrite the system (5.68) as

$$x'(t) = B(t)x + G(t, x). \tag{5.69}$$

The homogeneous ω-periodic system, associated with (5.68),

$$y'(t) = B(t)y, \tag{5.70}$$

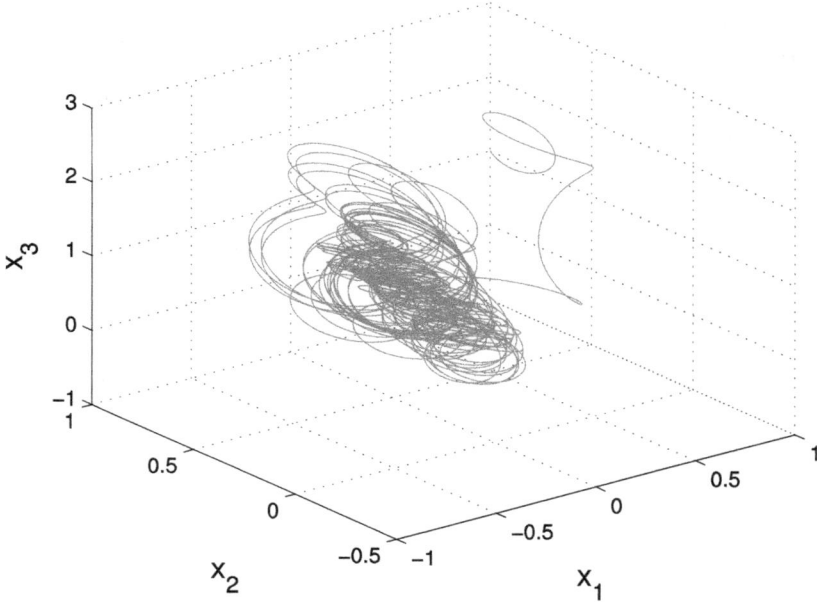

Figure 5.14. The trajectory of the solution $x(t)$, with $x_1(0) = 1$, $x_2(0) = 1$, $x_3(0) = 1$, which illustrates the Poisson stability of the system (5.67). Reproduced from [27]. CC BY 4.0.

has the fundamental matrix $Y(t)$, $Y(0) = I$, and the transition matrix $Y(t, s)$, $t, s \in \mathbb{R}$.

Assume that the following assumptions are valid.

(C9) The multipliers of the system (5.70) are in modulus less than one.

From the condition (C9) we have that there exist positive numbers $D \geqslant 1$ and β such that

$$\|Y(t, s)\| \leqslant De^{-\beta(t-s)}, \tag{5.71}$$

for $t \geqslant s$.

(C10) $D(L + d) < \beta$;

(C11) $\dfrac{D(mg + m_\phi + m_\psi)}{H} < \beta - Dd$,

where $d = \sup\limits_{t \in \mathbb{R}} \|D(t)\|$.

Theorem 5.3.3. [27] *If conditions (C2), (C3), (C5), (C6), and (C9) to (C11) hold, then system (5.68) admits a unique asymptotically stable Poisson stable solution.*

Proof. A bounded on the real axis function $z(t)$ is a solution of (5.69), if and only if it satisfies the equation

$$z(t) = \int_{-\infty}^{t} Y(t, s)G(s, z(s))ds, \; t \in \mathbb{R}. \tag{5.72}$$

Denote by $\mathcal{U}$ the Banach space of all Poisson stable functions $\nu(t) = (\nu_1, \nu_2, ..., \nu_n)$, $\nu_i \in \mathbb{R}$, $i = 1, 2, ..., n$, with common Poisson sequence t_k. The functions of space $\mathcal{U}$ satisfies the condition $\|\nu\|_1 < H$.

Introduce the operator Γ on $\mathcal{U}$ such that

$$\Gamma\nu(t) = \int_{-\infty}^{t} Y(t, s)G(s, \nu(s))ds, \ t \in \mathbb{R}. \qquad (5.73)$$

Let us show that the space $\mathcal{U}$ is invariant for the operator Γ. Fix a function $\nu(t)$ from $\mathcal{U}$. We have that

$$\|\Gamma\nu(t)\| \leqslant \int_{-\infty}^{t} \|Y(t, s)\|\|G(s, \nu(s))\|ds \leqslant \frac{D(dH + m_g + m_\phi + m_\psi)}{\beta}$$

for all $t \in \mathbb{R}$. Condition (C11) implies that $\|\Gamma\nu\|_1 < H$.

Next, we will use fixed positive number ε and an interval $[a, b]$, $-\infty < a < b < \infty$, and two numbers $c < a$, and $\xi > 0$ satisfying the following inequalities

$$\frac{4DK^2\xi}{\beta^3 e}(dH + m_g + m_\phi + m_\psi) < \frac{\varepsilon}{3}, \qquad (5.74)$$

$$\frac{2D}{\beta}(dH + m_g + m_\phi + m_\psi)e^{-\alpha(a-c)} < \frac{\varepsilon}{3}, \qquad (5.75)$$

and

$$\frac{D\xi}{\beta}[1 - e^{-\alpha(b-c)}] < \frac{\varepsilon}{3}. \qquad (5.76)$$

Using the condition (C9) and lemmas 5.2.4, 5.3.14, we obtain that $\|B(t + t_k) - B(t)\| < \xi$ for all $t \in \mathbb{R}$, and $\|G(t + t_k, \nu(t + t_k)) - G(t, \nu(t))\| < \xi$ for $t \in [c, b]$ and sufficiently large k. Then, applying the inequality (5.31), we obtain

$$\|\Gamma\nu(t + t_k) - \Gamma\nu(t)\| =$$

$$\left\|\int_{-\infty}^{t} Y(t + t_k, s + t_k)G(s + t_k, \nu(s + t_k))ds - \int_{-\infty}^{t} Y(t, s)G(s, \nu(s))ds\right\| \leqslant$$

$$\int_{-\infty}^{t} \|Y(t + t_k, s + t_k) - Y(t, s)\|\|G(s + t_k, \nu(s + t_k))\|ds +$$

$$\int_{-\infty}^{c} \|Y(t, s)\|\|G(s + t_k, \nu(s + t_k)) - G(t, s)\|ds +$$

$$\int_{c}^{t} \|Y(t, s)\|\|G(s + t_k, \nu(s + t_k)) - G(t, s)\|ds \leqslant$$

$$\int_{-\infty}^{t} \frac{2D^2\xi}{\beta^2 e} e^{-\frac{\beta}{2}(t-s)}(dH + m_g + m_\phi + m_\psi)ds +$$

$$\int_{-\infty}^{t} 2De^{-\beta(t-s)}(dH + m_g + m_\phi + m_\psi)ds + \int_{-\infty}^{t} De^{-\beta(t-s)}\xi ds \leqslant$$

$$\frac{4D\xi}{\beta^3 e}(dH + m_g + m_\phi + m_\psi) + \frac{2D}{\beta}(dH + m_g + m_\phi + m_\psi)e^{-\beta(a-c)} + \frac{D\xi}{\beta}[1 - e^{-\beta(b-c)}],$$

for all $t \in [a, b]$. Hence, the inequalities (5.74)–(5.76) give that $\|\Gamma\nu(t + t_k) - \Gamma\nu(t)\| < \varepsilon$ for $t \in [a, b]$. Therefore, the sequence $\Gamma\nu(t + t_k)$ uniformly converges to $\Gamma\nu(t)$ on the bounded interval of $\mathbb{R}$. Thus, we have shown that the operator Γ is invariant in $\mathcal{U}$.

Let us show that the operator $\Gamma: \mathcal{U} \to \mathcal{U}$ is contractive. Fix members $u(t)$ and $v(t)$ of $\mathcal{U}$. It is true that

$$\|\Gamma u(t) - \Gamma v(t)\| \leqslant \int_{-\infty}^{t} \|Y(t, s)\| \|G(s, u(s)) - G(s, v(s))\| ds \leqslant$$

$$\int_{-\infty}^{t} De^{-\beta(t-s)}(d + L)\|u(s) - v(s)\| ds \leqslant \frac{D(d + L)}{\beta}\|u(t) - v(t)\|_1,$$

for all $t \in \mathbb{R}$, and condition (C10) implies that the operator Γ is contractive.

Using the contraction mapping theorem, one can conclude that there exists a unique fixed point, $\bar{x}(t)$, of the operator Γ, which is the Poisson stable solution of the system (5.68). Let us investigate its stability.

If $x(t)$ is a solution of the equation (5.68), then

$$\bar{x}(t) - x(t) = Y(t, t_0)(\bar{x}(t_0) - x(t_0)) +$$

$$\int_{t_0}^{t} Y(t, s)(D(s)(\bar{x}(s) - x(s)) + (g(s, \bar{x}(s)) - g(s, x(s))))ds,$$

and

$$\|\bar{x}(t) - x(t)\| \leqslant \|Y(t, t_0)\| \|\bar{x}(t_0) - x(t_0)\| +$$

$$\int_{t_0}^{t} \|Y(t, s)\|(\|D(s)(\bar{x}(s) - x(s))\| + \|g(s, \bar{x}(s)) - g(s, x(s)\|)ds \leqslant$$

$$De^{-\beta(t-t_0)}\|\bar{x}(t_0) - x(t_0)\| + \int_{t_0}^{t} D(d + L)e^{-\alpha(t-s)}\|\bar{x}(s) - x(s)\| ds.$$

With the aid of the Gronwall–Bellman Lemma, one can verify that

$$\|\bar{x}(t) - x(t)\| \leqslant De^{-(\beta-D(d+L))(t-t_0)}\|\bar{x}(t_0) - x(t_0)\|, \ t \geqslant t_0. \tag{5.77}$$

Now, based on the condition (C10), we conclude that the Poisson stable solution $\bar{x}(t)$ of system (5.68) is asymptotically stable. The theorem is proved.

5.4 Notes

The present chapter is a significant part of our research design, issuing from the idea that differential equations are, first and foremost, missionaries for investigating the changing world and, consequently, the world entirely. This is why chaos research has to be a part of the *theory of dynamical equations*. That is, of ordinary, partial, functional-differential, and all other equations. We have already had results of the sort in papers on the replication of chaos [3, 11–18, 30–36], where an effective input-output mechanism of chaos extension is developed for the first time in literature. Now, we involve in the mechanism a more individual actor of the play—alpha unpredictable function. This chapter contains the results of our papers [24, 27, 37].

The proofs suggested in this chapter can serve for a wide variety of sciences. We have realized the method for neural networks. It is done in our articles [21, 22, 25, 26, 28]. We have introduced a new type of recurrence: a sum of two compartments, periodic and Poisson stable functions. We call it the *MPPS* function. Sufficient conditions for the dynamics to be Poisson stable have been determined. The novelty is not only convenient for theoretical analysis of differential and discrete equations of all types, but also holds significant implications for the study of chaotic dynamics. In the present work, we study quasi-linear ordinary differential equations. If one considers the periodic compartment in the Poisson stability and the achievements of the paper for simulations of the recurrence, then one can create new productive opportunities in the research of mechanical, electronic dynamics and neuroscience. The importance of these findings for theoretical analysis and simulations of chaotic dynamics cannot be overstated.

References

[1] Akhmet M 2020 A novel deterministic chaos and discrete random processes *ACM Int. Conf. Proc. Series* pp 53–6

[2] Akhmet M 2021 *Domain Structured Dynamics* (Bristol: IOP Publishing)

[3] Akhmet M, Akhmetova Z and Fen M O 2014 Chaos in economic models with exogenous shocks *J. Econ. Behav. Organ.* **106** 95–108

[4] Akhmet M, Arugaslan Cincin D, Tleubergenova M and Nugayeva Z 2021 Unpredictable oscillations for Hopfield-type neural networks with delayed and advanced arguments *Math. Open Access* **9** 1–19

[5] Akhmet M and Fen M O 2016 Poincaré chaos and unpredictable functions *Commun. Nonlinear Sci. Numer. Simul.* **48** 85–94

[6] Akhmet M and Fen M O 2016 Unpredictable points and chaos *Commun. Nonlinear Sci. Numer. Simul.* **40** 1–5

[7] Akhmet M and Fen M O 2017 Existence of unpredictable solutions and chaos *Turk. J. Math.* **41** 254–66

[8] Akhmet M and Fen M O 2018 Non-autonomous equations with unpredictable solutions *Commun. Nonlinear Sci. Numer. Simul.* **59** 657–70

[9] Akhmet M, Fen M O and Alejaily E M 2020 *Dynamics with Chaos and Fractals* (Berlin: Springer)

[10] Akhmet M, Fen M O, Tleubergenova M and Zhamanshin A A 2019 Unpredictable solutions of linear differential and discrete equations *Turk. J. Math.* **43** 2377–89

[11] Akhmet M and Fen M O 2012 Chaotic period-doubling and OGY control for the forced Duffing equation *Commun. Nonlinear Sci. Numer. Simulat.* **17** 1929–46

[12] Akhmet M and Fen M O 2013 Replication of chaos *Commun. Nonlinear Sci. Numer. Simulat.* **18** 2626–66

[13] Akhmet M and Fen M O 2014 Chaotification of impulsive systems by perturbations *Int. J. Bifurcation Chaos* **24** 1450078

[14] Akhmet M and Fen M O 2014 Entrainment by chaos *J. Nonlinear Sci.* **24** 411–39

[15] Akhmet M and Fen M O 2014 Generation of cyclic/toroidal chaos by Hopfield neural networks *Neurocomputing* **145** 230–9

[16] Akhmet M and Fen M O 2015 Attraction of Li–Yorke chaos by retarded SICNNs *Neurocomputing* **147** 330–42

[17] Akhmet M and Fen M O 2015 Li-Yorke chaos in hybrid systems on a time scale *Int. J. Bifurcation Chaos* **25** 1540024

[18] Akhmet M and Fen M O 2016 Homoclinic and heteroclinic motions in economic models with exogenous shocks *Appl. Math. Nonlinear Sci.* **1** 1–10

[19] Akhmet M, Fen M O, Tleubergenova M and Zhamanshin A 2019 Poincare chaos for a hyperbolic quasilinear system *Miskolc Math. Notes* **20** 33–44

[20] Akhmet M, Fen M O, Tleubergenova M and Zhamanshin A 2019 Unpredictable solutions of linear differential and discrete equations *Turk. J. Math.* **43** 2377–89

[21] Akhmet M, Seilova R D, Tleubergenova M and Zhamanshin A 2020 Shunting inhibitory cellular neural networks with strongly unpredictable oscillations *Commun. Nonlinear Sci. Numer. Simulat.* **89** 105287

[22] Akhmet M, Tleubergenova M and Nugayeva Z 2020 Strongly unpredictable oscillations of Hopfield-type neural networks *Math. Open Access* **8** 1–14

[23] Akhmet M, Tleubergenova M, Seilova R and Nugayeva Z 2022 Poisson stability in symmetrical impulsive shunting inhibitory cellular neural networks with generalized piecewise constant argument *Entropy* **14** 1754

[24] Akhmet M, Tleubergenova M and Zhamanshin A 2020 Quasilinear differential equations with strongly unpredictable solutions *Carpathian J. Math.* **36** 341–9

[25] Akhmet M, Tleubergenova M and Zhamanshin A 2019 Neural networks with Poincare chaos *ACM Int. Conf. Proc. Series* pp 34–7

[26] Akhmet M, Tleubergenova M and Zhamanshin A 2020 Inertial neural networks with unpredictable oscillations *Math. Open Access* **8** 1–11

[27] Akhmet M, Tleubergenova M and Zhamanshin A 2021 Modulo periodic poisson stable solutions of quasilinear differential equations *Entropy* **23** 1535

[28] Akhmet M, Tleubergenova M and Zhamanshin A 2024 Cohen–Grossberg neural networks with unpredictable and Poisson stable dynamics *Chaos Solitons Fractals* **178** 114307

[29] Akhmet M and Tola A 2020 Unpredictable strings *Kazakh Math.* J **20** 16–22

[30] Akhmet M U 2009 The complex dynamics of the cardiovascular system *Nonlinear Anal.* **71** e1922–31

[31] Akhmet M U 2009 Creating a chaos in a system with relay *Int. J. Qualit. Th. Diff Equ. Appl.* **3** 3–7

[32] Akhmet M U 2009 Devaney's chaos of a relay system *Commun. Nonlinear Sci. Numer. Simulat.* **14** 1486–93

[33] Akhmet M U 2009 Dynamical synthesis of quasi-minimal sets *Int. J. Bifurcation Chaos* **19** 2423–7

[34] Akhmet M U 2009 Li-yorke chaos in the system with impacts *J. Math. Anal. Appl.* **351** 804–10

[35] Akhmet M U 2009 Shadowing and dynamical synthesis *Int. J. Bifurcation Chaos* **19** 3339–46

[36] Akhmet M U 2010 Homoclinical structure of the chaotic attractor *Commun. Nonlinear Sci. Numer. Simulat.* **15** 1486–93

[37] Tleubergenova M, Zhamanshin A and Akhmet M 2023 Compartmental unpredictable functions *Mathematics* **11** 1069

[38] Birkhoff G D 1991 *Dynamical Systems* (Providence, RI: Colloquium Publications)

[39] Bohr H A 1947 *Almost Periodic Functions* (New York: Chelsea Publishing Company)

[40] Burton T A 1985 *Stability and Periodic Solutions of Ordinary and Functional Differential Equations* (Amsterdam: Elsevier Science)

[41] Cheban D and Liu Z 2020 Periodic, quasi-periodic, almost periodic, almost automorphic, birkhoff recurrent and Poisson stable solutions for stochastic differential equations *J. Differ. Equ.* **269** 3652–85

[42] Corduneanu C 2009 *Almost Periodic Oscillations and Waves* (New York: Springer)

[43] Driver R D 2012 *Ordinary and Delay Differential Equations* (New York: Springer)

[44] Farkas M 1994 *Periodic Motion* (New York: Springer)

[45] Fink A M 1974 *Almost Periodic Differential Equations* (New York: Springer)

[46] Haggarty R 1993 *Fundamentals of Mathematical Analysis* (Reading, MA: Addison-Wesley)

[47] Hartman P 2002 *Ordinary Differential Equations* (New York: SIAM)

[48] Levitan B and Zhikov V 1983 *Almost Periodic Functions and Differential Equations* (Cambridge: Cambridge University Press)

[49] Minorsky N 1947 *Introduction to Non-linear Mechanics: Topological Methods, Analytical Methods, Non-linear Resonance, Relaxation Oscillations* (Ann Arbor, MI: J. W. Edwards)

[50] Nemytskii V V and Stepanov V V 1960 *Qualitative Theory of Differential Equations* (Princeton, NJ: Princeton University Press)

[51] Poincaré H 1957 *New Methods of Celestial Mechanics Volumes I–III* (New York: Dover)

[52] Sell G R 1971 *Topological Dynamics and Ordinary Differential Equations* (London: Van Nostrand Reinhold Company)

[53] Shcherbakov B A 1969 Poisson stable solutions of differential equations, and topological dynamics (Russian) *Differ. Uravn.* **5** 2144–55

[54] Akhmet M, Tleubergenova M and Zhamanshin A 2024 Unpredictable solutions of compartmental quasilinear differential equations *Carpathian J. Math.* **40** 1–13

Part V

Numerical alpha unpredictability

Chapter 6

Ultra Poincaré chaos numerically

This chapter focuses on the numerical indicators of alpha unpredictability. It introduces a novel method for validating chaos in both continuous and discrete-time models, in addition to traditional techniques like Lyapunov exponents and bifurcation diagrams. The primary analytical tool is the sequential test, also known as the test for alpha unpredictability or the ultra Poincaré chaos test. Notably, it confirms the existence of chaos in systems that were previously deemed to exhibit strange non-chaotic attractors (SNCAs) due to their non-positive Lyapunov exponents. Additionally, the chapter showcases the confirmation of ultra Poincaré chaos through synchronization.

This chapter concerns numerical indications of alpha unpredictability. It reveals a new way to confirm the presence of chaos in continuous and discrete-time models, besides other techniques such as Lyapunov exponents and bifurcation diagrams. The primary instrument for analysis is *the sequential test* [5, 12, 13, 23, 29] which is also known as the *test for alpha unpredictability* or *ultra Poincaré chaos test*. We have confirmed that there exist models, which exhibit both Devaney and ultra Poincaré chaos. This also is true for Li–Yorke chaos, and those models investigated by Lyapunov exponent and bifurcation diagram analysis. One of the most exciting results is verifying alpha unpredictability for a model, which reveals intermittency. The peculiarity of an alpha unpredictability test lies in clarifying the divergence character of a single trajectory and the indication of just chaos. Potentially, it can be more effective than the conventional ways of indicating chaos. A section of the chapter is to confirm the presence of chaos in systems that have been previously accepted to have strange non-chaotic attractors [44, 49, 66] because of their non-positive Lyapunov exponents [34, 36, 37, 44, 55, 65, 66]. The results of [23] demonstrate that one can find alpha unpredictability, and consequently. ultra Poincaré chaos. The last part of the chapter is for ultra Poincaré chaos through synchronization. The idea is straightforward; the characteristics of generalized synchronization are suitable for claiming alpha unpredictability in both connected systems if they are correct for the driver one. Advanced results on synchronization of alpha unpredictability can be found in [12, 13, 29], where the *Delta synchronization* is introduced, and the ultra Poincaré test is developed for large scales. The results of the papers are not just for illustration, but are significant for applications in physics, biology, neural networks, and neuroscience. The theoretical achievements of the chapter are exemplified by systems satisfying definitions of different types of chaos or numerical methods. The analysis is supplemented with illustrative graphics.

6.1 Introduction and preliminaries

There are principally two opportunities for chaos detection in the dynamics of a model. It can be done either theoretically through the verification of ingredients of chaos [17, 32, 38, 54] or numerically by using bifurcation diagrams and Lyapunov exponents [40, 43, 52, 61, 67, 71, 72]. Numerical techniques are preferable for industry and science. For that reason, they are expected to be developed and new ones introduced. However, numerical observations of chaos generally are insufficient to determine the type of chaos that a system admits [72]. In the present chapter, we propose a numerical technique for the detection of alpha unpredictable solutions in continuous-time systems, and we call this technique *the sequential test*, the *alpha unpredictability test*, or *ultra Poincaré chaos test*. One of the novelties of this approach is that it can be used to determine the presence of chaos with a precise type which occurs due to an alpha unpredictable motion, i.e., ultra Poincaré chaos numerical characteristics [7, 8, 11, 16, 25, 26, 28]. Evaluation of Lyapunov exponents is widely used to indicate chaos since it is universally applicable. Nevertheless, there are examples of non-chaotic systems with positive Lyapunov exponents [19, 22, 24]. In contrast, analyses of period-doubling bifurcation diagrams confirm chaos in systems possessing unstable periodic solutions. The numerical method we introduce in this study, the sequential test, also applies to any continuous-time system without any restriction in dimension or periodic motions. Since the definition of alpha unpredictable function (as well as any definitions for other types of chaos) requests infinitely many iterations to indicate the presence of an unpredictable trajectory, like the other numerical techniques, we will apply the sequential test and embrace results for a finite number of iterations. It is known that there exist dynamics without mappings and some without, which are not differentiable [6–9, 21, 27]. Therefore, the bifurcation diagram technique may not be applicable in some cases, for instance, in symbolic dynamics. It cannot be considered also with the Lyapunov exponent approach. The alpha unpredictability test can be verified for all cases, which is a distinguishing feature. Consequently, it can be considered as a solid complementary instrument for chaos detection.

We demonstrate using the alpha unpredictability test to look for the existence of unpredictable solutions, and hence Poincaré chaos, in dynamics of Lorenz [32, 56] and Rössler [72] systems and the Hénon map. It is worth noting that even though the existence of chaos in these systems can also be deduced by other techniques such as Lyapunov exponents and bifurcation diagrams, they cannot demonstrate the existence of unpredictable solutions and a precise type of chaos. On the other hand, in [5], we showed the presence of an alpha unpredictable solution in the system introduced by Wei [79], whose largest Lyapunov exponent is zero. This result reveals an advantage of the sequential test compared to the technique of Lyapunov exponents.

Let us denote $\mathbb{N}$ the set of non-negative integers, and consider a metric (X, d) and a map $\pi: \mathbb{T}_+ \times X \longrightarrow X$, where $\mathbb{T}_+$ is either the set of non-negative real numbers or $\mathbb{N}$, or a semi-flow on X, i.e., $\pi(0, x) = x$ for all $x \in X$, $\pi(t, x)$ is continuous in the pair of variables t and x, and $\pi(t_1, \pi(t_2, x)) = \pi(t_1 + t_2, x)$ for all $t_1, t_2 \in \mathbb{T}_+, x \in X$.

A point $x \in X$ is called positively Poisson stable [73] if there exists a sequence $\{t_n\}$ satisfying $t_n \longrightarrow \infty$ such that $\pi(t_n, x) \longrightarrow x$, as $n \longrightarrow \infty$. For a given point $x \in X$, let Θ_x be the closure of trajectory $T(x) = \{\pi(t, x): t \in \mathbb{T}_+\}$.

The following definition is the instrument for the numerical analyses performed in this paper.

Definition 6.1.1. [17] *A point $p \in X$ and the trajectory through it are alpha unpredictable if there exist a positive number ε_0 (the unpredictability constant) and sequences $\{t_n\}$ and $\{s_n\}$, both of which diverge to infinity, such that $f(t_n, p) \to p$ as $n \longrightarrow \infty$ and $d[f(t_n + s_n, p), f(s_n, p)] > \varepsilon_0$ for each $n \in \mathbb{N}$.*

One can see that if a point is alpha unpredictable then it is Poisson stable.

The results in [8, 17, 28] reveal the presence of sensitivity and transitivity in a set Θ_p if p is an alpha unpredictable point. That is, they expose the appearance of ultra Poincaré chaos. In the course of the last definition, we suggest the alpha unpredictability test.

Many results for chaos were obtained by applying Lyapunov exponents. In what follows, we shall compare the achievements of the method with those obtained by alpha unpredictability analysis. The definition of the exponents follows.

Definition 6.1.2. [32]. *Let f be a smooth map on $\mathbb{R}^m$, and $\mathcal{J}_n = Df^n(v_0)$, where $Df^n(v_0)$ denote the first derivative matrix of the nth iterate of f, and for $k = 1, 2, \ldots, m$, let r_n^k be the length of the kth longest orthogonal axis of the ellipsoid $\mathcal{J}_n U$, where U is a unit sphere for an orbit with initial point v_0 during the first n iterations. The kth Lyapunov number of v_0 is defined by*

$$L_k = \lim_{n \to \infty} (r_k^n)^{1/n} \tag{6.1}$$

if this limit exists. The kth Lyapunov exponent of v_0 is $h_k = ln(L_k)$, $k = 1, 2, \ldots, m$.

6.2 The algorithm of the test and the application procedure

Let us consider the following autonomous system of differential equations,

$$x'(t) = f(x(t)), \tag{6.2}$$

where $f\colon \mathbb{R}^m \to \mathbb{R}^m$ is a continuous function. Let $x(t)$ be the solution of system (6.2) with initial condition $x(0) = x_0$, where x_0 is a given point in $\mathbb{R}^m$.

Due to definition 6.1.1, we say that the solution $x(t)$ satisfies the *alpha unpredictability test* if it is confirmed numerically that there exists a large natural number k and a positive number ε_0, large and increasing sequences $\{t_n\}$ and $\{s_n\}$ where $1 \leqslant n \leqslant k$, for the solution, such that $\|x(t_n) - x(0)\| = \delta_n$ is a decreasing sequence which becomes close to 0 and the inequality $\|x(t_n + s_n) - x(s_n)\| > \varepsilon_0$ is valid for every $1 \leqslant n \leqslant k$. In our numerical experiments, the largest value of t_k reached 1.9×10^6 and the smallest δ_k of order 10^{-3}, for continuous systems. In discrete models the largest values ζ_k are of order 10^8 and δ_k are small, of order 10^{-6}. The value of number k and the smallness of δ_n are closely related to the power and facilities a computer offers and the time interval length. It is obvious that in different calculations with different ε_0, δ_n or computers, similar results are not obtained. So the sequences $\{t_n\}$ and $\{s_n\}$ are not unique for a given solution. For convenience, we will call $\{t_n\}$ *the sequence of convergence* and $\{s_n\}$ *the sequence of separation*. In the light of definition 6.1.1, one may say that the solution is alpha unpredictable and the system (6.2) is ultra Poincaré chaotic if the alpha unpredictability test is satisfied.

We hypothesize that if the alpha unpredictability test works for a sufficiently large time interval, ultra Poincaré chaos is present—in other words, if the test conditions are confirmed for a sizeable experimental interval of time. We suppose that a theorem of this kind can be proved and suggest it as an open problem. The proof of such assertion may follow the arguments for the shadowing theorem [31, 33, 45, 64, 67].

Let us consider the solution of system (6.2), $x(t)$, with a proper initial condition, x_0, that satisfies the numerical method. While applying the alpha unpredictability test on it, there may exist some small-scale time series t_l in the sequence $\{t_n\}$, which satisfies the inequality $\|x(t_l + t) - x(t)\| > \varepsilon_0$ for small ε_0 for all $t \in [0, t_{\text{final}}]$, where t_{final} is a sufficiently large number. Hence, for system (6.2) to be chaotic, associated with a large positive number ε_0, we numerically evaluate the sequence $\{t_n\}$ on time interval $[t_{\text{fix}}, t_f]$, where t_{fix} is a fixed positive number such that $0 < t_{\text{fix}} < t_{\text{final}}$ and $t_{\text{fix}} > t_L$, wherein t_L is the greatest time value of the finite sequence $\{t_l\}$. Since δ_n is a decreasing sequence which becomes close to 0, then the inequality $\delta_n < \frac{1}{n}$ is valid for some n, where $1 \leqslant n \leqslant k$. We used this inequality for all autonomous systems considered in this paper, on which the alpha unpredictability test is applied. In order to obtain increasing sequences, $\{t_n\}$ and $\{s_n\}$, we set the condition

(C1) $t_{\xi+1} > t_\xi$ and $s_{\xi+1} > s_\xi$, $\xi = 1, 2, 3, \ldots.$

Succeeding, we will provide some description of the detailed steps which will be applied later to construct Matlab codes as algorithm 1.

Let $n = 1$ and $\tau_m = mh$, $m = 1, 2, \ldots.$

While $\|x(\tau_m) - x(0)\| < \frac{1}{1}$, fix $t_1 = \tau_{m_1}$ the first value of τ_ms satisfying the inequalities $\|x(\tau_m) - x(0)\| < 1$ and $\tau_{m_1} > t_{\text{fix}}$.

Set $\tau_i^* = ih$, $i = 1, 2, \ldots.$

While $\|x(t_1 + \tau_i^*) - x(\tau_i^*)\| > \varepsilon_0$, fix $s_1 = \tau_{i_1}^*$ the first value of τ_i^*s satisfying the inequality $\|x(t_1 + \tau_i^*) - x(\tau_i^*)\| > \varepsilon_0 f$.

Algorithm 1. Alpha unpredictability test for system (6.2).

1: Input t_{fix}
2: Set $l = t_{\text{fix}}$
3: Set $q = 0$
4: Input ε_0
5: Input *nspart*
6: Set $tmin = 0$
7: Set $dt = 0.01$
8: Find $tmax = nspart \cdot dt$
9: Input initial condition x_0
10: Find the numerical solution $x(t)$ of system (6.2) for the given interval.
11: **for** n = 1: k **do**
12:     **for** m = 1: nspart **do**
13:         **If** $\|x(\tau_m) - x(0)\| < \frac{1}{n}$ **then**
14:             **If** $l < \tau_m$ **then**
15:                 $l = \tau_m$
16:                 A(n) = l ▷ the matrix A(n) collects τ_ms, which satisfy lines 13 and 14 for each n
17:                 **break**   ▷ reckon the first τ_m satisfying lines 13 and 14 for each n
18:             **end if**
19:         **end if**
20:     **end for**
21: **end for**
22: **for** n = 1: k **do**
23: **for** i = 1: nspart **do**
24:     **If** $\|x(A(n) + \tau_i^*) - x(\tau_i^*)\| > \varepsilon_0$ **then**
25:         **If** $q < \tau_i^*$ **then**
26:             q = τ_i^*
27:                 B(n) = q ▷ the matrix B(n) collects τ_i^*, which satisfy lines 24 and 25 for each n
28:                 Display matrices $A(n)$ and $B(n)$
29:                 **break**   ▷ reckon the first τ_i^* satisfying lines 24 and 25 for each n
30:             **end if**
31:         **end if**
32:     **end for**
33: **end for**

Let $n = 2$.

While $\|x(\tau_m) - x(0)\| < \frac{1}{2}$, fix $t_2 = \tau_{m_2}$ the first value of τ_ms satisfying the inequalities $\|x(\tau_m) - x(0)\| < \frac{1}{2}$ and $\tau_{m_2} > t_1$.

While $\|x(t_2 + \tau_i^*) - x(\tau_i^*)\| > \varepsilon_0$, fix $s_2 = \tau_{i_2}^*$ the first value of τ_i^*s satisfying the inequalities $\|x(t_2 + \tau_i^*) - x(\tau_i^*)\| > \varepsilon_0$ and $\tau_{i_2}^* > s_1$

Let $n = N$, where $1 \leqslant N \leqslant k$.

While $\|x(\tau_m) - x(0)\| < \frac{1}{N}$, fix $t_N = \tau_{m_N}$ the first value of τ_ms satisfying the inequalities $\|x(\tau_m) - x(0)\| < \frac{1}{N}$ and $\tau_{m_N} > t_{N-1}$.

While $\|x(t_N + \tau_i^*) - x(\tau_i^*)\| > \varepsilon_0$, fix $s_N = \tau_{i_N}^*$ the first value of τ_i^*s satisfying the inequalities $\|x(t_N + \tau_i^*) - x(\tau_i^*)\| > \varepsilon_0$ and $\tau_{i_N}^* > s_{N-1}$.

In what follows, the Matlab code based on the above description will be as follows.

The alpha unpredictability test is also applicable on solutions of discrete systems. For this purpose, consider the autonomous discrete system

$$x(i + 1) = f(x(i)), \tag{6.3}$$

where $f\colon \mathbb{R} \to \mathbb{R}^m$ is a continuous function. Let $x(i) = x_i$ be the solution of system (6.3) with initial condition $x(0) = x_0$, where x_0 is a given point in $\mathbb{R}^m$.

Due to definition 6.1.1, we say that the solution x_i satisfies the alpha unpredictability test if it is confirmed numerically that there exist a large natural number k and a positive number ε_0, increasing sequences of natural numbers $\{\zeta_n\}$ and $\{\eta_n\}$, where $1 \leqslant n \leqslant k$, for the solution, such that $\|x_{\zeta_n} - x_0\| < \frac{1}{n}$ is a decreasing sequence which approaches to 0 and the inequality $\|x_{\zeta_n + \eta_n} - x_{\eta_n}\| > \varepsilon_0$ is valid for every $1 \leqslant n \leqslant k$. Similarly, as in the case of autonomous systems of differential equations, the sequences $\{\zeta_n\}$ and $\{\eta_n\}$ are not unique for a given solution. For convenience, we will call $\{\zeta_n\}$ *the sequence of convergence* and $\{\eta_n\}$ *the sequence of separation*. Setting this side by side with definition 6.1.1, one may say that the solution is alpha unpredictable and the system (6.3) is ultra Poincaré chaotic if the alpha unpredictability test is satisfied. By displacing the time series with indexes in algorithm 1, we derive a proper algorithm for system (6.3).

In this paper, we will apply the algorithm 1 to construct the sequences of convergence, $\{t_n\}$ or $\{\zeta_n\}$, and the sequence of divergence $\{s_n\}$ or $\{\eta_n\}$, respectively, for concrete models that satisfy the alpha unpredictability test. In other words, the algorithms are the basis of the test. Likewise, algorithm 1 is suitable for non-autonomous continuous and discrete systems.

6.3 Ultra Poincaré chaos for revisited models

In this part of the chapter, systems and irregular phenomena known for sophisticated dynamics are reconsidered from the alpha unpredictability point of view. This is realized for Devaney and Li–Yorke chaos, intermittency, and period doubling cascade. The Rössler model and Ikeda map were previously analyzed by the Lyapunov exponent method. Thus, not only has the irregularity been confirmed, but new numerical characteristics are discovered to clarify complexity.

6.3.1 Devaney's chaos subdued to the sequential test

One of the definitions of chaos was provided by Devaney [38] in 1976. To present this definition let us consider the autonomous discrete system

$$x_{i+1} = G(x_i), \tag{6.4}$$

where $G\colon J \longrightarrow J$, J is the solution space, is continuous. A point $p \in J$ is a periodic point if $G^n(p) = p$, for some $n \geqslant 1$, and $G^k(p) \neq p$, for $1 \leqslant k < n$. $G\colon J \longrightarrow J$ is said to be topologically transitive if there is a point $x_0 \in J$ such that the orbit of x_0 is dense in J. The mapping $G\colon J \longrightarrow J$ is said to have sensitive dependence on initial conditions if there exists $\varepsilon > 0$ such that, for any $x \in J$ and every $\delta > 0$, there exists $y \in J$ and $n \geqslant 0$ such that $\|x - y\| < \delta, \|G^n(x) - G^n(y)\| > \varepsilon$.

Definition 6.3.1. [38] *The map G is said to be Devaney chaotic if the following conditions are valid*:
- *It has sensitive dependence on initial conditions.*
- *It is topologically transitive.*
- *Periodic points of the mapping are dense in J.*

One of the most known chaotic equations is the Hénon map. It was introduced on 1975 by the french astronomer M. Hénon [32]. Also you can see in [32] that this map has positive Lyapunov exponents. In the book [38] it was proved that the Hénon map was Devaney chaotic. We will analyze whether the dynamics is ultra Poincaré chaotic using the alpha unpredictability test. Let us consider the following discrete model

$$\begin{aligned} x_{n+1} &= 1 - 1, 4x_n^2 + y_n \\ y_{n+1} &= 0.3x_n. \end{aligned} \tag{6.5}$$

For this system we took the initial values $[-0.275\,185\,753\,099\,546\,79,$ $-0.325\,156\,520\,338\,396\,54]$. Figure 6.1(a) shows the trajectory of system (6.5) within the initial values and figure 6.1(b) represents the solution graphs of each coordinate with respect to index i.

We will implement the alpha unpredictability test through algorithm 1 to system (6.5) with the fixed initial conditions and $\varepsilon_0 = 2.1$. The index i starts at 0 and continues till 10^8. As a result, we obtained 4219 terms for each sequence $\{\zeta_k\}$ and $\{\eta_k\}$, such that inequalities $\|x_{\zeta_k} - x_0\| < \frac{1}{k}$ and $\|x_{\zeta_k + \eta_k} - x_{\eta_k}\| > 2.1$ are satisfied for each $1 \leqslant k \leqslant 4219$.

Table 6.1 presents 11 elements of each sequence.

Next, the results achieved by algorithm 1 can be displayed graphically. For each element, say ζ_γ, within the sequence of convergence $\{\zeta_n\}$, can be drawn a particular graph of solutions of system (6.3) with initial conditions x_0 and $x(\zeta_\gamma)$. We will denote $x_{\text{shift}}(i) = x(i + \zeta_\gamma)$ the solution of the system within $x_0 = x(\zeta_\gamma)$. In these graphs the closeness at 0 will be visible, and the separation bigger than ε_0 between the two solution curves at the corresponding element of the sequence of separation η_n, η_γ. We will use this representation on any result obtained by employing algorithm 1.

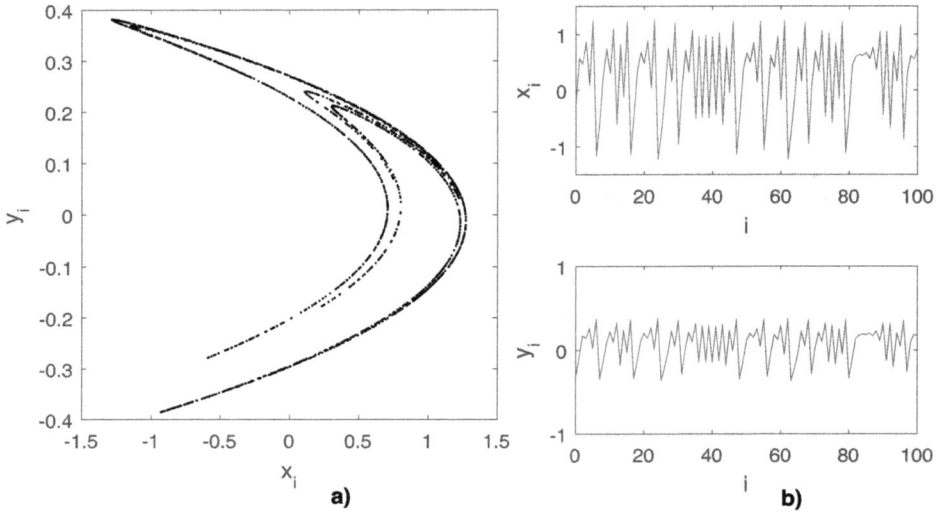

Figure 6.1. Simulations for the solution of system (6.5): (a) trajectory of the solution; (b) the graphs of each coordinate with respect to the index i.

Table 6.1. Selected elements from convergence and separation sequences obtained by algorithm 1 for system (6.5).

m	k	$1/k$	ζ_k	η_k
1	1	1	2	180
2	2	0.5	7	181
3	619	0.001 616	1 527 239	15 364
4	1084	0.000 923	6 504 212	26 361
5	1904	0.000 525	28 566 825	48 105
6	2337	0.000 428	46 003 272	58 728
7	2708	0.000 369	65 836 673	68 727
8	3008	0.000 332	12 056 560	75 921
9	3436	0.000 291	113 770 045	86 963
10	3710	0.000 27	139 217 264	93 357
11	4219	0.000 237	199 876 783	105 711

Following, we will represent an individual graph associated with $\zeta_\gamma = \zeta_2 = 7$. Since it is difficult to analize the two-dimensional graph, we will show the graph of system (6.5) for one dimension, where the distance between $\omega_i = \omega(i)$ and $\omega_{\text{shift}}(i)$, $\omega = x, y$ is bigger than the other dimension at point $i = \eta_\gamma = \eta_2 = 181$. The distance at $\eta_2 = 181$ is bigger in x dimension. In figure 6.2, the blue curve shows the graph of the solution of (6.5), $x_i = x(i)$, where the initial condition is $X_0 = X(0)$, while the red curve is the solution where the initial value is $X_0 = X(7)$, $x_{\text{shift}}(i) = x(i + 7)$, where

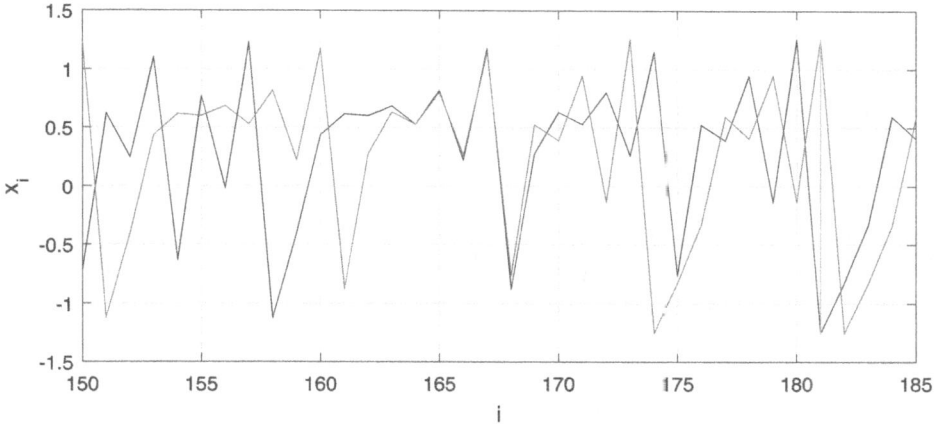

Figure 6.2. The blue curve is a graph of solution $x(i)$ for system (6.5), while the red curve is for $x_{\text{shift}}(i)$. The green line segment connects the points $(181, x(181))$ and $(181, x_{\text{shift}}(181))$ and is used for the distance evaluation between the solutions at $i = 181$.

$X(i) = (x(i), y(i))$ and $X_{\text{shift}}(i) = (x_{\text{shift}}(i), y_{\text{shift}}(i))$. The green line segment connects the points $(181, x(181))$ and $(181, x_{\text{shift}}(181))$.

The length of the green line segment is $|x(181) - x_{\text{shift}}(181)| = 2.512\,604\,8968 > \varepsilon_0$. In our calculation, we noticed that the distances,

$|x(16) - x_{\text{shift}}(16)| = 2.405\,695\,6028,$
$|x(23) - x_{\text{shift}}(23)| = 2.227\,305\,2694,$
$|x(47) - x_{\text{shift}}(47)| = 2.364\,476\,0759,$
$|x(61) - x_{\text{shift}}(61)| = 2.200\,589\,4834,$
$|x(119) - x_{\text{shift}}(119)| = 2.267\,728\,785,$
$|x(126) - x_{\text{shift}}(126)| = 2.160\,138\,8248,$
$|x(174) - x_{\text{shift}}(174)| = 2.402\,419\,8634,$

are bigger than ε_0, while the index values are smaller than η_2. Except for these indexes, it is also evident that two-dimensional distances $\| X(40) - X_{\text{shift}}(40) \| = 2.157\,836\,143$ and $\| X(72) - X_{\text{shift}}(72) \| = 2.192\,999\,1986$. The two-dimensional distance between the two solution curves at $i = 181$ is $\| X(181) - X_{\text{shift}}(181) \| = 2.547\,281\,0502$. Succeeding, let us consider the closeness of solutions $X(i)$ and $X_{\text{shift}}(i)$ of system (6.5) on the interval $[340, 420]$. Figure 6.3 presents the graph of solutions $y(i)$ and $y_{\text{shift}}(i)$ on $[340, 420]$, where the solution curves are close.

It is seen from figure 6.3, that the solutions $y(i)$ and $y_{\text{shift}}(i)$ are close to each other on the closed intervals $[343, 353]$, $[361, 368]$, $[384, 386]$, $[395, 400]$ and $[414, 417]$. The greatest distance between the two solution curves on these intervals is $0.108\,833\,6401$. If we consider the two-dimensional graph, the solutions $X(i)$ and $X_{\text{shift}}(i)$ are close on the closed intervals $[344, 349]$, $[362, 364]$, $[384, 385]$, $[393, 399]$ and $[414, 415]$. The greatest two-dimensional distance between the two solution curves on these intervals is $0.106\,939\,0621$.

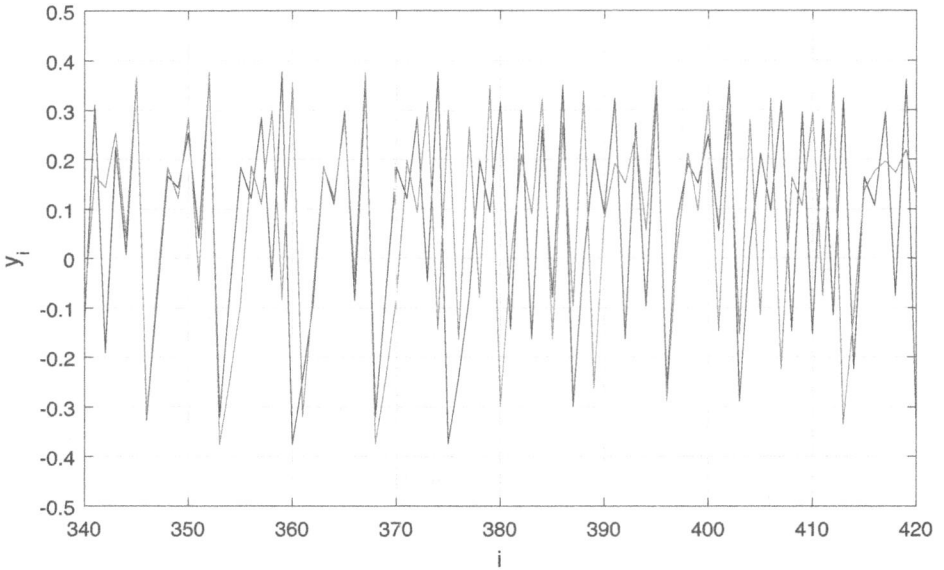

Figure 6.3. The blue and red curves present solutions $y(i)$ and $y_{\text{shift}}(i)$ of system (6.5) on the interval [340, 420].

6.3.2 Considering Li–Yorke chaos

Li–Yorke chaos was introduced in 1975 in paper [54]. To present this definition let J be an interval and consider the autonomous discrete system

$$x_{i+1} = F(x_i), \tag{6.6}$$

where $F: J \longrightarrow J$ is continuous.

Definition 6.3.2. [54] *The function F is said to be chaotic in the sense of Li–Yorke if the following conditions are fulfilled:*
- *For every $k = 1, 2, 3, ...,$ there is a periodic point in J having period k;*
- *There is an uncountable set $S \in J$ (containing noperiodic points), which satisfies the following conditions:*
  A) For every $p, q \in S$ with $p \neq q$

$$\lim_{n \to \infty} \sup |F^n(p) - F^n(q)| > 0 \tag{6.7}$$

and

$$\lim_{n \to \infty} \sup |F^n(p) - F^n(q)| = 0. \tag{6.8}$$

B) For every $p \in S$ and periodic point $q \in J$,

$$\lim_{n \to \infty} \sup |F^n(p) - F^n(q)| > 0. \tag{6.9}$$

In paper [54] it is also proved that the equation,

$$x_{i+1} = 3.9x_i(1 - x_i) \tag{6.10}$$

is Li–Yorke chaotic and shown that the solution with initial condition $x(0) = 0.5$ behaves irregularly. This is seen in figure 6.4.

We will apply the alpha unpredictability test on the system (6.10) within the set initial condition. The index i starts from 0 and continues to 10^8. Let $\varepsilon_0 = 0.7$. As a result, we obtained 17562 terms for the sequence of convergence and the sequence of separation, such that $\|x_{\zeta_k} - x_0\| < \frac{1}{k}$ and $\|x_{\zeta_k + \eta_k} - x_{\tau_k}\| > 0.7$ are satisfied for each $1 \leqslant k \leqslant 17\,562$. Table 6.2 shows 11 elements of each sequence.

Succeeding, let us consider a particular graph associated with one element within the sequence of convergence presented in table 6.2. Let $\zeta_\gamma = \zeta_{10} = 40$. So, consider figure 6.5. The blue curve shows the graph of the solution for (6.10) with the initial condition $x_0 = x(0)$, while the red curve is for the solution if the initial value is $x_0 = x(40)$. The green line segment connects the points $(45, x(45))$ and $(45, x_{\text{shift}}(45))$ and illustrates the distance between the two solutions.

The length of the green line segment is $|x(45) - x_{\text{shift}}(45)| = 0.751\,580\,2112 > \varepsilon_0$. In our calculations, we noticed that

$|x(7) - x_{\text{shift}}(7)| = 0.742\,961\,4716,$
$|x(19) - x_{\text{shift}}(19)| = 0.787\,039\,4876,$
$|x(24) - x_{\text{shift}}(24)| = 0.804\,523\,1135,$
$|x(31) - x_{\text{shift}}(31)| = 0.791\,589\,1888,$
$|x(41) - x_{\text{shift}}(41)| = 0.710\,576\,0726,$

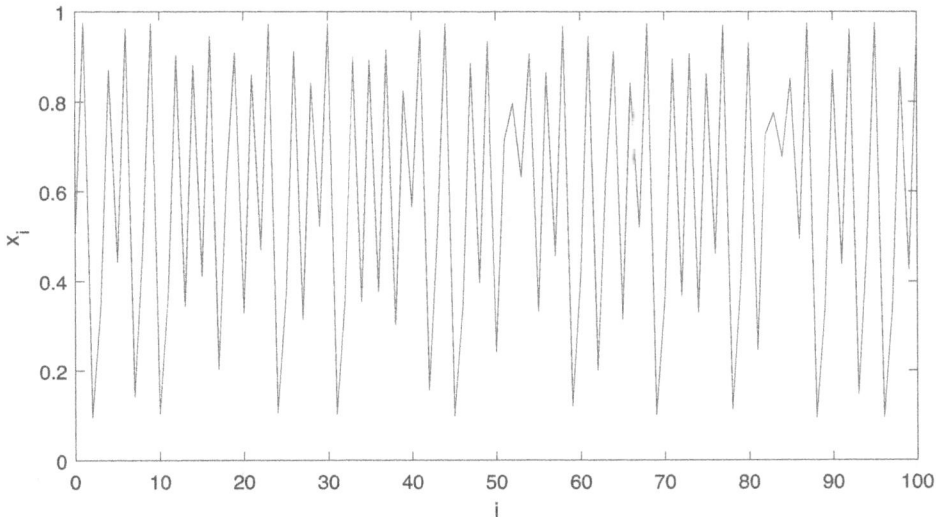

Figure 6.4. The solution of system (6.10) with the given initial condition.

Table 6.2. Selected elements from the convergence and separation sequences obtained for system (6.10) by algorithm 1.

m	k	$1/k$	ζ_k	η_k
1	1	1	2	2
2	10	0.1	40	45
3	1905	0.000 525	1 190 408	15 836
4	3764	0.000 266	4 619 782	30 909
5	5251	0.000 190	9 081 712	42 939
6	6997	0.000 143	16 316 573	57 253
7	8298	0.000 121	22 776 341	67 995
8	9549	0.000 105	29 930 023	78 430
9	11 026	0.000 091	39 821 708	90 551
10	12 460	0.000 080	50 743 495	10 207
11	17 562	0.000 057	99 977 681	16 027

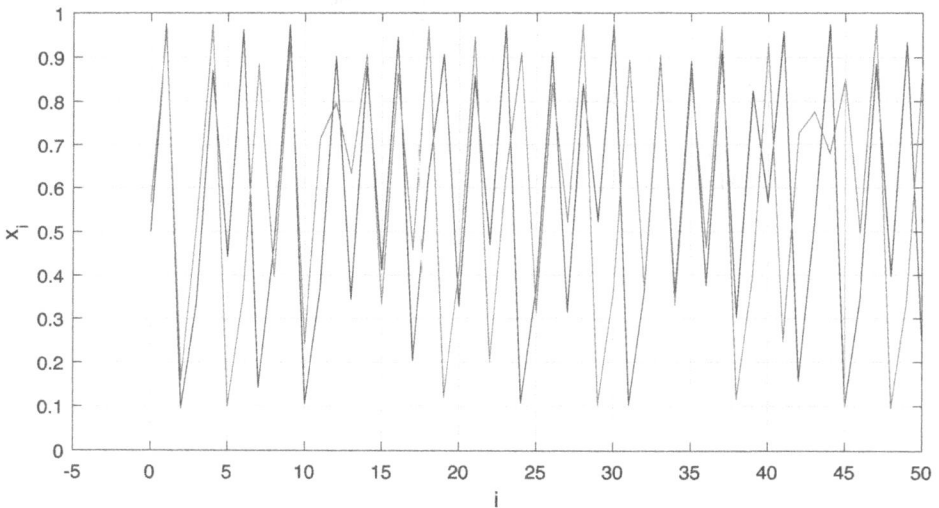

Figure 6.5. The blue curve is a graph of the solution of the system, $x(i)$, where the initial condition is $x_0 = x(0)$, while the red curve is for the solution with initial condition $x_0 = x(40)$, $x_{\text{shift}}(i)$. The vertical green line segment connects the points $(45, x(45))$ and $(45, x_{\text{shift}}(45))$ and illustrates the distance between the solutions at index $i = 45$.

where the index values are smaller than η_{10}. Next, let us consider the closeness of solutions $x(i)$ and $x_{\text{shift}}(i)$ of the system (6.10) on the interval [30, 100]. Figure 6.6 displays the graph of solutions on [30, 100]. It is seen from figure 6.6 that the solutions $x(i)$ and $x_{\text{shift}}(i)$ are close to each other on the closed intervals [32, 37], [57, 58], [65, 66], [79, 80], and [94, 98]. The greatest distance between the two solution curves on these intervals is 0.118 790 7046.

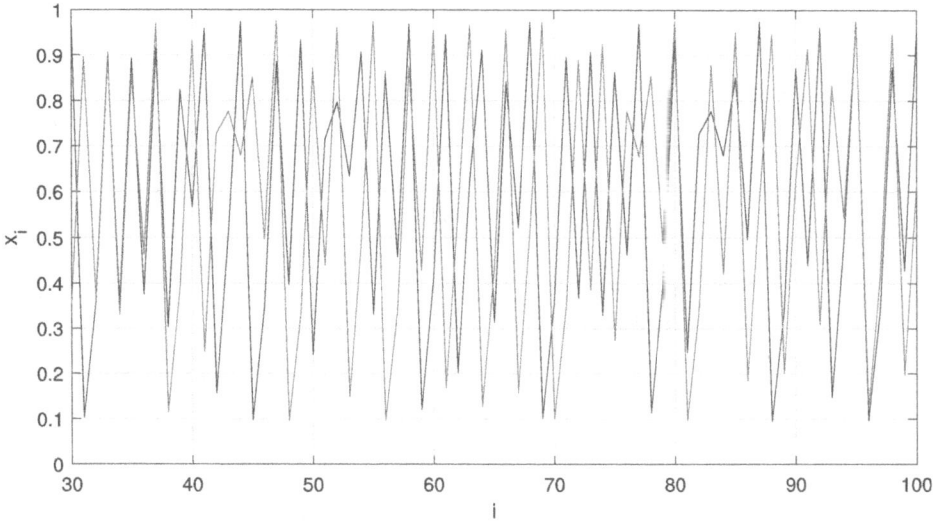

Figure 6.6. The blue and red curve present the solutions of $x(i)$ and $x_{\text{shift}}(i)$ of system (6.10) on the interval [30, 100].

6.3.3 Explaining results of bifurcation diagrams analysis

Systems chaotic by bifurcation diagram analysis possess periodic solutions [31]. In [14] one can find system (6.11) and confirm it is period-doubling route chaotic,

$$
\begin{aligned}
x_1' &= 10(x_2 - x_1) \\
x_2' &= 99.51x_1 - x_1x_3 - x_2 \\
x_3' &= x_1x_2 - \frac{8}{3}x_3.
\end{aligned}
\tag{6.11}
$$

The initial data, [23.319 088 231 571 342, −15.117 252 730 042 82, 130.763 839 152 679 31],

were used in the research for the illustration of irregularity. Figure 6.7(a) presents the trajectory of system (6.11) within the initial conditions, while figure 6.7(b) presents the solution graphs of each chaotic coordinate.

We will execute the alpha unpredictability test through algorithm 1 to system (6.11) with the fixed initial conditions. The time series starts at 0 and continues to 1.3×10^6, partitioned into pieces with distance 0.01 and $\varepsilon_0 = 45$. To ensure that the system (6.11) will satisfy the alpha unpredictability test for $\varepsilon_0 = 45$, we skipped $t_{\text{fix}} = 631.36$ while evaluating the sequence of convergence $\{t_k\}$. Within the given conditions and time interval, we obtained 347 terms for each sequence $\{t_k\}$ and $\{s_k\}$, such that $\|x_{t_k} - x_0\| < \frac{1}{k}$ and $\|x_{t_k + s_k} - x_{s_k}\| > 45$ are satisfied for each $1 \leqslant k \leqslant 347$. Table 6.3 provides 10 elements of each sequence.

Succeeding, the results achieved by algorithm 1 can be displayed graphically. For each element, say t_γ, within the sequence of convergence $\{t_n\}$, can be drawn a particular graph of solutions of the system (6.2) with initial conditions x_0 and $x(t_\gamma)$.

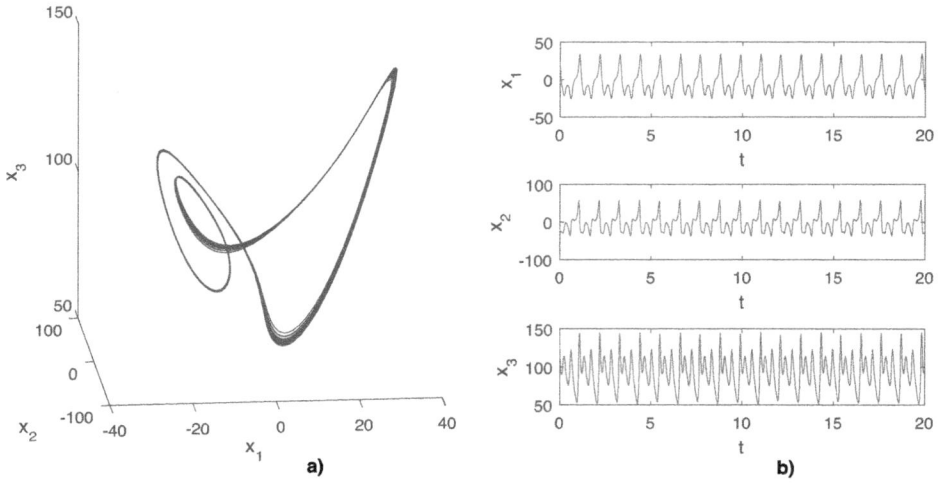

Figure 6.7. Simulations for the solution of system (6.11): (a) the trajectory of the solution; (b) the solution graphs of each coordinate with respect to time t. Reproduced with permission from [5]. Copyright © 2011–2024 L & H Scientific Publishing. All rights reserved.

Table 6.3. Selected elements from the convergence and separation sequences obtained by algorithm 1 for system (6.11).

m	k	$1/k$	t_k	s_k
1	1	1	640.35	117.87
2	38	0.012 195	13 076.12	137.74
3	79	0.006 711	68 393.97	160.89
4	111	0.003 937	118 970.29	180.69
5	146	0.003 311	234 621.96	200.54
6	175	0.002 519	280 456.75	217.08
7	207	0.002 079	331 033.07	233.63
8	258	0.001 706	592 366.1	263.38
9	310	0.001 464	846 071.02	293.15
10	347	0.001 186	1 025 217.53	313

We will denote $x_{\text{shift}}(t) = x(t + t_\gamma)$ the solution of the system within $x_0 = x(t_\gamma)$. In these graphs the closeness at 0 will be visible, and the separation bigger than ε_0 between the two solution curves at the corresponding element of the sequence of separation $\{s_n\}$, s_γ. We will use this representation on any result obtained by employing algorithm 1.

Following this description, we will draw a particular graph using $t_\gamma = t_1 = 640.35$. Since it is difficult to analize the three-dimensional graph, we will show the graph of the solution for system (6.11) coordinate-wise with respect to

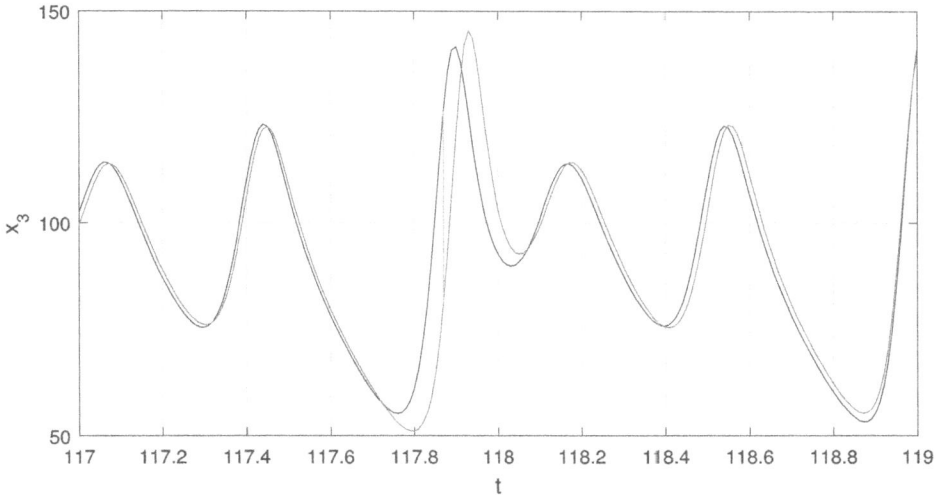

Figure 6.8. The blue curve shows a graph of the third coordinate, $x_3(t)$, while the red curve is for $x_{3_{\text{shift}}}(t)$. The vertical green line segment connects the points $(117.87, x_3(117.87))$ and $(117.87, x_{3_{\text{shift}}}(117.87))$ and illustrates the distance between the solutions at moment $t = 117.87$.

time, where distance between $x_\omega(t)$ and $x_{\omega_{\text{shift}}}(t)$, $\omega = 1, 2, 3$, is bigger than for other coordinates at $t = s_\gamma = s_1 = 117.87$. In figure 6.8, the blue curve shows graph of the third coordinate $x_3(t)$, for the solution of system (6.11), where the initial condition is $x_0 = x(0)$, while the red curve is the coordinate $x_{3_{\text{shift}}}(t) = x_3(640.35 + t)$, of the solution with $x_0 = x(640.35)$, such that $x(t) = (x_1(t), x_2(t), x_3(t))$ and $x_{\text{shift}}(t) = (x_{1_{\text{shift}}}(t), x_{2_{\text{shift}}}(t), x_{3_{\text{shift}}}(t))$. The green line segment connects the points $(117.87, x_3(117.87))$ and $(117.87, x_{3_{\text{shift}}}(117.87))$. The length of the green line segment is $|x_3(117.87) - x_{3_{\text{shift}}}(117.87)| = 44.934\ 797\ 66$, which is the greatest length between the solution curves until this time in x_3 coordinate. For x_1, the greatest length between the solution curves $|x_1(t) - x_{1_{\text{shift}}}(t)| = 12.827\ 509\ 11$ occurs at $t = 48.51$, while for x_2 $|x_2(t) - x_{2_{\text{shift}}}(t)| = 42.033\ 488\ 27$ occurs at $t = 48.47$. The space length at $t = 117.87$ is $\|x(117.87) - x_{\text{shift}}(117.87)\| = 47.036\ 037\ 13$, which is the longest until this time value.

If someone proceeds to further calculate the distance $\|x(t) - x_{\text{shift}}(t)\|$, they will notice that all found s_n satisfy the inequality $\|x(s_n) - x_{\text{shift}}(s_n)\| > 45$. Going on, we will show that this is true here and for others t_ns. For this purpose we chose four t_k values from table 6.3, and in table 6.4 we presented the lengths $\|x(t) - x_{\text{shift}_l}(t)\|$, where $x_{\text{shift}_l}(t) = x(t_l + t)$, for every s_k value of table 6.3.

Each of the distances shown above is greater than $\varepsilon_0 = 45$, confirming our claim that *every element of the sequence of separations satisfies the inequality* $\|x(s_n) - x_{\text{shift}_l}(s_n)\| > \varepsilon_0$, *for every* $1 \leqslant l, n \leqslant k$. Possibly, the alpha unpredictability test can be applied in this way to recognize or to be at least an additional method for analysis of chaos with multiple periods. Following this result, let us consider the closeness of the first coordinates for solutions $x(t)$ and $x_{\text{shift}}(t)$ of the system (6.11) on the interval $[0, 2]$. Figure 6.9 presents graphs of coordinates $x_1(t)$ and $x_{1_{\text{shift}}}(t)$ on

Table 6.4. The distance $\|x(t) - x_{\text{shift}_l}(t)\|$ for every s_k value of table 6.3, for $l = 1, 79, 175, 347$.

t	$\|x(t) - x_{\text{shift}_1}(t)\|$	$\|x(t) - x_{\text{shift}_{79}}(t)\|$	$\|x(t) - x_{\text{shift}_{175}}(t)\|$	$\|x(t) - x_{\text{shift}_{347}}(t)\|$
$s_1 = 117.87$	47.036 087 13	47.134 699 56	47.138 039 11	47.134 9988
$s_{38} = 137.74$	48.969 878 39	49.090 023 07	49.094 092 86	49.090 3839
$s_{79} = 160.89$	48.263 5579	48.384 666 58	48.388 769 17	48.385 0257
$s_{111} = 180.69$	46.367 003 21	46.461 766 46	46.464 975 64	46.462 044 34
$s_{146} = 200.54$	48.856 564 57	48.961 846 87	48.965 412 34	48.962 152 21
$s_{175} = 217.08$	49.687 216 77	49.798 472 11	49.802 240 05	49.798 791 75
$s_{207} = 233.63$	48.778 364 21	48.898 843 99	48.902 925 12	48.899 186 92
$s_{258} = 263.38$	49.702 384 09	49.819 121 38	49.823 075 17	49.819 447 91
$s_{310} = 293.15$	48.307 696 63	48.428 763 36	48.432 8644	48.429 096 14
$s_{347} = 313$	45.860 562 92	45.983 393 55	45.987 554 75	45.983 727 28

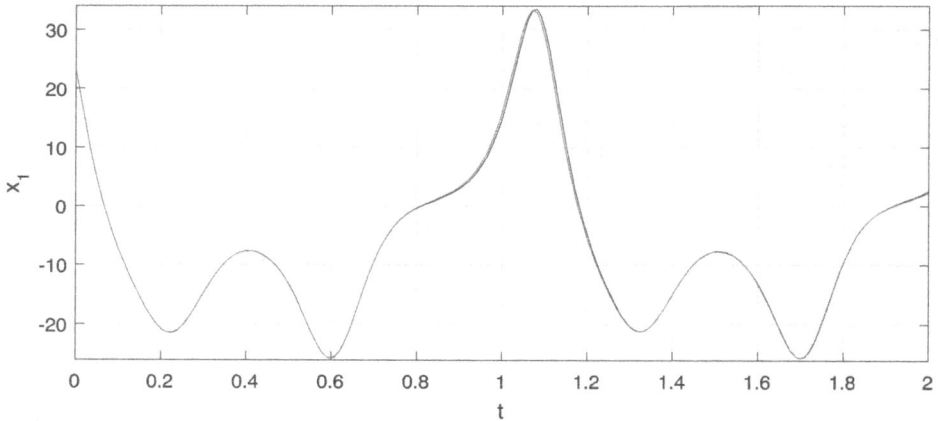

Figure 6.9. The blue and red curve present coordinates $x_1(t)$ and $x_{1_{\text{shift}}}(t)$ of system (6.11) on the interval $[0, 2]$. Since these curves are close to each other, in the picture they are seen as a single curve (in red).

$[0, 2]$, where the solution curves seen very near. One can notice from figure 6.9, that the solutions $x_1(t)$ and $x_{1_{\text{shift}}}(t)$ are close to each other on the closed intervals $[0, 0.82]$, $[1.3, 1.57]$, $[1.67, 1.71]$, and $[1.81, 1.94]$. The greatest distance between the two curves on these intervals is 0.092 787 609. If we consider the three-dimensional graph, the solutions $x(t)$ and $x_{\text{shift}}(t)$ are close on the interval $[0.03, 0.49]$. From our calculations, the greatest three-dimensional distance between the two solutions on this interval is 0.096 167 732.

6.3.4 Rössler System analyzed by its Lyapunov exponents

The Rössler system was first introduced by Otto E. Rössler in his paper written in 1976 [68], where several arguments were provided that showed it to be chaotic. The system that we used, which has positive Lyapunov exponents [75], is the following,

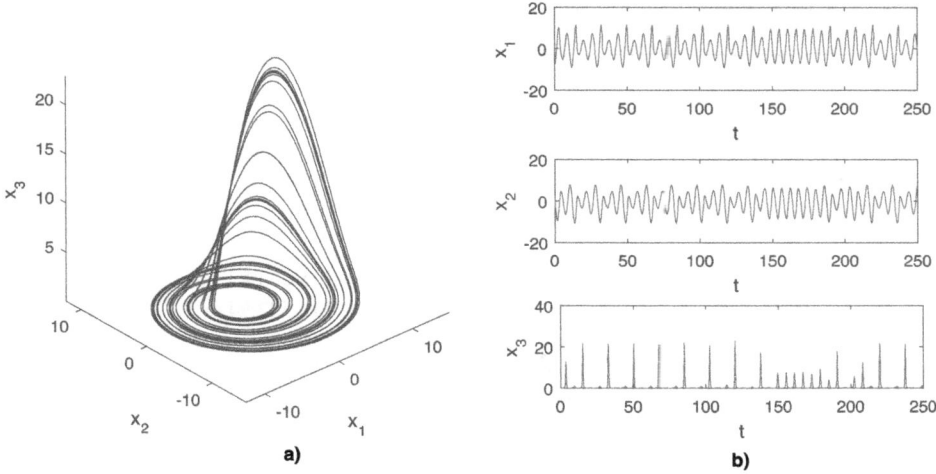

Figure 6.10. Simulations for the solution of system (6.12): (a) the trajectory of the solution; (b) the graphs of each coordinate.

$$
\begin{aligned}
x_1' &= -x_2 - x_3 \\
x_2' &= x_1 + 0.2x_2 \\
x_3' &= 0.2 + x_1 x_3 - 5.7x_3
\end{aligned}
\tag{6.12}
$$

with the point $[-7.920\,855\,070\,468\,1606, -0.322\,131\,574\,105\,066\,99, 0.014\,707\,110\,762\,462\,17]$ as an initial one. Figure 6.10(a) shows the trajectory of system (6.12), where the fixed initial conditions and figure 6.10(b) represents the solution graphs of each coordinate. We will implement the alpha unpredictability test through algorithm 1 to system (6.12) with the chosen initial condition on interval $[0, 1.9 \times 10^6]$ divided into pieces with distance 0.01 and $\varepsilon_0 = 22$. In order that the system (6.12) satisfy the alpha unpredictability test for $\varepsilon_0 = 22$, we skipped $t_{\text{fix}} = 35$ while finding the sequence of convergence $\{t_k\}$. For this system, we obtained 157 terms for each sequence $\{t_k\}$ and $\{s_k\}$, such that $\|x_{t_k} - x_0\| < \frac{1}{k}$ and $\|x_{t_k + s_k} - x_{s_k}\| > 22$ are satisfied for each $1 \leqslant k \leqslant 157$. Table 6.5 provides 10 elements of each sequence.

After we acquire the results, let us present the graph for $t_1 = 64.38$, which will be our selected t_γ. Since it is difficult to analize the three-dimensional graph, we will show graph of solution for system (6.12) coordinate-wise concerning time, where distance between $x_\omega(t)$ and $x_{\omega_{\text{shift}}}(t)$, $\omega = 1, 2, 3$, is bigger than the other coordinates at the moment $t = s_\gamma = s_1 = 20.85$. From our results, the distance at $s_1 = 20.85$ is bigger in x_3 coordinate. In figure 6.11, the blue curve shows a graph of the third coordinate for (6.12), $x_3(t)$, with the initial condition $x_0 = x(0)$, while the red curve is for the third coordinate, if $x_0 = x(64.38)$, $x_{3_{\text{shift}}}(t) = x_3(64.38 + t)$, where $x(t) = (x_1(t), x_2(t), x_3(t))$ and $x_{\text{shift}}(t) = (x_{1_{\text{shift}}}(t), x_{2_{\text{shift}}}(t), x_{3_{\text{shift}}}(t))$. The vertical green line segment connects the points $(20.85, x_2(20.85))$ and $(20.85, x_{2_{\text{shift}}}(20.85))$. If someone would calculate length of the green segment would have found that

Table 6.5. Selected elements from the convergence and separation sequences obtained by algorithm 1 for system (6.12).

m	k	$1/k$	t_k	s_k
1	1	1	64.38	20.85
2	16	0.0625	2253.47	512.43
3	30	0.033 333	19 922.92	916.43
4	56	0.017 857	97 107.19	2251.43
5	77	0.012 987	270 473.75	2796.21
6	89	0.011 236	405 263.82	3024.03
7	111	0.009 009	707 522.94	3620.64
8	126	0.007 937	984 335.56	3754.99
9	146	0.006 849	1503 902.36	3848.64
10	157	0.006 369	1878 637.91	3848.75

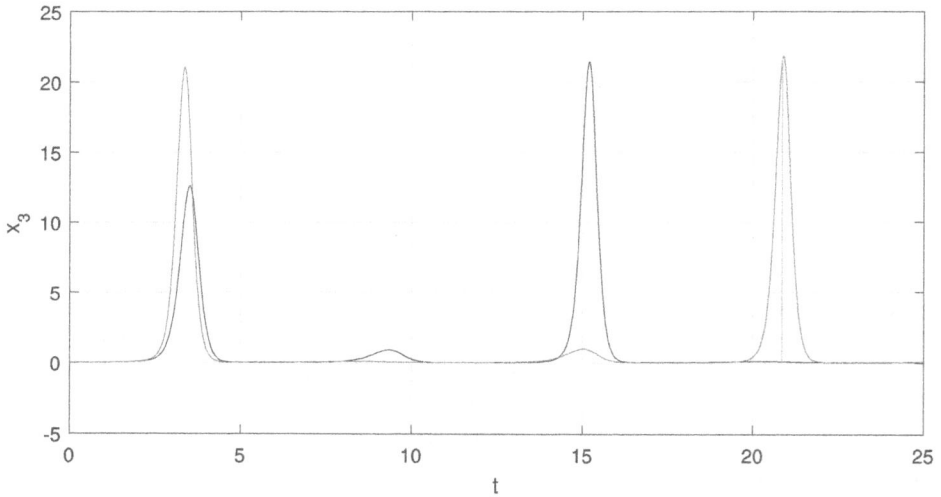

Figure 6.11. The blue curve shows the graph of the coordinate $x_3(t)$, while the red curve is for $x_{3_{\text{shift}}}(t)$. The vertical green line segment connects the points $(20.85, x_2(20.85))$ and $(20.85, x_{2_{\text{shift}}}(20.85))$ and illustrates the distance between solutions at the moment $t = 20.85$.

$|x_3(20.85) - x_{3_{\text{shift}}}(20.85)| = 21.571\,828\,63$, which is the greatest length between the solution curves until this time in x_3 coordinate. In x_1 the greatest distance between the solutions, $|x_1(t) - x_{1_{\text{shift}}}(t)| = 7.887\,399\,774$, occurs at $t = 20.46$, while in the second coordinate, $|x_2(t) - x_{2_{\text{shift}}}(t)| = 8.506\,374\,252$, occurs at $t = 19.49$. The space length at $t = 20.85$ is $\|x(20.85) - x_{\text{shift}}(20.85)\| = 22.082\,370\,84$, and is the largest until this time value.

Following this result, let us consider the closeness of solutions $x(t)$ and $x_{\text{shift}}(t)$ of the system (6.12) on the interval [822, 870]. Figure 6.12 presents the graph of solutions $x_3(t)$ and $x_{3_{\text{shift}}}(t)$ on [822, 870], where the solution curves are near.

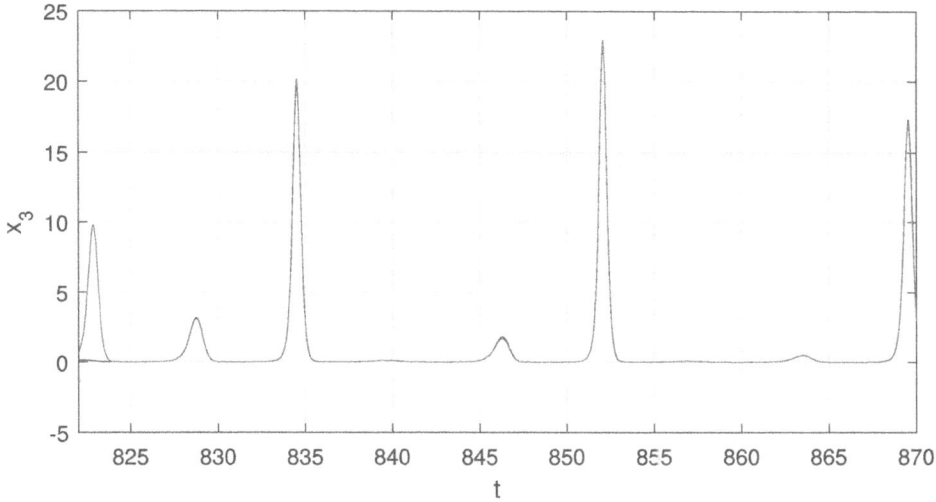

Figure 6.12. The blue and red curves present the coordinates $x_3(t)$ and $x_{3_{\text{shift}}}(t)$ of system (6.12) on the interval [822, 870].

It is seen from the figure, that the coordinates $x_3(t)$ and $x_{3_{\text{shift}}}(t)$ are near to each other on the closed interval [823.79, 869.51]. The greatest distance between the two solution curves on this interval is 0.099 134 999. If we consider the three-dimensional graph, the solutions $x(t)$ and $x_{\text{shift}}(t)$ are close on the closed interval [823.8, 865.06]. The greatest three-dimensional distance between the two solutions on this interval is 0.099 931 352.

6.3.5 Ikeda map with positive Lyapunov exponents

In this subsection, we consider the following equation taken from the book [32], where it was proven that equation (6.13) has positive Lyapunov exponents,

$$
\begin{aligned}
x_{n+1} &= 1 + 0.9(x_n \cos(\tau_n) - y_n \sin(\tau_n)) \\
y_{n+1} &= 0.9(x_n \sin(\tau_n) + y_n \cos(\tau_n)),
\end{aligned}
\tag{6.13}
$$

where $\tau_n = 0.4 - \left(\dfrac{6}{1 + x_n^2 + y_n^2}\right)$

For this system we took [0, 0] as initial conditions. Figure 6.13(a) shows the trajectory of the system within initial conditions and figure 6.13(b) represents the graphs of each coordinate with respect to index i. We will perform the alpha unpredictability test through algorithm 1 to system (6.13) with the chosen initial condition and $\varepsilon_0 = 2$. The index interval starts from 0 and continues to 3×10^6. As a result, we have obtained 754 terms for each sequence $\{\zeta_k\}$ and $\{\eta_k\}$, such that inequalities $\|x_{\zeta_k} - x_0\| < \frac{1}{k}$ and $\|x_{\zeta_k + \eta_k} - x_{\eta_k}\| > 2$ are satisfied for each $1 \leqslant k \leqslant 754$. Table 6.6 provides 11 elements of the sequences. After acquiring the results, we can sketch the graph for one of them, say the graph associated with $\zeta_\gamma = \zeta_{16} = 1193$. Since it is difficult to analyze the two-dimensional graph, we will

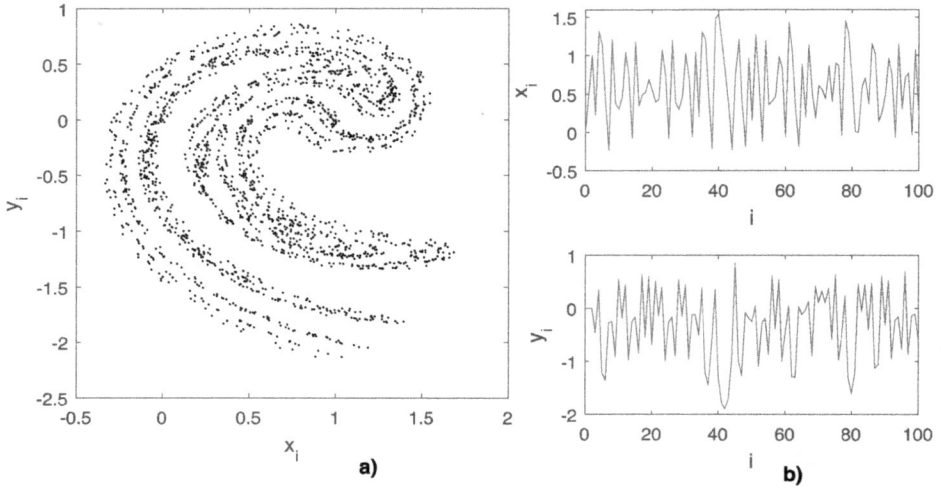

Figure 6.13. Simulations for the solution of system (6.13): (a) the trajectory; (b) the graphs of each coordinate with respect to index i.

Table 6.6. Selected elements from the convergence and separation sequences obtained by algorithm 1 for system (6.13).

m	k	$1/k$	t_k	s_k
1	1	1	2	34
2	16	0.0625	1193	384
3	95	0.010 526	204 042	1175
4	160	0.006 25	1 236 723	2221
5	201	0.004 975	2 394 484	2920
6	289	0.003 46	8 255 283	4048
7	354	0.002 825	14 874 258	4947
8	407	0.002 457	22 484 165	5858
9	520	0.001 923	51 130 431	7508
10	693	0.001 443	161 661 433	9873
11	754	0.001 326	295 347 460	10 735

show the graph of system (6.13) for one dimension, where the distance between $\omega_i = \omega(i)$ and $\omega_{\text{shift}}(i)$, $\omega = x, y$ is bigger than the other dimensions at index $i = \eta_\gamma = \eta_{16} = 384$. The distance at $\eta_{16} = 384$ is bigger in y dimension. In figure 6.14, the blue curve shows the graph of solution of (6.13), $y_i = y(i)$, where the initial condition is $X_0 = X(0)$, while the red curve is for the solution where the initial value is $X_0 = X(1193)$, $y_{\text{shift}}(i) = y(1193 + i)$, where $X(i) = (x(i), y(i))$ and $X_{\text{shift}}(i) = (x_{\text{shift}}(i), y_{\text{shift}}(i))$. The green line segment connects the points $(384, y(384))$ and $(384, y_{\text{shift}}(384))$.

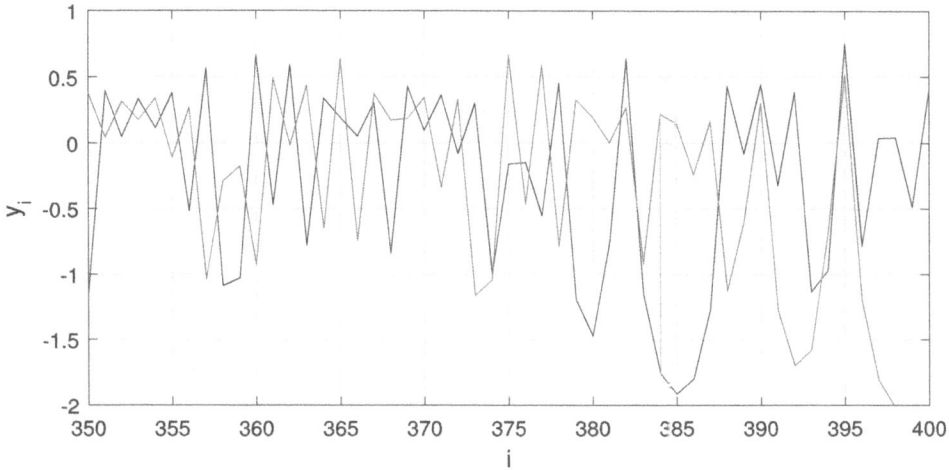

Figure 6.14. The graph for system (6.13) with respect to y_i and index i. The blue line is a graph for the solution $y(i)$, while the red one is for $y_{shift}(i)$. The vertical green line segment connects the points $(384, y(384))$ and $(384, y_{shift}(384))$ and illustrates the distance between the solutions at index $i = 384$.

From our calculations, we note that the length of the green segment $|y(384) - y_{shift}(384)| = 1.980\,393\,236 < \varepsilon_0$ reaches its maximum at index $i = 384$. Someone can find that

$$|y(22) - y_{shift}(22)| = 2.350\,274\,265,$$
$$|y(38) - y_{shift}(38)| = 2.085\,833\,749,$$
$$|y(40) - y_{shift}(40)| = 2.261\,857\,315,$$
$$|y(42) - y_{shift}(42)| = 2.175\,361\,502,$$
$$|y(55) - y_{shift}(55)| = 2.143\,415\,044,$$
$$|y(95) - y_{shift}(95)| = 1.983\,853\,779,$$
$$|y(106) - y_{shift}(106)| = 2.239\,737\,095,$$
$$|y(128) - y_{shift}(128)| = 1.991\,610\,248,$$
$$|y(129) - y_{shift}(129)| = 2.243\,506\,045,$$
$$|y(131) - y_{shift}(131)| = 2.434\,072\,116,$$
$$|y(190) - y_{shift}(190)| = 2.145\,400\,124,$$
$$|y(192) - y_{shift}(192)| = 2.212\,754\,503,$$
$$|y(212) - y_{shift}(212)| = 2.348\,997\,502,$$
$$|y(239) - y_{shift}(239)| = 2.014\,821\,546,$$
$$|y(242) - y_{shift}(242)| = 2.167\,259\,729,$$
$$|y(282) - y_{shift}(282)| = 2.281\,796\,984,$$
$$|y(288) - y_{shift}(288)| = 2.500\,601\,744.$$

Investigating these results, except the distance at $i = 95$, all other distances are bigger than ε_0. Except for these indexes, this is also evident at two-dimensional distances

$$\|X(20) - X_{shift}(20)\| = 2.112\,599\,494,$$
$$\|X(95) - X_{shift}(95)\| = 2.007\,853\,134,$$

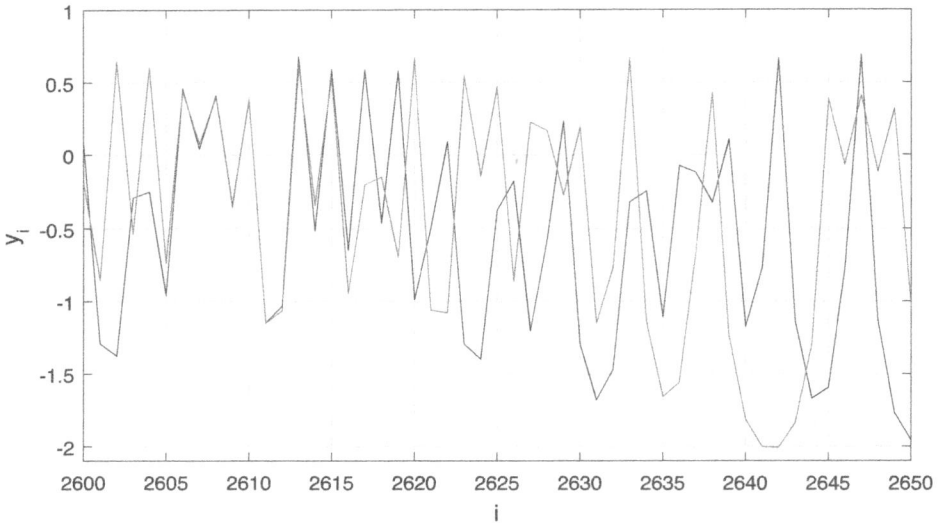

Figure 6.15. The blue and red curves present the solutions $y(i)$ and $y_{\text{shift}}(i)$ of system (6.13) on the interval [2600, 2650].

$$\|X(120) - X_{\text{shift}}(120)\| = 2.012\ 945\ 907,$$
$$\|X(123) - X_{\text{shift}}(123)\| = 2.211\ 804\ 315,$$
$$\|X(125) - X_{\text{shift}}(125)\| = 2.147\ 501\ 466,$$
$$\|X(126) - X_{\text{shift}}(126)\| = 2.219\ 332\ 021,$$
$$\|X(184) - X_{\text{shift}}(184)\| = 2.033\ 164\ 228,$$
$$\|X(189) - X_{\text{shift}}(189)\| = 2.084\ 321\ 694,$$
$$\|X(239) - X_{\text{shift}}(239)\| = 2.015\ 012\ 267,$$
$$\|X(244) - X_{\text{shift}}(244)\| = 2.304\ 006\ 889,$$
$$\|X(284) - X_{\text{shift}}(284)\| = 2.151\ 226\ 213,$$
$$\|X(285) - X_{\text{shift}}(285)\| = 2.025\ 867\ 605,$$

The two-dimensional distance at $i = 384$ is $\|X(384) - X_{\text{shift}}(384)\| = 2.293\ 721\ 422$. Next, let us consider the closeness of solutions $X(i)$ and $X_{\text{shift}}(i)$ of the system (6.13) on the interval [2600, 2650]. Figure 6.15 presents the graph of solutions $y(i)$ and $y_{\text{shift}}(i)$ on [2600, 2650], where the solution curves are close. From figure 6.15, note that the solutions $y(i)$ and $y_{\text{shift}}(i)$ are close to each other on the closed interval [2606, 2615]. The greatest distance between the two solution curves on this interval is 0.054 321 108. If we consider the two-dimensional graph, the solutions $X(i)$ and $X_{\text{shift}}(i)$ are close on the closed interval [2607, 2612]. The greatest two-dimensional distance between the two solution curves on this interval is 0.041 486 722.

6.3.6 Intermittency through the ultra Poincaré test

Intermittency or intermittent chaos is a periodic motion where at some specific time chaotic motions burst [53]. Consider the most well-known intermittent system [2], which has positive Lyapunov exponents,

$$
\begin{aligned}
x_1' &= 10(x_2 - x_1) \\
x_2' &= -x_1 x_3 + 166.29 x_1 - x_2 \\
x_3' &= x_1 x_2 - \frac{8}{3} x_3
\end{aligned}
\tag{6.14}
$$

with the initial point $[-6.902\,710\,153\,782\,7207, 6.121\,428\,586\,861\,6205, 146.734\,814\,043\,078\,05]$. Figure 6.16(a) shows the trajectory of system (6.14) having the fixed initial conditions, while figure 6.16(b) represents the graphs of each coordinate with respect to time t.

We will apply algorithm 1 on system (6.14) with the chosen initial condition and $\varepsilon_0 = 200$. The time interval starts from 0 and continues to 1.5×10^6, partitioned into pieces with distance 0.01. For the system (6.14) to satisfy the alpha unpredictability test for $\varepsilon_0 = 200$, we skip $t_{\text{fix}} = 11.64$ while finding the sequence $\{t_k\}$. Within the given conditions and time interval, we obtained 91 terms for each sequence $\{t_k\}$ and $\{s_k\}$, such that $\|x_{t_k} - x_0\| < \frac{1}{k}$ and $\|x_{t_k + s_k} - x_{s_k}\| > 200$ are satisfied for each $1 \leqslant k \leqslant 91$. Table 6.7 exhibits 10 elements of the sequence of convergence and the sequence of separation. Succeeding, we will sketch the graph of one element of the sequence of convergence exhibited in table 6.8, say for $t_\gamma = t_1 = 15.43$. Since it is

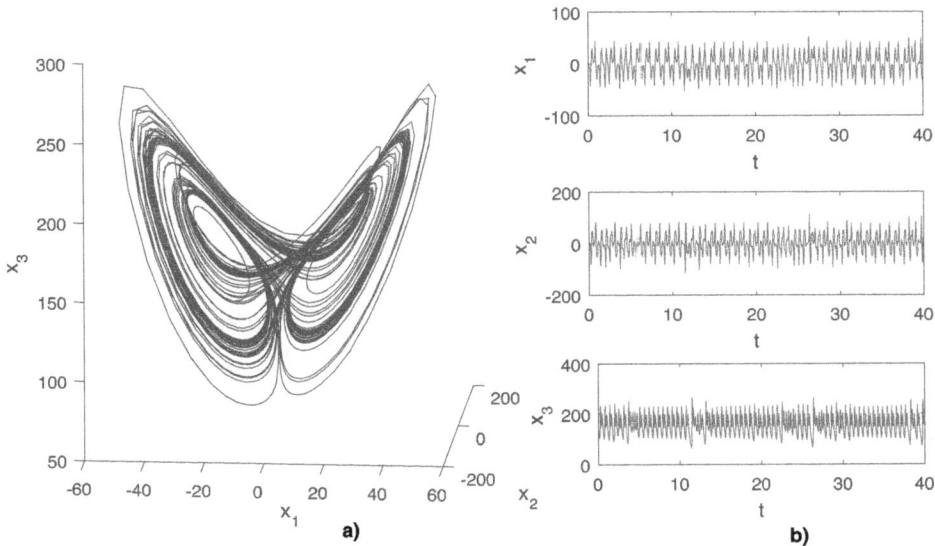

Figure 6.16. Simulations for the solution of system (6.14): (a) the trajectory: (b) the graphs of each coordinate with respect to time t.

Table 6.7. Selected elements from the convergence and separation sequences obtained by algorithm 1 for system (6.14).

m	k	$1/k$	t_k	s_k
1	1	1	15.43	96.98
2	12	0.083 333	2994.48	2322.64
3	23	0.043 478	12 029.93	4237.28
4	31	0.032 258	51 983.29	6852.58
5	45	0.022 222	162 216.36	11 141.28
6	56	0.017 857	401 509.63	14 399.56
7	63	0.015 873	580 535.22	16 311.43
8	70	0.014 286	802 996.63	18 649.65
9	79	0.012 658	1 016 186.84	20 581.15
10	91	0.010 989	1 420 323.21	24 758.91

Table 6.8. The result of the alpha unpredictability test applied to system (6.15). The time interval $[0, 1.2 \times 10^6]$ and the values $m_0 = 14$, $\sigma_0 = 0.15$, $\varepsilon_0 = 2.8$ are used. According to the result of the alpha unpredictability test shown in the table, the test is successful for system (6.15), and therefore, the system admits an alpha unpredictable solution.

n	$1/(n + m_0)$	t_n	s_n
1	0.066 666 667	115.97	23.55
69	0.012 048 193	5632.99	4694.84
125	0.007 194 245	19 260.76	15 392.37
174	0.005 319 149	53 962.87	29 485.77
228	0.004 132 231	107 313.81	44 615.45
276	0.003 448 276	191 797.09	64 090.49
338	0.002 840 909	348 109.94	92 989.3
387	0.002 493 766	528 701.98	119 583.17
430	0.002 252 252	718 208.65	146 308.61
493	0.001 972 387	1 007 781.9	191 617.32

difficult to analize the three-dimensional graph, we will graph the solution for system (6.14) for one dimension with respect to time, where the distance between $x_\omega(t)$ and $x_{\omega_{shift}}(t)$, $\omega = 1, 2, 3$ is bigger than the other dimensions at time $t = s_\gamma = s_1 = 96.98$, which occurs in x_2 dimension. In figure 6.17, the blue curve graphs the second coordinate of the solution for (6.14), with initial condition $x_0 = x(0)$, while the red curve is for the second coordinate of the solution, with the initial condition $x_0 = x(15.14)$, $x_{2_{shift}}(t) = x_2(15.43 + t)$, where $x(t) = (x_1(t), x_2(t), x_3(t))$ and $x_{shift}(t) = (x_{1_{shift}}(t), x_{2_{shift}}(t), x_{3_{shift}}(t))$. The vertical green line segment connects the points $(96.98, x_2(96.98))$ and $(96.98, x_{2_{shift}}(96.98))$. The length of the vertical green line segment is equal to $|x_2(96.98) - x_{2_{shift}}(96.98)| = 192.443\,61$, and is the greatest

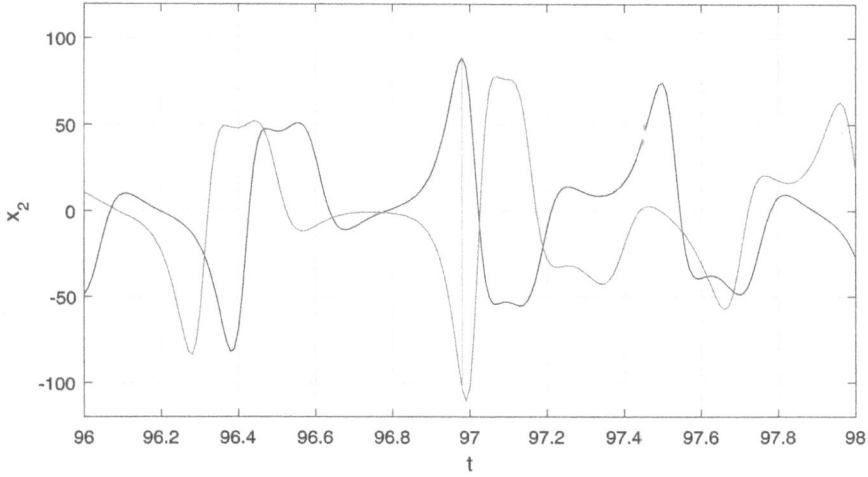

Figure 6.17. The blue curve is the graph of the second coordinate, $x_2(t)$, of the solution for system (6.14), while the red curve is for the second coordinate $x_{2_{shift}}(t)$. The vertical green line segment connects the points $(96.98, x_2(96.98))$ and $(96.98, x_{2_{shift}}(96.98))$ and illustrates the distance between the solutions at the moment $t = 96.98$.

length between the solutions until this time in the second coordinate. In x_1 the greatest length between the solution curves $|x_1(t) - x_{1_{shift}}(t)| = 84.048\,97$ occurs at $t = 25.1$, while in x_3 dimension $|x_3(t) - x_{3_{shift}}(t)| = 173.937\,22$ occurs at $t = 47.51$. The three-dimensional distance at $t = 96.98$ is $\|x(96.98) - x_{shift}(96.98)\| = 207.960\,4947$, which is the biggest until this point.

Following this result, let us consider the closeness of solutions $x(t)$ and $x_{shift}(t)$ of the system (6.14) on the interval [4772, 4786]. Figure 6.18 presents the graph of coordinates $x_2(t)$ and $x_{2_{shift}}(t)$ on [4772, 4786], where the solution curves are near. It is seen from figure 6.18, that $x_2(t)$ and $x_{2_{shift}}(t)$ are close to each other on the closed intervals [4775.88, 4776.07] and [4776.53, 4776.56]. The greatest distance between the two curves on these intervals is $0.469\,810\,895$. If we consider the space graph, the solutions $x(t)$ and $x_{shift}(t)$ are close on the closed interval [4775.88, 4776.07]. The greatest three-dimensional distance between the two solutions on this interval is $0.527\,310\,061$.

6.4 Alpha unpredictability test is positive for known strange non-chaotic attractors

This section demonstrates the power of the ultra Poincaré chaos test, proving that strange attractors, shown non-chaotic by conservative tests [44, 49], admit alpha unpredictability. Thus, continuous and discrete time systems, which are known as possessing strange non-chaotic attractors, are in the focus of the analysis. It is demonstrated that alpha unpredictable trajectories exist in the dynamics, which are accepted as not having chaos previously. In particular, it has been shown that the

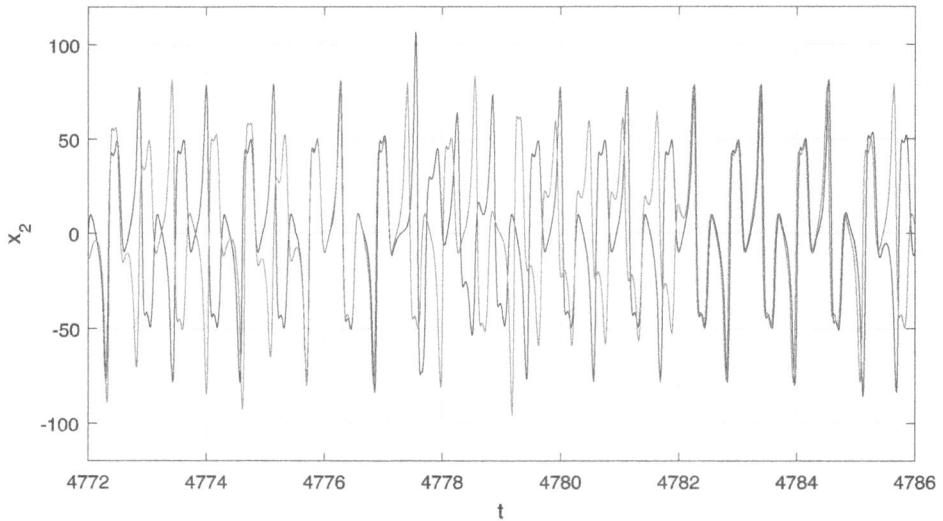

Figure 6.18. The blue and red curves present the solution of $x_2(t)$ and $x_{2_{\text{shift}}}(t)$ of system (6.14) on the interval [4772, 4786].

absence of positive Lyapunov exponents is insufficient for the decision. Even though we have shown below that the term *strange non-chaotic attractors* is not appropriate for models in [44, 49], we use it in this section to clarify the historical and methodological circumstances of the research.

6.4.1 Introduction and preliminaries

Strange non-chaotic attractors are either non-regular or non-chaotic [44, 49, 66]. Ruelle and Takens introduced them [69] in 1970, and Grebogi *et al* [44] provided a rigorous definition for such attractors in 1984. Strange non-chaotic attractors have irregular geometrical shapes and have non-positive Lyapunov exponents [34, 36, 37, 44, 55, 65, 66]. The latter property has led many researchers to conclude that strange non-chaotic attractors do not possess sensitive dependence on initial conditions, arguing that they are non-chaotic [34, 44, 65, 66]. Nevertheless, other studies show that strange non-chaotic attractors have sensitive dependence on initial conditions, but their measure of separation does not grow exponentially [42, 55, 63]. Even though the sensitive dependence on the initial conditions was shown for strange non-chaotic attractors, some examples do not satisfy the definition of chaos in the sense of Devaney [42].

A novel type of chaos called ultra Poincaré chaos was introduced in paper [17]. One of its distinguishing features compared to chaos in the sense of Devaney [38] and Li–Yorke [54] is that its definition is based only on the existence of a particular type of oscillation, which is called alpha unpredictable trajectory. It was rigorously proved in [8, 17] that an alpha unpredictable trajectory in a flow or semi-flow necessarily implies the presence of sensitive dependence on initial conditions. Some examples of dynamics possessing ultra Poincaré chaos are symbolic dynamics, logistic map, and Hénon map [16, 17].

Research for the presence of alpha unpredictable motions, and hence ultra Poincaré chaos, in systems of differential equations has begun with the study [18]. In this paper, an example of a continuous alpha unpredictable function was provided, and it was demonstrated that systems of differential equations with alpha unpredictable perturbations generate alpha unpredictable solutions. Additionally, the paper [19] is concerned with the theoretical investigation of alpha unpredictable solutions in dynamics of retarded differential equations and discrete-time systems. In order to investigate continuous-time systems for the presence of alpha unpredictable motions from the numerical point of view, a technique called the alpha unpredictability test was introduced in paper [5]. The alpha unpredictability test can be used in continuous-time and discrete-time systems with arbitrary high dimensions.

The primary purpose of the present study is to demonstrate that alpha unpredictable trajectories can exist in the dynamics of continuous-time and discrete-time systems with strange non-chaotic attractors. The alpha unpredictability test [5] is used to solve the problem. In particular, our results confirm that the systems under investigation admit the Lorenz sensitivity. The present results show using a numerical method [5, 12, 23, 29] that strange non-chaotic attractors are ultra Poincaré chaotic. They can be considered for comparison of ultra Poincaré and Devaney chaos [38], where the latter is suggested to be a good definition of chaotic dynamics [35, 41, 42]. Let us consider a metric space (X, d), where X is a non-empty set and $d: X \times X \to [0, \infty)$ is a metric, and let $\mathbb{T}_+$ be either the set of non-negative real numbers or non-negative integers. Suppose that $\pi: \mathbb{T}_+ \times X \to X$ is a semi-flow on X, i.e., $\pi(0, x) = x$ for all $x \in X$, $\pi(t, x)$ is continuous in the pair of variables t and x, and $\pi(t_1, \pi(t_2, x)) = \pi(t_1 + t_2, x)$ for all $t_1, t_2 \in \mathbb{T}_+$, $x \in X$. A point $x \in X$ is called positively Poisson stable (stable P^+) if there exists a sequence $\{t_n\}$ which diverges to infinity such that $\pi(t_n, x) \to x$ as $n \to \infty$ [73].

For a given point $x \in X$, let us note the closure of the trajectory $T(x) = \{\pi(t, x): t \in \mathbb{T}_+\}$ by Θ_x, i.e., $\Theta_x = \overline{T(x)}$. Θ_x is said to be a quasi-minimal set if the point x is stable P^+ and $T(x)$ is contained in a compact subset of X [73]. A point $x \in X$ and the trajectory through it are called alpha unpredictable if there exist a positive number ε_0 (the unpredictability constant) and sequences $\{t_n\}$ and $\{s_n\}$, both of which diverge to infinity, such that $\pi(t_n, x) \to x$ as $n \to \infty$ and $d[\pi(t_n + s_n, x), \pi(s_n, x)] > \varepsilon_0$ for each $n \in \mathbb{N}$ [17]. It is clear that if a point $x \in X$ is unpredictable, then it is stable P^+. The paper [17] reveals the presence of sensitivity in the quasi-minimal set Θ_x if x is an alpha unpredictable point in X. The presence of an alpha unpredictable point in X exposes ultra Poincaré chaos in the dynamics on the quasi-minimal set Θ_x [17].

The definitions of an alpha unpredictable function and an alpha unpredictable sequence are as follows.

Definition 6.4.1. [19] *A continuous and bounded function* $\vartheta: \mathbb{R} \to \mathbb{R}^p$ *is alpha unpredictable if there exist a positive number* ε_0 *and sequences* $\{t_n\}$, $\{s_n\}$, *both of* which diverge to infinity such that $\|\vartheta(t_n) - \vartheta(0)\| \to 0$ as $n \to \infty$ and $\|\vartheta(t_n + s_n) - \vartheta(s_n)\| \geqslant \varepsilon_0$ for each $n \in \mathbb{N}$.

Definition 6.4.2. [19] *A bounded sequence* $\{\kappa_i\}$, $i \in \mathbb{N}$, *in* $\mathbb{R}^p$ *is called alpha unpredictable if there exist a positive number* ε_0 *and sequences* $\{\zeta_n\}$, $\{\eta_n\}$, $n \in \mathbb{N}$, *of positive integers both of which diverge to infinity such that* $\|\kappa_{\zeta_n} - \kappa_0\| \longrightarrow 0$ *as* $n \longrightarrow \infty$ *and* $\|\kappa_{\zeta_n + \eta_n} - \kappa_{\eta_n}\| \geqslant \varepsilon_0$ *for each* $n \in \mathbb{N}$.

Definitions 6.4.1 and 6.4.2 are essential tools for formulating the alpha unpredictability test definition for continuous and discrete systems, respectively. The theory concerning the alpha unpredictability test for both continuous-time and discrete dynamics associated with examples of continuous-time and discrete strange non-chaotic attractors, where we show that they are ultra Poincaré chaotic, are provided in sections 6.4.2 and 6.4.3, respectively.

The rest of the section is organized as follows. In section 6.4.2 we consider a continuous-time system with a strange non-chaotic attractor, and show that this system admits an alpha unpredictable solution by using the alpha unpredictability test. Section 6.4.3, on the other hand, is devoted to a discrete-time system admitting a strange non-chaotic attractor. Utilizing the discrete counterpart of the alpha unpredictability test, the presence of an alpha unpredictable orbit is confirmed.

6.4.2 A continuous-time model

Let us consider the system

$$\begin{aligned} x'_1 &= x_2 \\ x'_2 &= -0.1x_2 + (1 + 0.34\cos t)x_1 - x_1^3. \end{aligned} \tag{6.15}$$

It was demonstrated in paper [49] that system (6.15) admits a strange non-chaotic attractor. Figure 6.19 shows the Lyapunov exponents graph of system (6.15), which points are non-positive.

To show the presence of an alpha unpredictable solution in the dynamics of the system (6.15), the ultra Poincaré chaos test will be utilized. For that purpose, we solve the system numerically in an equally partitioned time interval $[0, \sigma]$ with step size Δt, where $\sigma = \lambda \Delta t$ for some fixed sufficiently large natural number λ. A solution $x(t) = (x_1(t), x_2(t))$ of (6.15) with $x(0) = x_0$ is said to satisfy the alpha unpredictability test, if there exist a large natural number k, a positive number ε_0, strictly increasing finite sequences t_n and s_n, $1 \leqslant n \leqslant k$, of positive real numbers such that the inequalities

$$\|x(t_n) - x_0\| < 1/n \tag{6.16}$$

and

$$\|x(t_n + s_n) - x(s_n)\| > \varepsilon_0 \tag{6.17}$$

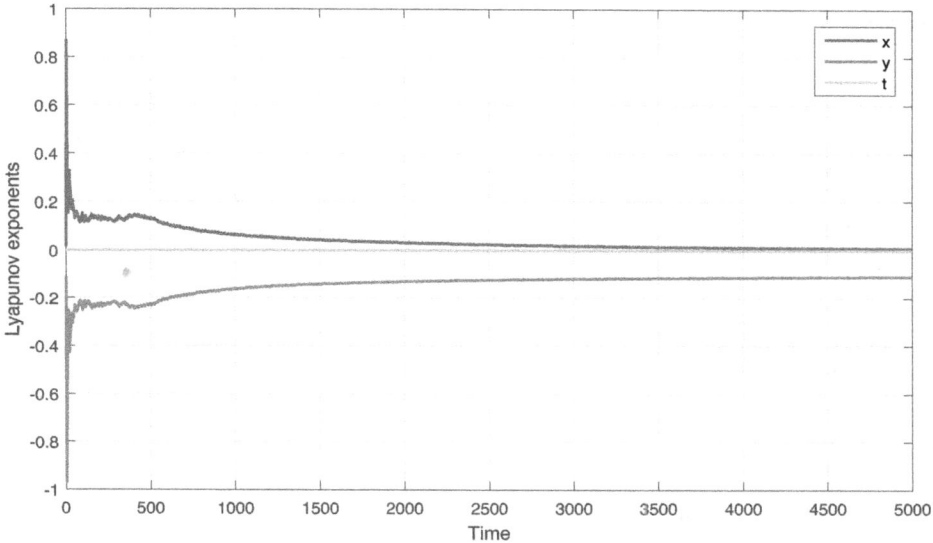

Figure 6.19. The Lyapunov exponents of system (6.15). Reproduced with permission from [4]. Copyright © 2011–2024 L & H Scientific Publishing. All rights reserved.

are valid for each $1 \leqslant n \leqslant k$, and $t_k + s_k$ is sufficiently close to σ. If the alpha unpredictability test is satisfied for system (6.15) then one can conclude that the system possesses an alpha unpredictable solution, and consequently, ultra Poincaré chaos.

In applications of the test, it is not an easy task to find an appropriate initial data x_0 such that the inequality (6.17) holds and $t_k + s_k$ is sufficiently close to σ. For that reason, we start with an arbitrary point y_0 near the attractor of system (6.15). Let us denote by $\varphi(t)$ the bounded solution of (6.15) satisfying $\varphi(0) = y_0$, which is considered in the equally partitioned time interval $[0, \sigma]$. First of all we numerically find positive numbers ω, which are not close to σ, such that there exist a large natural number k_ω and a strictly increasing finite sequence η_m^ω, $1 \leqslant m \leqslant k_\omega$, satisfying the inequality

$$\|\varphi(\eta_m^\omega + \omega) - \varphi(\omega)\| < 1/m \tag{6.18}$$

and $\eta_{k_\omega}^\omega$ is close to σ. Suppose that Ω is the set of all such numbers ω. Next, we fix a small positive number ε_1 which is greater than the error of the numerical technique used to solve system (6.15), and then find a subset Λ of Ω such that for each $\omega \in \Lambda$ there exist a positive large number $r_\omega \leqslant k_\omega$ and a strictly increasing finite sequence ρ_m^ω, $1 \leqslant m \leqslant r_\omega$, satisfying

$$\|\varphi(\eta_m^\omega + \rho_m^\omega) - \varphi(\rho_m^\omega)\| \geqslant \varepsilon_1 \tag{6.19}$$

for each $1 \leqslant m \leqslant r_\omega$ and $\eta_{r_\omega}^\omega + \rho_{r_\omega}^\omega$ is close to σ.

We set $x_0 = \varphi(\nu)$ for some ν in Λ, provided that Λ is nonempty, and choose a subsequence t_n of the finite sequence η_m^ν by omitting its first m_0 terms. In this case we

Algorithm 2. Alpha unpredictability test for system (6.15).

1:    Input m_0

2:    Set $l = \sigma_0$

3:    Set $q = 0$

4:    Input ε_0

5:    Input λ ▷ number of iterations

6:    Set $tmin = 0$

7:    Input Δt

8:    Find $\sigma = \lambda \cdot \Delta t$

9:    Input initial condition x_0

10:   Find the numerical solution $x(t)$ of system (6.15) for the given interval.

11: **for** $n = 1: k$ **do**

12:      **for** $j = 0: \lambda$ **do**

13:        Set $\tau_j = j\Delta t$

14:        **If** $\|x(\tau_j) - x(0)\| < \frac{1}{n+m_0}$ **thne**

15:          if $l < \tau_j$ then

16:            $l = \tau_j$

17:            $A(n) = l$ ▷ the matrix $A(n)$ collects τ_js, which satisfy lines 13 and 14 for each n

18:            **break**   ▷ reckon the first τ_j satisfying lines 14 and 15 for each n, which are the elements of the sequence of convergence t_n

19:          end if

20:        end if

21:      end for

22: end for

23: **for** $n = 1: k$ **do**

24:      **for** $j = 0: \lambda$ **do**

25:        Set $\tau_j^* = j\Delta t$

26:        if $\|x(A(n) + \tau_j^*) - x(\tau_j^*)\| > \varepsilon_0$ then

27:          **If** $q < \tau_j^*$ then

28:            $q = \tau_j^*$

29:            $B(n) = q$ ▷ the matrix $B(n)$ collects τ_j^*s, which satisfy lines 26 and 27 for each n

30:            **break**   ▷ reckon the first τ_j^* satisfying lines 24 and 25 for each n, which are the elements of the sequence of separation s_n

31:            Display matrices $A(n)$ and $B(n)$

32:          end if

33:        end if

34:      end for

35: end for

---

have $t_n = \eta^{\nu}_{n+m_0}$, $1 \leqslant n \leqslant k$. The application of the alpha unpredictability test is finalized by choosing an arbitrary large number ε_0 greater than ε_1 and finding a strictly increasing finite sequence s_n, $1 \leqslant n \leqslant k$, such that the inequality (6.17) holds for each $1 \leqslant n \leqslant k$ in which $x(t) = (x_1(t), x_2(t))$ is the solution of (6.15) satisfying $x(0) = x_0$ and $k \leqslant q$ is a sufficiently large natural number.

In algorithm 2, which is concerned with finding the finite sequences t_n and s_n, we set $\sigma_0 = \eta^{\nu}_{m_0}$ and $t_1 > \sigma_0$. The algorithm is as follows.

Taking $\Delta t = 0.01$, $\lambda = 1.2 \times 10^8$, and $y_0 = (0.1, 0)$ in the alpha unpredictability test, we obtained $\nu = 10$ and

$$x_0 = (0.246\ 429\ 004\ 871\ 484\ 89, -0.339\ 843\ 267\ 161\ 431\ 37).$$

Let $x(t) = (x_1(t), x_2(t))$ be the solution of system (6.15) satisfying $x(0) = x_0$. This solution is depicted in figure 6.20. Figure 6.20(a) represents the irregular trajectory while figure 6.20(b) shows the corresponding time series of each coordinate of $x(t)$. Figure 6.20 confirms the solution $x(t)$ behaves irregularly.

Applying algorithm 2 on system (6.15) with $\varepsilon_0 = 2.8$, $m_0 = 14$, and $\sigma_0 = 0.15$, we obtained 493 terms of each of the finite sequences t_n and s_n such that $\|x(t_n) - x_0\| < 1/(n + m_0)$ and $\|x(t_n + s_n) - x(s_n)\| > \varepsilon_0$ for $1 \leqslant n \leqslant k$, where $k = 493$. The time interval $[0, \sigma]$, where $\sigma = 1.2 \times 10^5$ is utilized for the solution $x(t)$. Table 6.8 shows the result of the alpha unpredictability test for system (6.15) such that 10 terms of each of the finite sequences t_n and s_n, as well as the values for $1/(n + m_0)$, are represented in the table. According to the results mentioned in table 6.8 we conclude that the alpha unpredictability test is satisfied. Therefore, there

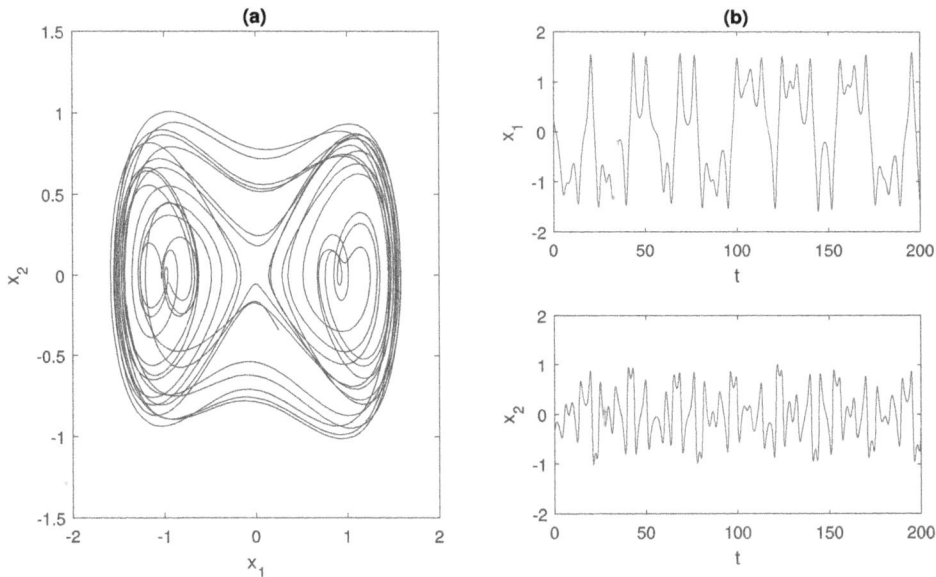

Figure 6.20. Irregular dynamics of system (6.15). The pictures shown in (a) and (b) both confirm that the system admits chaos. Reproduced with permission from [4]. Copyright © 2011–2024 L & H Scientific Publishing. All rights reserved.

exists an alpha unpredictable solution of system (6.15). Moreover, system (6.15) possesses sensitive dependence on initial conditions.

6.4.3 A discrete-time model

The main object of investigation of the present subsection is the discrete-time system

$$x_{i+1} = \frac{\tau}{1 + x_i^2 + y_i^2}(x_i \cos \theta_i + y_i \sin \theta_i)$$

$$y_{i+1} = \frac{\tau}{1 + x_i^2 + y_i^2}(-x_i \gamma \sin \theta_i + y_i \gamma \cos \theta_i) \qquad (6.20)$$

$$\theta_{i+1} = \theta_i + 2\pi\omega (\text{mod } 2\pi),$$

where $\tau = 2$, $\gamma = 0.5$, and $\omega = \frac{\sqrt{5}-1}{2}$. It was shown in paper [44] that system (6.20) possesses a strange non-chaotic attractor. Moreover, according to the results of the study [42], system (6.20) is not chaotic in the sense of Devaney. Figure 6.21 depicts the Lyapunov exponents of system (6.20). It is seen in figure 6.21 that the Lyapunov exponents are non-positive.

We will investigate the presence of an alpha unpredictable orbit in the dynamics of system (6.20) by applying the discrete counterpart of the alpha unpredictability test mentioned in the previous section.

Let us consider a solution $X_i = (x_i, y_i, \theta_i)$ of system (6.20) for $i = 0, 1, \ldots, \lambda$, where λ is a sufficiently large natural number. If there exist a large natural number k, a positive number ε_0, strictly increasing finite sequences ζ_n and η_n, $1 \leqslant n \leqslant k$, such that the inequalities

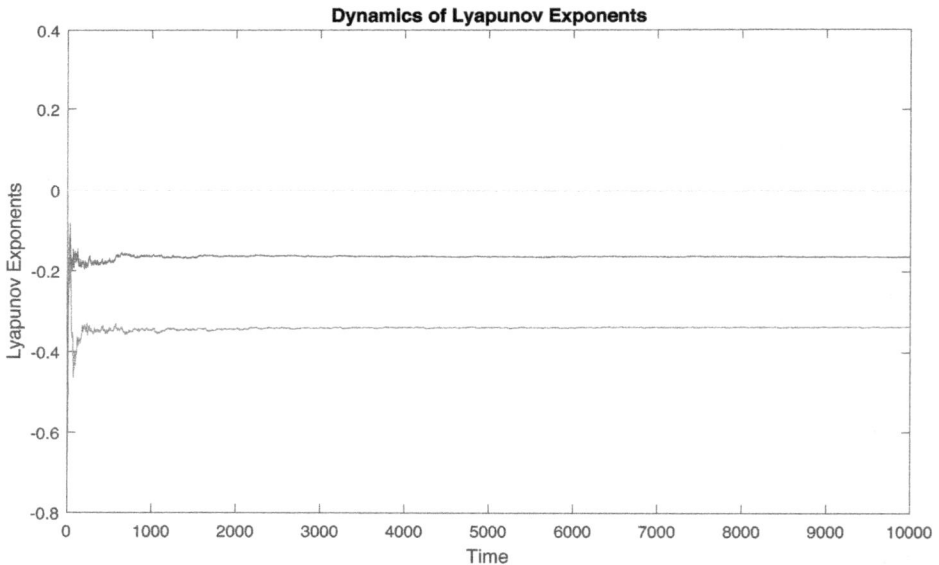

Figure 6.21. Lyapunov exponents of the discrete-time system (6.20). Reproduced with permission from [4].

$$\left\| X_{\zeta_n} - X_0 \right\| < 1/n \tag{6.21}$$

and

$$\left\| X_{\zeta_n + \eta_n} - X_{\eta_n} \right\| > \varepsilon_0 \tag{6.22}$$

are valid for each $1 \leqslant n \leqslant k$ and $\zeta_k + \eta_k$ is sufficiently close to λ, then we say that the alpha unpredictability test is satisfied for system (6.20), and hence, the system possesses an alpha unpredictable orbit.

Suppose that λ is a fixed sufficiently large natural number. Our first purpose is to choose an appropriate initial condition α for system (6.20) such that $\zeta_k + \eta_k$ is sufficiently close to λ. Let β be a point near the attractor of system (6.20), and consider a bounded solution ϕ_i of (6.20) with $\phi_0 = \beta$, where $0 \leqslant i \leqslant \lambda$. We find a positive number ν, which is not close to λ, such that there exist large natural numbers k_ν, s_ν with $s_\nu < k_\nu$, a small positive number ε_1, and strictly increasing finite sequences $\gamma_m^\nu, 1 \leqslant m \leqslant k_\nu$, and $v_m^\nu, 1 \leqslant m \leqslant m_\nu$, satisfying the inequalities

$$\left\| \phi_{\gamma_m^\nu + \nu} - \phi_\nu \right\| < 1/m, \tag{6.23}$$

for each $1 \leqslant m \leqslant k_\nu$, and

$$\left\| \phi_{\gamma_m^\nu + v_m^\nu} - \phi_{v_m^\nu} \right\| \geqslant \varepsilon_1 \tag{6.24}$$

for each $1 \leqslant m \leqslant s_\nu$, with $\gamma_{k_\nu}^\nu$ and $\gamma_{s_\nu}^\nu + v_{s_\nu}^\nu$ are close to λ. If such a positive number ν exists, then set $\alpha = \phi_\nu$.

Now, let $X_i = (x_i, y_i, \theta_i)$ be the solution of (6.20) with $X_0 = \alpha$. We will find finite sequences ζ_n, η_n, and a sufficiently large positive number ε_0 satisfying the inequality (6.22) for the solution z_i. For each $m = 1, 2, \ldots, k_\nu$ and each $j = 0, 1, \ldots, \lambda$, we set $\mu_m^j = \| X_{\eta_m^\nu + j} - X_j \|$ and let $\bar{\mu}_m = \max_{0 \leqslant j \leqslant \lambda} \mu_m^j$.

We choose a subsequence ζ_n of the finite sequence γ_m^ν by omitting its first m_0 terms, i.e., $\zeta_n = \gamma_{n+m_0}^\nu, 1 \leqslant n \leqslant k$. The alpha unpredictability test is finalized by fixing an arbitrary large number ε_0 between ε_1 and $\min_{1 \leqslant n \leqslant k'} \bar{\mu}_{m_n}$, and finding a strictly increasing finite sequence $\eta_n, 1 \leqslant n \leqslant k$, such that the inequality (6.17) holds for each $1 \leqslant n \leqslant k$, where $k \leqslant k'$ is a sufficiently large natural number with $\zeta_k + \eta_k$ is close to λ.

In algorithm 3, which is concerned with finding the finite sequences ζ_n and η_n, we set $\lambda_0 = \gamma_{m_0}^\nu$ and $\zeta_1 > \lambda_0$. The algorithm is as follows.

To show the existence of an alpha unpredictable orbit in the dynamics of system (6.20), we apply the alpha unpredictability test by taking $\beta = (0.1, 0, 0)$ and $\lambda = 10^8$. It is found that $\nu = 100$ and $\alpha = (x_0, y_0, \theta_0)$, where $x_0 = -0.121\,976\,909\,234\,366\,69$, $y_0 = 0.493\,032\,138\,757\,075\,91$ and $\theta_0 = 5.047\,904\,007\,138\,5182$. Denote by $X_i = (x_i, y_i, \theta_i)$ the solution of system (6.20), where $i = 0, 1, 2, \ldots, \lambda$, with $X_0 = \alpha$. Figure 6.22 shows the orbit X_i of (6.20). Figure 6.22 reveals the irregular dynamics of system (6.20).

Applying algorithm 3 with $\varepsilon_0 = 5.5$ to system (6.20), we obtained 1525 terms for each of the finite sequences ζ_n and η_n such that $\left\| X_{\zeta_n} - X_0 \right\| < \frac{1}{n}$ and

Algorithm 3. The alpha unpredictability test for equation (6.20).

1:     Input m_0
2:     Set $l = \lambda_0$
3:     Set $q = 0$
4:     Input ε_0
5:     Input λ ▷ number of iterations
6:     Input initial condition α
7:     Find the numerical solution X_i of system (6.20) satisfying $X_0 = \alpha$ for $0 \leqslant i \leqslant \lambda$.
8:     **for** $n = 1: k$ **do**
9:         **for** $j = 0: \lambda$ **do**
10:             **If** $\|X_j - X_0\| < \frac{1}{n + m_0}$ **then**
11:                 **If** $l < j$ **then**
12:                     $l = j$
13:                     $A(n) = 1$ ▷ the matrix A(n) collects js, which satisfy lines 10 and 11 for each n
14:                     **break**   ▷ reckon the first j satisfying lines 10 and 11 for each n, which are the elements of the sequence ζ_n
15:                 **end if**
16:             **end if**
17:         **end for**
18:     **end for**
19:     **for** $n = 1: k$ **do**
20:         **for** $r = 0: \lambda$ **do**
21:             **If** $\|X_{A(n)+r} - X_r\| > \varepsilon_0$ **then**
22:                 **If** $q < r$ **then**
23:                     $q = r$
24:                     $B(n) = q$ ▷ the matrix B(n) collects rs, which satisfy lines 21 and 22 for each n
25:                     **break**   ▷ reckon the first r satisfying lines 21 and 22 for each n, which are the elements of the sequence η_n
26:                     Display matrices $A(n)$ and $B(n)$
27:                 **end if**
28:             **end if**
29:         **end for**
30:     **end for**

$\|X_{\zeta_n + \eta_n} - X_{\eta_n}\| > 5.5$ for $1 \leqslant n \leqslant 1525$. Table 6.9 represents 10 terms of each of these finite sequences. The results of table 6.9 manifest that system (6.20) admits an alpha unpredictable orbit, and therefore, its dynamics is ultra Poincaré chaotic.

6.5 The generalized synchronization as a proof of alpha unpredictability

When generalized synchronization occurs, we demonstrate the extension of alpha unpredictable motions in coupled autonomous systems with skew product structure. We provide sufficient conditions for alpha unpredictable motions in the dynamics of

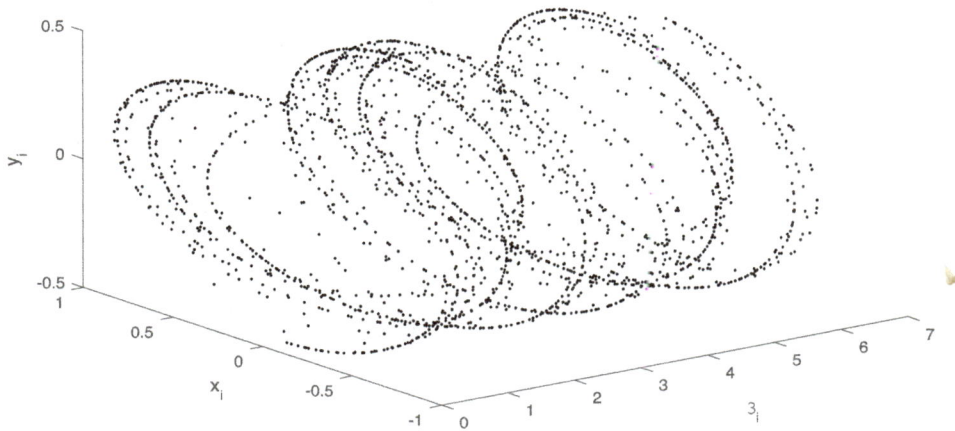

Figure 6.22. Irregular orbit of system (6.20). The initial data $x_0 = -0.121\,976\,909\,234\,366\,69$, $y_0 = 0.493\,032\,138\,757\,075\,91$ and $\theta_0 = 5.047\,904\,007\,138\,5182$ is used in the simulation. Reproduced with permission from [4]. Copyright © 2011–2024 L & H Scientific Publishing. All rights reserved.

Table 6.9. The result of the alpha unpredictability test applied to equation (6.20). The index interval is $[0, 10^8]$ and the value $\varepsilon_0 = 5.5$ are used. According to the result of the alpha unpredictability test shown in the table, one can confirm that the test is successful for system (6.20) such that it admits an alpha unpredictable solution.

n	$1/n$	ζ_n	η_n
1	1	5	10
266	0.003 759	1 138 905	1 842 687
436	0.002 294	3 074 428	4 974 429
697	0.001 435	8 059 903	12 966 072
871	0.001 148	12 462 889	20 165 278
980	0.001 020	15 808 760	25 579 011
1199	0.000 834	23 550 852	37 909 561
1343	0.000 745	29 785 679	47 876 330
1417	0.000 706	33 160 207	53 654 242
1525	0.000 656	38 134 736	61 703 199

the response system. The theoretical results are exemplified for coupled autonomous systems in which the drive is a hybrid dynamical system and the response is a Lorenz system. The auxiliary system approach and conditional Lyapunov exponents are utilized to detect the presence of generalized synchronization.

6.5.1 Introduction and preliminaries

A particular type of Poisson stable trajectory, alpha unpredictable, was introduced in paper [17]. An alpha unpredictable trajectory leads to Poincaré chaos in the associated quasi-minimal set. Such trajectories occur in symbolic dynamics, logistic and Hénon maps, and the Smale horseshoe [16, 17]. One of the essential features of Poincaré chaos is that a single alpha unpredictable trajectory in the dynamics can trigger it. This feature is the main difference between Poincaré chaos compared to chaos in the sense of Devaney [38] and Li–Yorke [54] since a collection of motions is required to define these chaos types.

Interesting results concerning alpha unpredictable motions as well as Poincaré chaos in topological spaces provided in papers [57, 58, 76] generalized the notion of alpha unpredictable points to the case of semiflows with arbitrary acting abelian topological monoids. In contrast, Thakur and Das [76] demonstrated that at least one of the factors is Poincaré chaotic provided that the same is valid for finite or countably infinite products of semiflows. Additionally, differential equations with hyperbolic linear parts exhibiting alpha unpredictable solutions were studied in [18–20], and the existence of alpha unpredictable outputs in cellular neural networks can be found in papers [10, 30].

In order to provide a larger class of differential equations possessing alpha unpredictable trajectories, in this study we take into account coupled systems in which generalized synchronization [70] takes place. More precisely, we consider the systems

$$x'(t) = F(x(t)), \tag{6.25}$$

and

$$y'(t) = G(x(t), y(t)), \tag{6.26}$$

where $F: \mathbb{R}^p \to \mathbb{R}^p$ and $G: \mathbb{R}^p \times \mathbb{R}^q \to \mathbb{R}^q$ are continuous functions. Systems (6.25) and (6.26) are respectively called the drive and response. It is worth noting that the coupled system (6.25)–(6.26) has a skew product structure. Our purpose is to rigorously prove that if the drive system (6.25) admits an alpha unpredictable solution, then the same is true for the response system (6.26) when they are synchronized in the generalized sense. Sufficient conditions to prove the unpredictability are given in section 6.5.2.

Pecora and Carroll initiated the concept of synchronization in coupled chaotic systems [62] for those that are identical, and it is generalized for non-identical systems by Rulkov *et al* [70]. Generalized synchronization characterizes the state of the response system when the drive drives the output of another system. Kocarev and Parlitz proved [51] that generalized synchronization occurs in coupled systems of the form (6.25)–(6.26) if and only if the response system (6.26) is asymptotically stable for all initial values in a neighborhood of the chaotic attractor. When generalized synchronization occurs, a functional relation exists between the states of the drive and response systems [51, 70]. For that reason, generalized synchronization allows us to predict the dynamics of the response system by the dynamics of the drive [1, 43]. Lyapunov exponent-based conditions, which imply that the response state is a smooth function of the drive state, were provided by Hunt *et al* [48].

In this study, we require that the state of the response (6.26) is a continuous function of the state of the drive (6.25). This criterion can be guaranteed by the auxiliary system approach, which was suggested by Abarbanel *et al* [1]. In this approach one needs to use another response system that is identical to the original one but independent of it, and monitor the final states after the transitories have died off when different initial data within the basin of attraction are utilized. Generalized synchronization occurs in the dynamics if the final states of the two response systems are identical [1, 43]. Mutual false nearest neighbors, conditional Lyapunov exponents, and Lyapunov functions are other techniques which can be used to determine generalized synchronization [43, 46, 51, 70].

The novelty of the present study is the verification of unpredictability in coupled systems under the presence of generalized synchronization. The concept of generalized synchronization was not taken into account in the studies [16, 18, 19]. For that reason, a different proof technique is applied. The recent results are essential not only from the theoretical point of view but also for applications since generalized synchronization can occur in various real world problems concerning image encryption, secure communication, lasers, electronic circuits, and neural networks [47]–[59].

Throughout the section, we make use of the Euclidean norm for vectors. We suppose that systems (6.25) and (6.26) admit compact invariant sets $\Lambda_x \subset \mathbb{R}^p$ and $\Lambda_y \subset \mathbb{R}^q$, respectively. In this case, a solution of the coupled system (6.25) and (6.26) with initial condition from the set $\Lambda_x \times \Lambda_y$ remains in that set.

In what follows, we consider a class of generalized synchronization in which the state of the response (6.26) is a continuous function of the state of the drive (6.25). More precisely, we say that generalized synchronization occurs in the dynamics of the coupled system (6.25) and (6.26) if there is a continuous transformation ψ such that for each $(x_0, y_0) \in \Lambda_x \times \Lambda_y$ the relation

$$\lim_{t \to \infty} \|y(t) - \psi(x(t))\| = 0$$

holds, where $x(t)$ and $y(t)$ are respectively the solutions of (6.25) and (6.26) with $x(0) = x_0$ and $y(0) = y_0$. The reader is referred to the paper [1] for further information concerning continuity of the transformation ψ.

The definition of an alpha unpredictable function which is utilized in the present study is as follows.

Definition 6.5.1. [19] *A uniformly continuous function $h: \mathbb{R} \to \Lambda$, where Λ is a compact subset of $\mathbb{R}^p$, is called alpha unpredictable if there exist positive numbers ε_0, r and sequences $\{\mu_n\}_{n \in \mathbb{N}}$ and $\{\nu_n\}_{n \in \mathbb{N}}$, both of which diverge to infinity such that $\|h(t + \mu_n) - h(t)\| \to 0$ as $n \to \infty$ uniformly on compact subsets of $\mathbb{R}$ and $\|h(t + \mu_n) - h(t)\| \geqslant \varepsilon_0$ for each $t \in [\nu_n - r, \nu_n + r]$ and $n \in \mathbb{N}$.*

The number ε_0 in definition 6.5.1 is called the unpredictability constant of the function $h(t)$ [19].

6.5.2 Ultra Poincaré chaos of the response system

The following assumptions on the response system (6.26) are required.

(A1)
There exists a positive number L_1 such that
$\|G(x_1, y) - G(x_2, y)\| \geqslant L_1\|x_1 - x_2\|$ for each $x_1, x_2 \in \Lambda_x$ and $y \in \Lambda_y$.

(A2)
There exists a positive number L_2 such that
$\|G(x, y_1) - G(x, y_2)\| \leqslant L_2\|y_1 - y_2\|$ for each $x \in \Lambda_x$ and $y_1, y_2 \in \Lambda_y$.

The following theorem provides a novel criterion for the existence of an alpha unpredictable solution in the dynamics of the response system (6.26).

Theorem 6.5.1. [77] *Suppose that the assumptions (A1) and (A2) are fulfilled. If the drive system (6.25) possesses an alpha unpredictable solution and generalized synchronization takes place in the dynamics of the coupled system (6.25) and (6.26), then the response system (6.26) also possesses an alpha unpredictable solution.*

Proof. Let $x(t)$ be an alpha unpredictable solution of the drive system (6.25). Since generalized synchronization occurs in the dynamics of the coupled system (6.25) and (6.26), there is a continuous transformation ψ and a point y_0 in Λ_y such that the equation

$$\lim_{t\to\infty} \|y(t) - \psi(x(t))\| = 0 \qquad (6.27)$$

is fulfilled, where $y(t)$ is the solution of the response system (6.26) with $y(0) = y_0$.

Let $\mathscr{C}$ be a compact subset of $\mathbb{R}$, and fix a positive number ε. One can find real numbers a and b such that $\mathscr{C} \subseteq [a, b]$. According to (6.27) there is a positive number T such that

$$\|y(t) - \psi(x(t))\| < \frac{\varepsilon}{3}$$

whenever $t \geqslant a + T$.

It is worth noting that

$$z(t) = (\tilde{x}(t), \tilde{y}(t)), \qquad (6.28)$$

where $\tilde{x}(t) = x(t + T)$ and $\tilde{y}(t) = y(t + T)$, is a solution of the coupled system (6.25) and (6.26). In the proof we will show that $\tilde{y}(t)$ is alpha unpredictable.

Because the transformation ψ is continuous there is a positive number δ such that if $\|x_1 - x_2\| < \delta$, then

$$\left\| \psi(x_1) - \psi(x_2) \right\| < \frac{\varepsilon}{3}. \tag{6.29}$$

Owing to the unpredictability of $x(t)$ there exist positive numbers ε_0, r and sequences $\{\mu_n\}_{n \in \mathbb{N}}$ and $\{\nu_n\}_{n \in \mathbb{N}}$, both of which diverge to infinity such that $\left\| x(t + \mu_n) - x(t) \right\| \to 0$ as $n \to \infty$ uniformly on compact subsets of $\mathbb{R}$ and $\left\| x(t + \mu_n) - x(t) \right\| \geqslant \varepsilon_0$ for each $t \in [\nu_n - r, \nu_n + r]$ and $n \in \mathbb{N}$. Assume without loss of generality that $\mu_n \geqslant 0$ for each n.

There is a natural number n_0 such that for $n \geqslant n_0$ and $t \in [a + T, b + T]$ the inequality

$$\left\| x(t + \mu_n) - x(t) \right\| < \delta$$

holds. Hence, the inequality

$$\left\| \psi(\tilde{x}(t + \mu_n)) - \psi(\tilde{x}(t)) \right\| < \frac{\varepsilon}{3}$$

is fulfilled for each $t \in [a, b]$ by means of (6.29). Accordingly, for $n \geqslant n_0$ and $t \in \mathscr{C}$, it can be verified that

$$\left\| \tilde{y}(t + \mu_n) - \tilde{y}(t) \right\| \leqslant \left\| \tilde{y}(t + \mu_n) - \psi(\tilde{x}(t + \mu_n)) \right\| + \left\| \psi(\tilde{x}(t + \mu_n)) - \psi(\tilde{x}(t)) \right\| + \left\| \psi(\tilde{x}(t)) - \tilde{y}(t) \right\| < \varepsilon.$$

For that reason, $\left\| \tilde{y}(t + \mu_n) - \tilde{y}(t) \right\| \to 0$ as $n \to \infty$ uniformly on compact subsets of $\mathbb{R}$.

In the remaining part of the proof, we will show the existence of numbers $\varepsilon_1 > 0$, $r_1 > 0$ and a sequence $\{\theta_n\}$ with $\theta_n \to \infty$ as $n \to \infty$ such that $\left\| \tilde{y}(t + \mu_n) - \tilde{y}(t) \right\| \geqslant \varepsilon_1$ for each $t \in [\theta_n - r_1, \theta_n + r_1]$ and $n \in \mathbb{N}$.

For each natural number n, we define $\omega_n = \nu_n - T$. Then,

$$\left\| \tilde{x}(t + \mu_n) - \tilde{x}(t) \right\| \geqslant \varepsilon_0$$

for each $t \in [\omega_n - r, \omega_n + r]$ and $n \in \mathbb{N}$.

Next, let us denote $G(x, y) = (G_1(x, y), G_2(x, y), \dots, G_q(x, y))$, where $G_i(x, y)$ is a real valued function for each $1 \leqslant i \leqslant q$. Because the function $G(x, y)$ is uniformly continuous on the compact region $\Lambda_x \times \Lambda_y$ one can find a positive number M_G such that $\left\| G(x, y) \right\| \leqslant M_G$ for each $x \in \Lambda_x, y \in \Lambda_y$. Since $\sup_{t \in \mathbb{R}} \left\| \tilde{y}'(t) \right\| \leqslant M_G$, the solution $\tilde{y}(t)$ of the response system (6.26) is uniformly continuous. Due to the uniform continuity of $\tilde{x}(t)$, there exists a positive number r_0 such that

$$\left\| G(\tilde{x}(t + \mu_n), \tilde{y}(t)) - G(\tilde{x}(\mu_n + \omega_n), \tilde{y}(\omega_n)) \right\| \leqslant \frac{L_1 \varepsilon_0}{4\sqrt{q}} \tag{6.30}$$

and

$$\left\| G(\tilde{x}(t), \tilde{y}(t)) - G(\tilde{x}(\omega_n), \tilde{y}(\omega_n)) \right\| \leqslant \frac{L_1 \varepsilon_0}{4\sqrt{q}} \tag{6.31}$$

for each $t \in [\omega_n - r_0, \omega_n + r_0]$ and $n \in \mathbb{N}$.

According to assumption $(A1)$, for each natural number n there is an integer i_n with $1 \leqslant i_n \leqslant q$ such that the inequality

$$|G_{i_n}(\bar{x}(\mu_n + \omega_n), \bar{y}(\omega_n)) - G_{i_n}(\bar{x}(\omega_n), \bar{y}(\omega_n))| \geqslant \frac{L_1}{\sqrt{q}} \|\bar{x}(\mu_n + \omega_n) - \bar{x}(\omega_n)\| \geqslant \frac{L_1 \varepsilon_0}{\sqrt{q}} \quad (6.32)$$

is valid.

Fix a natural number n. For each $t \in [\omega_n - r_0, \omega_n + r_0]$, one can attain by means of the inequalities (6.30), (6.31), and (6.32) that

$$\begin{aligned}
\left| G_{i_n}(\bar{x}(t + \mu_n), \bar{y}(t)) - G_{i_n}(\bar{x}(t), \bar{y}(t)) \right| &\geqslant \left| G_{i_n}(\bar{x}(\mu_n + \omega_n), \bar{y}(\omega_n)) - G_{i_n}(\bar{x}(\omega_n), \bar{y}(\omega_n)) \right| \\
&\quad - \left| G_{i_n}(\bar{x}(\mu_n + \omega_n), \bar{y}(\omega_n)) - G_{i_n}(\bar{x}(t + \mu_n), \bar{y}(t)) \right| \\
&\quad - \left| G_{i_n}(\bar{x}(t), \bar{y}(t)) - G_{i_n}(\bar{x}(\omega_n), \bar{y}(\omega_n)) \right| \\
&\geqslant \frac{L_1 \varepsilon_0}{2\sqrt{q}}.
\end{aligned}$$

There exist points $s_1, s_2, \ldots, s_q \in [\omega_n - r_0, \omega_n + r_0]$ such that

$$\begin{aligned}
\left\| \int_{\omega_n - r_0}^{\omega_n + r_0} [G(\bar{x}(s + \mu_n), \bar{y}(s)) - G(\bar{x}(s), \bar{y}(s))]ds \right\| &\geqslant 2r_0 \left| G_{i_n}(\bar{x}(s_{i_n} + \mu_n), \bar{y}(s_{i_n})) - G_{i_n}(\bar{x}(s_{i_n}), \bar{y}(s_{i_n})) \right| \\
&\geqslant \frac{L_1 r_0 \varepsilon_0}{\sqrt{q}}.
\end{aligned}$$

Therefore,

$$\begin{aligned}
\left\| \bar{y}(\mu_n + \omega_n + r_0) - \bar{y}(\omega_n + r_0) \right\| &\geqslant \left\| \int_{\omega_n - r_0}^{\omega_n + r_0} [G(\bar{x}(s + \mu_n), \bar{y}(s)) - G(\bar{x}(s), \bar{y}(s))]ds \right\| \\
&\quad - \left\| \bar{y}(\mu_n + \omega_n - r_0) - \bar{y}(\omega_n - r_0) \right\| \\
&\quad - \int_{\omega_n - r_0}^{\omega_n + r_0} \left\| G(\bar{x}(s + \mu_n), \bar{y}(s + \mu_n)) - G(\bar{x}(s + \mu_n), \bar{y}(s)) \right\| ds \\
&\geqslant \frac{L_1 r_0 \varepsilon_0}{\sqrt{q}} - \left\| \bar{y}(\mu_n + \omega_n - r_0) - \bar{y}(\omega_n - r_0) \right\| \\
&\quad - \int_{\omega_n - r}^{\omega_n + r} L_2 \left\| \bar{y}(s + \mu_n) - \bar{y}(s) \right\| ds.
\end{aligned}$$

The last inequality implies that

$$\max_{t \in [\omega_n - r_0, \omega_n + r_0]} \left\| \bar{y}(t + \mu_n) - \bar{y}(t) \right\| \geqslant \frac{L_1 r_0 \varepsilon_0}{2\sqrt{q}(r_0 L_2 + 1)}.$$

Suppose that

$$\max_{t \in [\omega_n - r_0, \omega_n + r_0]} \left\| \bar{y}(t + \mu_n) - \bar{y}(t) \right\| = \left\| \bar{y}(\mu_n + \zeta_n) - \bar{y}(\zeta_n) \right\|$$

for some $\zeta_n \in [\omega_n - r_0, \omega_n + r_0]$. Let us denote

$$\bar{r} = \frac{L_1 r_0 \varepsilon_0}{8 M_G \sqrt{q}(r_0 L_2 + 1)}.$$

It can be confirmed for $t \in [\zeta_n - \tilde{r}, \zeta_n + \tilde{r}]$ that

$$\left\| \tilde{y}(t + \mu_n) - \tilde{y}(t) \right\| \geq \left\| \tilde{y}(\mu_n + \zeta_n) - \tilde{y}(\zeta_n) \right\| - \left| \int_{\zeta_n}^{t} \| G(\tilde{x}(s + \mu_n), \tilde{y}(s + \mu_n)) - G(\tilde{x}(s), \tilde{y}(s)) \| ds \right|$$

$$\geq \frac{L_1 r_0 \varepsilon_0}{2\sqrt{q}(r_0 L_2 + 1)} - 2\tilde{r} M_G$$

$$= \frac{L_1 r_0 \varepsilon_0}{4\sqrt{q}(r_0 L_2 + 1)}.$$

We define $\theta_n = \zeta_n + \tilde{r}/2$ if $\zeta_n \in [\omega_n - r_0, \omega_n]$ and $\theta_n = \zeta_n - \tilde{r}/2$ if $\zeta_n \in (\omega_n, \omega_n + r_0]$.

The inequality $\left\| \tilde{y}(t + \mu_n) - \tilde{y}(t) \right\| \geq \varepsilon_1$ holds for each $t \in [\theta_n - r_1, \theta_n + r_1]$ and $n \in \mathbb{N}$, where

$$\varepsilon_1 = \frac{L_1 r_0 \varepsilon_0}{4\sqrt{q}(r_0 L_2 + 1)}$$

and $r_1 = \tilde{r}/2$. The sequence $\{\theta_n\}_{n \in \mathbb{N}}$ diverges to infinity since the same is true for the sequence $\{\omega_n\}_{n \in \mathbb{N}}$. Thus, $\tilde{y}(t)$ is alpha unpredictable. $\square$

Remark 6.5.1. *According to the proof of theorem 6.5.2, for each $n \in \mathbb{N}$ the interval $[\theta_n - r_1, \theta_n + r_1]$ is a subset of the interval $[\omega_n - r, \omega_n + r]$. Therefore, the solution $z(t) = (\tilde{x}(t), \tilde{y}(t))$ of the coupled system (6.25) and (6.26) defined by (6.28) is alpha unpredictable such that $\left\| z(t + \mu_n) - z(t) \right\| \to 0$ as $n \to \infty$ uniformly on compact subsets of $\mathbb{R}$ and $\left\| z(t + \mu_n) - z(t) \right\| \geq (\varepsilon_0^2 + \varepsilon_1^2)^{1/2}$ for each $t \in [\theta_n - r_1, \theta_n + r_1]$ and $n \in \mathbb{N}$, in which ε_0 and ε_1 are respectively the unpredictability constants of $\tilde{x}(t)$ and $\tilde{y}(t)$. Hence, one can conclude under the assumptions of theorem 6.5.1 that the coupled system (6.25)–(6.26) possesses an alpha unpredictable solution.*

In the next example we demonstrate the extension of alpha unpredictable solutions among coupled systems in which the drive is an autonomous hybrid system and the response is a Lorenz system.

6.5.3 An example with numerical simulations

According to the results of paper [16], the logistic map

$$\eta_{i+1} = \mu \eta_i (1 - \eta_i) \tag{6.33}$$

possesses an alpha unpredictable orbit for the values of the parameter μ between $3 + (2/3)^{1/2}$ and 4. Moreover, for such values of μ the unit interval $[0, 1]$ is invariant under the iterations of (6.33) [39].

Let $\{\eta_i^*\}_{i \in \mathbb{Z}}$ be an alpha unpredictable orbit of (6.33) with $\mu = 3.94$, which belongs to the unit interval $[0, 1]$, and suppose that $\gamma \colon \mathbb{R} \to [0, 1]$ is the piecewise constant function defined by $\gamma(t) = \eta_i^*$ for $t \in (i, i + 1]$, $i \in \mathbb{Z}$. The function $\gamma(t)$ is the solution of the impulsive system

$$\gamma'(t) = 0, \quad \Delta\gamma \mid_{t=i} = \eta_i^* - \eta_{i-1}^* \tag{6.34}$$

satisfying the initial condition $\gamma(0) = \eta_{-1}^*$, where $\Delta\gamma \mid_{t=i} = \gamma(i+) - \gamma(i)$ and $\gamma(i+) = \lim_{t \to i^+} \gamma(t)$. The impulse moments of (6.34) coincide with the ones of the solution of the discontinuous dynamical system

$$s'(t) = -1, \quad \Delta s \mid_{s=0} = 1 \tag{6.35}$$

with $s(0) = 0$.

It was demonstrated in study [18] that the differential equation

$$\phi'(t) = -\phi(t) + \gamma(t), \tag{6.36}$$

admits the unique uniformly continuous alpha unpredictable solution

$$\phi(t) = \int_{-\infty}^{t} e^{-(t-s)}\gamma(s)ds,$$

which is globally exponentially stable. Theorem 5.2 [16], on the other hand, implies that the function

$$\tilde{\phi}(t) = (2\phi(t) + 0.1\sin(\phi(t)), 3\phi(t), 2.5\phi(t) + 0.2\cos(\phi(t)))$$

is also alpha unpredictable. For that reason, the system

$$\begin{aligned}
x'_1(t) &= -0.5x_1(t) + 2\phi(t) + 0.1\sin(\phi(t)) \\
x'_2(t) &= -0.3x_2(t) + 0.2\arctan(x_1(t)) + 3\phi(t) \\
x'_3(t) &= -0.4x_3(t) + 2.5\phi(t) + 0.2\cos(\phi(t))
\end{aligned} \tag{6.37}$$

possesses a unique alpha unpredictable solution by theorem 4.1 [18]. Accordingly, the autonomous hybrid system (6.33)-(6.34)-(6.35)-(6.36)-(6.37) has a unique alpha unpredictable solution.

Figure 6.23 shows the time-series of the x_1, x_2, and x_3-coordinates of the solution of this hybrid system corresponding to the initial data $\zeta_0 = 0.76$, $\phi(0) = 0.41$, $x_1(0) = 2.26$, $x_2(0) = 6.48$, $x_3(0) = 3.89$, and value of the parameter $\mu = 3.94$. The irregularity of each of the time-series confirms the presence of an alpha unpredictable solution.

Next, let us consider the Lorenz system [56]

$$\begin{aligned}
y'_1(t) &= -20y_1(t) + 20y_2(t) \\
y'_2(t) &= -y_1(t)y_3(t) + 41.05y_1(t) - y_2(t) \\
y'_3(t) &= y_1(t)y_2(t) - 3y_3(t).
\end{aligned} \tag{6.38}$$

We use system (6.37) as the drive, and establish unidirectional coupling between (6.37) and (6.38) by setting up the response system

$$\begin{aligned}
y'_1(t) &= -20y_1(t) + 20y_2(t) + 2.9x_1(t) \\
y'_2(t) &= -y_1(t)y_3(t) + 41.05y_1(t) - y_2(t) + 2.6x_2(t) \\
y'_3(t) &= y_1(t)y_2(t) - 3y_3(t) + 2.4x_3(t),
\end{aligned} \tag{6.39}$$

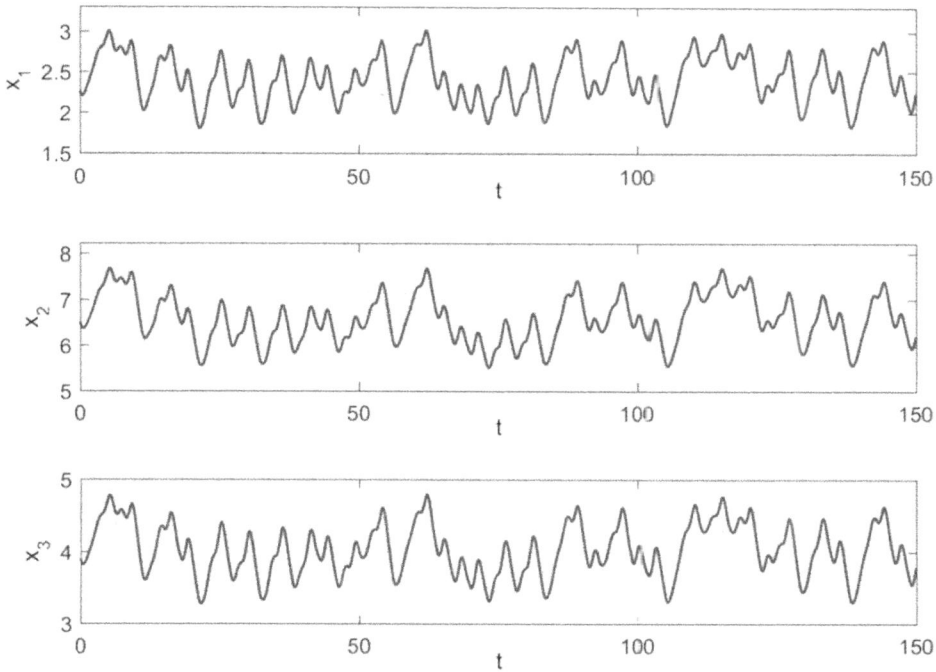

Figure 6.23. Time-series of the x_1, x_2, and x_3-coordinates of the autonomous hybrid system (6.33)-(6.34)-(6.35)-(6.36)-(6.37). Each time-series is irregular, and this confirms the presence of an alpha unpredictable solution in the dynamics. Reproduced with permission from [76].

where $x(t) = (x_1(t), x_2(t), x_3(t))$ is a solution of (6.37). It can be verified that the assumptions $(A1)$ and $(A2)$ hold for system (6.39).

Using the solution of the drive system (6.37), whose coordinates are depicted in figure 6.23, the trajectory of the response system (6.39) shown in figure 6.24 is obtained. In the simulation, the initial data $y_1(0) = 10.91$, $y_2(0) = 10.58$, and $y_3(0) = 41.62$ are utilized. The irregular behavior of this trajectory reveals the presence of an alpha unpredictable solution.

Now, we will make use of the auxiliary system approach [1] to show the presence of generalized synchronization. We take into account the auxiliary system

$$z'_1(t) = -20z_1(t) + 20z_2(t) + 2.9x_1(t)$$
$$z'_2(t) = -z_1(t)z_3(t) + 41.05z_1(t) - z_2(t) + 2.6x_2(t) \qquad (6.40)$$
$$z'_3(t) = z_1(t)z_2(t) - 3z_3(t) + 2.4x_3(t),$$

which is an identical copy of the response system (6.39). Again using the solution of (6.37) shown in figure 6.23 and the initial data $y_1(0) = 10.91$, $y_2(0) = 10.58$, $y_3(0) = 41.62$, $z_1(0) = 10.52$, $z_2(0) = 10.03$, and $z_3(0) = 41.96$, we represent in figure 6.25 the projection of the stroboscopic plot of the hybrid system (6.33)-(6.34)-(6.35)-(6.36)-(6.37)-(6.39)-(6.40) on the $y_1 - z_1$ plane. The plot is obtained by omitting the first 200 iterations in order to eliminate the transients. Because it takes

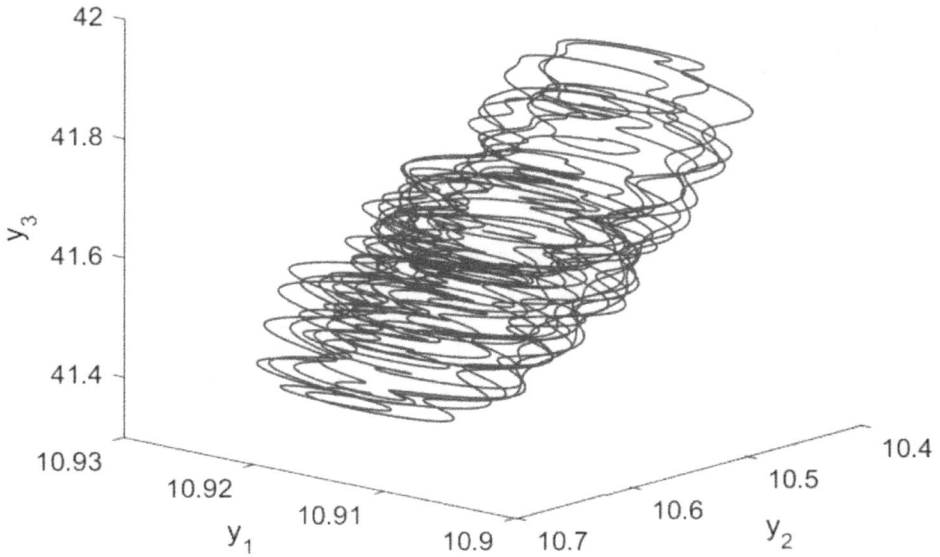

Figure 6.24. The projection of the trajectory of autonomous hybrid system (6.33)-(6.34)-(6.35)-(6.36)-(6.37)-(6.39) on the $y_1 - y_2 - y_3$ space. The irregularity observed in the figure proves that system (6.39) admits an alpha unpredictable solution. Reproduced with permission from [76].

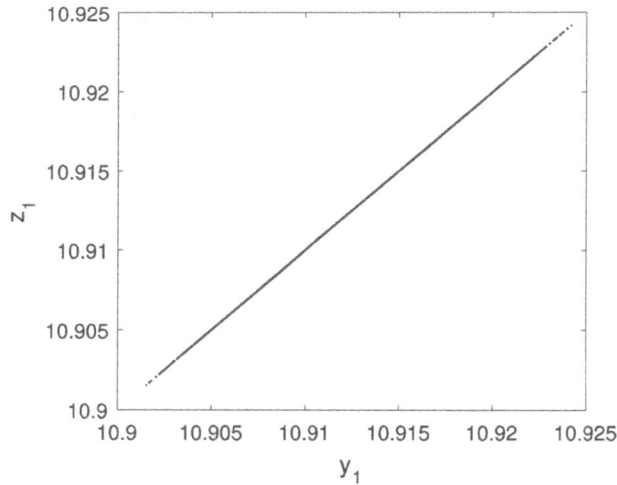

Figure 6.25. The result of the auxiliary system approach applied to the coupled system (6.37)-(6.39). The figure confirms that (6.37) and (6.39) are synchronized and there is an alpha unpredictable solution of the response system (6.39). Reproduced with permission from [76].

place on the line $z_1 = y_1$, one can confirm that generalized synchronization occurs, and therefore, there exists a continuous transformation ψ, which has no explicit time dependence, such that the equation

$$\lim_{t \to \infty} \|y(t) - \psi(x(t))\| = 0$$

is fulfilled, where $x(t) = (x_1(t), x_2(t), x_3(t))$ and $y(t) = (y_1(t), y_2(t), y_3(t))$ are respectively the states of (6.37) and (6.39) [1, 43]. Hence, in accordance with our theoretical results, the response system (6.39) admits an alpha unpredictable solution. Moreover, an alpha unpredictable motion takes place also in the dynamics of the coupled system (6.37)-(6.39) in accordance with remark 6.5.1.

To prove the presence of generalized synchronization one more time, let us evaluate the conditional Lyapunov exponents of the response (6.39). For that purpose, we take into account the corresponding variational system

$$
\begin{aligned}
w'_1(t) &= -20w_1(t) + 20w_2(t) \\
w'_2(t) &= (-y_3(t) + 41.05)w_1(t) - w_2(t) - y_1(t)w_3(t) \qquad (6.41) \\
w'_3(t) &= y_2(t)w_1(t) + y_1(t)w_2(t) - 3w_3(t).
\end{aligned}
$$

When the solution $y(t) = (y_1(t), y_2(t), y_3(t))$ of (6.39) with $\zeta_0 = 0.76$, $\phi(0) = 0.41$, $x_1(0) = 2.26$, $x_2(0) = 6.48$, $x_3(0) = 3.89$, $y_1(0) = 10.91$, $y_2(0) = 10.58$, $y_3(0) = 41.62$, and $\mu = 3.94$ is utilized, the largest Lyapunov exponent of system (6.41) is calculated as -0.2371, i.e., all conditional Lyapunov exponents of system (6.39) are negative. This confirms that (6.37) and (6.39) are synchronized in the generalized sense [43, 51], and accordingly, the response system (6.39) as well as the coupled systems (6.37)–(6.39) possess alpha unpredictable solutions.

6.6 Notes

This chapter reveals a novel numerical approach, the alpha unpredictability test, which proves chaos through observing sequences of numbers. We examined systems that were either Devaney or Li–Yorke chaotic or satisfied the bifurcation diagram analysis or Lyapunov exponents method. These dynamics subdued the alpha unpredictability test under its conditions. For each system, we provided some elements from the sequences of convergence and separation. Famous models such as Rössler, Ikeda, and one with intermittency were subdued to the test examination. The results of the first section are from [3]. Additionally, they are supplied with illustrative graphics associated and corresponding explanations.

The second section of the chapter is concerned with the presence of alpha unpredictable solutions in systems with strange non-chaotic attractors (SNCAs). Examples of continuous-time and discrete-time systems are provided. The alpha unpredictability test [5, 23] is utilized to demonstrate the existence of alpha unpredictable solutions in the dynamics under investigation. It was proved in paper [17] that the existence of an alpha unpredictable motion in dynamics implies the presence of sensitive dependence on initial conditions. Thus, the results of the present study also reveal that systems with SNCAs can possess Lorenz-sensitive dependence on initial conditions even though their Lyapunov exponents are non-positive. The main results of the section are from the papers [3, 5, 23, 77].

Synchronization is one of the phenomena that can occur in coupled chaotic systems [43]. This phenomenon can be observed in various fields such as image encryption, secure communication, lasers, electronic circuits, and neural networks [47, 50, 59, 74, 78]. An alpha unpredictable trajectory makes the corresponding dynamics exhibit chaotic behavior [8, 17]. In the third section of the chapter, it is rigorously proved that if a drive system possesses an alpha unpredictable solution, then the response also has the same property under the conditions mentioned in section 6.5.2. The content of the section was published in [77] The proposed technique makes obtaining high-dimensional systems possessing alpha unpredictable trajectories possible. In the future, our approach can be used to detect alpha unpredictable trajectories in time-delayed and mutually coupled systems [60]. More detailed and sophisticated analysis of synchronization can be found in papers [12, 13, 29], where the sequential test has been applied for both actors of the unison dynamics, and a novel, delta synchronization, based on the convergence and separation sequences analysis was introduced.

References

[1] Abarbanel H D I, Rulkov N F and Sushchik M M 1996 Generalized synchronization of chaos: the auxiliary system approach *Phys. Rev.* E **53** 309–19

[2] Aizawa Y 1983 Symbolic dynamics approach to intermittent chaos *Prog. Theor. Phys.* **144** 1249–63

[3] Akhmet M, Fen M O and Tola A 2019 The sequential test for chaos *arXiv* arXiv: 1904.09127

[4] Akhmet M, Fen M O and Tola A 2022 Strange non-chaotic attractors with unpredictable trajectories *J. Vib. Test. Syst. Dyn.* **6** 317–27

[5] Akhmet M, Fen M O and Tola A 2023 A numerical analysis of Poincaré chaos *Discontinuity Nonlinearity Complexity* **12** 183–95

[6] Akhmet M 2019 Domain-structured chaos in discrete random processes arXiv preprint arXiv:1912.10478

[7] Akhmet M 2020 A novel deterministic chaos and discrete random processes *ACM Int. Conf. Proc. Series* pp 53–6

[8] Akhmet M 2021 *Domain Structured Dynamics* (Bristol: IOP Publishing)

[9] Akhmet M 2022 Unpredictablity in Markov chains *Carpathian J. Math.* **38** 13–9

[10] Akhmet M, Akhmetova Z and Fen M O 2014 Chaos in economic models with exogenous shocks *J. Econ. Behav. Organ.* **106** 95–108

[11] Akhmet M, Arugaslan Cincin D, Tleubergenova M and Nugayeva Z 2021 Unpredictable oscillations for Hopfield-type neural networks with delayed and advanced arguments *Math. Open Access* **9** 1–19

[12] Akhmet M, Baskan K and Yesil C 2022 Delta synchronization of Poincaré chaos in gas discharge-semiconductor systems *Chaos* **32** 083137

[13] Akhmet M, Baskan K and Yesil C 2023 Revealing chaos synchronization below the threshold in coupled Mackey-glass systems *Math. Open Access* **11** 3197

[14] Akhmet M and Fen M O 2016 *Replication of Chaos in Neural Networks, Economics and Physics* (Berlin: Springer)

[15] Tokmak F, Fen M O and Akhmet M 2023 A novel criterion for unpredictable motions *Filomat* **37** 6151–60

[16] Akhmet M and Fen M O 2016 Poincaré chaos and unpredictable functions *Commun. Nonlinear Sci. Numer. Simul.* **48** 85–94

[17] Akhmet M and Fen M O 2016 Unpredictable points and chaos *Commun. Nonlinear Sci. Numer. Simul.* **40** 1–5

[18] Akhmet M and Fen M O 2017 Existence of unpredictable solutions and chaos *Turk. J. Math.* **41** 254–66

[19] Akhmet M and Fen M O 2018 Non-autonomous equations with unpredictable solutions *Commun. Nonlinear Sci. Numer. Simul.* **59** 657–70

[20] Akhmet M, Fen M O, Tleubergenova M and Zhamanshin A A 2019 Unpredictable solutions of linear differential and discrete equations *Turk. J. Math.* **43** 2377–89

[21] Akhmet M, Fen M O and Alejaily E M 2020 A randomly determined unpredictable function *Kazakh Math. J.* **20** 30–6

[22] Akhmet M, Fen M O, Tleubergenova M and Zhamanshin A 2019 Unpredictable solutions of linear differential and discrete equations *Turk. J. Math.* **43** 2377–89

[23] Akhmet M, Fen M O and Tola A 2022 Strange non-chaotic attractors with unpredictable trajectories *J. Vib. Test. Syst. Dyn.* **6** 317–27

[24] Akhmet M, Tleubergenova M and Zhamanshin A 2020 Quasilinear differential equations with strongly unpredictable solutions *Carpathian J. Math.* **36** 341–9

[25] Akhmet M, Tleubergenova M and Zhamanshin A 2019 Neural networks with Poincare chaos *ACM Int. Conf. Proc. Series* pp 34–7

[26] Akhmet M, Tleubergenova M and Zhamanshin A 2020 Inertial neural networks with unpredictable oscillations *Math. Open Access* **8** 1–11

[27] Akhmet M, Tleubergenova M and Zhamanshin A 2023 Unpredictable solutions of duffing type equations with Markov coefficients arXiv: 2303.17336arXiv

[28] Akhmet M, Tleubergenova M and Zhamanshin A 2024 Cohen-Grossberg neural networks with unpredictable and Poisson stable dynamics *Chaos Solitons Fractals* **178** 114307

[29] Akhmet M, Yesil C and Baskan K 2023 Synchronization of chaos in semiconductor gas discharge model with local mean energy approximation *Chaos Solitons Fractals* **167** 113035

[30] Akhmet M and Yilmaz E 2013 *Neural Networks with Discontinuous/Impact Activations* (New york: Springer)

[31] Akhmet M U 2009 Shadowing and dynamical synthesis *Int. J. Bifurcation Chaos* **19** 3339–46

[32] Sauer T D, Yorke J A and Alligood K T 1996 *Brain Theory and Neural Networks* (New York: Springer)

[33] Anosov D 1967 Geodesic flows on closed Riemannian manifolds with negative curvature *Proc. Steklov Inst. Math.* **90** 1–235

[34] Badard R 2008 A lot of strange attractors: chaotic or not? *Chaos* **18** 023127

[35] Blanchard F 2009 Topological chaos: what may this mean? *J. Diff Equ. Appl.* **15** 23–46

[36] Chua L O and Brown R 1998 Clarifying chaos ii: Bernoulli chaos, zero lyapunov exponents and strange attractors *Int. J. Bifurcation Chaos* **8** 1–32

[37] de Moura A P S 2007 Strange nonchaotic repellers *Phys. Rev. E* **76** 1–4

[38] Devaney R L 1987 *An Introduction to Chaotic Dynamical Systems* (Menlo Park, CA: Addison-Wesley)

[39] Devaney R L and Nitecki Z 1979 Shift automorphisms in the Henon mapping *Math. Phys.* **67** 137–46

[40] Feigenbaum M J 1980 Universal behavior in nonlinear systems *Los Alamos Sci./Summer* **1** 4–27

[41] Glasner B and Weiss E 1993 Sensitive dependence on initial conditions *Nonlinearity* **6** 1067–75

[42] Jäger T H, Keller G and Glendinning P 2006 Sensitive dependence on initial conditions *Nonlinearity* **19** 2005–22

[43] Gonzáles-Miranda J M 2004 *Synchronization and Control of Chaos* (London: Imperial College Press)

[44] Ott E, Pelikan S, Yorke J A and Grebogi C 1984 Strange attractors that are not chaotic *Physica* D **13** 261–8

[45] Yorke J A, Grebogi C and Hammel S M 1987 Do numerical orbits of chaotic dynamical processe represent true orbits? *J. Complex* **3** 136–45

[46] He R and Vaidya P G 1992 Analysis and synthesis of synchronous periodic and chaotic systems *Phys. Rev.* A **46** 7387–92

[47] Huang X and Cao J 2006 Generalized synchronization for delayed chaotic neural networks: a novel coupling scheme *Nonlinearity* **19** 2797–811

[48] Hunt B R, Ott E and Yorke J A 1997 Differentiable generalized synchronization of chaos *Phys. Rev.* E **55** 4029–34

[49] Kapitaniak T 1993 Generating strange nonchaotic trajectories *Phys. Rev.* E **47** 1408–10

[50] Kinzel W, Englert A and Kanter I 2010 On chaos synchronization and secure communication *Phil. Trans. R. Soc.* A **368** 379–89

[51] Kocarev L and Parlitz U 1996 Generalized synchronization, predictability, and equivalence of unidirectionally coupled dynamical systems *Phys. Rev. Lett.* **76** 1816–9

[52] Koçak K J and Palmer H 2010 Lyapunov exponents and sensitive dependence *J. Dyn. Diff. Equ.* **22** 381–98

[53] Kohyama T and Aizawa Y 1984 Theory of the intermittent chaos *Prog. Theor. Phys.* **71** 917–29

[54] Li T Y and Yorke J A 1975 Period three implies chaos *Am. Math. Mon.* **82** 985–92

[55] Yue Y, Xie J, Grebogi C and Li G 2013 Strange nonchaotic attractors in a nonsmoothdynamical system *Commun. Nonlinear Sci. Numer. Simulat.* **78** 1–10

[56] Lorenz E N 1963 Deterministic non-periodic flows *J. Atmos. Sci.* **20** 130–41

[57] Miller A 2019 Unpredictable points and stronger versions of Ruelle-Takens and Auslander-Yorke chaos *Topology Appl.* **253** 7–16

[58] Miller V H and Jansen B H 2008 Oscillatory neural network for pattern recognition: trajectory based classification and supervised learning *Biol. Cybern.* **99** 459–71

[59] Moon S, Baik J J and Seo J M 2021 Chaos synchronization in generalized Lorenz and an application to image encryption *Commun. Nonlinear Sci. Numer. Simulat.* **96** 105708

[60] Moskalenko O I, Koronovskii A A and Plotnikova A D 2021 Pecularities of generalized synchronization in unidirectionally and mutually coupled time-delayed systems *Chaos Solit. Fract.* **148** 111031

[61] Naife 1960 *Qualitative Theory of Differential Equations* (Princeton, NJ: Princeton University Press)

[62] Pecora L M and Carroll T L 1990 Synchronization in chaotic systems *Phys. Rev. Lett.* **64** 821–5

[63] Pikovsky A and Feudel U 1995 Characterizing strange nonchaotic attractors *Chaos* **5** 253–260

[64] Pilugin S 1999 *Shadowing in Dynamical Systems* (Berlin: Springer)

[65] Prasad A, Mehra V and Ramakrishna R 1998 Strange nonchaotic attractors in the quasiperiodically forced logistic map *Phys. Rev. E* **57** 1576–84

[66] Prasad A, Nandi A and Ramaswamy R 2007 Aperiodic nonchaotic attractors, strange and otherwise *Int. J. Bifurcation Chaos* **17** 3397–407

[67] Robinson C 2012 *An Introduction to Dynamical Systems: Continuous and Discrete* (Providence, RI: American Mathematical Society)

[68] Rössler O E 1976 An equation for continuous chaos *Phys. Lett.* **57A** 397–8

[69] Ruelle D 1978 An inequality of the entropy of differentiable maps *Bol. Soc. Bras. Mat.— Bull./Brazilian Math. Soc.* **9** 83–7

[70] Rulkov N, Sushchik L, Tsimring M and Abarbanel H 1995 Generalized synchronization of chaos in directionally coupled chaotic systems *Phys. Rev. E* **51** 980

[71] Sander E and Yorke J A 2011 Period-doubling cascades galore *Ergod. Theor. Dynam. Syst.* **31** 1249–67

[72] Schuster H G and Just W 2005 *Deterministic Chaos: An Introduction* (Weinheim: Wiley-VCH)

[73] Sell G R 1971 *Topological Dynamics and Ordinary Differential Equations* (London: Van Nostrand Reinhold Company)

[74] Silva I G D, Buldú J M, Mirasso C R and García-Ojalvo J 2006 Synchronization by dynamical relaying in electronic circuit array *Chaos* **16** 043113

[75] Sprott J C 2017 Asymmetric bistability in the Rössler system *Acta Physica Polonica* B **48** 97–107

[76] Thakur R and Das R 2020 Strongly Ruelle–Takens, strongly Auslander–Yorke and Poincaré chaos on semiflows *Commun. Nonlinear Sci. Numer. Simulat.* **81** 105018

[77] Tokmak F, Fen M O and Akhmet M 2023 A novel criterion for unpredictable motions *Filomat* **37** 6151–60

[78] Uchida A, Higa K, Shiba T, Yoshimori S, Kuwashima H and Iwasawa F 2003 Generalized synchronization of chaos in He-Ne lasers *Phys. Rev. E.* **68**

[79] Wei Z 2011 Dynamical behaviors of a chaotic system with no equilibria *Phys. Lett.* A **376** 102–8

Part VI

Randomness and alpha labeling

Chapter 7

Alpha labeled randomness

Numerous papers have meticulously examined the various antagonistic forms of irregular motions, including deterministic chaos and stochastic processes. This book explores the relationships between stochastic processes and deterministic chaos as an investigation into the essential characteristics of dynamics influenced by randomness. This chapter demonstrates alpha unpredictability in Bernoulli schemes, Markov chains, and modular unbounded stochastic processes. These findings are applied to the random construction of alpha unpredictable functions and contribute to advancing chaos studies in differential equations and the development of new types of stochastic differential equations.

Unsurprisingly, the two antagonistic sorts of irregular motions, deterministic chaos and stochastic processes, are the subject of scrupulous research in many papers. There are different tasks in the approach. First, one wants to find common statistical characteristics for dynamics which are essentially different. Another is to randomize deterministic chaos and find out what kinds of features are invariant under perturbations. And the third task is to find isomorphic dynamics. It does not matter if they are deterministic or stochastic, if there exists isomorphism, they can be considered abstractly the same.

Our approach to the relations between stochastic processes and deterministic chaos can be understood clearly if one remembers that the theory of probability seeks to identify features in chance dynamics. In other words, one must agree that there are no purely deterministic and random processes—the ideal necessary and chance dynamics occur only theoretically. Moreover, necessarily elements are found even in theoretical analysis for processes with chance. Accordingly, our research has shown that ingredients of chaos exist in stochastic processes, such as Bernoulli schemes and Markov chains.

In this study, the main role is given to alpha unpredictability. The alpha unpredictable motion is distinct because it exhibits all the features needed for chaos.

Consequently, one can examine the chaotic nature of an isolated realization of stochastic dynamics. This ability is inapplicable for other types of deterministic chaos, since they are based on sensitivity, which is about collective behavior and consequently, randomness is a dominant factor. One must be clear that when speaking of chaos in stochastic dynamics, we mean the ingredients used to describe chaos. That is, agree that the ingredients of chaos necessarily appear in processes with chance. For example, the appearance of a finite realization is a necessary event for a random process. Next, the occurrence of alpha unpredictability is necessary for certain random dynamics such as the Bernoulli scheme.

We have obtained several results already. One of the most significant is the alpha unpredictable strings and the two laws for them, which were introduced and proved in [16]. The second law is analogous to the Bernoulli theorem. Next, we find that any finite realization of the Bernoulli scheme and Markov chains with finite state space is an arc of an unpredictable alpha trajectory. The method of searching chaotic ingredients was extended for continuous time random processes. Randomly determined unpredictable functions is an important contribution to functional analysis. Thus, the functions class needed for chaos research and applications has been significantly enlarged. Generally speaking, we have moved the study to probability research in many senses. Finally, construction of unpredictable functions as realizations of random processes will create a new line of research for stochastic differential equations. The first steps in the study have been done already in our papers. We investigate the existence of unpredictable solutions of equations, which perturbations are alpha unpredictable functions generated randomly. This will be described in detail in chapter 8. Another interesting problem to research in the future is statistical characteristics for concrete random processes with discrete and continuous time based on numerical parameters provided in ultra Poincaré chaos investigation.

We will start with two ideas supporting what this chapter is doing. They are from the list of ideas in [96], and papers [91–94]. Authors of the book recognized that complementary to the stochastic dynamics being investigated in terms of probability distribution and their various moments, 'or many purposes especially illuminating approach, is the study of individual outcomes of the stochastic process of interest' [96]. This view appreciates our choice of alpha unpredictable trajectories, which can be searched individually in random processes. Another argument is that B. Ornstein demonstrated that isomorphism exists for some random and deterministic processes, such as symbolic dynamics and logistic maps.

Recently, we have proven that realizations of stochastic processes with finite state space are alpha unpredictable functions [4–7, 16]. There are many observations that Markov chains are relative to symbolic dynamics. Our papers also proved that Markov chains and the Bernoulli scheme with finite state space are a rich source of alpha unpredictability, which can be used in functional differential equations. At the same time, the models have chaotic inputs and are proven to have ingredients such as sensitivity, alpha unpredictability, and others.

The results of this chapter are not surprising if one considers the following simple implication. The theory of stochastic processes consists of a list of necessities that have been found for the dynamics of chance. Mathematical chaos consists of ingredients that must confirm to these dynamics. This is why we may look for the ingredients of chaos in random dynamics. If they occur and are proven to be there, one can say that chaos is present in stochastic processes. Be assured that we do not refer to determinism in processes with chance, but non-trivial ingredients, which are common for deterministic chaos and random dynamics. This time, we have chosen alpha unpredictable sequences, the main ingredient of ultra Poincaré chaos, for the discussion. Advanced characteristics support the choice of the concept; namely, Poisson stability and the alpha unpredictability. The sequences are proposed for specific occurrences in random experiments of Bernoulli schemes and Markov chains. We apply alpha labeling for dynamics with orbits whirling among infinitely many modules. Thus, the conservative conditions of chaos existence are weakened to model irregular dynamics. The research proves that ultra Poincaré chaos is of exceptional use for analysis of stochastic processes. To confirm ultra Poincaré chaos in a random process, we can simulate an alpha unpredictable function, which is defined through the realization of a Bernoulli scheme. We have contributed to the probability theory by shaping the characteristics of alpha unpredictable strings. New laws are formulated and verified for Bernoulli schemes. Thus, we have proposed new necessary events for probabilistic analysis. We hope that they are present in the processes sufficiently intensively to be applied, let's say, in stochastic differential equations theory.

7.1 Alpha unpredictability in Bernoulli schemes

Fix the alphabet 1, 2, ... , N, and consider the set $\Omega = \{1, 2, ... , N\}^{\mathbb{N}}$, where $\mathbb{N}$ is the set of all natural numbers, as a sample space. Elements ω of Ω are infinite sequences $(\omega_1, \omega_2, ...)$, where ω_i, $i \in \mathbb{N}$, are members of the alphabet, and probabilities of them are positive. The sum of probabilities is equal to the unit. If the random variables are independent then we are dealing with a Bernoulli scheme or shift. Additionally, we will consider finite strings $(\omega_1, ... , \omega_n)$, $n \in \mathbb{N}$ to be members of the cylindrical sets $\Omega^n = \{1, 2, ... , N\}^n$. An element of the space Ω happening in the process will be called an infinite or finite realization depending on the length of a string. They are orbits of the dynamic. In the case $N = 2$, the dynamics is said to be the Bernoulli process [40, 104]. Symbolic dynamics, a natural source of Bernoulli schemes, is important in studying dynamical systems. It is known [48] that there are dynamics, Axiom A systems among them, which exhibit a product of Cantor set and a smooth manifold, and the dynamics on the Cantor set are isomorphic to the Bernoulli shift. The theorem [91–93] of Ornstain proves that the shifts are isomorphic provided that their entropy is equal.

Assign for each member of the alphabet infinitely many alpha labels $\mathscr{F}_{i_j...}$ where $i_j = 1, ... , N$, $j = 2, ...$, and assume that in the set, a metric d is determined. Thus, one can consider the alpha set, $\mathscr{F} = \{\mathscr{F}_{i_1 i_2...}, i_j = 1, ... , N, j = 1, ...\}$. It is easy to verify that the diameter and separation conditions are valid for the set $\mathscr{F}$. That is, the set is a chaotic structure. Alpha chaos is valid for the dynamics $(\mathscr{F}, \varphi, \delta)$, where φ is the abstract similarity map. The alpha-distance $\delta(\mathscr{F}_{i i_2...}, \mathscr{F}_{j j_2...})$ is equal to the metric d for labeled elements of the alphabet. In the light of chapter 2 results, this means that the Poincaré, Li–Yorke, and Devaney ingredients are present for the triple and in the random process. In what follows, we will pay attention only to ultra-Poincaré chaos. More precisely, to the alpha unpredictable sequence. This notion is the strict evidence for the ultra-Poincaré chaos and vise versa [4, 19, 20].

Definition 7.1.1. [19, 20] *A bounded sequence κ_i, $i \geqslant 1$, in $\mathbb{R}^p$ is called alpha unpredictable if there exist a positive number ε_0 and sequences ζ_n, η_n, $n \geqslant 1$, of positive integers, both of which diverge to infinity such that $\left\|\kappa_{i+\zeta_n} - \kappa_i\right\| \to 0$ as $n \to \infty$ for each i in bounded intervals of integers and $\left\|\kappa_{\zeta_n + \eta_n} - \kappa_{\eta_n}\right\| > \varepsilon_0$ for each natural number n.*

For a finite metric space (S, d), the last definition has the next form.

Definition 7.1.2. *A bounded sequence $s_i \in S$, $i \geqslant 0$, is called alpha unpredictable if there exist a positive number ε_0 and sequences ζ_n, η_n, $n \geqslant 0$ of positive integers, both of which diverge to infinity such that $s_{i+\zeta_n} = s_i$ for each bounded interval of integers, if n is sufficiently large, and $d(s_{\zeta_n + \eta_n}, s_{\eta_n}) > \varepsilon_0$ for each natural number n.*

The closure of the alpha unpredictable orbit of the abstract similarity chaos is the quasi-minimal set, see chapter 2 and chapter 3. It contains an uncountable set of alpha unpredictable trajectories. Let us fix one of them. It is also an alpha unpredictable infinite realization of a Bernoulli scheme. Adding to the definition 7.1.2, the following assertion is valid.

Theorem 7.1.1. *Each finite realization of the Bernoulli scheme coincides with an arc of each alpha unpredictable realization. That is, an alpha unpredictable realization happens in each experiment of the scheme, and is a particular event.*

Thus, one has to recognize that each experiment in the Bernoulli scheme generates an arc of the alpha unpredictable orbit. It is in full accordance with the principle of ergodic theory [111] that a single trajectory proves behavior of the whole dynamics and all other trajectories. In other words, the random process is a reproduction of that exhibiting chaos with probability.

Now is the moment to explain why we have chosen the alpha unpredictable trajectory, and hence, the ultra Poincaré chaos, for the discussion, as opposed to any other types of chaos. One may consider the transitive orbit of Devaney chaos. It is, like the alpha unpredictable orbit, dense in the domain of chaos. However, the alpha unpredictable sequence is preferable because it admits more constructive characteristics than the transitive orbit. They are the sequence of moments, when the values of a motion recurrently visit the initial state and those that guarantee the divergence. The result is helpful for research and applications of random processes such as coin tossing, dice rolling, traffic lights, tetrahedron dice rolling, and five city entrances.

7.2 Alpha unpredictable functions randomly

A special type of Poisson stable orbit, the alpha unpredictable trajectory, which leads to the ultra Poincaré chaos, was introduced in paper [20] and is carefully described in chapter 2. Ensuing papers [4, 19, 22] were concerned with the alpha unpredictable functions of continuous and discrete arguments. This is why, to confirm the ultra Poincaré chaos in a random process with probability one, we can simulate an alpha unpredictable function, which is defined through the realization of a Bernoulli scheme.

Definition 7.2.1. [22] *A uniformly continuous and bounded function* $v: \mathbb{R} \to \mathbb{R}^p$ *is alpha unpredictable if there exist positive numbers* ε_0, δ *and sequences* t_n, u_n *all of which diverge to infinity such that* $v(t + t_n) \to v(t)$ *as* $n \to \infty$ *uniformly on compact subsets of* $\mathbb{R}$ *and* $\|v(t + t_n) - v(t)\| \geqslant \varepsilon_0$ *for each* $t \in [u_n - \delta, u_n + \delta]$ *and* $n \in \mathbb{N}$.

In [22], it is shown that each simulation of the equation

$$x' = -2x + \pi(t) \tag{7.1}$$

asymptotically approaches the solution of the equation (7.1), which is an alpha unpredictable function in the sense of the definition 7.2.1 if the perturbation $\pi(t)$ is based on an alpha unpredictable sequence in the way described below. In what follows, we perform the simulation by using a random alpha unpredictable sequence.

Consider the Bernoulli process, that is, the discrete time random process with $N = 2$, $\omega_1 = 1$, $\omega_2 = 2$, and $d(s_1, s_2) = |1 - 2| = 1$, and take into account the corresponding dynamics of the similarity map such that label $\mathscr{S}_{1 i_2 \dots, i_k, \dots}$ is assigned for 1 and $\mathscr{S}_{2 i_2 \dots}$ is assigned for 2, where $i_j = 1, 2, j = 2, \dots$. We assume that $\delta(\mathscr{S}_{i_1 i_2 \dots}, \mathscr{S}_{j_1 j_2 \dots}) = d(\omega_{i_1}, \omega_{j_1})$ if $i_1, j_1 = 1, 2$, to obtain the chaotic structure $\mathscr{S} = \{\mathscr{S}_{i_1 i_2 \dots}, i_j = 1, 2, j = 1, \dots\}$. By theorem 7.1.3, each finite realization is an arc of an alpha unpredictable sequence. Let us fix a finite realization ω_k, $i_k = 1, 2$, $k \geqslant 1$, of the Bernoulli process. Consider the piecewise constant function, $\pi(t)$, defined on the positive real semi-axis through the equation $\pi(t) = \omega_k$ for $t \in [\frac{k}{100}, \frac{k+1}{100})$, $k \in \mathbb{Z}$. It satisfies conditions for discontinuous alpha unpredictable functions in section 4.2.

Consider the function

$$\Theta(t) = \int_{-\infty}^{t} e^{-2(t-s)} \pi(s) ds, \tag{7.2}$$

It is worth noting that $\Theta(t)$ is the unique bounded on the whole real axis globally exponentially stable solution of the differential equation (7.1).

One can confirm that the function $\Theta(t)$ is uniformly continuous since its derivative is bounded. Because the sequence $\{i_k\}$ is alpha unpredictable, there exist a positive number ε_0 and sequences $\{\zeta_n\}$, $\{\eta_n\}$ both of which diverge to infinity such that

$|i_{k+\zeta_n} - i_k| \to 0$ as $n \to \infty$ for each k in bounded intervals of integers and $|i_{\zeta_n + \eta_n} - i_{\eta_n}| \geqslant \varepsilon_0$ for each $n \in \mathbb{N}$. Fix an arbitrary positive number ε and arbitrary real numbers α, β with $\beta > \alpha$. Let N be a sufficiently large natural number satisfying

$$N \geqslant \ln\left(\frac{2}{3\varepsilon}\right).$$

There exists a natural number n_0 such that for each $n \geqslant n_0$ the inequality

$$|i_{k+\zeta_n} - i_k| < \frac{\varepsilon}{2}$$

holds for $k = 100(\lfloor \alpha \rfloor - N)$, $100(\lfloor \alpha \rfloor - N) + 1, \ldots, 100\lfloor \beta \rfloor$, where $\lfloor \alpha \rfloor$ and $\lfloor \beta \rfloor$ respectively denote the largest integers which are not greater than α and β. Accordingly, if $n \geqslant n_0$, then we have

$$|\pi(t + \zeta_n/100) - \pi(t)| < \frac{\varepsilon}{2} \tag{7.3}$$

for $t \in [\lfloor \alpha \rfloor - N, \lfloor \beta \rfloor + 1)$.

The last inequality implies that $|\Theta(t + \zeta_n/100) - \Theta(t)| < \varepsilon$ for $t \in [\lfloor \alpha \rfloor, \lfloor \beta \rfloor + 1]$. Hence, $|\Theta(t + \zeta_n/100) - \Theta(t)| \to 0$ as $n \to \infty$ uniformly on the interval $[\alpha, \beta]$.

On the other hand, one can confirm for each $n \in \mathbb{N}$ that $|\pi(t + \zeta_n/100) - \pi(t)| \geqslant \varepsilon_0$ for $t \in [\eta_n/100, (\eta_n + 1)/100)$. For fixed $n \in \mathbb{N}$, using the equation

$$\Theta(t + \zeta_n/100) - \Theta(t) = \Theta(\zeta_n/100 + \eta_n/100) - \Theta(\eta_n/100) - 2\int_{\eta_n/100}^{t} (\Theta(s + \zeta_n/100) - \Theta(s))ds$$

$$+ \int_{\eta_n/100}^{t} (\pi(s + \zeta_n/100) - \pi(s))ds$$

we attain that

$$|\Theta(\zeta_n/100 + \eta_n/100 + 1/100) - \Theta(\eta_n/100 + 1/100)| \geqslant \left| \int_{\eta_n/100}^{\eta_n/100 + 1/100} (\pi(s + \zeta_n/100) - \pi(s))ds \right| -$$

$$|\Theta(\zeta_n/100 + \eta_n/100) - \Theta(\eta_n/100)| - 2 \left| \int_{\eta_n/100}^{\eta_n/100 + 1/100} (\Theta(s + \zeta_n/100) - \Theta(s))ds \right|.$$

Therefore, one can verify that

$$\sup_{t \in [\eta_n/100, (\eta_n + 1)/100]} |\Theta(t + \zeta_n/100) - \Theta(t)| \geqslant \frac{\varepsilon_0}{202}.$$

Thus, there exists a sequence $\{u_n\}$ with $\eta_n/100 \leqslant u_n \leqslant (\eta_n + 1)/100$, $n \in \mathbb{N}$, such that

$$|\Theta(u_n + \zeta_n/100) - \Theta(u_n)| \geqslant \frac{\varepsilon_0}{202}.$$

For $t \in [u_n - \delta, u_n + \delta]$, where $\delta = \varepsilon_0/2424$, we have that

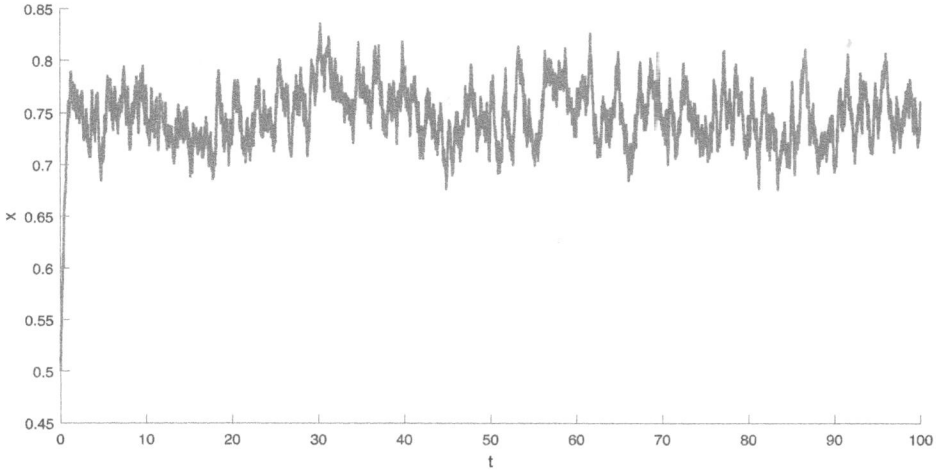

Figure 7.1. The solution $\Psi(t)$, which approximates asymptotically $\Theta(t)$.

$$|\Theta(t + \zeta_n/100) - \Theta(t)| \geqslant \left| |\Theta(u_n + \zeta_n/100) - \Theta(u_n)| - \frac{5}{2} \left| \int_{u_n}^{t} |\Theta(s + \zeta_n/100) - \Theta(s)| ds \right| \right.$$

$$\left. - \left| \int_{u_n}^{t} |\pi(s + \zeta_n/100) - \pi(s)| ds \right| \geqslant \frac{\varepsilon_0}{404}.$$

It is clear that $u_n \to \infty$ as $n \to \infty$. Thus, the function $\Theta(t)$ is alpha unpredictable. This is the first case, when a random process has determined deterministic alpha unpredictable function. To visualize the function, we shall apply the asymptotic stability of the alpha unpredictable solution. It means that one can simulate any solution of the equation, and it will ultimately approach the alpha unpredictable function. Accordingly, we depict in figure 7.1 the solution $\Psi(t)$, $\Psi(0) = 0.5$, of the equation (7.1), which approximates the alpha unpredictable solution $\Theta(t)$.

7.3 Alpha unpredictable strings and statistical laws

The Bernoulli process, and more broadly the Bernoulli scheme, are fundamental and compelling concepts in probability theory, statistics, random processes, and dynamical processes related to chaos [1, 34, 39, 44, 54, 67, 69, 70, 76, 90–94, 104, 112]. In this section, we propose introducing a new type of event in addition to traditional probabilistic events, called an alpha unpredictable string. This new concept offers intriguing opportunities for expanding existing theories and exploring the valuable characteristics of deterministic ultra Poincaré chaos within stochastic dynamics. We define infinite sequences that include alpha unpredictable strings. The first and second laws of large (alpha unpredictable) strings are discussed theoretically. The findings of section 7.1 are particularly useful for comparison, as they are more general in certain respects.

7.3.1 Basic dynamics of strings

Next, we introduce the main concept of this section, alpha unpredictable strings, and utilize it to determine alpha unpredictable sequences.

Let a_i, $i = 0, 1, 2, \ldots$, be an infinite sequence of symbols.

Definition 7.3.1. [16] *A finite array* $(a_s, a_{s+1}, \ldots, a_{s+k})$, *where s and k are positive integers, is said to be an alpha unpredictable string of length k if* $a_i = a_{s+i}$, *for* $i = 0, 1, 2, \ldots, k - 1$, *and* $a_k \neq a_{s+k}$.

Definition 7.3.2. [16] *The sequence* a_i, $i \geqslant 0$, *is alpha unpredictable if it admits alpha unpredictable strings with arbitrary large lengths.*

Definition 7.3.3. [16] *The sequence* a_i, $i \geqslant 0$, *is alpha unpredictable if there exist sequences* ζ_n, η_n *of positive integers, both of which diverge to infinity such that* $a_{\zeta_n + l} = a_l$, $l = 0, 1, 2, \ldots, \eta_n - 1$, *and* $a_{\zeta_n + \eta_n} \neq a_{\eta_n}$, *for each* $n \in \mathbb{N}$.

Theorem 7.3.1. *The definitions 7.3.2 and 7.3.3 are equivalent.*

Proof. Let sequence a_i be alpha unpredictable in the sense of definition 7.3.3. Then the finite arrays $(a_{\zeta_n}, a_{\zeta_n + 1}, \ldots, a_{\zeta_n + \eta_n})$ are alpha unpredictable strings of length η_n, for each natural n. Thus, the sequence admits alpha unpredictable strings with arbitrary large lengths. Conversely, let a_i be a sequence that admits alpha unpredictable strings of arbitrary large lengths, i. e., there is a sequence i_n, $n = 1, 2, 3, \ldots$, such that the finite arrays $(a_{i_n}, a_{i_n + 1}, \ldots, a_{i_n + k})$ are alpha unpredictable strings. By setting $\zeta_n = i_n$ and $\eta_n = i_n + k$ we deduce that the sequence a_i is alpha unpredictable in light of definition 7.3.3.

Fix a positive integer k and denote by S_k the sets of all indexes s such that the strings $(a_s, a_{s+1}, \ldots, a_{s+k})$ are alpha unpredictable within the sequence a_i, $i = 1, 2, \ldots$, which is not necessarily alpha unpredictable.

Theorem 7.3.2. [16] *The sets S_l and S_q do not intersect if $l < q$.*

Proof. Assume, on the contrary, that sets S_l and S_q admit a common element s. Then, we have that $a_l \neq a_{s+l}$ if $s \in S_l$ and $a_l = a_{s+l}$ if $s \in S_q$. This contradiction completes the proof.

Theorem 7.3.3. [16] *Assume that a_i, $i \geqslant 0$, is an alpha unpredictable sequence. Then each a_j with positive j is the first element of an alpha unpredictable string, if $a_j = a_0$.*

Proof. Assume the opposite. Then one can show that the sequence α is periodic one, that it is not an alpha unpredictable sequence.

7.3.2 Laws of large strings

In this part, we consider a discrete-time random process with the finite state space of N different symbols 1, 2, ... , N.

It admits values i with positive probabilities p_i, $i = 1, 2, \ldots , N$, which sum is equal to the unit, and the variables are independent. A realization ω of the process is the sequence ω_i, $i = 1, 2, \ldots$, and a finite realization ω_m is the array ω_i, $i = 1, 2, \ldots , m$. We claim that stochastic processes with discrete-time and finite-state spaces satisfy the following theorem.

Theorem 7.3.4. [16] *(the first law of large strings) A discrete time random process with a finite state space admits an uncountable set of realizations, which are alpha unpredictable sequences in the sense of definition 7.3.2.*

Proof. Let us consider the space Σ_N of infinite sequences of a finite set of symbols s_1, s_2, ... , s_r, with the metric

$$d(\xi, \zeta) = \Sigma_{k=0}^{\infty} \frac{|\xi_k - \zeta_k|}{2^k}, \tag{7.4}$$

where $\xi = (\xi_0 \xi_1 \xi_2 \ldots)$, $\zeta = (\zeta_0 \zeta_1 \zeta_2 \ldots)$. The Bernoulli shift σ on Σ_N is defined as $\sigma(\xi_0 \xi_1 \xi_2 \ldots) = (\xi_1 \xi_2 \xi_3 \ldots)$. The map is continuous and Σ_N is a compact metric space [112].

The set of all realizations of the random dynamics coincides with the set of all sequences of the symbolic dynamics on Σ_N. According to the result in [19], the symbolic dynamics admits an alpha unpredictable point, i^*, a sequence from the set Σ_N. There is the uncountable set of points, which are alpha unpredictable sequences in the sense of definition 7.3.2.

Each realization must be in the closure for the alpha unpredictable orbit. The density is considered in the shift dynamics sense. The property of the metric implies that each arc of any sequence in space coincides with some arc of the alpha unpredictable sequence.

Denote by n(m) the number of elements equal to ω_0 in a finite string. The limit $E[\omega_0] = \lim_{m \to \infty} n(m)/m$ is said to be the expected value such that $E[\omega_0] = p_i$, if $\omega_0 = i$, $i = 1, \ldots , N$ [40].

Theorem 7.3.8 implies the equality $N(m) = n(m)$, where $N(m)$ is the number of alpha unpredictable strings, which intersect the array. Hence, the following proposition is correct, which can be helpful in applications.

Theorem 7.3.5. [16] *If a realization ω is an alpha unpredictable sequence, then the relation*

$$\lim_{m \to \infty} \frac{N(m)}{m} = E[a_0] \tag{7.5}$$

is valid.

Theorem 7.3.6. [16] *(the second law of large strings) If the discrete time random process admits a finite state space, then the relation*

$$\lim_{m \to \infty} P\left(\left| \frac{N(m)}{m} - E[\omega_0] \right| < \varepsilon \right) = 1 \tag{7.6}$$

holds for any $\varepsilon > 0$.

Proof. Since of theorem 7.3.4, the Bernoulli scheme is ultra-Poincaré chaotic. Consequently, there is an alpha unpredictable realization, which is dense in the set of all other realizations with topology of uniform convergence on each bounded interval of integers. Because of the finite space set, an arc of the alpha unpredictable sequence exists for each finite realization, which coincides with the realization. Now, taking into account the time-uniform property, we can conclude that each of the finite realization is, in fact, an arc of an alpha unpredictable orbit and consequently, by the last theorem the proof is complete.

The last result is in the full accordance with the theorem 7.1.3, which says that any finite realization is an arc of a fixed alpha unpredictable realization common for all experiments.

7.4 Modular chaos in random processes

We will apply the method of alpha-chaos for a dynamics with orbits whirling among infinitely many modules. Consequently, the conservative conditions of chaos existence will be weakened to model irregular dynamics. The research proves that the ultra-Poincaré chaos is of exceptional use for analysis of stochastic processes. Examples, illustrating the results are provided. Section 2.6 clarifies the chapter's chaotic basis.

7.4.1 Chaos and randomness

The paradigm of chaos relies on the two distance antagonists, proximality and separation. They utilize metrical or topological closeness and divergence in the description considering either couples of orbits [42, 78] or a single motion [4, 19, 20]. In the present research, we have renounced the idea that transitivity and density of periodic motions of a class of motions have to be described metrically, and even topologically. It is suggested, first time in literature, that the closeness can be considered not in the metrical sense, but through *indexing*, although separation still entirely rely on the distance. The antagonists are still present, but not in the way of conventional theories. The process of indexing is adjacent with the similarity map. More precisely, the indexing is the instrument for the mapping to be chaotic. Trajectories move through different sets, and even various metric spaces when time increases, and each of the sets admits its chaotic structure. This is why we say that the *modular chaos* is in the focus of the research or, more precisely, *modular alpha chaos*.

The method of alpha-chaos is initiated in [3, 8, 9]. Results were obtained, which rely on dynamics in a single metrical or topological space, fractals and bounded random processes such as Bernoulli schemes [3, 9, 14]. This time the method has been enriched in two senses. Firstly, we consider infinite multiples of chaotically structured modules. Secondly, the conservative chaos conditions are weakened by considering indexing for closeness. This helps us determine unbounded deterministic dynamics chaos and find chaotic structures for continuous-time random processes, bounded and unbounded. We have accepted chaos presence in discrete time stochastic dynamics if realizations coincide with orbits of the modular similarity map and explain its presence for continuous-time random processes. One must say that the concepts of ultra Poincaré chaos and alpha unpredictable sequences [4, 16, 19, 20] are principal for the exploration.

The approach will be developed further, by considering multidimensional indexing and infinitely dimensional dynamics. Additionally, the definitions can be considered for processes, which are only partially stochastic for hybrid dynamics, where intervals of deterministic and/or stochastic behavior are combined.

Let us highlight the probabilistic essence of the chaos and the random process relations. Considering the modular ultra Poincaré chaos, one can see a modular alpha unpredictable orbit exists such that any finite realization is an arc of that one. Consequently, we have to conclude that a modular alpha unpredictable orbit is a certain event in any experiment we observe.

In the basis of our investigation are the effective methods, which have been developed for chaos investigation, *alpha dynamics and alpha chaos*.

The research on similarity goes back to G Leibniz, who introduced the notions of recursive self-similarity [125]. The idea of a self-similar set was first considered by Moran,[87]. He gave a mathematical definition of a geometric construction as a collection of sets satisfying specific conditions. Further, definitions of self-similarity and related problems were discussed in papers [36, 55, 62–64, 67, 77, 89, 95, 105, 106]. Our research is based on a point-set structure in metric spaces and this is why we call it *abstract self-similarity*[9]. It does not rely on special functions, and the *similarity map* is naturally combined with the concept. The next extension of the present results can be obtained based on papers [36, 55, 62–64, 77, 89, 105].

The potential applications of modular chaos are vast and practical. They naturally relate to problems of control and synchronization [49], enriching the theory of stochastic differential equations [90] and Markov chains [47]. By involving the results of modular chaos investigation in these studies, we can utilize methods of replication of chaos and alpha unpredictable dynamics [6, 10, 14, 19], opening up new avenues for research and application.

It is of intense interest to compare deterministic and random dynamics. More precisely, to find features of chaos in stochastic processes and recognize probabilistic features in deterministic motions. If the first one is considered, for instance, in papers [92, 93], in the paper [65] arguments of statistics are provided to show deterministic features in random processes. The present research is in the second stream, when the random processes can be assigned with properties of chaos. Our results rely not on the statistical parameters, but on the concept, which has been introduced and developed during the last four years, ultra-Poincaré chaos [3, 4, 15, 19, 20].

7.4.2 Modular chaotic random processes

Let us describe the type of the stochastic processes [44] that are in the focus of this section. Consider a random process with continuous time or discrete time and with continuous or discrete state space as a family of random variables $\mathbf{X}(t)$ indexed by the parameter t with the range $\mathscr{I}$, which is either an interval of the real line or an infinite set of integers bounded from below. Realizations of the dynamics are not necessarily continuous if the time is continuous and they must not be bounded functions for both sorts of time. Denote by $S = \bigcup_{t \in \mathscr{I}} S^t$ the state space of the process, and consider it with a distance d. We assume positive probability for all members of the state space, and for each fixed t one of the members of the state space must necessarily happen.

Suppose that for each fixed $t \in \mathscr{I}$ there exists an alpha labeling,

$$\mathscr{S}^t = \left\{ \mathscr{S}^t_{i_1 i_2 \dots i_n \dots} : i_k = 1, 2, \dots, m, \quad k = 1, 2, \dots \right\}, \tag{7.7}$$

where m is a fixed natural number, common for all $t \in \mathscr{I}$, larger than or equal to two, such that each element of the set S^t is alpha labeled with a member of the set $\mathscr{S}^t$. The corresponding to the distance d function δ is common for all $t \in \mathscr{I}$.

Consider all possible infinite strictly increasing sequences $t_\alpha = \{t_i^\alpha\}$, $i = 1, 2, \dots$, in $\mathscr{I}$, with $\mathscr{A}$, the set of all indexes α.

We shall say that the random process $\mathbf{X}(t)$ is *(strongly) modular chaotic*, if $\mathscr{S}^\alpha = \bigcup_i \mathscr{S}^{t_i^\alpha}$ is a (strong) modular chaotic structure for $S^\alpha = \bigcup_i S^{t_i^\alpha}$ with each $\alpha \in \mathscr{A}$, and there exists a positive number ε_0, which is not larger than all separation constants ε_0^α, $\alpha \in \mathscr{A}$.

The choice of modular chaos for the random dynamics description is approved because the set of all realizations for dynamics $\mathbf{X}(t_\alpha)$ with fixed $\alpha \in \mathscr{A}$ coincides with the set of all trajectories of the affine similarity map φ on the set $\mathscr{S}^\alpha$, and the dynamics of the affine map is similar to the symbolic dynamics. To argue this, let us turn the discussion to the lower indexes analysis. Consider the space $\Sigma_m = \{(i_1 i_2 \dots) \,|\, i_k = 1, 2, \dots, m\}$ of infinite sequences on finite number of symbols with the metric

$$d(s, t) = \sum_{k=1}^{\infty} \frac{|i_k - j_k|}{2^{k-1}},$$

where $(i_1 i_2 \ldots)$, and $(j_1 j_2 \ldots)$ are elements of Σ_m. The Bernoulli shift $\sigma\colon \Sigma_m \to \Sigma_m$ is defined as $\sigma(s_0 s_1 s_2 \ldots) = (s_1 s_2 s_3 \ldots)$. The map σ is continuous and (Σ_r, d) is a compact metric space [42, 112]. According to the result in [19], the symbolic dynamics admits an alpha unpredictable point, i^*, an element of the set [20]. It is important for us that Σ_m is a closure for the alpha unpredictable orbit. The peculiarity of the metric implies that each arc of any sequence in the space coincides with some arc of the alpha unpredictable sequence with a shift precision. In other words, because of the finite number of indexes involved in the construction, one has the complete coincidence with a part of the alpha unpredictable orbit. Moreover, the coincidence does not last forever, because of the alpha unpredictability. This is why each iteration of the Bernoulli shift σ results in what is done by the Bernoulli trial, and we accept the random dynamics as modular chaotic. Consequently, the following assertion is valid for a fixed alpha unpredictable realization of the random process.

Theorem 7.4.1. [5] *Each finite orbit of the dynamics* $\mathbf{X}(t_\alpha)$ *coincides with an arc of the alpha modular realization, which is a certain event.*

Let us give more arguments for chaos presence based on our research, considering the sensitivity of the dynamics. Fix an infinite sequence. Consider it as a realization. The probability that the realization will happen as the prescribed sequence is equal to zero. Hence, the complementary event, that there is a finite moment such that the realization changes its value from the fixed sequence value, admits the probability 1. It is easy to see that it is the probability of the sensitivity. More precisely, if we start two experiments with an identical state, then there exists a moment, with probability 1, such that they diverge. Thus, we can say the sensitivity presents in the random dynamics with probability 1. One can call the property a *random sensitivity*.

The final conclusion of the discussion is that each alpha unpredictable realization of the chain appears at each experiment of the chain as a certain event, and the sensitivity of the dynamics occurs with probability 1. So, we can not claim that the chaos appears as a deterministic phenomenon in the random process, but what we can say two things. The first is that chaos occurs with probability 1, and secondly, one can look for ingredients of deterministic chaos, which are present in the dynamics as certain events. The problem of searching for the ingredients can be beneficial for applications, and looks attractive for theoretical issues.

The closure of the alpha unpredictable realization is said to be the quasi-minimal set [103]. It contains an uncountable set of alpha unpredictable realizations. One can say that the result of the present research is the proof that the union of all infinite realizations of the random process is a quasi-minimal set, which is a certain event. And this is another reason we call the process *chaotic*.

It is clear that for the discrete-time random process, it is sufficient to consider only the sequence related to the set of indexes $\mathscr{I}$ itself. We do not exclude that the method of modular chaos may be of use even if the modular chaotic structure exists not for all α from $\mathscr{A}$, but for some of them.

Based on the last discussion, one can say that stochastic processes are more 'irregular' than deterministic chaos. Nevertheless, we have strong confidence that by developing the present results, one can reduce the difference between the two concepts, if not diminish it altogether.

Consider several scalar real-valued functions $f_j(t), j = 1, \ldots, m$, defined on a time interval, $\mathscr{I}$, of real numbers. Assume that the ranges of the functions are disjointed and the minimal distance between elements of distinct ranges is larger than a positive number ε_0. Define the random process $\mathbf{X}(t)$ such that for each fixed $t \in \mathscr{I}$, it admits a value of one of the functions f_j with positive probability $p_j, j = 1, \ldots, m$. The probabilities do not depend on t, and the sum of all probabilities is equal to the unit if t is fixed.

We have that the state space is $S = \bigcup_{t \in \mathscr{I}} S^t$, where $S^t = \{f_j(t) | j = 1, \ldots, m, t \in \mathscr{I}\}$. Let us construct the alpha set,

$$\mathscr{S}^t = \left\{ \mathscr{S}^t_{i_1 i_2 \ldots i_n \ldots} : i_k = 1, 2, \ldots, m, \ \ k = 1, 2, \ldots \right\}, \tag{7.8}$$

where $\mathscr{S}^t_{i_1 i_2 \ldots i_n \ldots} = f_{i_1}(t)$, $i_j = 1, 2, \ldots, m$, $j = 2, \ldots$. One can check that the union, $\mathscr{S}$, of sets $\mathscr{S}^t$, $t \in \mathscr{I}$, is an alpha set for the set S, and consequently the random process is modular alpha chaotic.

7.4.3 Examples with numerical simulations

Next, we will provide three examples of modular alpha chaotic random processes.

Example 7.4.1. Let us consider the random process $\mathbf{X}(t)$, which is equal to t or $-t$ for each fixed $t \in [1, 4]$, with probability 1/2. It is clear that the dynamics is a continuous-time random process with a continuous state space. Fix the parameter value t and obtain that $S^t = \{t, -t\}$ consists of two elements, and the distance between the two points is equal $|t - (-t)| = 2t \geqslant \varepsilon_0 = 2$. Let us construct a chaotic structure for the set S^t. Denote $S^t_{i_1 i_2 \ldots} = t$ with $i_1 = 1$, and i_j is equal to 1 or 2 for all $j \geqslant 2$. Similarly, $S^t_{i_1 i_2 \ldots} = -t$ with $i_1 = 2$ and i_j is equal to 1 or 2 for all $j \geqslant 2$. According to discussion in the last example, the random process is modular chaotic. To illustrate the dynamics, we consider the sequence $t_i = i/100$, $i = 100, 2, \ldots, 400$, randomly determine values $\mathbf{X}(t_i)$ and draw the graph of the piece-wise constant function φ equal to $\mathbf{X}(t_i)$ on the interval $[i/100, (i + 1)/100), i = 100, \ldots, 399$. The graph is seen in figure 7.2 and it approximates one of the realizations of the random dynamics.

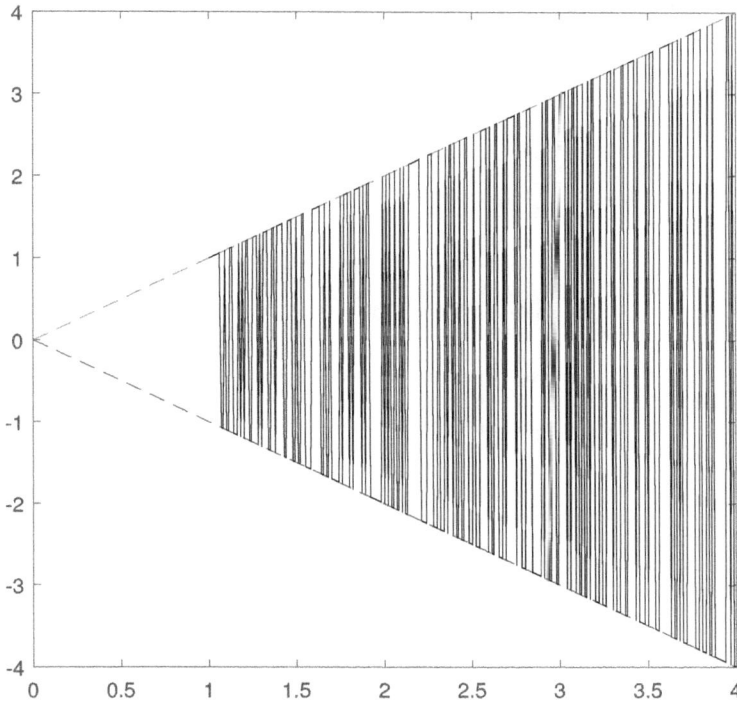

Figure 7.2. The graph of the function φ in interval $[1, 4]$. The red dashed line is the graph of the identity function t and the blue dashed line is the graph of function $-t$. Reproduced with permission from [5].

Example 7.4.2. Let us determine functions $f_i(t) = t$ and $g_i(t) = -t$ on the interval $[i/10, (i + 1)/10)$ for each $i = 10, 11, \ldots$. Construct the metric space of continuous functions, $S^i = \{f_i(t), g_i(t)\}$, $t \in [i/10, (i + 1)/10)$, for each $i = 10, 11, \ldots$, with the distance $d(f_i(t), g_i(t)) = \max_t |t - (-t)|$. Introduce the discrete time random process $\mathbf{X}(i)$, $i = 10, 11, \ldots$, which is equal to the function $f_i(t)$ or $g_i(t)$ with probability $1/2$ on the interval $[i/10, (i + 1)/10)$. One can easily see that the random dynamics are modular chaotic, since one can construct appropriate modular chaotic structure. The graph of a realization, ψ, on the interval $[1, 4]$ is seen in figure 7.3. For better visibility of the dynamics, the vertical lines connecting pieces of the graph are drawn.

7.4.4 Notes

The chapter consists of results that bring the two types, deterministic and indeterministic complexities, much closer together. We simply did this so that the Bernoulli schemes and processes are considered, but we strongly believe that the approach of alpha-dynamics has to be applied for more advanced stochastic and statistical phenomena rather than being accomplished in this book.

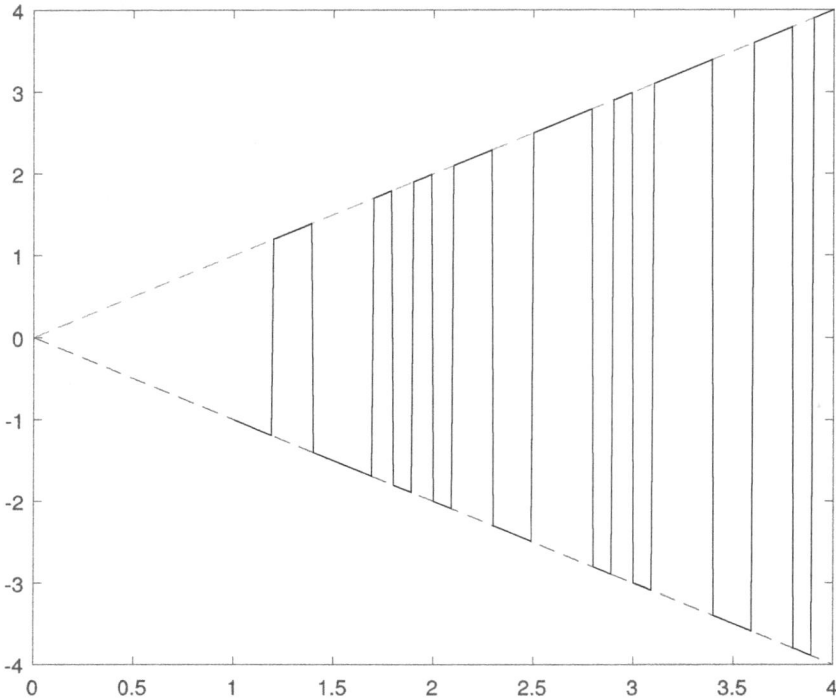

Figure 7.3. The graph of function ψ over interval [1, 4]. The red dashed line is the graph of the identity function t and the blue dashed line is the graph of function $-t$. The vertical lines connecting pieces of the graph are drawn for better visibility. Reproduced with permission from [22].

We have pursued two goals in the present research. The first is to reveal signs of deterministic chaos in random phenomena, and the second, to maximize their presence. New definitions of chaos to fulfill the two tasks have been developed. All the three are considered in this research for random processes with discrete and continuous time. First, the ultra Poincaré chaos ingredients are recognized for the Bernoulli schemes, and they are visualized for such a popular process as coin tossing. The approach is utilized for randomly constructed alpha unpredictable functions.

Surprisingly, we have discovered that classical dynamics with the simplest statistical laws of large numbers can yield new features that can be quantitatively characterized. For this, alpha unpredictable strings in the realizations, considered infinite sequences, were determined. Then, the laws of large strings, which are strongly related to the laws of large numbers, were proved. A new gate for further analysis of chaotic dynamics in probability and random processes has been opened.

Applying the alpha modular chaos concept to random dynamics analysis is sensible, since the set of all realizations for a random dynamics coincides with the set of all chaotic trajectories. To approve the phenomenon, the domain structured chaos notion, which has been considered in our previous papers, is extended. The new type of chaos has been developed such that the notions of transitivity and density of periodic orbits are free of metrical or topological features, and are based on the comprehension that two points are close if they belong to the same set by indexing. Moreover, trajectories move among different sets. When the time increases, each of the sets admits its chaotic structure. It is important that this concept helped us find a chaotic structure for continuous-time random processes, bounded and unbounded. We accept the phenomenon as a sign of deterministic chaos in the dynamics, since realizations of the corresponding discrete time random process are the same as trajectories of the deterministic affine similarity map introduced in section 2.6. The concept can be considered for processes, which are only partiall stochastic in time, that is hybrid dynamics, through combinations of intervals, where a dynamics is deterministic or stochastic.

Hopefully, this may generate new approaches for non-stationary processes and Markov chains inside the stochastic dynamics theory [90]. Our research results may help develop chaos in non-equilibrium processes [96]. The methods of chaotic dynamics, e.g., synchronization and control [49], can be extended to dynamics with probability. Moreover, one can expect that results for symbolic dynamics considered in the course of the present research can give new effects for entropy [1, 125], harmonic analysis, discrete mathematics, probability, and operator algebras [67].

References

[1] Adler R, Konheim A and McAndrew M 1965 Topological entropy *Trans. Am. Math. Soc.* **114** 309–19

[2] Aihara K, Takabe T and Toyoda M 1990 Chaotic neural networks *Phys. Lett.* A **144** 333–40

[3] Akhmet M 2019 Domain-structured chaos in discrete random processes arXiv preprint arXiv:1912.10478

[4] Akhmet M 2021 *Domain Structured Dynamics* (Bristol: IOP Publishing)

[5] Akhmet M 2022 Modular chaos for random processes *Discontinuity Nonlinearity Complexity* **11** 191–201

[6] Akhmet M 2022 Unpredictablity in Markov chains *Carpathian J. Math.* **38** 13–9

[7] Akhmet M, Tleubergenova M and Zhamanshin A 2023 Unpredictable solutions of Duffing type equations with Markov coefficients *arXiv* arXiv: 2303.17336

[8] Akhmet M and Alejaily E M 2019 Domain-structured chaos in a Hopfield neural network *Int. J. Bifurcation Chaos* **29** 195–205

[9] Akhmet M and Alejaily E M 2021 Abstract similarity, chaos and fractals *Discrete Continuous Dyn. Syst. Ser.* B **26** 2479–97

[10] Akhmet M and Fen M O 2016 *Replication of Chaos in Neural Networks, Economics and Physics* (Berlin: Springer)

[11] Akhmet M and Fen M O 2016 Poincaré chaos and unpredictable functions *Commun. Nonlinear Sci. Numer. Simul.* **48** 85–94

[12] Akhmet M and Fen M O 2016 Unpredictable points and chaos *Commun. Nonlinear Sci. Numer. Simul.* **40** 1–5

[13] Akhmet M and Fen M 2018 Non-autonomous equations with unpredictable solutions *Commun. Nonlinear Sci. Numer. Simul.* **59** 657–70

[14] Akhmet M, Fen M O and Alejaily E M 2020 *Dynamics with Chaos and Fractals* (Berlin: Springer)

[15] Akhmet M, Fen M and Alejaily E M 2020 A randomly determined unpredictable function *Kazakh Math. J.* **20** 30–6

[16] Akhmet M and Tola A 2020 Unpredictable strings *Kazakh. Math. J.* **20** 16–22

[17] Akhmet M, Arugaslan D and Yilmaz E 2010 Stability analysis of recurrent neural networks with piecewise constant argument of generalized type *Neural Netw.* **23** 305–11

[18] Akhmet M, Arugaslan Cincin D, Tleubergenova M and Nugayeva Z 2021 Unpredictable oscillations for Hopfield-type neural networks with delayed and advanced arguments *Math. Open Access* **9** 1–19

[19] Akhmet M and Fen M O 2016 Poincaré chaos and unpredictable functions *Commun. Nonlinear Sci. Numer. Simul.* **48** 85–94

[20] Akhmet M and Fen M O 2016 Unpredictable points and chaos *Commun. Nonlinear Sci. Numer. Simul.* **40** 1–5

[21] Akhmet M and Fen M O 2017 Existence of unpredictable solutions and chaos *Turk. J. Math.* **41** 254–66

[22] Akhmet M and Fen M O 2018 Non-autonomous equations with unpredictable solutions *Commun. Nonlinear Sci. Numer. Simul.* **59** 657–70

[23] Akhmet M, Fen M O and Alejaily E M 2018 Dynamics motivated by Sierpinski fractals ArXiv e-prints

[24] Akhmet M and Fen M O 2015 Attraction of Li-Yorke chaos by retarded SICNNs *Neurocomputing* **147** 330–42

[25] Akhmet M, Fen M O and Kirane M 2015 Almost periodic solutions of retarded SICNNs with functional response on piecewise constant argument *Neural Comput. Appl.* **13** 1–10

[26] Akhmet M, Fen M O, Tleubergenova M and Zhamanshin A 2019 Poincare chaos for a hyperbolic quasilinear system *Miskolc Math. Notes* **20** 33–44

[27] Akhmet M, Fen M O, Tleubergenova M and Zhamanshin A 2019 Unpredictable solutions of linear differential and discrete equations *Turk. J. Math.* **43** 2377–89

[28] Akhmet M, Seilova R D, Tleubergenova M and Zhamanshin A 2020 Shunting inhibitory cellular neural networks with strongly unpredictable oscillations *Commun. Nonlinear Sci. Numer. Simulat.* **89** 105287

[29] Akhmet M, Tleubergenova M and Nugayeva Z 2020 Strongly unpredictable oscillations of Hopfield-type neural networks *Math. Open Access* **8** 1–14

[30] Akhmet M, Tleubergenova M and Zhamanshin A 2019 Neural networks with Poincare chaos *ACM Int. Conf. Proc. Series* pp 34–7

[31] Akhmet M, Tleubergenova M and Zhamanshin A 2020 Inertial neural networks with unpredictable oscillations *Math. Open Access* **8** 1–11

[32] Akhmet M and Yilmaz E 2013 *Neural Networks with Discontinuous/Impact Activations* (New York: Springer)

[33] Arbib M A 1988 *Brain Theory and Neural Networks* (Cambridge, MA: MIT Press)

[34] Arnold L 1998 *Random Dynamical Systems* (New York: Springer)

[35] Babcock K L and Westervelt R M 1986 Stability and dynamics of simple electronic neural networks with added inertia *Physica D* **23** 464–9

[36] Bandt C and Graf S 1992 Self-similar sets 7. A characterization of self-similar fractals with positive Hausdorff measure *Proc. Am. Math. Soc.* **114** 995–1001

[37] Bel A and Rotstein H G 2019 Membrane potential resonance in non-oscillatory neurons interacts with synaptic connectivity to produce network oscillations *J. Comput. Neurosci.* **46** 169–95

[38] Bouzerdoum A and Pinter R B 1993 Shunting inhibitory cellular neural networks: derivation and stability analysis *IEEE Trans. Circuits Systems I Fund. Theory and Appl.* **40** 215–21

[39] Bowen R 1970 Markov partitions for axiom a diffeomorphisms *Am. J. Math.* **92** 725–47

[40] Castaneda L B, Arunachalam V and Dharmarajs S 2012 *Introduction to Probability and Stochastic Processes with Applications* (Hoboken, NJ: Wiley)

[41] Chen L and Zhao H 2008 Global stability of almost periodic solution of shunting inhibitory cellular neural networks with variable coefficients Chaos Solitons Fractals 35 351-7; Derrida B and Meir R 1988 Chaotic behavior of a layered neural network *Phys. Rev. A* **38** 3116–9

[42] Devaney R L 1987 *An Introduction to Chaotic Dynamical Systems* (Menlo Park, CA: Addison-Wesley)

[43] Ding H S, Liang J and Xiao T J 2008 Existence of almost periodic solutions for SICNNs with time-varying delays *Phys. Lett.* A **372** 5411–6

[44] Doob J L 1953 *Stochastic Processes* (New York: Wiley)

[45] Dong Q, Matsui K and Huang X 2002 Existence and stability of periodic solutions for Hopfield neural network equations with periodic input *Nonlinear Anal.* **49** 471–9

[46] Driver R D 2012 *Ordinary and Delay Differential Equations* (Berlin: Springer Science + Business Media) p 20

[47] Gagniuc P A 2017 *Markov Chains: From Theory to Implementation and Experimentation* (Hoboken, NJ: Wiley)

[48] Gaspard P 1998 *Chaos, Scattering and Statistical Mechanics* (Cambridge: Cambridge University Press)

[49] Gonzáles-Miranda J M 2004 *Synchronization and Control of Chaos* (London: Imperial College Press)

[50] Gray C M, Konig P, Engel A K and Singer W 1989 Oscillatory responses in cat visual cortex exhibit inter-columnar synchronization which reflects global stimulus properties *Nature* **338** 334–7

[51] Guevara M R, Glass L, Mackey M C and Shrier A 1983 Chaos in neurobiology *IEEE Trans. Syst. Man. Cybern.* **13** 790–8

[52] Gui Z and Ge W 2006 Periodic solution and chaotic strange attractor for shunting inhibitory cellular neural networks with impulses *Chaos* **16** 033116

[53] Hartman P 2002 *Ordinary Differential Equations* (New York: SIAM)

[54] Hajek B 2015 *Random Processes for Engineers* (Cambridge: Cambridge University Press)

[55] Hata M 1985 On the structure of self-similar sets Japan *J. Appl. Math.* **2** 381–414

[56] Hayashi Y 1994 Oscillatory neural network and learning of continuously transformed patterns *Neural Netw.* **7** 219–31

[57] He G, Shrimali M D and Aihara K 2008 Threshold control of chaotic neural network *Neural Netw.* **21** 114–21

[58] He Y, Wu M and She J H 2006 Delay-dependent exponential stability of delayed neural networks with time-varying delay *IEEE Trans. Circuits Syst. II Exp. Briefs* **53** 553–7

[59] Hoppensteadt F C and Izhekevich E M 1999 Oscillatory neurocomputers with dynamic connectivity *Phys. Rev. Lett.* **82** 2983–6

[60] Huang X and Cao J 2003 Almost periodic solution of shunting inhibitory cellular neural networks with time-varying delay *Phys. Lett.* A **314** 222–3

[61] Hutcheon B and Yarom Y 2000 Resonance, oscillation and the intrinsic frequency preferences of neurons *Trends Neurosci.* **23** 216–22

[62] Hutchinson J 1981 Fractals and self-similarity *Indiana Univ. Math. J.* **30** 713–47

[63] Falconer K J 1995 Sub-self-similar sets *Trans. Am. Math. Soc.* **347** 3121–9

[64] Edgar G A 1990 *Measure, Topology, and Fractal Geometry* (New York: Springer)

[65] Eisencraft M, Monteiro L H A and Soriano D C 2017 White Gaussian chaos *IEEE Commun. Lett.* **21** 17–9

[66] Jiang H, Zhang L and Teng Z 2005 Existence and global exponential stability of almost periodic solution for cellular neural networks with variable coefficients and time-varying delays *IEEE Trans. Neural Netw.* **16** 1340–51

[67] Jorgensen P E T 2006 *Analysis and Probability: Wavelets, Signals, Fractals* **vol 234** (New York: Springer)

[68] Juhong G E and Jing X U 2012 Weak resonant double Hopf bifurcations in an inertial four-neuron model with time delay *Int. J. Neural Syst.* **22** 22–63

[69] Karlin S and Taylor H E 2012 *A First Course in Stochastic Processes* (New York: Academic)

[70] Klenke A 2006 *Probability Theory* (Berlin: Springer)

[71] Ke Y Q and Miao C F 2011 Stability analysis of BAM neural networks with inertial term and time delay *WSEAS. Trans. Syst.* **10** 425–38

[72] Ke Y Q and Miao C F 2013 Stability analysis of inertial Cohen-Grossberg-type neural networks with time delays *Neurocomputing* **117** 196–205

[73] Ke Y Q and Miao C F 2013 Stability and existence of periodic solutions in inertial BAM neural networks with time delay *Neural Comput. Appl.* **23** 1089–99

[74] Korn H and Faure P 2003 Is there chaos in the brain? II. Experimental evidence and related models *Neurosci. C. R. Biol.* **326** 787–840

[75] Landau I D and Sompolinsky H 2018 Coherent chaos in a recurrent neural network with structured connectivity *PLoS Comput Biol.* **14** e1006309

[76] Lasota A and Mackey M C 1985 *Probabilistic Properties of Deterministic Systems* (Cambridge: Cambridge University Press)

[77] Lau K S, Ngai S M and Rao H 2001 Iterated function systems with overlaps and the self-similar measures *J. London Math. Soc.* **63** 99–115

[78] Li T Y and Yorke J A 1975 Period three implies chaos *Am. Math. Mon.* **82** 985–92

[79] Li R and Zhou X 2013 A note on chaos in product maps *Turk. J. Math.* **37** 665–75

[80] Li X and Song S 2013 Impulsive control for existence, uniqueness, and global stability of periodic solutions of recurrent neural networks with discrete and continuously distributed delays *IEEE Trans. Neural. Netw. Learn. Syst.* **24** 868–77

[81] Li Y and Meng X 2018 Impulsive control for existence, uniqueness, and global stability of periodic solutions of recurrent neural networks with discrete and continuously distributed delays *Int. J. Syst. Sci.* **49** 2490–505

[82] Liao G, Chen X and Sanchez E 2002 Delay-dependent exponential stability analysis of delayed neural networks: an LMI approach *Neural Netw.* **15** 855–66

[83] Liu B and Huang L 2008 The intrinsic electrophysiological properties of mammalian neurons: insights into central nervous system function *Nonlinear Anal. Real World Appl.* **3** 830841

[84] Liu Z and Liao L 2004 Existence and global exponential stability of periodic solution of cellular neural networks with time-varying delays *J. Math. Anal. Appl.* **290** 247–62

[85] Llinas R 1988 The intrinsic electrophysiological properties of mammalian neurons: insights into central nervous system function *Science* **242** 1654–64

[86] Miller V H and Jansen B H 2008 Oscillatory neural network for pattern recognition: trajectory based classification and supervised learning *Biol. Cybern.* **99** 459–71

[87] Moran P A P 1946 Additive functions of intervals and Hausdorff measure *Proc. Cambridge Philos.* pp 15–23 Soc. 42

[88] Muscinelli S P, Gerstner W and Schwalger T 2018 How single neuron properties shape chaotic dynamics and signal transmission in random neural networks *PLoS Comput. Biol.* **15** e1007122

[89] Ngai S M and Wang Y 2001 Hausdorff dimension of overlapping self-similar sets *J. London Math. Soc.* **63** 665–72

[90] Oksendal B K 2003 Stochastic *Differential Equations: An Introduction with Applications* (Berlin: Springer)

[91] Ornstein D 1970 Bernoulli shifts with the same entropy are isomorphic *Adv. Math.* **4** 337–52

[92] Ornstein D 1995 In what sense can a deterministic system be random? *Chaos Solitons Fractals* **59** 139–41

[93] Ornstein D 2004 Kolmogorov, random processes, and Newtonian dynamics *Russian Math. Surveys* **59** 313–7

[94] Ornstein D and Weiss B 1998 On the Bernoulli nature of systems with some hyperbolic structure *Ergod. Theor. Dynam. Syst.* **18** 139–41

[95] Pesin Y and Weiss H 1996 On the dimension of deterministic and random cantor-like sets, symbolic dynamics, and the Eckmann-Ruelle conjecture *Commun. Math. Phys.* **182** 105–53

[96] Nicolis G and Prigojin I 1989 *Exploring Complexity* (New York: Freeman and Company)

[97] Raiko T and Valpola H 2011 Oscillatory neural network for image segmentation with biased competition for attention *Adv. Exp. Med. Biol.* **718** 175–85

[98] Rakkiyappan R, Gayathri D, Velmurugan G and Cao J 2019 Exponential synchronization of inertial memristor-based neural networks with time delay using average impulsive interval approach *Neural Process. Lett.* **50** 2053–71

[99] Rakkiyappan R, Premalatha S, Chandrasekar A and Cao J 2016 Stability and synchronization analysis of inertial memristive neural networks with time delays *Cogn. Neurodyn.* **10** 437–51

[100] Schmidt H, Avitabile D, Montbrió E and Roxin A 2018 Network mechanisms underlying the role of oscillations in cognitive tasks *PLoS Comput Biol.* **14** e1006430

[101] Selverston A I and Moulins M 1985 Oscillatory neural networks *Annu. Rev. Physiol.* **47** 29–48

[102] Skarda C A and Freeman W J 1987 How brains make chaos in order to make sense of the world? *Behav. Brain Sci.* **10** 161–95

[103] Sell G R 1971 *Topological Dynamics and Ordinary Differential equations* (London: Van Nostrand Reinhold Company)

[104] Shields P 1973 *The Theory of Bernoulli Shifts* (Chicago, IL: University of Chicago Press)

[105] Spear D W 1992 Measure and self-similarity *Adv. Math.* **91** 143–57

[106] Stella S 1992 On Hausdorff dimension of recurrent net fractals *Proc. Amer. Math. Soc.* **116** 389–400

[107] Thakur R and Das R 2020 Strongly Ruelle-Takens, strongly Auslander-Yorke and Poincaré chaos on semiflows *Commun. Nonlinear Sci. Numer Simulat.* **81** 105018

[108] Qi J, Li C and Huang T 2015 Stability of inertial BAM neural network with time-varying delay via impulsive control *Neurocomputing* **161** 162–7

[109] Qin S, Gu L and Pan X 2020 Exponential stability of periodic solution for a memristor-based inertial neural network whith time delays *Neural Comput. Appl.* **32** 3265–81

[110] Qu J, Wang R, Yan C and Du Y 2014 Oscillations and synchrony in a cortical neural network *Cogn. Neurodyn.* **8** 157–66

[111] Walters P 1982 *An Introduction to Ergodic Theory* (New York: Springer)

[112] Wiggins S 1988 *Global Bifurcation and Chaos Analytical Methods* (New York: Springer)

[113] Wanatabe M, Aihara K and Kondo S 1997 Self-organization dynamics in chaotic neural networks *Control and Chaos Model* **8** 320–33

[114] Wang L 2010 Existence and global attractivity of almost periodic solutions for delayed high-ordered neural networks *Neurocomputing* **73** 802–8

[115] Wang L and Ross J 1990 Osillations and chaos in neural networks: an exactly solvable model *Proc. Natl. Acad. Sci.* **87** 9467–71

[116] Wen S, Huang T, Zeng Z, Chen Y and Li P 2015 Circuit design and exponential stabilization of memristive neural networks *Neural Netw.* **63** 48–56

[117] Winfree A 1980 *The Geometry of Biological Time* (New York: Springer)

[118] Xia Y, Cao J and Huang Z 2007 Existence and exponential stability of almost periodic solution for shunting inhibitory cellular neural networks with impulses *Chaos Solitons Fractals* **34** 1599–607

[119] Yi Z, Heng P A and Fu A W C 1999 Estimate of exponential convergence rate and exponential stability for neural networks *IEEE Trans. Neural Netw.* **50** 1487–93

[120] Yu B and Liu W 2004 Fractal analysis of permeabilities for porous media *AIChE J.* **50** 46–57

[121] Zhang W, Huang T, Li C and Yang J 2018 Robust stability of inertial BAM neural networks with time delays and uncertainties via impulsive effect *Neural Process. Lett.* **44** 245–56

[122] Zhang Z and Quan Z 2015 Global exponential stability via inequality technique for inertial BAM neural networks with time delays *Neurocomputing* **151** 1316–26

[123] Zhao Y, Cai Y and Fan G 2016 Dynamical behavior for fractional-order shunting inhibitory cellular neural networks *J. Nonlinear Sci. App.* **9** 4589–99

[124] Zhou Q and Zhao C 2010 Exponential convergence of shunting inhibitory cellular neural networks with continuously distributed delays *Annales Polonici Mathematici* **98** 241–50

[125] Zmeskal O, Dzik P and Vesely M 2013 Entropy of fractal systems *Comput. Math. Appl.* **66** 135–46

Part VII

Markov chains and differential equations

Chapter 8

Markov chains and stochastic differential equations

This chapter delves into quasi-linear differential equations perturbed by alpha unpredictable functions generated via Markov chains. Specifically, it examines a stochastic differential equation of Duffing type with Markov components. The existence, uniqueness, and exponential stability of alpha unpredictable solutions are established. The novel method of included intervals is utilized.

Two foundational concepts supporting the chapter's narrative are drawn from perspectives in [49] and papers [50–53]. Authors from the book determined that, in studying stochastic dynamics, analyzing individual outcomes of interest complements investigations into probability distribution and various moments, offering an illuminating approach for many purposes [49]. Furthermore, B. Ornstein illustrated the existence of isomorphism between certain random and deterministic processes such as symbolic dynamics and logistic maps.

Recent studies have demonstrated that realizations of various stochastic processes with finite state spaces are indeed alpha unpredictable functions [2, 4, 5, 22, 23], detailed in section 7.2. Notably, observations suggest a connection between Markov chains and symbolic dynamics. The research showcased that Markov chains and Bernoulli shifts with finite state spaces serve as rich sources of alpha unpredictability, terms that find applications in differential equations. These models, incorporating chaotic inputs and crucial elements like sensitivity and alpha unpredictability, are validated. The chapter progresses by intensively exploring the verification of alpha unpredictable solutions, building upon prior studies in differential equations [2, 11–13, 19] and merging these insights with methods for establishing and ensuring the stability of recurrent solutions, including those that are periodic and almost periodic [1, 30, 33, 38, 45, 47].

The chapter's structure unfolds as follows: section 8.1 presents demonstrations validating that Markov chains with finite state spaces exhibit alpha chaos, where infinite realizations of dynamics manifest as alpha unpredictable sequences. This leads to the design of an additional algorithmic mechanism generating unpredictable alpha functions, with an illustrative example featuring a random walk to reinforce the theoretical results. These proposed constructions are then applied in subsequent sections. Section 8.2 delves into a stochastic differential equation of Duffing type featuring Markov components that are designated as alpha unpredictable. This chapter affirms the existence, uniqueness, and exponential stability of alpha unpredictable solutions within this context. The synchronization of divergence and convergence sequences of coefficients and solutions is emphasized, accompanied by numerical examples that substantiate the theoretical assertions.

8.1 Alpha chaos in Markov chains

In this section, we have formalized the realizations of Markov chains as systematically constructed sequences and elucidated the rationale behind why random dynamics exhibit alpha unpredictability within the realm of ultra-Poincaré chaos. We present an illustrative example supported by thorough numerical simulations.

The primary objective of Markov research [41] was to demonstrate that random processes involving dependent events could exhibit behavior akin to those with independent events. This led to the creation of simple yet highly effective models with numerous applications. The pivotal role of Markov chains in shaping the theory of random dynamics and its practical applications cannot be overstated. For instance, the ergodic theorem was rigorously verified for the first time. Observations suggest a profound link between Markov chains, symbolic dynamics, and the Bernoulli scheme. Donald Ornstein's seminal work [50–53] underscored the isomorphism between B-automorphisms like subshifts of finite type and Markov shifts, Anosov flows, Sinai's billiards, ergodic automorphisms of the n-torus [55], and continued fraction transformations. The shared characteristics across these sources not only facilitate the description of various random processes in terms of chaos but also indicate their fundamental alignment. Researchers have explored both avenues, seeking chaos within random dynamics and identifying stochastic elements in deterministic motions [25–27]. Therefore, the discussion on chaos in Markov chains within this section is part of a broader, more comprehensive project. The concept of the alpha unpredictable orbit [2, 11], characterized as a singular isolated movement representing ultra Poincaré chaos [2, 10] and identified within the chains as a specific event, serves as a focal point in this chapter segment. The crux of this paper lies in establishing the existence of a realization—an infinite alpha unpredictable sequence within a finite state space—where the closure, in terms of convergence on bounded intervals, encompasses all infinite realizations. It is essential to note that this convergence aligns with the alpha unpredictable realization for each finite-length orbit, implying that every dynamics simulation mirrors an aspect of the alpha unpredictable realization within the Markov chain. The certainty of an event emerges when it consistently occurs in every experiment iteration.

One can apply finite [25, 27] or countable [29] partitioning to find randomness in deterministic motions. However, we utilize uncountable partitioning [2, 4, 5, 7, 14, 17].

While finite [25, 27] or countable [29] partitioning aids in unveiling randomness within deterministic motions, we deploy uncountable partitioning [2, 4, 5, 7, 14, 17] to establish robust connections between random processes and ultra Poincaré chaos. Chaos generation is formalized through specially structured sets subjected to the similarity map, emphasizing chaotic behavior inherent in Bernoulli scheme dynamics and extending to various random and deterministic processes across discrete and continuous time [3, 4, 15, 23]. The exploration progresses to showcase how chaos manifests within Markov chains, particularly focusing on time-homogeneous chains with finite state spaces and memory considerations. The methodology delves into investigating the dynamics, homing in on individual motion descriptions to enhance our understanding of the underlying geometry. The research highlights the harmonious behavior between an isolated realization in a Markov chain and a thoughtfully selected realization within a Bernoulli scheme, underscoring their dynamic equivalence facilitated through the path of the similarity map in a chosen space. The closure of the alpha unpredictable realization encapsulates what is termed as the quasi-minimal set [57], housing an uncountable array of alpha unpredictable realizations. Our findings affirm that the quasi-minimal set, representing the union of all infinite realizations within the Markov chain, stands as a certain event where each experimental run yields an orbit within this set, serving as an alternate expression of the core result. Our research underscores that the Bernoulli scheme [15, 23], Markov chains, and abstract hyperbolic dynamics [3] all embody a similar type of chaos. Notably, ultra Poincaré chaos aligns seamlessly with these motions, presenting new and exceptional opportunities, particularly within stochastic processes. We envision further research avenues extending this analysis to other Bernoulli automorphisms as studied by D. Ornstein [52], and expanding these results to encompass the majority of stochastic processes provided appropriately structured domains are established.

8.1.1 Markov chains with alpha dynamics

A Markov chain functions as a stochastic model describing a series of potential events where the likelihood of each event is predicated solely on the prior state achieved [35, 37]. These chains find myriad applications serving as statistical models for real-world processes like analyzing queues at airports, monitoring customer traffic, evaluating currency exchange rate fluctuations, managing cruise control systems in vehicles, and gauging animal population dynamics [42]. Consider a discrete-time stochastic process $X_n, n \geqslant 0$, operating on a countable set S. Essentially, this entails a cohort of random variables denoting a probability space (Ω, F, P), where P signifies a probability measure over a range of events, designated by F, within a space of events known as Ω. Here, the set S represents the process's state space, where $X_n \in S$ indicates the state of the process at a given time point, n.

The Markov chain is a stochastic process characterized by the property, which mandates that for all $s_i, s_j \in S$ and $n \geqslant 0$, the transition probability $P\{X_{n+1} = s_j | X_0, \ldots, X_n\} = P\{X_{n+1} = s_j | X_n\}$ holds true. Furthermore, the property dictates that $P\{X_{n+1} = s_j | X_n = s_i\} = p_{ij}$, where p_{ij} denotes the probability of transitioning from state i to state j. This property stipulates that at any specific time n, the subsequent state X_{n+1} is conditionally independent of the past states $X_0, \ldots, X_{n-1}$, given the present state X_n. In essence, this signifies that the transition probabilities remain unaltered by the time parameter n, highlighting the time-homogeneous nature of the chain. Should the transition probabilities vary with time, the process would evolve into a non-time-homogeneous Markov chain.

Consider a finite state space $S = \{s_1, \ldots, s_m\}$, where m is a natural number, not smaller than two, and a metric d for the space. Denote $p_{ij} = p_j(s_i)$ the Markov probability for s_j such that $\Sigma_{j=1}^m p_{ij} = 1$ for all $i = 1, \ldots, m$. In what follows, we shall investigate the problem of chaos considering that all the probabilities p_{ij}, $i, j = 1, \ldots, m$, are positive.

If f_{ij} is the event, which consists of two elementary events, s_i and s_j, happen successively, then an infinite realization of the Markov chain X_n, $n \geqslant 0$ can be formalized as the infinite sequence $f_{i_1 j_1} f_{i_2 j_2} \cdots f_{i_n j_n} \cdots$, with $i_k = j_{k-1}$ for all $k = 2, 3, \ldots$. We have that $p(f_{ij}) = p_{ij}$. The formalization does not give advantages, if one considers the chains without memory, but it makes easier the discussion of the processes with memory, in what follows.

Present the last sequence as the element $\mathscr{F}_{i_1 i_2 \ldots}$, $i_k = 1, 2, \ldots, m$, $k = 1, 2, \ldots$, of the space $\mathscr{F}$ with the metric $\delta(\mathscr{F}_{i_1 i_2 \ldots}, \mathscr{F}_{l_1 l_2 \ldots}) = \Sigma_{k=1}^\infty d(s_{i_k}, s_{l_k})/2^k$, where $\mathscr{F}_{l_1 l_2 \ldots}$, $l_k = 1, 2, \ldots, m$, $k = 1, 2, \ldots$ is another member of the space.

Next, we will consider Markov chains with memory of a non-zero length, besides the zero length memory. We start with the length equal to two such that the element f_{ijk} presents three elementary events s_i, s_j, and s_k to happen successively, and the probability for s_k is equal to $p_{ijk} = p_k(s_i, s_j)$. Then the Markov chain with memory has the formal presentation $f_{i_1 j_1 k_1} f_{i_2 j_2 k_2} \cdots f_{i_n j_n k_n} \cdots$, where $j_l = k_{l-1}$, $i_l = j_{l-1}$ for all $l = 2, 3, \ldots$. Accepting the last sequence as the element $\mathscr{F}_{i_1 i_2 \ldots}$, $i_k = 1, 2, \ldots, m$, $k = 1, 2, \ldots$, of the space $\mathscr{F}$ with the metric $\delta(\mathscr{F}_{i_1 i_2 \ldots}, \mathscr{F}_{l_1 l_2 \ldots}) = \Sigma_{k=1}^\infty d(s_{i_k}, s_{l_k})/2^k$ we attain the basis, common with that for the chain without memory.

At last, consider the Markov process with the memory of the length equal to arbitrary natural number n. Then we formalize the discussion with the elements $f_{i^1, \ldots, i^n}$, which consist of successive elementary events $s_{i^1}, \ldots, s_{i^n}$, such that the sequence $f_{i_l^1, \ldots, i^n} \cdots, f_{i_l^1, \ldots, i^n} \cdots$ with $i_l^j = i_{l-1}^{j+1}$, $j = 1, \ldots, n-1$, $l = 2, 3, \ldots$, is the formalization of the chain. We obtain the structure for the dynamics research if we accept the last sequence as an element $\mathscr{F}_{i_1^1 i_2^1 \ldots}$, $i_k^1 = 1, 2, \ldots, m$, $k = 1, 2, \ldots$, of the space $\mathscr{F}$, making stress on the events with the indexes i_k^1, $k = 1, 2, \ldots$.

It is clear that for all cases, regardless are they with memory or without memory, we have constructed one and the same space $\mathscr{F}$ of elements $\mathscr{F}_{i_1 i_2 \ldots}$, $i_k = 1, 2, \ldots, m$, $k = 1, 2, \ldots$, with the distance $\delta(\mathscr{F}_{i_1 i_2 \ldots}, \mathscr{F}_{l_1 l_2 \ldots}) = \Sigma_{k=1}^\infty d(s_{i_k}, s_{l_k})/2^k$. Opposite to that in chapter 2, the distance δ is now a metric. This is why the space is the main object of analysis for chaos presence in the random process, in what follows.

Complete the dynamics with the special mapping $\varphi \colon \mathscr{F} \to \mathscr{F}$ such that

$$\varphi(\mathscr{F}_{i_1 i_2 \ldots i_n \ldots}) = \mathscr{F}_{i_2 i_3 \ldots i_n \ldots}, \tag{8.1}$$

for each element of the set. The map φ is in the paradigm of the Bernoulli shift [56, 60], known for its symbolic dynamics. It is said to be the *abstract similarity map* [7] or *alpha map* as it is convenient to describe fractals, which are determined through abstract self-similarity. The abstract similarity map surpasses the Bernoulli shift in many senses. This is detailed in chapter 2.

The definitions imply that the appearance of values on the map is the same as for the members of the state space, which are ordered successively. Consequently, if one proves that the map is of a chaotic type, then we have to recognize that the chaos is proper for the stochastic dynamics if appropriate probability arguments are provided.

The following alpha sets,

$$\mathscr{F}_{i_1 i_2 \ldots i_n} = \bigcup_{j_k = 1,2,\ldots,m} \mathscr{F}_{i_1 i_2 \ldots i_n j_1 j_2 \ldots}, \tag{8.2}$$

where indices $i_1, i_2, \ldots, i_n$, are fixed, were introduced in [6, 7]. They are described also in chapter 2.

It is clear that

$$\mathscr{F} \supseteq \mathscr{F}_{i_1} \supseteq \mathscr{F}_{i_1 i_2} \supseteq \cdots \supseteq \mathscr{F}_{i_1 i_2 \ldots i_n} \supseteq \mathscr{F}_{i_1 i_2 \ldots i_n i_{n+1}} \cdots, \quad i_k = 1, 2, \ldots, m, \quad k = 1, 2, \ldots,$$

that is, the sets form a nested sequence.

Since one can verify that

$$\varphi^n(\mathscr{F}_{i_1 i_2 \ldots i_n}) = \mathscr{F}, \tag{8.3}$$

for arbitrary natural number n and $i_k = 1, 2, \ldots, m, \quad k = 1, 2, \ldots$, there is a reason to call φ a *similarity map*.

We will say that for the sets $\mathscr{F}_{i_1 i_2 \ldots i_n}$ the *diameter condition* is valid, if

$$\max_{i_k = 1,2,\ldots,m} \operatorname{diam} \mathscr{F}_{i_1 i_2 \ldots i_n} \to 0 \quad \text{as} \quad n \to \infty, \tag{8.4}$$

where $\operatorname{diam}(A) = \sup\{\delta(\mathbf{x}, \mathbf{y}) : \mathbf{x}, \mathbf{y} \in A\}$, for a set A in $\mathscr{F}$.

Define the function $\delta(A, B) = \inf\{\delta(\mathbf{x}, \mathbf{y}) : \mathbf{x} \in A, \mathbf{y} \in B\}$, for two nonempty sets A and B in $\mathscr{F}$. Set $\mathscr{F}$ satisfies the *separation condition* of degree n if there exist a positive number ε_0 and a natural number n such that for arbitrary indices $i_1 i_2 \ldots i_n$ one can find indices $j_1 j_2 \ldots j_n$ such that

$$\delta(\mathscr{F}_{i_1 i_2 \ldots i_n}, \mathscr{F}_{j_1 j_2 \ldots j_n}) \geqslant \varepsilon_0. \tag{8.5}$$

Next, we will formulate two theorems on ultra-Poincaré and Devanye chaos. Verification of the assertions is given in [7] and chapter 2.

Theorem 8.1.1. [7] *If the diameter and separation conditions are valid, then the dynamics $(\mathscr{F}, \delta, \phi)$ is chaotic in the sense of Devaney.*

The last theorem claims that the similarity map φ possesses the three ingredients of Devaney chaos, namely density of periodic points, transitivity and sensitivity. A point $\mathcal{F}_{i_1 i_2 i_3 \ldots} \in \mathcal{F}$ is periodic with period n if its index consists of endless repetitions of a block of n terms.

In [2, 10, 11], Poisson stable motion is utilized to distinguish chaotic behavior from periodic motions in Devaney and Li–Yorke types. The dynamics are called Poincarè chaos. The next theorem shows that the ultra-Poincaré chaos is valid for the similarity mapping. The domain structured chaos unites models, which admit all types of theoretical chaos such as U-ultra-Poincaré Li–Yorke and Devaney [6, 7]. Still, the first is most convenient and best suited to confirm that chaos is proper for random dynamics.

Let us give the definition of the alpha unpredictability in the sense of the dynamics $(\mathcal{F}, \delta)$.

Definition 8.1.1. [10] *A point $\mathcal{F}_{i_1 i_2 \ldots}$ from the set $\mathcal{F}$ is unpredictable if there exist a positive number ε_0 and sequences t_m, s_m, of natural numbers, both of which diverge to infinity such that $\varphi^{t_m}(\mathcal{F}_{i_1 i_2 \ldots i_k \ldots}) \to \mathcal{F}_{i_1 i_2 \ldots i_k \ldots}$ as $m \to \infty$ and $\delta(\varphi^{t_m + s_m}(\mathcal{F}_{i_1 i_2 \ldots i_k \ldots}), \varphi^{t_m}(\mathcal{F}_{i_1 i_2 \ldots i_k \ldots})) \geqslant \varepsilon_0$ for each natural number m.*

Theorem 8.1.2. [6, 7] *If the diameter and separation conditions are valid, then the alpha map possesses ultra-Poincaré chaos.*

Proof. Let us fix an alpha unpredictable point of the dynamics as a member $\mathcal{F}_{i_1 i_2 \ldots i_k \ldots}$ of $\mathcal{F}$, which index sequence $i_1 i_2 \ldots i_k \ldots$ consists of subsequently chosen all strings of length one, then all possible strings of length two, etc, to infinity. The sample of this construction is seen for the two symbolic alphabet cases in the next diagram,

$$(\underbrace{0 \; 1}_{1 \; blocks} \, | \underbrace{00 \; 01 \; 10 \; 11}_{2 \; blocks} | \underbrace{000 \; 001 \; 010 \; 011 \; \ldots}_{3 \; blocks} | \ldots).$$

It is easily seen by the diameter condition that the point is Poisson stable. That is, there exists a sequence of integers, t_m, divergent to infinity such that $\varphi^{t_m}(\mathcal{F}_{i_1 i_2 \ldots i_k \ldots}) \to \mathcal{F}_{i_1 i_2 \ldots i_k \ldots}$ as $n \to \infty$. Consider the alpha unpredictability property. We will show that there is a sequence s_m divergent to infinity such that $\delta(\varphi^{t_m + s_m}(\mathcal{F}_{i_1 i_2 \ldots i_k \ldots}), \varphi^{t_m}(\mathcal{F}_{i_1 i_2 \ldots i_k \ldots})) \geqslant \varepsilon_0$, where ε_0 is the separation constant for the space $(\mathcal{F}, \delta)$. Assume, on the contrary, that there is no such sequence. Then, $\delta(\varphi^{t_m + i}(\mathcal{F}_{i_1 i_2 \ldots i_k \ldots}), \varphi^{t_m}(\mathcal{F}_{i_1 i_2 \ldots i_k \ldots})) < \varepsilon_0$, for all natural i. This contradicts the diameter condition and the choice of the initial point, which means an absence of the initial string of length n in the index sequence of the point $\varphi^{t_m}(\mathcal{F}_{i_1 i_2 \ldots i_k \ldots})$.

It is easily seen that for a finite metric space (S, d), the definition 8.1.2 has the next form, which is most convenient for the present research.

Definition 8.1.2. [5] *A bounded sequence $s_i \in S$, $i = 0, 1, \ldots$, is called alpha unpredictable if there exist a positive number ε_0 and sequences t_n, s_n, $n = 1, 2, \ldots$, of positive integers, both of which diverge to infinity such that $s_{i+t_n} = s_i$ for each bounded interval of integers, if n is sufficiently large, and $d(s_{t_n + s_n}, s_{t_n}) \geqslant \varepsilon_0$ for each natural number n.*

Theorem 8.1.3. [5] *Each finite realization of the Markov chain corresponds to an arc of s^*. This indicates that the alpha-unpredictable realization occurs in every experiment of the chain and constitutes a specific event.*

In the following part, we will establish a fixed alpha-unpredictable realization of the chain, which can be defined as follows.

We begin by selecting an alpha-unpredictable point $\mathscr{F}^*_{i_1 i_2 \ldots}$ from the mapping φ, along with the sequence $s^* = s^*_{i_k k}$, corresponding to the alpha-unpredictable realization of the Markov chain. Based on definition 8.1.5 and theorem 8.1.3, the following statement holds:

Theorem 8.1.4. [5] *For every finite realization of the Markov chain, there exists a corresponding one identical to the realization arc of s^*.*

It is important to acknowledge that every experiment involving a Markov chain, regardless of whether it possesses memory, produces an arc of an alpha-unpredictable orbit. This finding constitutes a significant contribution of this research. It aligns with a key principle of ergodic theory [59] that a single trajectory can represent the overall dynamics and behavior of all other trajectories. Therefore, we can conclude that this represents a fixed alpha-unpredictable orbit, effectively capturing chaos as a specific event. This property is what accounts for the irregularity seen in each finite sample path of the chain.

Now, let us consider the random dynamics for deterministic sensitivity. For the discussion, let us fix an infinite sequence of elements of the state space. The probability that a simulated realization will coincide with the sequence forever equals zero. Consequently, the complimentary event that an experiment with the same start state will diverge from the sequence at some finite time equals one. Because of the metric it means that the sensitivity is present for the random dynamics with probability one.

To finalize the relationship of the stochastic dynamics and deterministic chaos, we conclude that there are elements of the deterministic chaos, which are certain events such as appearance of an alpha unpredictable sequence as a finite realization, and there are ingredients fo chaos such as the sensitivity, which are most probable, but not deterministic, not certain events. It is useful to say that chaos is proper for the random processes with probability one altogether.

8.1.2 An example: random walk

Consider the following Markov chain as an example. Let the real valued scalar dynamics $X_{n+1} = X_n + Y_n$, $n \geqslant 0$, be given such that $Y_n = \{-1, 1\}$ is a random variable. Since we expect for the chaotic dynamics realizations to be bounded, the special chain with boundaries is constructed. That is, the probability distribution $P(1) = P(-1) = 1/2$, if $X_n \neq 1, 4$, and certain events $Y_n = -1$ if $X_n = 4$, and $Y_n = 1$

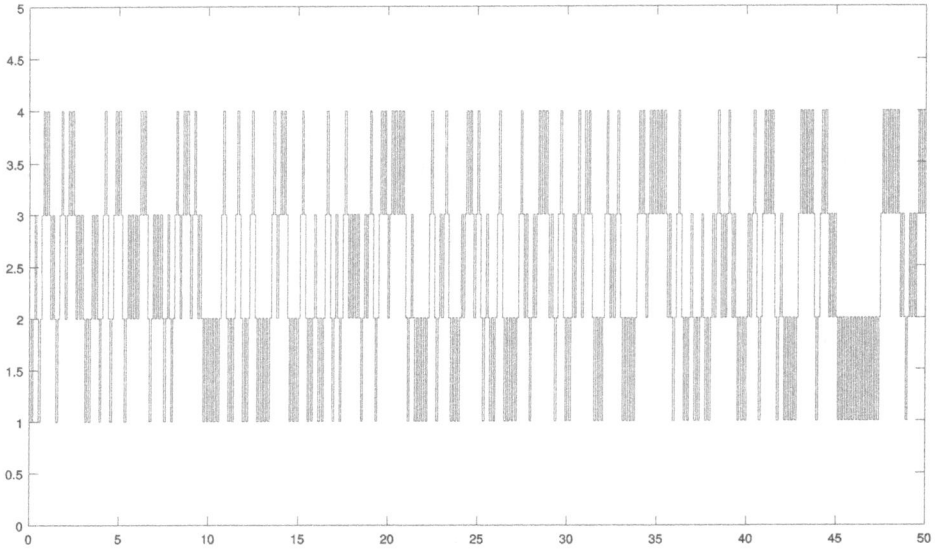

Figure 8.1. The graph of function $\varphi(t)$ as geometry of the alpha unpredictable realization of the chain X_n, $n \geqslant 0$. The vertical lines are drawn for better visibility. Reproduced with permission from [5].

if $X_n = 1$. To satisfy the construction of the present research, we will make the following agreements. First of all, denote $s_0 = 1$, $s_1 = 2$, $s_2 = 3$, $s_3 = 4$. Consider the state space $S = \{s_1, s_2\}$. Introduce the following events, $f_{11} = \{s_1, s_0, s_1\}$, $f_{12} = \{s_1, s_2\}, f_{21} = \{s_2, s_1\}, f_{22} = \{s_2, s_3, s_2\}$. It is clear that $p_{i1} + p_{i2} = 1$, $i = 1, 2$, and all the probabilities are equal to the half. That is, despite the differences with the theoretical chain construction in the main body of the paper, we have the sequences $f_{i_1 j_1} f_{i_2 j_2} \ldots f_{i_n j_n} \ldots$, with $i_k = j_{k-1}$ for all $k = 2, 3, \ldots$, and we see the circumstances of the domain structured chaos. Consequently, there is alpha unpredictability. To visualize an alpha unpredictable realization, we will draw the graph of the function $\varphi(t) = X_n$, if $t \in [n/10, (n + 1)/10)$, $0 \leqslant t \leqslant 60$. According to theorem 8.1.7, it is an arc of an alpha unpredictable sequence. The graph of the function is seen in figure 8.1. It illustrates an alpha unpredictable sequence, the sample path of the random walk.

8.2 Duffing type equations with Markov coefficients

Studying the Duffing equation with variable coefficients is of significant importance. As highlighted in Moon's book [46], particular attention is given to cases where the coefficients display irregular behavior, making the model particularly intriguing. This raises the question: What happens when the perturbations are random and take the form of noise, especially when this noise is linked to asymptotic properties of divergence and convergence? Two primary issues arise from this scenario. The first concerns how to incorporate deviations in the coefficients while ensuring accurate evaluations. The second focuses on the connection between stochastic processes and what we typically understand as deterministic chaos. This does not imply that a random process can be regarded as a deterministic phenomenon; rather, the goal is to demonstrate that significant characteristics associated with chaos can emerge in dynamics generated by processes like Markov events, which occur with certain probabilities. This article addresses these two questions, building upon results obtained in our earlier research. In [10, 11], we introduced a new concept known as the alpha-unpredictable point, which is defined when the state space is functional. Furthermore, our research in [5] established that any infinite time realization of a Markov process with a finite state space and without memory represents an alpha unpredictable sequence.

Stochastic differential equations typically involve one or more stochastic processes as terms, with their solutions also being stochastic processes. Generally, a stochastic differential equation can be viewed as an ordinary differential equation perturbed by a term that depends on a white noise variable, derived from Brownian motion or the Wiener process. These equations are often seen as continuous-time counterparts to stochastic difference equations.

In our research, we align with the suggestion to view realizations of random dynamics both as functions in general and as sequences in particular. As noted in the book [49], 'We have described stochastic dynamics in terms of probability distributions and their various moments. A complimentary, and for many purposes especially illuminating approach, is the study of individual outcomes of the stochastic process of interest'. We concur with this viewpoint and believe that individual outcomes should be considered both for applications and as perturbations for various theoretical models. In the current paper, we individualize both the inputs —coefficients of the Duffing-type equation—and the outputs—the solutions of the equation.

In this study, stochastic processes take on multiple roles. The first is the discrete Markov chain with a finite state space and without memory; memory-adept processes may be explored in future extensions. A realization of this chain serves as an input for a dissipative stochastic inhomogeneous equation. The random solutions of these equations are then utilized as coefficients and inputs for the stochastic Duffing equation. Ultimately, we demonstrate that the solution to the Duffing-type equation is a continuous alpha unpredictable function. The framework established in this study can be applied to many other theoretical challenges and practical applications.

The concept of alpha unpredictability was introduced in [10, 11] and has since been applied to various fields such as differential equations, neural networks, and gas discharge-semiconductor systems [5, 8, 9, 19, 20, 24]. This concept serves as a powerful tool for identifying chaos [2, 14, 17, 40, 43, 44, 58].

Research on Markov chains [41] has demonstrated that random processes, although composed of dependent events, can exhibit behaviors characteristic of independent events. This led to the development of simple dynamics that have proven to be highly effective in numerous applications. The significance of Markov processes in the evolution of random dynamics theory and its applications cannot be overstated. For instance, the ergodic theorem was initially rigorously established for these dynamics. Numerous findings indicate that Markov chains are closely linked to symbolic dynamics and the Bernoulli scheme. A pivotal contribution in understanding this connection was made by Donald Ornstein, who showed that (B)-automorphisms—such as subshifts of finite type, Markov shifts, Anosov flows, Sinai's billiards, and ergodic automorphisms of the (n)-torus—are isomorphic [50–53]. Given these results, it is essential to demonstrate that various random processes can be framed in terms of chaos and exhibit analogous characteristics. Researchers have explored both chaos in random dynamics and stochastic features within deterministic motions [26–28]. Therefore, the issue of chaos in Markov chains, which is the focus of our interest, is part of a broader and significant endeavor.

The alpha-unpredictable orbit [11], considered as a single isolated motion that exemplifies ultra Poincaré chaos [10], has been identified as a specific event within the context of Markov chains [5]. Thus, our current results are not surprising when viewed in light of the research findings from [50] and [5].

The Duffing equation is expressed as [31]:

$$x'' + ax' + bx + cx^3 = F_0 \cos(\lambda t), \tag{8.6}$$

where a represents the damping coefficient, b and c are the stiffness (restoring) coefficients, F_0 is the excitation coefficient, λ denotes the frequency of excitation, and t is time. The majority of studies concerning this equation assume that the coefficients a, b, c and F_0 are constant [39, 61]. However, when considering the original model, it is plausible to attribute mechanical reasons for the variability of these coefficients. For instance, damping and driving forces may not remain constant [32].

The primary focus of this section is the following stochastic differential equation:

$$x''(t) + (p_0(t) + p_1(t))x'(t) + (q_0(t) + q_1(t))x(t) + (r_0(t) + r_1(t))x^3(t) = (F_0(t) + F_1(t))\cos(\lambda t), \tag{8.7}$$

where t, $x \in \mathbb{R}$, λ is a real constant, and $p_0(t)$, $q_0(t)$, $r_0(t)$ and $F_0(t)$ are continuous periodic functions. The coefficients $p_1(t)$, $q_1(t)$, $r_1(t)$ and $F_1(t)$ are derived from realizations of Markov processes, which is why we refer to them as Markovian coefficients. It is also worth noting that the right-hand side of the equation is treated as a coefficient. While the periodic components of the coefficients are introduced to ensure the stability of the solutions to equation (8.7), the Markovian components introduce irregularities in the solutions.

It is important to emphasize that the main goal of this research is not to establish chaos in the model but rather to demonstrate the existence of a stochastic output for the equation that exhibits the property of alpha unpredictability, and to show that it is synchronized with the asymptotic characteristics of the stochastic perturbations present in the model.

8.2.1 Elements of alpha unpredictable functions

The following definitions are basic in the theory of alpha unpredictable points, orbits, and functions introduced [10, 11] and developed further in papers [2, 14, 16, 17, 19, 20, 40, 43, 44, 58].

Definition 8.2.1. [10] *A uniformly continuous and bounded function* $\psi \colon \mathbb{R} \to \mathbb{R}$ *is alpha unpredictable if there exist positive numbers* ε_0, σ *and sequences* t_n, s_n *both of which diverge to infinity such that* $|\psi(t + t_n) - \psi(t)| \to 0$ *as* $n \to \infty$ *uniformly on compact subsets of* $\mathbb{R}$ *and* $|\psi(t + t_n) - \psi(t)| \geqslant \varepsilon_0$ *for each* $t \in [s_n - \sigma, s_n + \sigma]$ *and* $n \in \mathbb{N}$.

In what follows, we shall call t_n and s_n as convergence and divergence sequences, respectively. The presence of the convergence sequence is the argument that any alpha unpredictable function is Poisson stable [11, 12, 57], but not vice versa.

Definition 8.2.2. [48, 57] *A function* $\phi(t) \colon \mathbb{R} \to \mathbb{R}$, *bounded and continuous, is said to be Poisson stable if there is a sequence of moments* t_n, $t_n \to \infty$ *as* $n \to \infty$, *such that the sequence* $\phi(t + t_n)$ *uniformly converges to* $\phi(t)$ *on each bounded interval of the real axis.*

The discrete version of the definition 8.2.1 is as follows.

Definition 8.2.3. [10] *A bounded sequence* $\{k_i\} \in \mathbb{R}$, $i \in \mathbb{Z}$, *is called alpha unpredictable if there exist a positive number* ε_0 *and the sequences* $\{\zeta_n\}$, $\{\eta_n\}$, $n \in \mathbb{N}$, *of positive integers, both of which diverge to infinity such that* $|k_{i+\eta_n} - k_i| \to 0$ *as* $n \to \infty$ *for each* i *in bounded intervals of integers and* $|k_{\zeta_n + \eta_n} - k_{\eta_n}| \geqslant \varepsilon_0$ *for each* $n \in \mathbb{N}$.

In this paper, we shall consider alpha unpredictable sequences with non-negative arguments and call them also alpha unpredictable sequences [2].

Let us give examples of alpha unpredictable functions. Using an alpha unpredictable sequence, k_i, one can construct a piecewise constant function $\phi(t)$, such that $\phi(t) = k_i$ on intervals $t \in [hi, h(i + 1)]$, where h is a real number. In paper [15], the function $\phi(t)$ is determined through the solution of the logistic map and the Bernoulli process is used. Another alpha unpredictable function, $W(t)$, is a continuous solution of differential equation $W'(t) = \alpha W(t) + \phi(t)$, where α is a negative number. In figure 8.2(a) the graph of function $\phi(t) = \lambda_i$, for $t \in [i, i + 1)$, $i = 0, 1, 2, \ldots$, is shown, where λ_i is the alpha unpredictable solution [2, 11] of the logistic map, $\lambda_{i+1} = \mu \lambda_i (1 - \lambda_i)$, $i \in \mathbb{Z}$, with $\lambda_0 = 0.4$, $\mu = 3.9$. Figure 8.2(b) depicts the graph of the solution, $w(t)$, of the equation with $w(0) = 0.6$ and $\alpha = -2$, which exponentially approaches the unique alpha unpredictable solution, $W(t)$, of the non-homogeneous equation. This is why the red line can be considered [18] for $t > 40$ as the graph of an alpha unpredictable function. In the present paper, the coefficients of the stochastic differential equation (8.7) are determined by applying the algorithm for $W(t)$, but randomly such that a Markov chain is used instead of the logistic equation.

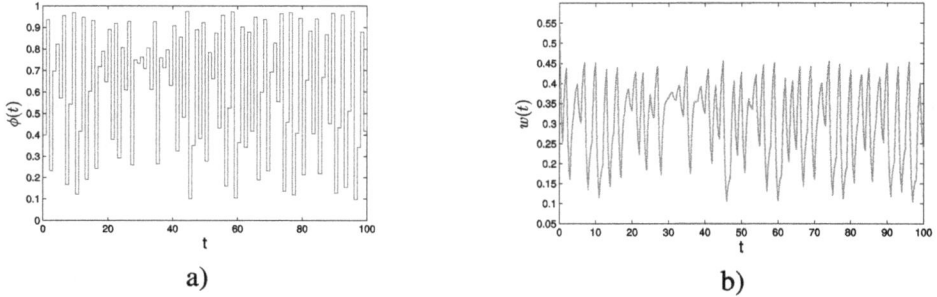

a)   b)

Figure 8.2. The graphs of the discontinuous and continuous functions, $\phi(t)$ and $w(t)$. Reproduced with permission from [22].

8.2.2 Markovian coefficients

In this part of the paper, we demonstrate algorithms to construct Markovian coefficients for Duffing type equation (8.7).

A Markov chain is a stochastic model, which describes a sequence of possible events such that the probability of each event depends only on the state attained in the previous one [35, 37, 42].

Since we expect the chaotic dynamics realizations to be bounded, the special Markov chain with boundaries is constructed below. Let the real valued scalar dynamics

$$X_{n+1} = X_n + Y_n, n \geqslant 0, \qquad (8.8)$$

be given such that Y_n is a random variable with values in $\{-2, 2\}$ with probability distribution $P(2) = P(-2) = 1/2$, if $X_n \neq -4, 6$, and certain events $Y_n = 2$, if $X_n = -4$, and $Y_n = -2$, if $X_n = 6$. To satisfy the construction of the present research, we will make the following agreements. First of all, denote $s_0 = -4$, $s_1 = -2$, $s_2 = 0$, $s_3 = 2$, $s_4 = 4$, $s_5 = 6$. Consider the state space of the process $S = \{s_0, s_1, s_2, s_3, s_4, s_5\}$, and the value $X_n \in S$ is the state of the process at time n. The Markov chain is a random process which satisfies property $P\{X_{n+1} = s_j | X_0, \ldots, X_n\} = P\{X_{n+1} = s_j | X_n\}$ for all $s_i, s_j \in S$ and $n \geqslant 0$, and, moreover, $P\{X_{n+1} = s_j | X_n = s_i\} = p_{ij}$, where p_{ij} is the transition probability that the chain jumps from state i to state j. It is clear that $\sum_{j=0}^{5} p_{ij} = 1$ for all $i = 0, \ldots, 5$. The alpha unpredictability of infinite realizations of the dynamics is proved by theorem 2.2 [20].

Next, we shall need the ρ-type piecewise constant unpredictable functions, which are defined through the Markov chain such that $\rho(t) = X_n$, if $t \in [hn, h(n + 1))$. To visualize the ρ-type functions in figure 8.3(a) the graph of the function $\rho(t) = X_n$, if $t \in [hn, h(n + 1))$, where $h = 0.5$, $0 \leqslant t \leqslant 100$ is drawn.

On the basis of the ρ-type functions, we introduce the σ-type piecewise constant alpha unpredictable functions such that

$$\sigma(t) = \Sigma(\rho(t)), \qquad (8.9)$$

where $\Sigma(s)$ is a continuous function, which satisfies the inverse Lipschitz condition. It can be shown that σ-type functions are discontinuous alpha unpredictable [2].

8-12

a)

b)

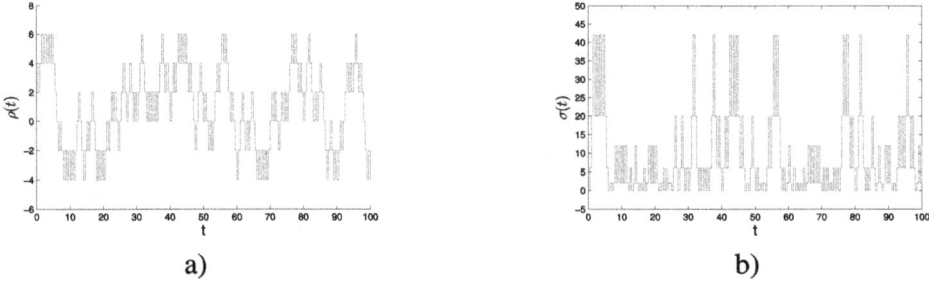

Figure 8.3. The piecewise constant functions $\rho(t)$ and $\sigma(t)$. The vertical lines are drawn for better visibility. Reproduced with permission from [22].

Figure 8.3(b) depicts the graph of piecewise constant alpha unpredictable function $\sigma(t) = \rho^2(t) + \rho(t)$.

Now, let us define another type of function to finalize construction of continuous alpha unpredictable functions through the Markov process. Consider ordinary differential equation

$$x'(t) = \alpha x(t) + \sigma(t), \qquad (8.10)$$

where α is a negative number. The equation (8.10) admits a unique exponentially stable alpha unpredictable solution [12]. We say that the solution of the equation (8.10) is Θ-type alpha unpredictable function. It is impossible to specify the initial value of the solution, but applying the property of exponential stability one can consider any solution as arbitrary close. In figure 8.4, the graph of the solution, $x(t)$, $x(0) = 0.6$ of equation (8.10), where the parameter α is equal to -3, and $\sigma(t) = \rho^2(t) + \rho(t)$ is shown. The solution exponentially approaches an alpha unpredictable function $\Theta(t)$. Thus, the algorithm for three types of alpha unpredictable functions, which will be applied to build the Markovian coefficients, has been finalized. Next, we shall apply it for each of the coefficients in the stochastic differential equation (8.7).

Let us consider the following dissipative equations

$$p_1'(t) = \alpha_p p_1(t) + p(t), \qquad (8.11)$$

$$q_1'(t) = \alpha_q q_1(t) + q(t), \qquad (8.12)$$

$$r_1'(t) = \alpha_r r_1(t) + r(t), \qquad (8.13)$$

$$F_1'(t) = \alpha_F F_1(t) + f(t), \qquad (8.14)$$

where α_p, α_q, α_r and α_F are negative real numbers, and $p(t)$, $q(t)$, $r(t)$ and $f(t)$ are alpha unpredictable functions of σ-type. That is, $p(t) = P(\rho(t))$, $q(t) = Q(\rho(t))$, $r(t) = R(\rho(t))$ and $f(t) = F(\rho(t))$, where $P(s)$, $Q(s)$, $R(s)$, and $F(s)$, are continuous functions with inverse Lipschitz property and the function $\rho(t)$ is determined above. The exponentially stable and bounded solutions $p_1(t)$, $q_1(t)$, $r_1(t)$ and $F_1(t)$ of

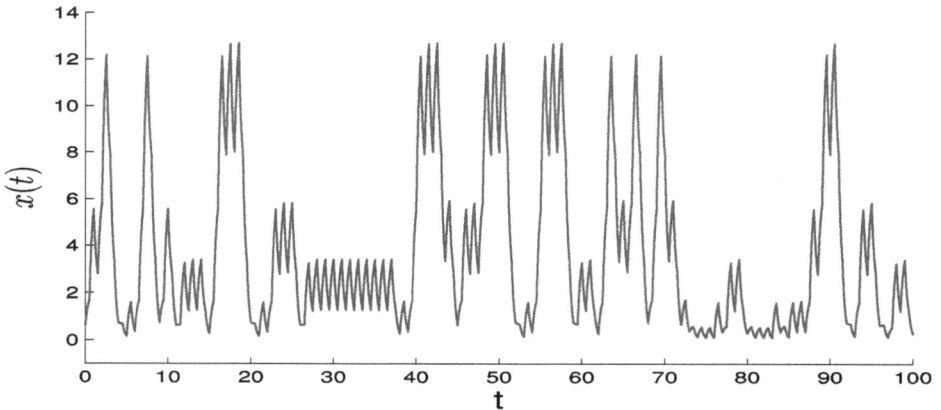

Figure 8.4. The solution $x(t)$ of equation (8.10) with initial value $x(0) = 0.6$ exponentially approaches the alpha unpredictable Markovian function. Reproduced with permission from [22].

equations (8.11)–(8.14) are Θ-type functions. The functions are considered Markovian components of the coefficients in the Duffing type equation (8.7).

In this paper, we utilize Markov chains without memory for the coefficients, but it is clear that one can consider chains with memories of arbitrary finite length in future studies.

8.2.3 Random processes

In the present part of the section, under certain conditions, it is rigorously proved that an exponentially stable alpha unpredictable solution takes place in the dynamics of the stochastic differential equation with Markovian coefficients.

We will make use of the norm $\|v\| = \max(|v_1|, |v_2|)$ for a two-dimensional vector $v = (v_1, v_2)$, and a corresponding norm for square matrices. For stochastic differential equation (8.7) it is provided that a solution $x(t)$ and its derivative $x'(t)$ are bounded such that $\sup_{t\in\mathbb{R}} |x(t)| < H$, $\sup_{t\in\mathbb{R}} |x'(t)| < H$, where H is a fixed positive number.

Assume that stochastic differential equation (8.7) satisfies the following conditions:

(C1)
the functions $p_0(t)$, $q_0(t)$, $r_0(t)$, and $F_0(t)$ are continuous periodic with common positive period ω such that $\lambda = \frac{2\pi}{\omega}$;

(C2)
the Markovian components $p_1(t)$, $q_1(t)$, $r_1(t)$ $F_1(t)$ are of Θ-type with common sequences of convergence t_n, and divergence s_n such that there exist positive numbers σ, ε_0, which satisfy $|p_1(t + t_n) - p_1(t)| \geqslant \varepsilon_0$, $|q_1(t + t_n) - q_1(t)| \geqslant \varepsilon_0$, $|r_1(t + t_n) - r_1(t)| \geqslant \varepsilon_0$, $|F_1(t + t_n) - F_1(t)| \geqslant \varepsilon_0$, for all $t \in [s_n - \sigma; s_n + \sigma]$;

(C3)
$t_n \to 0 \ (mod \ \omega)$ as $n \to \infty$;

(C4)
$s_n \to 0 \ (mod \ \frac{\omega}{2})$ as $n \to \infty$.

The equation (8.7) can be written as the system

$$x_1'(t) = x_2(t),$$
$$x_2'(t) = -q_0(t)x_1(t) - p_0(t)x_2(t) - q_1(t)x_1(t) - p_1(t)x_2(t) - \qquad (8.15)$$
$$(r_0(t) + r_1(t))x_1^3(t) + (F_0(t) + F_1(t))\cos(\lambda t).$$

Consider the homogeneous system associated with (8.15):

$$x_1'(t) = x_2(t),$$
$$x_2'(t) = -q_0(t)x_1(t) - p_0(t)x_2(t). \qquad (8.16)$$

Let $X(t), t \in \mathbb{R}$ be the fundamental matrix of system (8.16) such that $X(0) = I$, and I is the 2×2 identical matrix. Moreover, $X(t, s) = X(t)X^{-1}(s)$ is the transition matrix of system (8.16) such that $X(t + \omega, s + \omega) = X(t, s)$ for all $t, s \in \mathbb{R}$.

The following assumption is needed,

(C5)
the multipliers of system (8.16) in modulus are less than one.

The last condition implies that there exist positive numbers $K > 1$ and μ, which satisfy

$$\|X(t, s)\| < Ke^{-\mu(t-s)}, \qquad (8.17)$$

for $t \geqslant s$ [36].

For convenience, let us introduce notations,

$q_0 = \sup_{t \in \mathbb{R}} |q_0(t)|, \ p_0 = \sup_{t \in \mathbb{R}} |p_0(t)|, \ p_1 = \sup_{t \in \mathbb{R}} |p_1(t)|, \ q_1 = \sup_{t \in \mathbb{R}} |q_1(t)|, \ r_0 = \sup_{t \in \mathbb{R}} |r_0(t)|,$

$$r_1 = \sup_{t \in \mathbb{R}} |r_1(t)|, \ F = \sup_{t \in \mathbb{R}} |F_0(t) + F_1(t)|.$$

Throughout the section, the following additional conditions are required:

(C6)
$\frac{K}{\mu}(H(p_1 + q_1) + (r_0 + r_1)H^3 + F) < H;$

(C7)
$\frac{K}{\mu}(p_1 + q_1 + 3(r_0 + r_1)H^2) < 1.$

We will consider the system (8.15) in the matrix form

$$y' = A(t)y + B(t)y + C(t, y) + G(t), \tag{8.18}$$

where $y(t) = column(y_1(t), y_2(t))$,

$$A = \begin{pmatrix} 0 & 1 \\ -q_0(t) & -p_0(t) \end{pmatrix}, \; B(t) = \begin{pmatrix} 0 & 0 \\ -q_1(t) & -p_1(t) \end{pmatrix},$$

$$C(t, y) = \begin{pmatrix} 0 \\ -(r_0(t) + r_1(t))y_1^3 \end{pmatrix}, \; G(t) = \begin{pmatrix} 0 \\ (F_0(t) + F_1(t))\cos(\lambda t) \end{pmatrix}.$$

Let us show the alpha unpredictability of the function $G(t)$. Moreover, that the convergence and divergence sequences of the function are common with those for Markovian components. Fix a positive number ε and a bounded interval $I \subset \mathbb{R}$. As with condition (C2), (C3), there exists a natural number n_1 such that

$$|\cos(\lambda(t + t_n)) - \cos(\lambda t)| < \frac{\varepsilon}{2F},$$

for all $t \in \mathbb{R}$ and $n > n_1$. Additionally, there exists a natural number n_2 such that

$$|F_0(t + t_n) + F_1(t + t_n) - F_0(t) - F_1(t)| < \frac{\varepsilon}{2}$$

for all $t \in I$ and $n > n_2$. Therefore, it is true that

$$\|G(t + t_n) - G(t)\| = |(F_0(t + t_n) + F_1(t + t_n))\cos(\lambda(t + t_n)) - (F_0(t) + F_1(t))\cos(\lambda t)| \leqslant$$
$$\|F_0(t + t_n) + F_1(t + t_n)\||\cos(\lambda(t + t_n)) - \cos(\lambda t)| +$$
$$|\cos(\lambda t)|\|F_0(t + t_n) + F_1(t + t_n) - F_0(t) - F_1(t)\| \leqslant F\frac{\varepsilon}{2F} + \frac{\varepsilon}{2} < \varepsilon,$$

for all $t \in I$ and $n > \max(n_1, n_2)$. On the other hand, there exist positive numbers ε_0, σ and sequence u_n such that $|F_1(t + t_n) - F_1(t)| \geqslant \varepsilon_0$ for each $t \in [s_n - \sigma, s_n + \sigma]$, $n \in \mathbb{N}$. Moreover, for sufficiently large number n one can attain that $|F_0(t + t_n) - F_0(t)| < \frac{\varepsilon_0}{8}$ and $|\cos(\lambda(t + t_n)) - \cos(\lambda t)| < \frac{\varepsilon_0}{8F}$, $t \in \mathbb{R}$. Applying conditions (C3), (C4), we obtain that $\cos(\lambda(s_n + t_n)) = 0 \; (mod \; \frac{\omega}{2})$ as $n \to \infty$. Hence, due to the uniform continuity of the cosine function, there exists positive number $\sigma_1 < \sigma$ such that $\min|\cos(\lambda(t + t_n))| > \frac{\varepsilon_0}{2}$ for $t \in [s_n - \sigma_1, s_n + \sigma_1]$ and $n \to \infty$. This is why, we get that

$$\|G(t + t_n) - G(t)\| = |(F_0(t + t_n) + F_1(t + t_n))\cos(\lambda(t + t_n)) - (F_0(t) + F_1(t))\cos(\lambda t)| =$$
$$|(F_1(t + t_n) - F_1(t))\cos(\lambda(t + t_n)) + (F_1(t) + F_0(t)(\cos(\lambda(t + t_n)) - \cos(\lambda t)) +$$
$$(F_0(t + t_n) - F_0(t))\cos(\lambda(t + t_n))| \geqslant |(F_1(t + t_n) - F_1(t))\cos(\lambda(t + t_n))| -$$
$$|(F_1(t) + F_0(t))(\cos(\lambda(t + t_n)) - \cos(\lambda t))| - |(F_0(t + t_n) - F_0(t))\cos(\lambda(t + t_n))| >$$
$$\varepsilon_0 \min|\cos(\lambda(t + t_n))| - F\frac{\varepsilon_0}{8F} - \frac{\varepsilon_0}{8} > \frac{\varepsilon_0}{4},$$

for $t \in [s_n - \sigma_1, s_n + \sigma_1]$. Thus, the function $G(t)$ is alpha unpredictable with sequences t_n, s_n, and positive numbers $\frac{\varepsilon_0}{4}$, σ_1.

Condition (C5) implies that a bounded on the real axis function $z(t)$ is a solution of system (8.18) if and only if it satisfies the equation

$$z(t) = \int_{-\infty}^{t} X(t, s)[B(s)z(s) + C(s, z(s)) + G(s)]ds, \ t \in \mathbb{R}. \tag{8.19}$$

Denote by U the set of bounded and uniform continuous functions $v(t) = column(v_1(t), v_2(t))$, with common convergence sequence t_n such that $\|v(t)\|_0 < H$, where $\|v(t)\|_0 = \sup_{\mathbb{R}} \|v(t)\|$.

Define on U the operator ϕ as

$$\phi v(t) = \int_{-\infty}^{t} X(t, s)(B(s)v(s) + C(s, v(s)) + G(s))ds. \tag{8.20}$$

Lemma 8.2.1. *The operator ϕ is invariant in U.*

Proof. Fix a function $v(t)$ that belongs to U. We have that

$$\|\phi v(t)\| \leqslant \int_{-\infty}^{t} \|X(t, s)\|(\|B(s)\|\|v(s)\|+$$

$$\|C(s, v(s))\| + \|G(s)\|)ds \leqslant \frac{K}{\mu}((p_1 + q_1)H + (r_0 + r_1)H^3 + F)$$

for all $t \in \mathbb{R}$. Therefore, by the condition (C6) it is true that $\|\phi v\|_0 < H$.

Next, the method of included intervals [19, 21] will be utilized to prove that Poisson stability is invariant in U. Let us show that $\|\phi v(t + t_n) - \phi v(t)\| \to 0$ on each bounded interval of $\mathbb{R}$. Fix an arbitrary positive number ε and a closed interval $[a, b]$, $-\infty < a < b < \infty$, of the real axis. Let us choose two numbers $c < a$, and $\xi > 0$ satisfying

$$\frac{K}{\mu}(H(p_1 + q_1) + (r_0 + r_1)H^3 + F)e^{-\mu(a-c)} < \frac{\varepsilon}{4}, \tag{8.21}$$

$$\frac{K}{\mu}\xi(p_1 + q_1 + H + 3(r_0 + r_1)H^2 + H^3 + 1)[1 - e^{-\mu(b-c)}] < \frac{\varepsilon}{2}. \tag{8.22}$$

Conditions (C3) and (C4) imply that for sufficiently large n, the following inequalities are valid: $\|B(t + t_n) - B(t)\| < \xi$, $\|G(t + t_n) - G(t)\| < \xi$, $|r_0(t + t_n) + r_1(t + t_n) - r_0(t) - r_1(t)| < \xi$, and $\|v(t + t_n) - v(t)\| < \xi$ for $t \in [c, b]$. We obtain that

$$\|\phi v(t + t_n) - \phi v(t)\| = \left\| \int_{-\infty}^{t} X(t, s)(B(s + t_n)v(s + t_n) + C(s + t_n, v(s + t_n)) + G(s + t_n))ds - \right.$$

$$\left. \int_{-\infty}^{t} X(t, s)(B(s)v(s) + C(s, v(s)) + G(s))ds \right\| \leqslant$$

$$\left\| \int_{-\infty}^{t} X(t, s)(B(s + t_n)v(s + t_n) - B(s)v(s) + C(s + t_n, v(s + t_n)) - C(s, v(s)) + \right.$$

$$G(s + t_n) - G(s))ds \right\| \leqslant \int_{-\infty}^{c} \|X(t, s)\|(\|B(s + t_n)v(s + t_n) - B(s)v(s)\| +$$

$$\|C(s + t_n, v(s + t_n)) - C(s, v(s))\| + \|G(s + t_n) - G(s)\|)ds +$$

$$\int_{c}^{t} \|X(t, s)\|(\|B(s + t_n)(v(s + t_n) - v(s))\| + \|v(s)(B(s + t_n) - B(s))\|)ds +$$

$$\int_{c}^{t} \|X(t, s)\|(\|C(s + t_n, v(s + t_n)) - C(s + t_n, v(s))\| +$$

$$\|C(s + t_n, v(s)) - C(s, v(s))\|)ds + \int_{c}^{t} \|X(t, s)\|\|G(s + t_n) - G(s)\|ds \leqslant$$

$$\frac{2K}{\mu}((p_1 + q_1)H + (r_0 + r_1)H^3 + F)e^{-\mu(a-c)} + \frac{K}{\mu}(\xi(p_1 + q_1) + H\xi)[1 - e^{-\mu(b-c)}] +$$

$$\frac{K}{\mu}(3\xi(r_0 + r_1)H^2 + \xi H^3)[1 - e^{-\mu(b-c)}] + \frac{K}{\mu}\xi[1 - e^{-\mu(b-c)}],$$

is correct for all $t \in [a, b]$. From inequalities (8.21) and (8.22) it follows that $\|\phi v(t + t_n) - \phi v(t)\| < \varepsilon$ for $t \in [a, b]$. Therefore, the sequence $\phi v(t + t_n)$ uniformly converges to $\phi v(t)$ on each bounded interval of $\mathbb{R}$.

The function $\phi v(t)$ is uniformly continuous, since its derivative is uniformly bounded on the real axis. Thus, the set U is invariant for the operator ϕ.

Theorem 8.2.1. [22] *The stochastic differential equation (8.7) with Markovian coefficients admits a unique exponentially stable alpha unpredictable solution provided that the conditions (C1)–(C7) are valid. Moreover, the divergence and convergence sequences of the output stochastic dynamics are common with those, t_n and s_n, of the stochastic components of the coefficients.*

Proof. Let us prove completeness of the set U. Consider a Cauchy sequence $\phi^k(t)$ in U, which converges to a limit function $\phi(t)$ on $\mathbb{R}$. Fix a closed and bounded interval $I \subset \mathbb{R}$. We get that

$$\|\phi(t + t_n) - \phi(t)\| \leqslant \|\phi(t + t_n) - \phi^k(t + t_n)\| + \tag{8.23}$$
$$\|\phi^k(t + t_n) - \phi^k(t)\| + \|\phi^k(t) - \phi(t)\|.$$

One can choose sufficiently large n and k, such that each term on the right side of (8.23) is smaller than $\frac{\varepsilon}{3}$ for an arbitrary $\varepsilon > 0$ and $t \in I$. Thus, we conclude that the sequence $\phi(t + t_n)$ is uniformly converging to $\phi(t)$ on I. That is, the set U is complete.

Next, we shall show that the operator $\phi: U \to U$ is a contraction. For any $\varphi(t)$, $\psi(t) \in U$, one can attain that

$$\|\phi\varphi(t) - \phi\psi(t)\| \leqslant \int_{-\infty}^{t} \|X(t, s)\|(\|B(s)\|\|\varphi(s) - \psi(s)\| + \|C(s, \varphi(s)) - C(s, \psi(s))\|)ds \leqslant$$

$$\frac{K}{\mu}\Big((p_1 + q_1)\|\varphi(t) - \psi(t)\|_0 + (r_0 + r_1)(|\varphi_1^2(t)| + |\varphi_1(t)||\psi_1(t)| + |\psi_1^2(t)|)\|\varphi(t) - \psi(t)\|_0\Big) <$$

$$\frac{K}{\mu}(p_1 + q_1 + 3(r_0 + r_1)H^2)\|\varphi(t) - \psi(t)\|_0.$$

Therefore, the inequality $\|\phi\varphi - \phi\psi\|_0 < \frac{K}{\mu}(p_1 + q_1 + 3(r_0 + r_1)H^2)\|\varphi - \psi\|_0$ holds, and according to the condition (C7) the operator $\Pi\colon U \to U$ is a contraction.

By the contraction mapping theorem there exists the unique fixed point, $z(t) \in U$ of the operator ϕ, which is the unique solution of stochastic differential equation (8.7). In what follows, we will show that the solution $z(t)$ is unpredictable.

Applying the relations

$$z(t) = z(s_n) + \int_{s_n}^{t} A(s)z(s)ds + \int_{s_n}^{t} B(s)z(s)ds + \int_{s_n}^{t} C(s, z(s))ds + \int_{s_n}^{t} G(s)ds$$

and

$$z(t + t_n) = z(s_n + t_n) + \int_{s_n}^{t} A(s + t_n)z(s + t_n)ds + \int_{s_1}^{t} B(s + t_n)z(s + t_n)ds +$$

$$\int_{s_n}^{t} C(s + t_n, z(s + t_n))ds + \int_{s_n}^{t} G(s + t_n)ds$$

we obtain that

$$z(t + t_n) - z(t) = z(s_n + t_n) - z(s_n) + \int_{s_n}^{t} (A(s + t_n)z(s + t_n) - A(s)z(s))ds +$$

$$\int_{s_n}^{t} (B(s + t_n)z(s + t_n) - B(s)z(s))ds +$$

$$\int_{s_n}^{t} (C(s + t_n, z(s + t_n)) - C(s, z(s)))ds + \int_{s_n}^{t} (G(s + t_n) - G(s))ds.$$

Using conditions (C3), (C4), and uniform continuity of the entries of the matrix $A(t)$, periodic function $r_0(t)$ and solution $z(t)$, one can find a positive number σ_2 and integers l, k, n_0 such that the following inequalities are satisfied

$$\sigma_2 < \sigma_1; \tag{8.24}$$

$$\|A(t + t_n) - A(t)\| < \varepsilon_0(\frac{1}{l} + \frac{2}{k}), \quad t \in \mathbb{R}, n > n_0; \tag{8.25}$$

$$|r_0(t + t_n) - r_0(t)| < \varepsilon_0(\frac{1}{l} + \frac{2}{k}), \quad t \in \mathbb{R}, n > n_0; \tag{8.26}$$

$$\frac{2l\sigma_2}{3}\left(\varepsilon_0\Big[\frac{1}{4} - (p_1 + q_1 + \max(q_0 + p_0, 1) + H + 3H^2(r_0 + r_1) + H^3)(\frac{1}{l} + \frac{2}{k})\Big] - \right. \tag{8.27}$$
$$2(p_1 + q_1)H - 2H^3r_1) \geqslant \varepsilon_0;$$

$$\|z(t+s) - z(t)\| < \varepsilon_0 \min(\frac{1}{k}, \frac{1}{4l}), \quad t \in \mathbb{R}, \ |s| < \sigma_2. \tag{8.28}$$

Let the numbers σ_2, l and k as well as numbers $n \in \mathbb{N}$, be fixed. Consider the following two alternatives: (i) $\|z(s_n + t_n) - z(s_n)\| < \varepsilon_0/l$; (ii) $\|z(s_n + t_n) - z(s_n)\| \geqslant \varepsilon_0/l$.

(i) Using (8.28) one can show that

$$\|z(t+t_n) - z(t_n)\| \leqslant \|z(t+t_n) - z(s_n+t_n)\| + \|z(s_n+t_n) - z(s_n)\| + \tag{8.29}$$

$$\|z(s_n) - z(t)\| < \frac{\varepsilon_0}{l} + \frac{\varepsilon_0}{k} + \frac{\varepsilon_0}{k} = \varepsilon_0(\frac{1}{l} + \frac{2}{k}), \tag{8.30}$$

if $t \in [s_n, s_n + \sigma_2]$.

Therefore, the inequalities (8.24)–(8.29) imply that

$$\|z(t+t_n) - z(t)\| \geqslant \int_{s_n}^t \|G(s+t_n) - G(s)\| ds - \|z(s_n+t_n) - z(s_n)\| -$$

$$\int_{s_n}^t \|B(s+t_n)z(s+t_n) - B(s)z(s)\| ds - \int_{s_n}^t \|A(s+t_n)z(s+t_n) - A(s)z(s)\| ds -$$

$$\int_{s_n}^t \|C(s+t_n, z(s+t_n)) - C(s, z(s))\| ds \geqslant \int_{s_n}^t \|G(s+t_n) - G(s)\| ds - \|z(s_n+t_n) - z(s_n)\| -$$

$$\int_{s_n}^t \|B(s+t_n) - B(s)\| \|z(s+t_n)\| ds - \int_{s_n}^t \|B(s)\| \|z(s+t_n) - z(s)\| ds -$$

$$\int_{s_n}^t \|A(s+t_n) - A(s)\| \|z(s+t_n)\| ds - \int_{s_n}^t \|A(s)\| \|z(s+t_n) - z(s)\| ds -$$

$$\int_{s_n}^t |r_0(s+t_n)z_1^3(s+t_n) - r_0(s+t_n)z_1^3(s))| ds - \int_{s_n}^t |r_0(s+t_n)z_1^3(s) - r_0(s)z_1^3(s))| ds -$$

$$\int_{s_n}^t |r_1(s+t_n)z_1^3(s+t_n)) - r_1(s+t_n)z_1^3(s))| ds - \int_{s_n}^t |r_1(s+t_n)z_1^3(s) - r_1(s)z_1^3(s)| ds \geqslant$$

$$\sigma_2\frac{\varepsilon_0}{4} - \frac{\varepsilon_0}{l} - 2\sigma_2(p_1+q_1)H - \sigma_2(p_1+q_1)\varepsilon_0(\frac{1}{l}+\frac{2}{k}) - \sigma_2\varepsilon_0(\frac{1}{l}+\frac{2}{k})H -$$

$$\sigma_2\max(q_0+p_0, 1)\varepsilon_0(\frac{1}{l}+\frac{2}{k}) - 3\sigma_2 r_0 H^2\varepsilon_0(\frac{1}{l}+\frac{2}{k}) - \sigma_2\varepsilon_0(\frac{1}{l}+\frac{2}{k})H^3 -$$

$$3\sigma_2 r_1 H^2\varepsilon_0(\frac{1}{l}+\frac{2}{k}) - 2\sigma_2 r_1 H^3 > \frac{\varepsilon_0}{2l}$$

for $t \in [s_n, s_n + \sigma_2]$.

(ii) If $|z(t_n + s_n) - z(s_n)| \geqslant \varepsilon_0/l$ it is not difficult to find that (8.28) implies

$$\|z(t+t_n) - z(t)\| \geqslant \|z(t_n+s_n) - z(s_n)\| - \|z(s_n) - z(t)\| - \tag{8.31}$$

$$\|z(t+t_n) - z(t_n+s_n)\| \geqslant \frac{\varepsilon_0}{k} - \frac{\varepsilon_0}{l} - \frac{\varepsilon_0}{4l} = \frac{\varepsilon_0}{2l}, \tag{8.32}$$

for $t \in [s_n - \sigma_2, s_n + \sigma_2]$ and $n \in \mathbb{N}$. Thus, it can be concluded that $z(t)$ is an alpha unpredictable solution with sequences t_n, s_n and positive numbers $\frac{\sigma_2}{2}, \frac{\varepsilon_0}{2l}$.

Finally, let us discuss the exponential stability of the solution $z(t)$. It is true that

$$z(t) = X(t, t_0)z(t_0) + \int_{t_0}^{t} X(t, s)(B(s)z(s) + C(s, z(s)) + G(s))ds.$$

Denote by $\bar{z}(t)$ another solution of stochastic differential equation (8.7) such that

$$\bar{z}(t) = X(t, t_0)\bar{z}(t_0) + \int_{t_0}^{t} X(t, s)(B(s)\bar{z}(s) + C(s, \bar{z}(s)) + G(s))ds.$$

Making use of the relation

$$\bar{z}(t) - z(t) = X(t, t_0)(\bar{z}(t_0) - z(t_0)) + \int_{t_0}^{t} X(t, s)(B(s)(\bar{z}(s) - z(s)) + C(s, \bar{z}(s)) - C(s, z(s)))ds,$$

one can obtain

$$\|\bar{z}(t) - z(t)\| \leqslant \|X(t, t_0)\|\|\bar{z}(t_0) - z(t_0)\| + \int_{t_0}^{t} \|X(t, s)\|(\|B(s)\|\|\bar{z}(s) - z(s)\| +$$

$$\|C(s, \bar{z}(s)) - C(s, z(s))\|)ds \leqslant Ke^{-\mu(t-t_0)}\|\bar{z}(t_0) - z(t_0)\| +$$

$$\int_{t_0}^{t} Ke^{-\mu(t-s)}((p_1 + q_1)\|\bar{z}(s) - z(s)\| + (r_0 + r_1)(|\bar{z}_1^2(s)| + |\bar{z}_1(s)||z_1(t)| + \tag{8.33}$$

$$|z_1^2(s)|)\|\bar{z}(s) - z(s)\|)ds \leqslant \frac{K}{\mu}(p_1 + q_1 + 3(r_0 + r_1)H^2)\|\bar{z}(t) - z(t)\|,$$

for $t \in \mathbb{R}$. With the aid of the Gronwall-Bellman Lemma, one can verify that

$$\|\bar{z}(t) - z(t)\| \leqslant Ke^{(K(p_1 + q_1 + 3(r_0 + r_1)H^2) - \mu)(t-t_0)}\|\bar{z}(t_0) - z(t_0)\|, \tag{8.34}$$

for all $t \geqslant t_0$, and condition (C7) implies that the alpha unpredictable solution, $z(t)$, is an exponentially stable solution of stochastic differential equation (8.7). The theorem is proved.

The following section provides an example to confirm the theoretical results by using numerical simulations. It illustrates various alpha unpredictable dynamics of the stochastic equation of Duffing type (8.7) for different contributions of periodic and non-periodic components of coefficients.

8.2.4 A numerical example and discussions

Below, to visualize the exponentially stable alpha unpredictable solution of Θ-type and determine dynamics of Markov coefficients, we shall apply solutions $\phi_p(t)$, $\phi_p(0) = 0.5$, $\phi_q(t)$, $\phi_q(0) = 0.6$, $\phi_r(t)$, $\phi_r(0) = 0.4$, and $\phi_F(t)$, $\phi_F(0) = 0.3$, of the dissipative equations (8.11)–(8.14), where $\alpha_p = -5$, $\alpha_q = -3$, $\alpha_r = -2$, $\alpha_F = -4$ and $p(t) = q(t) = r(t) = f(t) = \rho(t)$. The piecewise constant function $\rho(t)$, is constructed by a Markov chain with values over intervals $[hn, h(n + 1)]$, $n \in \mathbb{N}$, and described in section 8.2.2.

Consider the following stochastic Duffing equation

$$x''(t) + (p_0(t) + p_1(t))x'(t) + (q_0(t) + q_1(t)) + (r_0(t) + r_1(t))x^3(t) = \tag{8.35}$$
$$(F_0(t) + F_1(t))\cos(\lambda t),$$

where $\lambda = 1$, $p_0(t) = 2 - 0.3\sin(4t)$, $q_0(t) = 2 - 0.2\cos(2t)$, $r_0(t) = 0.04\cos(4t)$, $F_0(t) = 0.05\sin(8t)$, $p_1(t) = -0.4\phi_p(t)$, $q_1(t) = 0.1\phi_q(t)$, $r_1(t) = 0.02\phi_r(t)$ and $F_1(t) = 0.02\phi_F(t)$. The periodic functions $p_0(t)$, $q_0(t)$, $r_0(t)$ and $F_0(t)$ with common period $\omega = 2\pi$. All conditions from (C1) to (C7) hold with $K = 1.5$, $\mu = 2\pi$, $\bar{p}_0 = 2.3$, $\bar{q}_0 = 2.2$, $\bar{r}_0 = 0.04$, $\bar{p}_1 = 0.48$, $\bar{q}_1 = 0.2$, $\bar{r}_1 = 0.06$, $\bar{F} = 0.03$ and $\bar{H} = 0.025$. According to theorem 8.2.6, the equation (8.35) admits a unique exponentially stable alpha unpredictable solution. In figures 8.5, 8.6, and 8.7, the graphs of the coordinates and trajectories of solutions $x(t)$ for stochastic differential equation (8.35) with $h = 0.2\pi$, 2π, 8π, and initial values $x_1(0) = x_2(0) = 0$ are shown. The solutions $x(t)$ exponentially approach the alpha unpredictable solutions, $z(t)$, as time increases.

The theorem 8.2.6 can be interpreted as a result on response-driver synchronization [34] of the alpha unpredictability in the stochastic system (8.11)–(8.14) and the stochastic Duffing equation (8.7). That is, the theorem claims, in particular, that the alpha unpredictable solution $(p_1(t), q_1(t), r_1(t))$ of the system and the alpha unpredictable solution, $z(t)$, admit common sequences of convergence and diver-

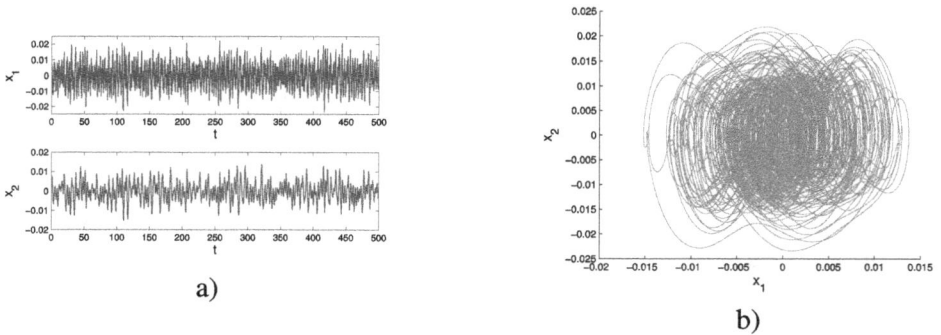

a)

b)

Figure 8.5. The time series of the coordinates and trajectory of the solution $x(t)$ of equation (8.35), with Markovian components obtained for $h = 0.1\pi$. The stochastic influence is strong because the time step of the Markov chain is smaller than the period. Reproduced with permission from [22].

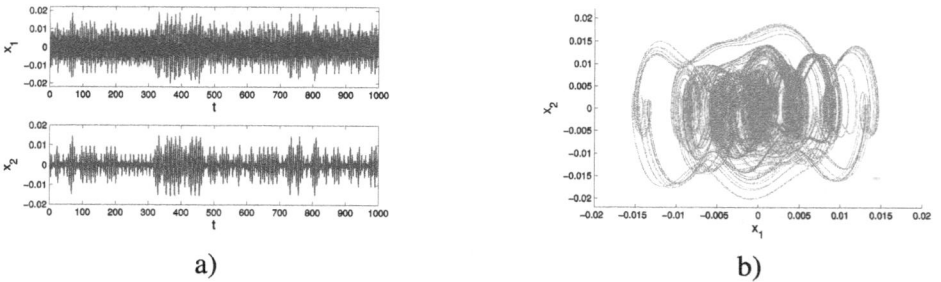

a)

b)

Figure 8.6. The x_1, x_2-coordinates and trajectory of the solution of (8.35), with Markovian components obtained for $h = 2\pi$. That is, the value of time steps is equal to the period 2π and our simulations show that the periodicity still cannot be seen clearly in this case. Reproduced with permission from [22].

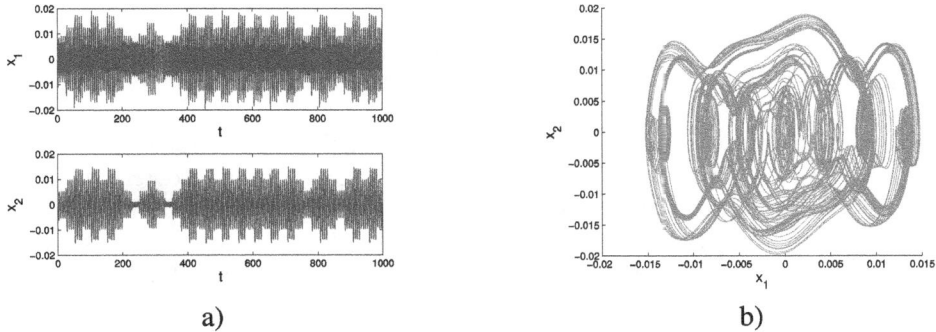

Figure 8.7. The graphs of coordinates and trajectory of the solution $x(t)$ of equation (8.35), where Markovian components obtained for $h = 8\pi$. One can see that several intervals of periodicity are placed in one step of the constancy.

gence. Delta synchronization of the alpha unpredictability for gas discharge-semi-conductor systems is considered in [8, 24].

We consider the various simulations for the model since they are with different steps h of the Markov function $\rho(t)$. The choice makes a qualitative difference in the stochastic dynamics. If the step is in the range $h \leqslant 2\pi$ the behavior is strongly irregular, and there is no indication of periodicity. For $h > 2\pi$, one can observe that periodicity is seen locally in time, and the phenomenon of intermittency [54] appears. Thus, our results demonstrate quantitative asymptotic characteristics and the possibility to learn reasons for different phenomena of chaos. Possibly, the simulations may shed light on the origins of intermittency. Additionally, for $h \leqslant 2\pi$ the effect of periodicity is seen in the 'symmetry' of the phase portraits, which is also explained by the state space's finite values. Any symmetry cannot be seen for the values of h less than 2π, since the stochastic dynamics dominates significantly.

8.3 Notes

The outcome of research in the first section is the existence of an alpha unpredictable sequence as a realization of the Markov chain, and the sequence appears as a finite realization of each experiment of the process. That is, the appearance of the sequence is a specific event for the stochastic dynamics. Simultaneously, the sensitivity appears not deterministically but with probability 1. From this point of view one can say that the deterministic chaos is an event for the stochastic dynamics with probability one. This result is true for many other discrete time random processes. For instance, the Bernoulli scheme. The significant use of the investigation is that one can unite methods of deterministic chaos with those for stochastic dynamics. Many other opportunities may appear. Among the methods are control and synchronization of chaos [34]. We have discussed that the sensitivity is present with probability one. It is valid for the Bernoulli scheme [23] and other dynamics, which can be proven for abstract similarity dynamics. Next, our study will relate Markov processes to continuous and unbounded time, which is also connected to many other random processes. Our results shed more light on the Markov chains as ergodic processes since we have shown an uncountable set of realizations: alpha unpredictable orbits of the abstract similarity dynamics. Each of them is dense in the set of all realizations [10].

The last section of the chapter is about stochastic differential equations of a particular type. One can interpret the Duffing equation model as a deterministic one with coefficients, which have been constructed randomly, then fixed and utilized as perturbations. We disagree with this view since the Markov chain is a noise source. Then it is transformed to continuous perturbations of the model, but the randomness of the construction is still present. This is why one can consider the equation (8.7) as a stochastic differential equation. The model is proven to having a solution, an alpha unpredictable function with the quantitative characteristics synchronized with those for the original Markov chain, considered as perturbations. Another interesting phenomenon is that we effectively apply a combination of regular and irregular phenomena in the process, assuming periodicity in components as a regular contribution to the external circumstances.

References

[1] Akhmet M 2020 *Almost Periodicity, Chaos and Asymptotic Equivalence* (Berlin: Springer)

[2] Akhmet M 2021 *Domain Structured Dynamics* (Bristol: IOP Publishing)

[3] Akhmet M 2022 Abstract hyperbolic chaos *Discontinuity Nonlinearity Complexity* **11** 133–8

[4] Akhmet M 2022 Modular chaos for random processes *Discontinuity Nonlinearity Complexity* **11** 191–201

[5] Akhmet M 2022 Unpredictablity in Markov chains *Carpathian J. Math.* **38** 13–9

[6] Akhmet M and Alejaily E M 2019 Domain-structured chaos in a Hopfield neural network *Int. J. Bifurcation Chaos* **29** 195–205

[7] Akhmet M and Alejaily E M 2021 Abstract similarity, chaos and fractals *Discrete Continuous Dyn. Syst. Ser. B* **26** 2479–97

[8] Akhmet M, Baskan K and Yesil C 2022 Delta synchronization of Poincaré chaos in gas discharge-semiconductor systems *Chaos* **32** 083137

[9] Akhmet M, Baskan K and Yesil C 2023 Revealing chaos synchronization below the threshold in coupled Mackey-glass systems *Math. Open Access* **11** 3197

[10] Akhmet M and Fen M O 2016 Poincaré chaos and unpredictable functions *Commun. Nonlinear Sci. Numer. Simul.* **48** 85–94

[11] Akhmet M and Fen M O 2016 Unpredictable points and chaos *Commun. Nonlinear Sci. Numer. Simul.* **40** 1–5

[12] Akhmet M and Fen M O 2018 Non-autonomous equations with unpredictable solutions *Commun. Nonlinear Sci. Numer. Simul.* **59** 657–70

[13] Akhmet M, Fen M O and Alejaily E M 2020 *Dynamics with Chaos and Fractals* (Berlin: Springer)

[14] Akhmet M, Fen M O and Alejaily E M 2021 Dynamics with fractals *Discontinuity Nonlinearity Complexity* **10** 173–84

[15] Akhmet M, Fen M O and Alejaily E M 2020 A randomly determined unpredictable function *Kazakh Math. J.* **20** 30–6

[16] Akhmet M, Fen M O, Tleubergenova M and Zhamanshin A 2019 Poincare chaos for a hyperbolic quasilinear system *Miskolc Math. Notes* **20** 33–44

[17] Akhmet M and Alejaily E M 2021 Abstract fractals *Discontinuity Nonlinearity and Complexity* **10** 135–42

[18] Akhmet M, Seilova R D, Tleubergenova M and Zhamanshin A 2020 Shunting inhibitory cellular neural networks with strongly unpredictable oscillations *Commun. Nonlinear Sci. Numer. Simulat.* **89** 105287

[19] Akhmet M, Tleubergenova M and Zhamanshin A 2020 Quasilinear differential equations with strongly unpredictable solutions *Carpathian J. Math.* **36** 341–9

[20] Akhmet M, Tleubergenova M and Zhamanshin A 2020 Inertial neural networks with unpredictable oscillations *Math. Open Access* **8** 1–11

[21] Akhmet M, Tleubergenova M and Zhamanshin A 2022 Compartmental Poisson stability in non-autonomous differential equations *Nonlinear Dynamics and Complexity* (Berlin: Springer) pp 1–23

[22] Akhmet M, Tleubergenova M and Zhamanshin A 2023 Unpredictable solutions of Duffing type equations with Markov coefficients *arXiv* 2303.17336

[23] Akhmet M and Tola A 2020 Unpredictable strings *Kazakh Math. J.* **20** 16–22

[24] Akhmet M, Yesil C and Baskan K 2023 Synchronization of chaos in semiconductor gas discharge model with local mean energy approximation *Chaos Solitons Fractals* **167** 113035

[25] Alekseev V M and Yakobson M V 1981 Symbolic dynamics and hyperbolic dynamic systems *Phys. Rep.* **75** 290–325

[26] Bowen R 1970 Markov partitions and minimal sets for axiom A diffeomorphisms *Am. J. Math.* **92** 907–18

[27] Bowen R 1970 Markov partitions for axiom a diffeomorphisms *Am. J. Math.* **92** 725–47

[28] Bowen R 1973 Symbolic dynamics for hyperbolic flows *Am. J. Math.* **95** 429–40

[29] Bunimovich L and Ya S 1980 Markov partitions for dispersed billiards *Commun. Math. Phys.* **78** 247–80

[30] Corduneanu C 2009 *Almost Periodic Oscillations and Waves* (New York: Springer)

[31] Duffing G 1918 *Erzwungene Schwingungen bei Veranderlicher Eigen–frequenz und ihre technische Bedeutung* (Braunschweig: F. Vieweg und Sohn)

[32] Estevez P G, Kuru S, Negro J and Nieto L M 2011 Solutions of a class of duffing oscillators with variable coefficients *Int. J. Theor. Phys.* **50** 2046–56

[33] Farkas M 1994 *Periodic Motion* (New York: Springer)

[34] Gonzáles-Miranda J M 2004 *Synchronization and Control of Chaos* (London: Imperial College Press)

[35] Hajek B 2015 *Random Processes for Engineers Cambridge* (Cambridge: Cambridge University Press)

[36] Hartman P 2002 *Ordinary Differential Equations* (New York: SIAM)

[37] Karlin S and Taylor H E 2012 *A First Course in Stochastic Processes* (New York: Academic)

[38] Levitan B and Zhikov V 1983 *Almost Periodic Functions and Differential Equations* (Cambridge: Cambridge University Press)

[39] Liu B W and Tunc C 2015 Pseudo almost periodic solutions for a class of nonlinear Duffing system with a deviating argument *J. Appl. Math. Comput.* **49** 233–42

[40] Mahajan A, Thakur R and Das R 2024 Sensitivity and unpredictability in semiflows on topological spaces *Commun. Nonlinear Sci. Numer. Simulat.* **133** 107949

[41] Markov A A 1971 *Extension of the Limit theorems of Probability Theory to a Sum of Variables Connected in a Chain* **vol 1** (New York: Wiley)

[42] Meyn S and Tweedie R L 2009 *Markov Chains and Stochastic Stability* (Cambridge: Cambridge University Press)

[43] Miller A 2019 Unpredictable points and stronger versions of Ruelle-Takens and Auslander-Yorke chaos *Topology Appl.* **253** 7–16

[44] Miller V H and Jansen B H 2008 Oscillatory neural network for pattern recognition: trajectory based classification and supervised learning *Biol. Cybern.* **99** 459–71

[45] Minorsky N 1947 *Introduction to Non-linear Mechanics: Topological Methods, Analytical Methods, Non-linear Resonance, Relaxation Oscillations* (Ann Arbor, MI: J.W. Edwards)

[46] Moon F 2004 *Chaotic Vibrations: An Introduction for Applied Scientists and Engineers* (Hoboken, NJ: Wiley)

[47] Naife A 1960 *Qualitative Theory of Differential Equations* (Princeton, NJ: Princeton University Press)

[48] Nemytskii V V and Stepanov V V 1960 *Qualitative Theory of Differential Equations* (Princeton, NJ: Princeton University Press)

[49] Nicolis G and Prigojin I 1989 *Exploring Complexity* (New York: Freeman)

[50] Ornstein D 1970 Bernoulli shifts with the same entropy are isomorphic *Adv. Math.* **4** 337–52

[51] Ornstein D 1995 In what sense can a deterministic system be random? *Chaos Solitons Fractals* **59** 139–41

[52] Ornstein D 2004 Kolmogorov, random processes, and Newtonian dynamics *Russian Math. Surv.* **59** 313–7

[53] Ornstein D and Weiss B 1998 On the Bernoulli nature of systems with some hyperbolic structure *Ergod. Theor. Dynam. Syst.* **18** 139–41

[54] Pomeau Y and Manneville P 1980 Intermittent transition to turbulence in dissipative dynamical systems *Commun. Math. Phys.* **74** 189–97

[55] Ponce G and Varao R 2019 *An Introduction to the Kolmogorov-Bernoulli Equivalence* (Switzerland: Springer)

[56] Robinson C 2012 *An Introduction to Dynamical Systems: Continuous and Discrete* (Providence, RI: American Mathematical Society)

[57] Sell G R 1971 *Topological Dynamics and Ordinary Differential equations* (London: Van Nostrand Reinhold Company)

[58] Thakur R and Das R 2020 Strongly Ruelle-Takens, strongly Auslander-Yorke and Poincaré chaos on semiflows *Commun. Nonlinear Sci. Numer. Simulat.* **81** 105018

[59] Walters P 1982 *An Introduction to Ergodic Theory* (New York: Springer)

[60] Wiggins S 1988 *Global Bifurcation and Chaos: Analytical Methods* (New York: Springer)

[61] Zeng W Y 1997 Almost periodic solutions for nonlinear Duffing equations *Acta Math. S* **13** 373–80

Part VIII

Fractals with alpha spaces

Chapter 9

Alpha induced dynamics in fractals and cubes

This chapter focuses on the challenge of generating chaotic dynamics within various geometries, exploring this issue from two fundamentally different perspectives. On one hand, it affirms the universality of fractals, recognized for their self-similar characteristics. On the other hand, it examines Euclidean cubes, where alpha labeling ensures chaos within dense subsets and across broad categories of shapes that are topologically equivalent to those under investigation. This approach is not intended to limit the boundaries of geometric classification but rather aims to broaden these complex perspectives through abstract construction.

9.1 Historical observations

In previous works [4, 6] and chapter 2 of ths book, a new method of chaos formation has been developed. This method relies more on the partition of domains rather than on mappings. The extension of this approach considers chaos as a generic dynamical phenomenon in various real-world, scientific, and industrial contexts. The phenomenon of chaos is viewed from two perspectives. The first aspect relates to the number of models that exhibit chaotic dynamics. This is a complex issue as the number of equations displaying chaos is still relatively small compared to those showing regular behavior, reflecting the evolution of the theory. The second aspect focuses on the density of chaotic points within the domain or state space of the dynamics. This aspect has not been extensively explored in existing literature. Chaotic domains are typically viewed as fractal objects with dimensions smaller than that of the state space, leading to the assumption that chaotic points are sparse and the domain has low density, especially in one-dimensional dynamics. To address these challenges, the study aims to establish multidimensional chaos for simple geometric objects like cubes. This approach involves analyzing the domain structure to create an invariant set under a chaotic map. This is achieved through a recursive division of the set into elements that meet specific criteria. The findings are applicable to all figures that are topologically similar to cubes, demonstrating the realization of multidimensional chaos through a structured approach.

The mathematical notion of similarity [13, 39, 77] underscores the essence of commonalities between objects, whether abstract or tangible. It denotes relationships among point sets, with self-similarity emerging as a pivotal attribute. The concept of self-similarity, pioneered by Leibniz in the 17th century [82], delineates figures whose constituents mirror the whole across all scales. Key landmarks in self-similarity include the Weierstrass function, the Cantor set, and space-filling curves. Fractals epitomize sets that showcase self-similarity across scales, featuring a Hausdorff dimension surpassing the topological dimension. Self-similarity and fractional dimensions stand out as elemental traits of fractals, with the former representing a cardinal method for sculpting sets with fractional dimensions [21].

Moran [61] introduced the concept of a self-similar set in 1946, providing a mathematical definition of a geometric construction as a collection of sets that meet specific conditions. Researchers have extensively studied and expanded Moran's construction, exploring various methods. Building on this work, researchers in [66, 67, 73] generalized the approach to create different classes of Moran-like geometric constructions using symbolic space. They utilized topological conjugacy to describe the symbolic representation of these constructions, known as Cantor-like sets. The papers [12, 25, 27, 28, 34, 37, 45, 62, 72] delve into exciting definitions of self-similarity, discussing dimension and measure-related problems. These manuscripts focus on sets in Euclidean space $\mathbb{R}^n$, defining a self-similar set as a union of its images under similarity transfunctions [28].

Fractal geometry serves as a mathematical tool for elucidating the structures found in nature that exhibit varying levels of self-similarity and for creating artificial structures. The late 19th century witnessed increased interest in self-similarity, notably with Karl Weierstrass's introduction of a function that is continuous but not differentiable everywhere, leading to the development of the Weierstrass function as a prototypical self-similar fractal. Georg Cantor's creation of the Cantor set in 1883 stands as a pivotal and influential fractal, exemplifying the principles and applications of fractals. The work of Helge von Koch in 1904 on a continuous yet non-differentiable curve further enriched the understanding of self-similar fractals. Waclaw Sierpinski's contributions to fractal mathematics included the introduction of the Sierpinski gasket in 1916, a triangular fractal generated through a recursive process of removing symmetrical components from an initial triangle. Sierpinski also developed the Sierpinski carpet, a square fractal following a similar recursive procedure. The discovery of Julia sets by Gaston Julia and Pierre Fatou in 1917–19 marked another advancement in fractal exploration, with these sets being constructed through iterative functions. The coinage of the term 'fractal' by Benoit Mandelbrot in 1975 encapsulates geometric structures characterized by self-similarity [51, 53]. Felix Hausdorff's establishment of fractional dimension during the same period became an invaluable tool for analyzing and defining the geometrical intricacies of fractals. Paul Lévy's investigations into self-similar curves and surfaces led to the delineation of the Lévy C curve, a novel fractal curve introduced in 1938.

Fractals and chaos are intricately linked concepts stemming from disparate mathematical realms: geometry and dynamics. Chaos domains represent assemblies of points where dynamic processes unfold, often revealing fractal attributes. These domains may manifest as self-similar sets or harbor fractional dimensions, exemplified by the renowned Cantor set attributed to the logistic map. The interplay of porosity, sensitivity, and continuity in the initial conditions significantly influences chaotic systems. Certain fractals, such as the Sierpinski carpet and triangle, the Koch Snowflake, and others, serve as apt domains for chaotic motion. Intriguingly, structured chaos can transcend to finite-dimensional Euclidean cubes, posing an unconventional departure from inherently fractal settings. The symbolic affinity shared between chaos and fractals is compelling and traces back to their historical evolution.

Several researchers have highlighted the close relationship between chaos and fractal geometry. This connection is evident in the dynamics of the Fatou–Julia iteration used to create Julia and Mandelbrot sets, where neighboring points near the boundary in the domain can exhibit vastly different behaviors, showcasing sensitivity in fractal structures. Chaos provides insights into irregularity and trajectory divergence based on dynamics, while the concept of fractals allows for the study of complex geometric formations. The correlation between chaos and fractals becomes more pronounced when the fractal dimension quantifies how a trajectory fills its phase space, indicating the presence of a strange attractor [60]. Research on the chaotic nature of fractals has primarily focused on specific types classified as disconnected fractals [14, 20, 23]. The concept of topological conjugacy has been employed to demonstrate the invariance of these fractal sets under specific chaotic maps. Limited studies have explored chaotic dynamical systems for fractals, with notable work conducted on the Sierpinski carpet, showing chaotic behavior in terms of topological mixing in the associated dynamical system defined on the Sierpinski carpet set [26].

The application of Bernoulli's shift projection to an infinite symbol sequence is a potent tool for investigating non-linearity. However, a key challenge arises in establishing this projection as an auxiliary mapping. We introduce a method for structuring state spaces that directly embodies the essential dynamical aspects between alpha-labeled elements. The concept originated in our exploration of dynamics within fractals and later proved to be applicable beyond fractal shapes. Surprisingly, conventional chaotic models such as the logistic map and Baker transformation exhibit clearer and more elegant interpretations when viewed through the lens of abstract similarity dynamics. Our approach to exploring dynamical features and organizing intricate motions is designed to be accessible to novice readers. The advantages include verifying chaos in a majority of self-similar fractals, generating unpredictable alpha functions, uncovering novel principles for random processes, and identifying chaotic attributes within discrete and continuous stochastic motions.

9.2 Alpha induced dynamics in metric spaces

In this part of the book, we shall specify knowledge of alpha dynamics of chapter 2 with two conditions. The first one is that the labeling is applied to geometries, which are initially not assigned with dynamics. Furthermore, the one-to-one nature of the alpha labeling operator, which is isometric, simplifies our discussion and enhances the clarity of our results. This unique characteristic allows us to see the outcomes more clearly and effectively than our intuition alone. To aid the reader, we have structured the information into three subsections: one on alpha labels, another on alpha-induced dynamics, and the third on the conditions of chaos existence, making it easier to navigate and understand.

9.2.1 Alpha labels and alpha map

Consider the following *alpha set*,

$$\mathscr{F} = \{\mathscr{F}_{i_1 i_2 \ldots i_n \ldots} : i_k = 1, 2, \ldots, m, \quad k = 1, 2, \ldots\}, \tag{9.1}$$

with fixed natural number m. The elements of $\mathscr{F}$ are indexed symbols and they are called *alphalabels* in the chapter.

To achieve the continuity of dynamics on an initial value in the research, we shall need the following subsets of alpha labels, *n-dimensional rectangles*,

$$\mathscr{F}_{i_1 i_2 \ldots i_n} = \bigcup_{j_k = 1, 2, \ldots, m} \mathscr{F}_{i_1 i_2 \ldots i_n j_1 j_2 \ldots}, \tag{9.2}$$

where indices $i_1, i_2, \ldots, i_n$, are fixed.

The rectangles are nested and satisfy the following relations,

$$\mathscr{F} \supseteq \mathscr{F}_{i_1} \supseteq \mathscr{F}_{i_1 i_2} \supseteq \cdots \supseteq \mathscr{F}_{i_1 i_2 \ldots i_n} \supseteq \mathscr{F}_{i_1 i_2 \ldots i_n i_{n+1}} \cdots, \quad i_k = 1, 2, \ldots, m, \quad k = 1, 2, \ldots.$$

Next, we shall determine the map $\varphi \colon \mathscr{F} \to \mathscr{F}$ such that

$$\varphi(\mathscr{F}_{i_1 i_2 \ldots i_n \ldots}) = \mathscr{F}_{i_2 i_3 \ldots i_n \ldots}. \tag{9.3}$$

One can see that for arbitrary natural number n and $i_k = 1, 2, \ldots, m, \quad k = 1, 2, \ldots, n$, iterations of the map result that

$$\varphi^n(\mathscr{F}_{i_1 i_2 \ldots i_n}) = \mathscr{F}. \tag{9.4}$$

The definition (9.3) and the equality (9.4) make it reasonable to call φ *abstract similarity map* or *similarity map*. We shall call it also *alpha-map*, and the number n *order of similarity*. Correspondingly, the *alpha set* is also called *abstract similarity set*.

To complete the definition of the dynamics on alpha labels, one needs a metric for the couple $(\mathscr{F}, \varphi)$. In the following subsection, we shall show how the metric can be inherited from the dynamics under investigation.

9.2.2 Alpha induced dynamics

In this part of the paper, we shall show the mechanism of alpha labeling, and correspondingly, we shall see how the alpha-metric δ for abstract similarity dynamics $(\mathscr{F}, \varphi, \delta)$ can be designed. That is, we shall show how the abstract couple $(\mathscr{F}, \varphi)$ can be intervened with the element of a dynamical model, namely the metric, such that the abstract object became ready for research applications. In the present paper, investigation is focused on chaos. The construction of abstract similarity dynamics (alpha dynamics) is strongly joined with the procedure of alpha labeling itself.

Consider a metric space (F, d) where d is the distance. It is assumed that set F is uncountable. We shall say that the abstract similarity set $\mathscr{F}$ is an *alpha label set* for F if each element of F is labeled exactly by one member of $\mathscr{F}$, such that every element in $\mathscr{F}$ serves as a label for F. In what follows, $\mathscr{L}: F \to \mathscr{F}$ denotes *alpha-operator* such that $\mathscr{F}_{i_1 i_2 \ldots i_n \ldots} = \mathscr{L}(f)$ if $\mathscr{F}_{i_1 i_2 \ldots i_n \ldots}$ is a label for f. The operator is a bijection. This is a restriction for the present fractal analysis, but it can be removed in future research. The general presentation of the dynamics can be found in chapter 2.

Definition 9.2.1. *A function $\delta: \mathscr{F} \times \mathscr{F} \to \mathbb{R}$, is said to be an alpha induced metric in the alpha set $\mathscr{F}$, if $\delta(\mathscr{F}_{i_1 i_2 \ldots i_n \ldots}, \mathscr{F}_{j_1 j_2 \ldots jn \ldots}) = d(f_1, f_2)$, where $\mathscr{F}_{i_1 i_2 \ldots i_n \ldots}$ and $\mathscr{F}_{j_1 j_2 \ldots jn \ldots}$ are alpha labels of elements f_1 and f_2 of F.*

Definition 9.2.2. *A map ψ is called an alpha induced map in the metric space (F, d), if for each element f of F, it is true that alpha labels of f and $\psi(f)$ are $\mathscr{F}_{i_1 i_2 \ldots i_n \ldots}$ and $\mathscr{F}_{i_2 \ldots i_n \ldots}$, respectively. That is, $\psi(f) = \mathscr{L}^{-1}(\varphi(\mathscr{L}(f)))$.*

In what follows, we shall consider the dynamics $(F, \psi, d) = (F, \mathscr{L}^{-1}\varphi\mathscr{L}, d)$. It will be called a *dynamics induced by alpha labeling* or simply *alpha induced dynamics* in what follows. Moreover, we shall consider the dynamical system $(\mathscr{F}, \varphi, \delta)$, where the abstract couple $(\mathscr{F}, \delta)$ is a metric space.

9.2.3 How to make chaotic geometry

Finally, in this subsection, we describe how to apply the dynamics induced by alpha labeling to research complexity in originally non-dynamical problems. For example, it can be the task of chaos arrangement in fractals. So, consider a metric space (F, d), which is not assigned with a mapping. Suppose that one can find an alpha label set $\mathscr{F}$ for the space and, correspondingly alpha space $(\mathscr{F}, \delta)$ with the similarity map φ such that the triple $(\mathscr{F}, \varphi, \delta)$ admits certain dynamical properties. In the present study, they are the separation and diameter conditions, such that the abstract dynamics is chaotic. Then, by introducing the alpha induced map $\psi = \mathscr{L}^{-1}\varphi\mathscr{L}$ in the metric space (F, d), one obtains that the alpha induced dynamics (F, ψ, d) is chaotic.

The following feature for alpha spaces, combined with the action of the alpha similarity map, is equivalent to what is said to be continuous dependence on the initial value for dynamics determined by differential or difference equations.

Definition 9.2.3. *It will be said that an alpha space $(\mathscr{F}, \delta)$ satisfies the diameter condition if for all rectangles $\mathscr{F}_{i_1 i_2 \ldots i_n}$ it is valid that*

$$\sup_{i_k=1,2,\ldots,n} \operatorname{diam} \mathscr{F}_{i_1 i_2 \ldots i_n} \to 0 \quad as \quad n \to \infty, \tag{9.5}$$

where $\operatorname{diam}(A) = \sup\{\delta(\mathbf{x}, \mathbf{y}): \mathbf{x}, \mathbf{y} \in A\}$ for a set A in $\mathscr{F}$.

The diameter condition implies that for arbitrary positive ε there exists a finite ε-net, and consequently a complete label space is compact. Finally, let us remark that the diameter condition implies the existence of finite ε-nets with arbitrary positive ε, and consequently the space is bounded. The topological properties of the alpha space will be utilized for chaotic dynamics, in what follows.

Next, we shall describe a feature for alpha dynamics, which is important for mathematics complexity. We shall call it the *separation condition*. It causes the alpha unpredictability of Poincaré chaos and sensitivity of Lorenz chaos, if they are results of alpha labeling.

Consider the distance between two nonempty bounded sets A and B in $\mathscr{F}$ as the function $\delta(A, B) = \inf\{\delta(\mathbf{x}, \mathbf{y}) : \mathbf{x} \in A, \mathbf{y} \in B\}$.

Definition 9.2.4. *The alpha set $\mathscr{F}$ satisfies the separation condition of degree n, if there exist a positive number ε_0 and a natural n such that for arbitrary $i_1 i_2 \ldots i_n$ one can find indices $j_1 j_2 \ldots j_n$ such that*

$$\delta(\mathscr{F}_{i_1 i_2 \ldots i_n}, \mathscr{F}_{j_1 j_2 \ldots j_n}) \geqslant \varepsilon_0. \tag{9.6}$$

The real number ε_0 is called the separation constant.

The separation condition combined with the action of the similarity map is an abstraction of the Lorenz sensitivity. Thus, the diameter and separation conditions make the circumstances for chaos, which for dynamical systems are caused by continuity in the initial value and sensitivity. The conditions are only what we need to guarantee chaos on alpha dynamics. Nevertheless, it is of strong interest to search for other criteria. We believe that the conditions can be elaborated further, as well as new structures similar to alpha spaces introduced for the research of complexity.

Next, we shall provide definitions of alpha unpredictability in the two dynamics under discussion.

Definition 9.2.5. *A point $\mathscr{F}_{i_1 i_2 \ldots}$ from alpha set $\mathscr{F}$ is alpha unpredictable if there exist a positive number ε_0 and sequences t_m, s_m, of natural numbers, both of which diverge to infinity such that*
- $\varphi^{t_m}(\mathscr{F}_{i_1 i_2 \ldots i_k \ldots}) \to \mathscr{F}_{i_1 i_2 \ldots i_k \ldots}$ *as* $m \to \infty$
- $\delta(\varphi^{t_m + s_m}(\mathscr{F}_{i_1 i_2 \ldots i_k \ldots}), \varphi^{s_m}(\mathscr{F}_{i_1 i_2 \ldots i_k \ldots})) \geqslant \varepsilon_0$ *for each natural number m.*

Definition 9.2.6. [8] *A point $f \in F$ and the trajectory through it are alpha unpredictable if there exist a positive number ε_0 (the alpha unpredictability constant) and sequences $\{\zeta_n\}$ and $\{\kappa_n\}$, of natural numbers, both of which diverge to infinity, such that*
- $\lim_{n\to\infty} \psi^{\zeta_n}(f) = f$;
- $d[\psi^{\zeta_n + \kappa_n}(f), \psi^{\zeta_n}(f)] \geqslant \varepsilon_0$ *for each natural number n.*

We call the dynamics, which exists because of the alpha unpredictability point, *ultra Poincaré chaos.* The existence of infinitely many alpha unpredictable Poisson stable trajectories that lie in a compact set meets all requirements of chaos. Based on this, chaos can appear in the dynamics on a quasi-minimal set, which is the closure of a Poisson stable trajectory. Therefore, the ultra Poincaré chaos is referred to as the dynamics on the quasi-minimal set of trajectory initiated from the alpha unpredictable point.

To validate the last discussions, we suggest the following assertion. The proof is provided in chapter 2.

Theorem 9.2.1. *If the diameter and separation conditions are valid, then alpha dynamics $(\mathscr{F}, \varphi, \delta)$ and $(F, \mathscr{L}^{-1}\varphi\mathscr{L}, d)$ possess alpha unpredictable points and ultra Poincaré chaos, provided that the space $(\mathscr{F}, \delta)$ is compact.*

The following theorem proves that the similarity map φ and the alpha induced map ψ possess the ingredients of Devaney chaos, density of periodic points, transitivity, and sensitivity. A point $\mathscr{F}_{i_1 i_2 i_3...}$ of $\mathscr{F}$ is periodic with period n if its index consists of endless repetitions of a block of n terms.

Theorem 9.2.2. *If the diameter and separation conditions hold, then the dynamics $(\mathscr{F}, \varphi, \delta)$ and $(F, \mathscr{L}^{-1}\varphi\mathscr{L}, d)$ are chaotic in the sense of Devaney.*

In addition to Devaney and Poincaré chaos, it can be shown that Li–Yorke chaos also takes place in the dynamics of the map φ. First, we shall introduce needed definitions.

Definition 9.2.7. *A couple $(\mathscr{F}_{i_1 i_2...}, \mathscr{F}_{j_1 j_2 \,...}\,)$ of elements from $\mathscr{F}$ is called proximal if*
$$\liminf_{n\to\infty}\delta(\varphi^n(\mathscr{F}_{i_1 i_2...}), \varphi^n(\mathscr{F}_{j_1 j_2 \,...}\,)) = 0.$$

Definition 9.2.8. *A couple $(\mathscr{F}_{i_1 i_2...}, \mathscr{F}_{j_1 j_2 \,...}\,)$ of elements from $\mathscr{F}$ is frequently separated if*
$$\limsup_{n\to\infty}\delta(\varphi^n(\mathscr{F}_{i_1 i_2...}), \varphi^n(\mathscr{F}_{j_1 j_2 \,...}\,)) > 0.$$

Definition 9.2.9. *A subset of $\mathscr{F}$ is called scrambled if each couple of it is proximal and frequently separated.*

Definition 9.2.10. *The dynamics $(\mathscr{F}, \varphi, \delta)$ is Li–Yorke chaotic, if*
- *there exists an n-periodic orbit for φ with arbitrary natural period;*
- *there is an uncountable scrambled subset;*
- *each element of the scrambled set makes a frequently separated couple with any periodic point.*

Theorem 9.2.3. *The alpha dynamics* $(\mathscr{F}, \varphi, \delta)$ *and alpha induced dynamics* $(F, \mathscr{L}^{-1}\varphi\mathscr{L}, d)$ *are Li–Yorke chaotic if the diameter and separation conditions hold.*

We call the chaos for dynamics $(\mathscr{F}, \phi, \delta)$ *abstract similarity chaos* or *alpha chaos*, and *alpha induced chaos* or *alpha chaos* for $(F, \mathscr{L}^{-1}\varphi\mathscr{L}, d)$. Obviously, it is a significant result for dynamical systems theory, since it confirms that all the three types, ultra Poincaré, Li–Yorke, and Devaney chaos, are present if a model exhibits alpha chaos.

9.3 Chaotic cubes

This section shows that chaotic behavior can be found on dense subsets of cubes with arbitrary finite dimensions. The method of alpha labeling is applied. Thus, we shall prove that a cube can be a quasi-minimal set—that is a closure of alpha unpredictable trajectory. Previously, this was known only for the unit section and the Devaney and Li–Yorke chaos.

Chaos has become a fundamental concept deeply integrated into many, if not all, fields of science such as physics, biology, medicine, engineering, culture, and human activities [17, 71]. The chaotic behavior of some physical and biological properties was formerly attributed to random or stochastic processes or uncontrolled forces [44, 50]. The appearance of chaos in deterministic systems drew the borderline between deterministic chaos and stochastic noise. The idea is manifested in the chaotic behavior of simple dynamical systems. However, the randomness theory of Kolmogorov-Martin-Löf still provides a deeper understanding of the origins of deterministic chaos [17]. The fundamental theoretical framework of chaos was developed in the last quarter of the 20th century. During that period, different types and definitions of chaos were formulated. In general, chaos can be defined as a periodic, long-term behavior in a deterministic system that exhibits sensitive dependence on initial conditions [43]. Devaney [23] and Li–Yorke [48] chaos are the most frequently used types, which are characterized by transitivity, sensitivity, frequent separation, and proximality. Another common type occurs through a period-doubling cascade, which is a sort of route to chaos through local bifurcations [29, 68, 69]. In the papers [7, 8], ultra Poincaré chaos was introduced through alpha unpredictable point concepts. Further, it was developed to alpha unpredictable functions and sequences.

Whoever researches in this field can discern from the literature that there is a scientific conception that chaos is everywhere. Realizing such an ideation requires mathematical tools to conceptualize all manifestations of the phenomenon. Strictly speaking, we should develop simple chaotic mechanisms that emulate complex behaviors. Investigating the fundamental aspects of high-dimensional chaotic states is necessary in this direction. Indeed, mathematical modeling of real-world problems shows that real life is very often a high-dimensional chaos, and even chaotic activities in our everyday lives are difficult to describe via low-dimensional systems [38].

Recently, in papers [4, 6], we have developed a new method of chaos formation that depends somewhat more on partition of the domain than on a map. That is, the map is a natural consequence of the chaotic domain structure. In the present study, we extend the approach of consideration of the methodological problem for chaos as a *generic* dynamical phenomenon in the real world, science, and industry. The phenomenon is understood in two aspects. The first is connected to the number of models which admit chaotic dynamics. This problem is complex to discuss since the number of differential, discrete, and other equations exhibiting chaos is still remarkably small if one compares it with the number of those admitting regular behavior. The second aspect concerns the density of chaotic points in the domain or the state space of the dynamics. This question has not been considered in detail (as far as we know) in literature. This is because the chaotic domains are usually considered fractal objects; their dimension is less than the dimension of the state space. Consequently, it is assumed without discussion that the points of chaos are sparse and the domain admits a small density only for one-dimensional unimodal dynamics, as is the case for the unit section where it was proven that all points of the set can be chaotic [23, 48]. As a first step towards the goal, we show how to establish multi-dimensional chaos for simple geometrical objects in the present chapter. We focus on constructing an invariant set under a chaotic map in the domain structure. Our approach is based on a recursive division of the set into infinite elements that satisfy specific conditions. A chaotic map can be defined on the set using infinite sequences to index the set points. Thus, in this section, we realize the alpha dynamics for simple geometrical objects. The results are valid for all figures, which are topologically equivalent to cubes.

9.3.1 Chaotic line segment

For the sake of comprehension, let us start with the line segment $F = (0, 1]$. Divide the set into 4 parts $F_1 = [0, \frac{1}{4}]$, $F_2 = (\frac{1}{4}, \frac{1}{2}]$, $F_3 = (\frac{1}{2}, \frac{3}{4}]$, $F_4 = (\frac{3}{4}, 1]$ (see figure 9.1 (b)). Divide each part F_{i_1}, $i_1 = 1, 2, 3, 4$ again into 4 equal parts and denote them by $F_{i_1 i_2}$, $i_2 = 1, 2, 3, 4$. Continue in this procedure such that, at the *kth* step of the

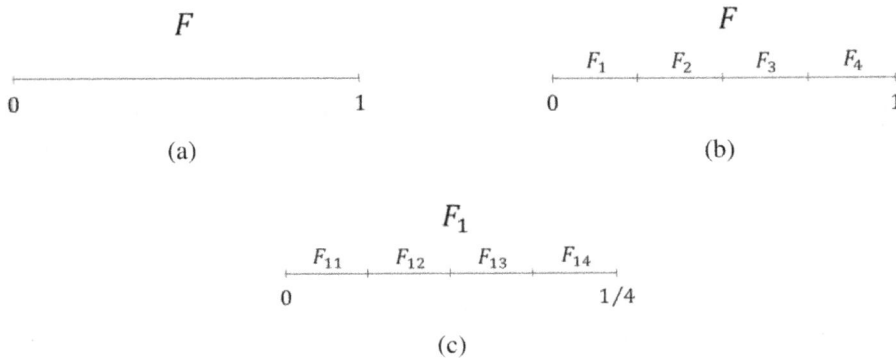

Figure 9.1. The first three alpha labeling steps of the unit section.

partition, each part $F_{i_1 i_2 \dots i_{k-1}}$ is divided into 4 equal parts denoted as $F_{i_1 i_2 \dots i_k}$, $i_p = 1, 2, 3, 4$, $p = 1, 2, \dots, k$. Figure 9.1(c) illustrates the second step of the partition of the part F_1.

Now, one can see that each point of the unit section is an intersection of a nested lines F_{i_1}, $F_{i_1 i_2}$, ..., and it is reasonable to alpha label the point as $\mathscr{F}_{i_1 i_2 \dots i_k \dots}$, such that

$$\mathscr{F} = \{\mathscr{F}_{i_1 i_2 \dots i_k \dots} \mid i_p = 1, 2, 3, 4, \ p = 1, 2, \dots\}. \tag{9.7}$$

is an *alpha label set* for the unit section in the sense of domain structured dynamics determined in the last chapter. It is sensible to consider the line with a Euclidean metric. Then the length of a section $F_{i_1 i_2 \dots i_k}$, $k = 1, 2, \dots$ is equal to $1/4k$, and the diameter condition is obviously valid. Similarly, we can see that the first degree separation condition is true with $\varepsilon_0 = 1/4$. Indeed, $d(F_1, F_3) = d(F_1, F_4) = d(F_2, F_4) = \frac{1}{4}$.

Let us now consider the similarity map $\varphi: \mathscr{F} \to \mathscr{F}$ defined by

$$\varphi(\mathscr{F}_{i_1 i_2 \dots i_k \dots}) = \mathscr{F}_{i_2 i_3 \dots i_k \dots}, \tag{9.8}$$

such that for fixed sequence $i_1 i_2 \dots i_k$, $\varphi(\mathscr{F}_{i_1 i_2 \dots i_k}) = \mathscr{F}_{i_2 i_3 \dots i_k}$ and $\varphi(\mathscr{F}_{i_1}) = \mathscr{F}$. We call each part $\mathscr{F}_{i_2 i_3 \dots i_k}$ *rectangle of order k*.

Considering the results in section 9.1, one can agree that the similarity map admits dynamics with Devaney, Li–Yorke, and ultra Poincaré chaos. Thus, we have shown that the line segment is a domain for chaos. This simple case is frankly pointed out in [23] for the Devaney chaos of logistic map '$f(x) = 4x(1 - x)$' on the interval $[0, 1]$.

To realize similarity map φ geometrically one should find proper alpha induced map ψ. We suggest to apply the following four-folded scalar Baker transformation,

$$\psi(x) = \begin{cases} 4x & 0 < x \leqslant \dfrac{1}{4}, \\[2mm] 4\left(x - \dfrac{1}{4}\right) & \dfrac{1}{4} < x \leqslant \dfrac{1}{2}, \\[2mm] 4\left(x - \dfrac{1}{2}\right) & \dfrac{1}{2} < x \leqslant \dfrac{3}{4}, \\[2mm] 4\left(x - \dfrac{3}{4}\right) & \dfrac{3}{4} < x \leqslant 1, \end{cases}$$

where the first order subsets are $F_1 = (0, \frac{1}{4}]$, $F_2 = (\frac{1}{4}, \frac{1}{2}]$, $F_3 = (\frac{1}{2}, \frac{3}{4}]$, $F_4 = (\frac{3}{4}, 1]$. For each $i = 1, 2, 3, 4$, we have that $\{\psi(x): x \in F_i\}$ is the unit section. Define, for each $j = 1, 2, 3, 4$, the second order subsets as $F_{ij} = \{x \in F_i: \psi(x) = F_j\}$. Generally, a kth order subset can be defined by

$$F_{i_1 i_2 \dots i_k} = \{x \in F_{i_1 i_2 \dots i_{k-1}}: \psi(x) = F_{i_2 i_3 \dots i_k}],$$

where $i_p = 1, 2, 3, 4$, $p = 1, 2, \dots, k$. One can see that sets of all order are the same as chosen above for the alpha labeling, and that the four-folded Baker map is an alpha induced map for the alpha dynamics. This is why it is the easiest way to conclude that the Baker transformation is alpha chaotic, and consequently, there is the ultra Poincaré, Li–Yorke, and Devaney chaos.

9.3.2 Chaotic square and 3D cube

On the basis of the above construction, more general cases of chaotic domains can be investigated. Consider the unit square (cube) F, $(0, 1]^2$, $((0, 1]^3)$. Similarly to the line case, we divide F into 16 squares (64 cubes) and denote them by F_{i_1}, $i_1 = 1, 2, \ldots, 16$ $(F_{i_1}, i_1 = 1, 2, \ldots, 64)$. Illustration of the first step construction for a square and cube are seen in figure 9.2. Again we divide each square (cube) F_i into 16 squares (64 cubes) and denote them by $F_{i_1 i_2}$, $i_2 = 1, 2, \ldots, 16$ $(F_{i_1 i_2}, i_2 = 1, 2, \ldots, 64)$. We continue in this procedure such that, at the kth step of partition, each part $F_{i_1 i_2 \ldots i_{k-1}}$ is divided into 16 squares (64 cubes) denoted as $F_{i_1 i_2 \ldots i_k}$, $i_p = 1, 2, \ldots, 16$, $p = 1, 2, \ldots, k$ $(F_{i_1 i_2 \ldots i_k}, i_p = 1, 2, \ldots, 64, p = 1, 2, \ldots, k)$. Likewise, an alpha label-set can be defined by

$$\mathscr{F} = \{\mathscr{F}_{i_1 i_2 \ldots i_k \ldots} \mid i_p = 1, 2, \ldots, m, \ p = 1, 2, \ldots\},$$

where m is 16 for the square and 64 for the cube, and each $\mathscr{F}_{i_1 i_2 \ldots i_k \ldots}$ is the corresponding alpha label for the intersection. Analogously, in these cases, the alpha map φ is defined, and it is chaotic in the sense of Devaney, Li–Yorke, and ultra Poincaré.

9.3.3 Chaos in multidimensional cubes

For a general case, consider the unit n-dimensional cube $F = (0, 1]^n$. The first step consists of dividing F into 4^n equal parts (n-dimensional sub-cubes) denoted as F_{i_1}, $i_1 = 1, 2, \ldots, 4^n$. In the second step, each part F_{i_1} is again divided into 4^n equal parts denoted as $F_{i_1 i_2}$, $i_2 = 1, 2, \ldots, 4^n$. Continue in this procedure such that, at the kth step of partition, each part $F_{i_1 i_2 \ldots i_{k-1}}$ is divided into 4^n equal sub-cubes denoted as

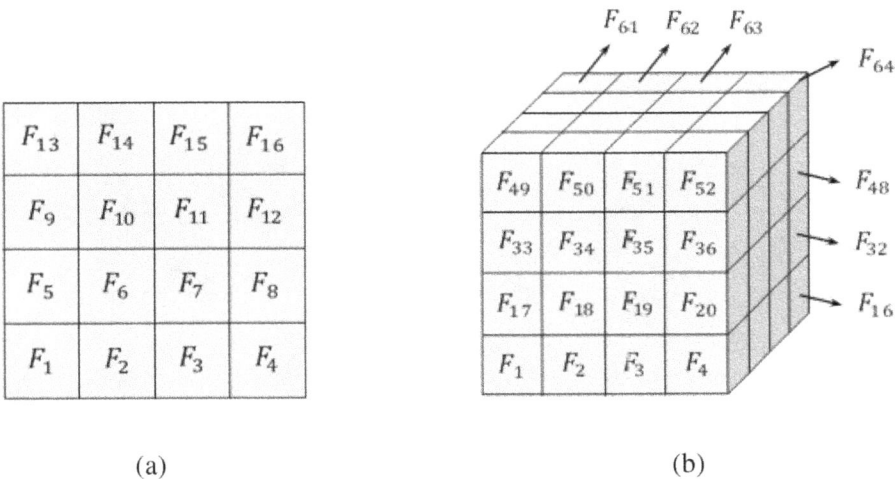

(a)   (b)

Figure 9.2. The partition procedure of a square and cube, which is identical with the alpha labeling.

$F_{i_1 i_2 \ldots i_k}$, $i_p = 1, 2, \ldots, 4^n$, $p = 1, 2, \ldots, k$. At an infinite iteration of this process, all points of the cube F can be represented as the collection of alpha labels, $\mathscr{F}_{i_1 i_2 \ldots i_k \ldots}$, i.e., the alpha label set is the following collection,

$$\mathscr{F} = \{\mathscr{F}_{i_1 i_2 \ldots i_k \ldots} \mid i_p = 1, 2, \ldots, 4^n, \ p = 1, 2, \ldots\}. \tag{9.9}$$

Let us now verify the diameter and separation properties stated in section 9.1. At any step k, the diagonal of the sub-cube $\mathscr{F}_{i_1 i_2 \ldots i_k}$ is given by $\frac{\sqrt{n}}{4^k}$, and since n is finite, the diameter property holds.

For the separation property, we will show that it is valid with a separation constant $\varepsilon_0 = \frac{\sqrt{n}}{4}$. Consider an edge of the cube as a projection of the cube on the dimension. The partition allows one to find several sub-cubes such that the distance between them and a sub-cube A is not less than $\frac{1}{4}$ in the projection. Consequently, there exists a sub-cube, say B, which is distanced from A in all projections not less than $\frac{1}{4}$. This is why the distance between A and B is greater than or equal to $\frac{\sqrt{n}}{4}$. Here, we point out that the separation constant has a direct relationship with the dimension of the cube. This simply means that the high dimensionality makes chaos stronger, and this comports with the concept that higher-state space dimensions allow for more complex attitudes, including chaotic behavior [35].

In figure 9.3, we provide simulation results to illustrate the chaotic behavior of trajectories initiated at points on the unit square and cube under the similarity map φ.

The proposed approach for generating chaotic dynamics on infinite sets sheds light on the rule of the domain of chaos. In classical theory, the requirements and properties of chaos are usually described through the map; then, the chaotic behavior is reflected in the image, characterized by quantitative measures such as fractional dimension, similarity, and scaling. An example of such a case is the Lorenz attractor, whose capacity dimension is 2.06 ± 0.01 [49]. The same could be said for an invariant subset of the domain, such as a Cantor set for the logistic map, which has the Hausdorff dimension $0.630\,93$ [53]. In the present research, we devote

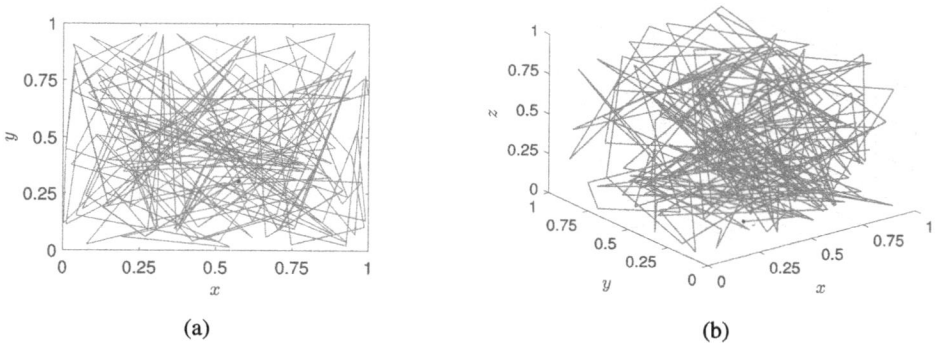

Figure 9.3. The chaotic behavior of trajectories: (a) initiated at the point $\approx(0.574\,633\,7359, 0.302\,773\,8565)$ on the unit square, and (b) initiated at the point $\approx(0.376\,643\,7951, 0.264\,931\,8230, 0.027\,599\,8135)$ on the unit cube.

more attention to the domain and less attention to the properties of the map. In particular, we impose specific conditions on the domain, namely, the diagonal and separation properties. On the other hand, the continuity and injectivity requirements for the motion are ignored since a chaotic map need not be continuous or injective [1, 24, 42, 47, 65]. We regard these properties of a chaotic map as important characteristics only from the analytical side; that is to say, they are handy for handling the map to prove the presence of chaos [20, 80]. However, it must be rigorously correct to consider them as intrinsic properties of chaos. Disregarding such properties helps us to organize a given infinite set as an invariant domain of the map, which leads to a chaotic regime with the domain, the map, and the image all having equal dimensions. This is different from the classical situation in continuous chaotic dynamics, where the dimension of the map is usually larger than that for the domain and image, as in the case of the above two examples, the Lorenz system and the logistic map.

The results presented in this section reveal that the points of an n-dimensional cube are chaotic under the defined map. That is, the cube is a quasi-minimal set. The approach can be extended for each infinite set that admits topological properties common to the cube. This reinforces the concept that the chaotic points are dense in the state space of dynamics of many real-world phenomena. Moreover, the results would be crucial for numerous relevant analyses and applications. For example, a good application can be performed for the Brownian motion [18, 78] where the random movement of particles may be considered in light of the present results.

9.4 Alpha chaos in Fatou–Julia iterations

The brilliant idea of famous scientists P. Fatou and G. Julia [41] is an approximation of infinite time dynamics in a bounded area with long orbits. The method does not have advantages for regular motions, but is very useful for complex behavior, and modern computational powers make this approach flourish for applications. Nowadays, the simple and robust algorithm is a unique way for constructing sophisticated geometries. In the present section, we use the Fatou–Julia iterations for alpha labeling.

9.4.1 Logistic maps with alpha chaos

Let $F = [0, 1]^n$ is the n-dimensional unit cube. Consider the logistic map $\psi = (\psi_1, \psi_2, ... \psi_n) \colon \mathbb{R}^n \to \mathbb{R}^n$ defined by

$$
\begin{aligned}
x_{k+1}^1 &= \psi_1(x_k^1) = r_1 x_k^1 (1 - x_k^1), \\
x_{k+1}^2 &= \psi_2(x_k^2) = r_2 x_k^2 (1 - x_k^2), \\
&\vdots \\
x_{k+1}^n &= \psi_n(x_k^n) = r_n x_k^n (1 - x_k^n),
\end{aligned}
\tag{9.10}
$$

where $r_i > 4$, $i = 1, 2, ... , n$ are parameters.

In the paper [6], we adopt the idea of Fatou–Julia iteration, sometimes called "Escape Time Algorithm" (ETA), and develop a scheme for constructing the

Sierpinski carpet. The scheme is based on the iterations of the modified planar tent map. The Sierpinski fractals are typically generated using iterated function system (IFS) [14, 15], which is defined as a collection of affine transformations. The technique of the FJI is based on detecting the points of a fractal set through the limits of their iterations under a specific map. Here, we shall extend the technique to include any possible criterion for grouping points in a given domain. That is, domain structuring will be realized. Proceeding upon the properties of the logistic map, ETA is applied for the map (9.10). We iterate the points in F under the map (9.10) such that in each iteration, we keep only the points whose images do not escape the domain F_0. The resulting points from the first iteration belong to the subsets F_i, $i = 1, 2, \ldots, 2^n$. In the second iteration, the non-escaped points belong to 2^{2n} subsets, and each subset is indexed as F_{ij}, $j = 1, 2, \ldots, 2^n$ such that $F_{ij} \subset F_i$ and $\psi(F_{ij}) = F_j$. Similarly, a subset resulting at the kth iteration is indexed as $F_{i_1 i_2 \ldots i_k}$ such that $F_{i_1 i_2 \ldots i_k} \subset F_{i_1 i_2 \ldots i_{k-1}}$ and $\psi(F_{i_1 i_2 \ldots i_k}) = F_{i_2 i_3 \ldots i_k}$.

Based on the algorithm, it is clear that the separation condition holds. Thus, we describe the points $F_{i_1 i_2 i_3 \ldots} = \lim_{k \to \infty} F_{i_1 i_2 \ldots i_k}$, and then the self-similar set F corresponding to the above algorithm, is defined as the collection of the points $F_{i_1 i_2 i_3 \ldots}$. It is clear that the map ψ and similarity map ϕ coincide on F. To agree with the alpha labeling theory, we assign a point notation $\mathscr{F}_{i_1 i_2 i_3 \ldots}$. It is clear that ψ is an alpha induced map on F. For r_i, $i = 1, 2, \ldots, n > 4$, the separation condition of the first degree and diameter condition are guaranteed to be valid for the set F, and therefore, theorems 9.2.8, 9.2.7 and 9.2.13 imply that the similarity map φ, and alpha induced map ψ are chaotic in the sense of ultra Poincaré, Li–Yorke, and Devaney.

For numerical simulation, let us consider the 2-dimensional system

$$\begin{aligned} x_{n+1} &= \psi_1(x_n) = r_1 x_n (1 - x_n), \\ y_{n+1} &= \psi_2(y_n) = r_2 y_n (1 - y_n), \end{aligned} \tag{9.11}$$

with $r_1 = 4.2$ and $r_2 = 4.3$. We fix $G = [0, 1] \times [0, 1]$ and apply ETA for (9.11). The first iteration will generate the sets F_i, $i = 1, 2, 3, 4$. In the second iteration, we get the sets F_{ij}, $j = 1, 2, 3, 4$, and so on. Figure 9.4 shows the subsets constructed in the first three iterations. The chaotic set of the system (9.11) corresponding to the alpha label-set is that resulting from an infinite iteration of this procedure. The set is a sort

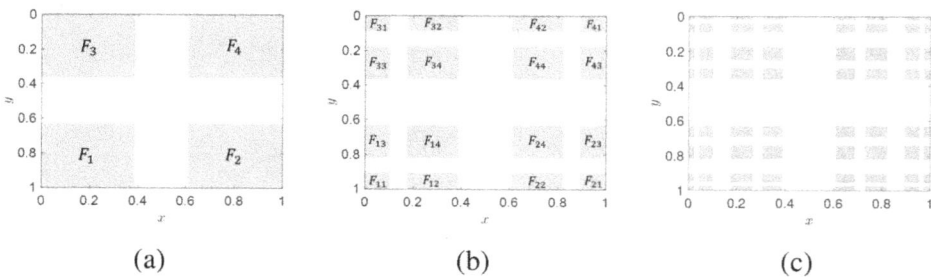

(a) (b) (c)

Figure 9.4. The subsets corresponding to the third order abstract rectangles. They were obtained by the first three iterations of the map (9.11). Reproduced with permission from [6].

of Cantor dust, which is the Cartesian product of two Cantor sets [46]. That is, the domain of the infinitely iterated logistic map is a fractal.

9.4.2 Perturbed logistic systems

This section delves into a significant investigation of systems of several logistic equations, each of which is perturbed to a slight degree. This research holds substantial importance in the field of differential equations and chaos theory. Considering differential equations with small parameters, we have proved and demonstrated that the models admit alpha chaos. The domain of dynamics can be considered as non-self-similar fractals. Moreover, they can be accepted as examples of non-trivial figures determined through abstract fractals ingredients of section 2.7.

Let F denote the initial set $[0, 1] \times [0, 1]$ and consider the 2-dimensional perturbed logistic map $\psi = (\psi_1, \psi_2)$: $\mathbb{R}^2 \to \mathbb{R}^2$ defined by

$$
\begin{aligned}
x_{n+1} &= \psi_1(x_n, y_n; \mu_1) = r_1 x_n(1 - x_n) + \mu_1 y_n, \\
y_{n+1} &= \psi_2(x_n, y_n; \mu_2) = r_1 y_n(1 - y_n) + \mu_2 x_n,
\end{aligned}
\tag{9.12}
$$

where r_1, r_2, μ_1 and μ_2 are parameters. The last term on the left-hand side of both equations in system (9.12) can be considered as a perturbation of a unimodal map. It is known that, if a unimodal map is regular [36, 81], then it is structurally stable, and therefore, any small perturbation does not affect the topological properties of the map [11]. Such an inference can be extended for high-dimensional unimodal maps. Numerical simulations can provide an adequate verification of the unimodal properties of the perturbed map.

Similarly to section 9.4.1, we apply ETA to the map (9.12). The points that do not escape F_0 in the first iteration belong to the subsets F_i, $i = 1, 2, 3, 4$. In the second iteration, the resulting points belong to the subsets indexed by F_{ij}, $j = 1, 2, 3, 4$ such that $F_{ij} \subset F_i$ and $\psi(F_{ij}) = F_j$. Similarly, a subset resulting at the nth iteration is indexed as $F_{i_1 i_2 \ldots i_n}$ such that $F_{i_1 i_2 \ldots i_n} \subset F_{i_1 i_2 \ldots i_{n-1}}$ and $\psi(F_{i_1 i_2 \ldots i_n}) = F_{i_2 i_3 \ldots i_n}$. Figure 9.5 shows the subsets constructed in the first three iterations with the parameters $r_1 = 4.2$, $r_2 = 4.5$, $\mu_1 = 0.03$, and $\mu_2 = -0.05$. We claim that the ultimate results of the iterations is a non-self-similar fractal. Obviously, one can show that it has a fractal dimension, but another way is to show that we have an abstract fractal presentation (see section 2.7) for the figure.

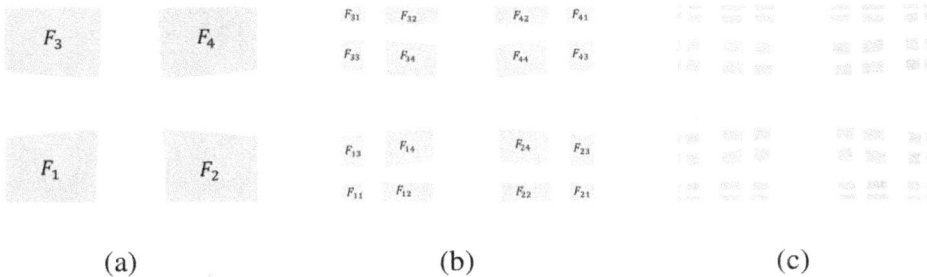

(a)　　　　　　　　　(b)　　　　　　　　　(c)

Figure 9.5. The first three iterations of the self-similar set construction by using the map (9.12). Reproduced with permission from [6].

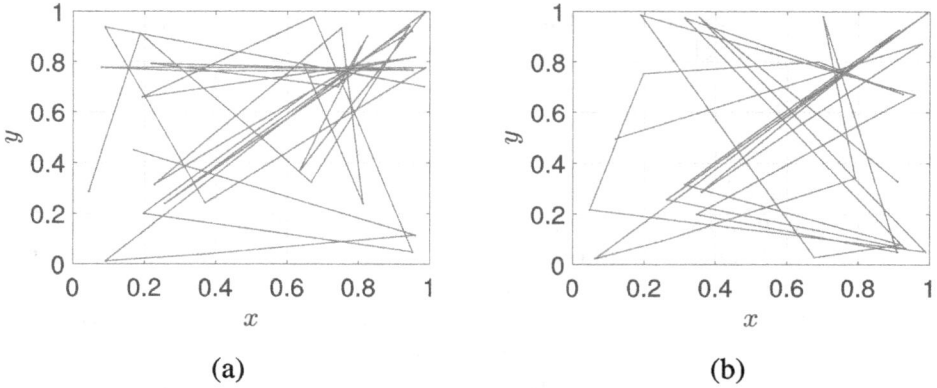

Figure 9.6. The two trajectories of the system (9.12). Reproduced with permission from [6].

Depending on the choice of the coefficients r_i and relying on the smallness of the coefficients μ_i, we have that the diameter and separation conditions for abstract similarity and chaos are fulfilled. Therefore, we could say that the alpha map φ agrees with (9.12) such that $\varphi = \mathscr{L}\psi\mathscr{L}^{-1}$ and is chaotic on the alpha set $\mathscr{F}$ as a union of alpha labels $\mathscr{F}_{i_2 i_3 \ldots}$. Obviously, it is true for the dynamics of the logistic multidimensional map. Figure 9.6 depicts two trajectories which start at the point (0.044 608 921 784 357, 0.287 657 531 506 301) and (0.910 182 036 407 281, 0.329 865 973 194 639). The irregular behavior of the trajectories reveals the presence of chaos in (9.12).

Next, we shall consider the system of an arbitrary fixed finite number of logistic maps. The discussion is of minimum formality, but the idea is clear and hopefully it can provide avenues for the new analysis of chaos and fractals.

Let F be a compact set in $\mathbb{R}^n$ such that it contains an open neighborhood of the n-dimensional unit cube and the set F to be sufficiently near to the cube. Consider the n-dimensional perturbed logistic map $\psi = (\psi_1, \psi_2, \ldots \psi_n): \mathbb{R}^n \to \mathbb{R}^n$ which is defined by

$$
\begin{aligned}
x_{k+1}^1 &= \psi_1(x_k^1, x_k^2, \ldots, x_k^n; \mu_1) = r_1 x_k^1(1 - x_k^1) + \mu_1 \chi_1(x_k^1, x_k^2, \ldots, x_k^n), \\
x_{k+1}^2 &= \psi_2(x_k^1, x_k^2, \ldots, x_k^n; \mu_2) = r_2 x_k^2(1 - x_k^2) + \mu_2 \chi_2(x_k^1, x_k^2, \ldots, x_k^n), \\
&\vdots \\
x_{k+1}^n &= \varphi_n(x_k^1, x_k^2, \ldots, x_k^n; \mu_n) = r_n x_k^n(1 - x_k^n) + \mu_n \chi_n(x_k^1, x_k^2, \ldots, x_k^n),
\end{aligned}
\quad (9.13)
$$

where μ_i, $i = 1, 2, \ldots, n$ are parameters and $\chi = (\chi_1, \chi_2, \ldots, \chi_n; \mu_i)$ is a continuous function. Due to the continuity of χ and for r_i, $i = 1, 2, \ldots, n > 4$ and sufficiently small μ_i, $i = 1, 2, \ldots, n$, one can show that an alpha label set can be constructed using (9.13), and thus, chaotic behavior in the sense of ultra Poincaré, Li–Yorke, and Devaney for an n-dimensional perturbed logistic map would be expected to appear. The domain for infinite iterations is a non-self-similar fractal. We obtain here another benefit of the result—fractals with dimension larger than three, but this has to be proven in future research. One of the methods for that can be the theory of abstract fractals, which basics are provided in section 2.7.

9.5 Notes

We anticipate that the concept of abstract similarity will introduce new foundational principles in chaos and fractal studies, serving as a valuable tool across diverse domains such as harmonic analysis, discrete mathematics, probability, and operator algebras [40]. It is worth noting that self-similarity often emerges in a multitude of problems spanning various fields like wavelets, fractals, and graph systems [40]. Our approach to chaos research places a distinct emphasis on the precedence of the domain over the mapping process [4]. To elaborate, the conventional methodology for chaos exploration typically commences with defining a map possessing specific attributes like unimodality, hyperbolicity, period three, and topological conjugacy to a standard chaotic map. Subsequently, chaotic phenomena are identified, leading to an analysis of the structure of the chaotic attractor. For instance, in scenarios like Li–Yorke chaos, the initial focus lies on defining a map with period three character- istics, followed by outlining specific properties of the chaos domain. A similar strategy applies to chaos in unimodal maps, with the chaos domain often represented by a Cantor set. In the context of the Smale horseshoe situation, the construction of both the map and domain are interdependent, with domain development commencing from an initial set and evolving step by step under the influence of a designated map. This mutual coordination between domain structure and map characteristics is a defining feature of our approach. In our proposed method, we initiate by constructing a domain within a metric space under specific conditions conducive to hosting chaotic manifestations. Subsequently, a similarity map is crafted based on the invariant and self-similarity traits of the domain to delineate an abstract motion. This tailored approach ensures that the map is apt for alpha sets and any fractal emanating from self-similarity, thereby enabling the demonstration of chaos within these fractal categories. Implementing this method- ology, we delve into various types of chaos—ultra Poincaré, Devaney, and Li– Yorke—across self-similar fractal frameworks encompassing structures like Sierpinski fractals, the Koch curve, and the Cantor set.

We firmly believe that the alpha dynamics method stands as a potent instrument for theoretical analysis, particularly in sophisticated dynamics scenarios where multifaceted models demand intricate structural dynamics assessment. This method can be harnessed across three key areas—structural dynamization, chaos certifi- cation, and stochastic process analysis—particularly relevant in exploring complex, irregular dynamics detailed in the introductory segment of this chapter. Exploring innovative structuring pathways presents an avenue for intriguing problem for- mulations. Presently, two key structuring tools are at our disposal. The first tool hinges on similarity-based structuring, while the second tool revolves around Fatou– Julia iterations, showcasing inherent interrelationships. The iteration method often takes precedence as the principal structuring mechanism, with alpha labeling serving as its foundation. While these methods are interconnected, their equivalency underscores their utility in dynamic systems modeling. Notably, structuring via similarity proves effective in delineating fractals characterized by self-similarity, extending even to non-fractal domains as exemplified through chaotic arrangements within finite-dimensional cubes within this chapter. The Fatou–Julia methodology in constructing abstract similarity sets is straightforward—starting with a bounded region, iteratively collecting points after n iterations to envelop an n-dimensional rectangle, eventually culminating in the structured domain defined by the envelope intersections. This method requires an initial first-order partition, exemplified by the logistic map iterations shaping the Cantor middle set. We envisage the practical application of this methodology in establishing abstract similarity chaos within figures like the Julia set and other sophisticated structures, furnishing algorithmic support for abstract constructions and dynamic equations.

References

[1] Addabbo T, Fort A, Rocchi S and Vignoli V 2011 Digitized chaoor pseudo-random number generation in cryptography *Chaos-based Cryptography: Theory, Algorithms and Applications* ed L Kocarev and S Lian (Berlin: Springer) pp 67–97

[2] Addison P S 1997 *Fractals and Chaos: An Illustrated Course* (Bristol: Institute of Physics Publishing)

[3] Akhmet M 2022 Unpredictablity in Markov chains *Carpathian J. Math.* **38** 13–9

[4] Akhmet M and Alejaily E M 2019 Domain-structured chaos in a Hopfield neural network *Int. J. Bifurcation Chaos* **29** 195–205

[5] Akhmet M and Alejaily E M 2020 Conservative-progressive growth method for fractal curves and functions (unpublished)

[6] Akhmet M and Alejaily E M 2021 Abstract similarity, chaos and fractals discrete continuous *Dyn. Syst. Ser.* B **26** 2479–97

[7] Akhmet M and Fen M O 2016 Poincaré chaos and unpredictable functions *Commun. Nonlinear Sci. Numer. Simul.* **48** 85–94

[8] Akhmet M and Fen M O 2016 Unpredictable points and chaos *Commun. Nonlinear Sci. Numer. Simul.* **40** 1–5

[9] Akhmet M, Fen M O and Alejaily E M 2018 *Dynamics motivated by Sierpinski fractals* ArXiv e-prints

[10] Anovitz L M and Cole D R 2015 Characterization and analysis of porosity and pore structures *Rev. Mineral. Geochem.* **80** 61–164

[11] Avila A and Moreira C G 2003 Bifurcations of unimodal maps arXiv preprint math/0306157 Dynamical Systems pp 1–22

[12] Bandt C and Graf S 1992 Self-similar sets 7. A characterization of self-similar fractals with positive Hausdorff measure *Proc. Am. Math. Soc.* **114** 995–1001

[13] Barabasi L 1995 *Fractal Concepts in Surface Growth* (Cambridge: Cambridge University Press)

[14] Barnsley M 1988 *Fractals Everywhere* (New York: Academic)

[15] Barnsley M F and Demko S 1985 Iterated function systems and the global construction of fractals *Proc. R. Soc. Ser.* A **399** 243–75

[16] Barnsley M F, Elton J, Hardin D P and Massopust P R 1989 Hidden variable fractal interpolation functions *SIAM J. Math. Anal.* **20** 1218–42

[17] Bolotin Y, Tur A and Yanovsky V 2013 *Chaos: Concept, Control and Constructive Use* (Berlin: Springer)

[18] Brown R 1928 A brief account of microscopical observations made in the months of June, July and August, 1927, on the particles contained in the pollen of plants; and on the general existence of active molecules in organic and inorganic bodies *Philos. Mag. Ann. Philos. New Ser.* **4** 161–78

[19] Carbone A, Gromov M and Prusinkiewicz P 2000 *Pattern Formation in Biology, Vision and Dynamics* (Singapore: World Scientific)

[20] Chen G and Huang Y 2011 *Chaotic Maps: Dynamics, Fractals and Rapid Fluctuations, Synthesis Lectures on Mathematics and Statistics* (San Rafael, CA: Morgan & Claypool Publishers)

[21] Crownover R M 1995 *Introduction to Fractals and Chaos* (Boston, MA: Jones and Bartlett)

[22] Davis H T 1989 On the fractal character of the porosity of natural sandstone *Europhys. Lett.* **8** 629–32

[23] Devaney R L 1987 *An Introduction to Chaotic Dynamical Systems* (Menlo Park, CA: Addison-Wesley)

[24] Değirmenci N and Koçak Ş 2010 Chaos in product maps *Turk. J. Math.* **34** 593–600

[25] Edgar G A 1990 *Measure, Topology, and Fractal Geometry* (New York: Springer)

[26] Ercai C 1997 Chaos for the Sierpinski carpet *J. Stat. Phys.* **88** 979–84

[27] Falconer K J 1985 *The Geometry of Fractal Sets* (Cambridge: Cambridge University Press)

[28] Falconer K J 1995 Sub-self-similar sets *Trans. Am. Math. Soc.* **347** 3121–9

[29] Feigenbaum M J 1980 Universal behavior in nonlinear systems *Los Alamos Sci./Summer* **1** 4–27

[30] Feldman D P 2012 *Chaos and Fractals: An Elementary Introduction* (Oxford: Oxford University Press)

[31] Ganji D D and Kachapi S H H 2015 *Application of Nonlinear Systems in Nanomechanics and Nanofluids: Analytical Methods and Applications* (New York: Elsevier)

[32] Hahn H K, Georg M and Peitgen H O 2005 Fractal aspects of three-dimensional vascular constructive optimization *Fractals in Biology and Medicine* (Berlin: Birkhäuser) pp 55–66

[33] Hardin D P and Massopust P R 1986 The capacity for a class of functions *Cormm. Math. Phys.* **105** 455–60

[34] Hata M 1985 On the structure of self-similar sets *Japan J. Appl. Math.* **2** 381–414

[35] Hilborn R C 2000 *Chaos and Nonlinear Dynamics: An Introduction for Scientists and Engineers* (New York: Oxford University Press)

[36] Hunt B R and Kaloshin V Y 2010 *Handbook of Dynamical Systems* ed A Katok and B Hasselblatt (Amsterdam: Elsevier Science) pp 43–87 Broer

[37] Hutchinson J 1981 Fractals and self-similarity *Indiana Univ. Math. J.* **30** 713–47

[38] Ivancevic V G and Ivancevic T T 2007 *High-Dimensional Chaotic and Attractor Systems* (Dordrecht: Springer)

[39] Janahmadov A K and Javadov M 2018 *Fractal Approach to Tribology of Elastomers* (Switzerland: Springer)

[40] Jorgensen P E T 2006 Analysis and probability: wavelets *Analysis and Probability: Wavelets, Signals, Fractals* **vol 234** (New York: Springer)

[41] Julia G 1918 Mémoire suv i'itération des fonctions rationelles *J. Math. Pure Appl.* **8** 47–245

[42] Keener J P 1980 Chaotic behavior in piecewise continuous difference equations *Trans. Am. Math. Soc.* **261** 589–604

[43] Kellert S H 1994 *In the Wake of Chaos: Unpredictable Order in Dynamical Systems* (Chicago, IL: University of Chicago Press)

[44] Korsch H J, Jodl H J and Hartmann T 2008 *Chaos: A Program Collection for the PC* (Berlin: Springer)

[45] Lau K S, Ngai S M and Rao H 2001 Iterated function systems with overlaps and the self-similar measures *J. London Math. Soc.* **63** 99–115

[46] Layek G C 2015 *An Introduction to Dynamical Systems and Chaos* (New Delhi: Springer)

[47] Li R and Zhou X 2013 A note on chaos in product maps *Turk. J. Math.* **37** 665–75

[48] Li T Y and Yorke J A 1975 Period three implies chaos *Am. Math. Mon.* **82** 985–92

[49] Lorenz E N 1963 Deterministic non-periodic flows *J. Atmos. Sci.* **20** 130–41

[50] Lynch S 2017 *Dynamical Systems with Applications using Mathematica* 2nd edn (Cham: Springer International Publishing)

[51] Mandelbrot B B 1975 *Les Objets Fractals: Forme, Hasard, et Dimension* (Paris: Flammarion)

[52] Mandelbrot B B 1977 *Fractals: Form, Chance, and Dimension* (San Francisco, CA: W. H. Freeman)

[53] Mandelbrot B B 1983 *The Fractal Geometry of Nature* (New York: Freeman)

[54] Mandelbrot B B 2004 *Fractals and Chaos: The Mandelbrot Set and Beyond* (New York: Springer)

[55] Masoller C, Schifino A C and Sicardi R L 1995 Characterization of strange attractors of Lorenz model of general circulation of the atmosphere *Chaos, Solit. Fract.* **6** 357–66

[56] Massopust P R 1987 Dynamical systems, fractal functions and dimension *Topology Proc.* **12** 93–110

[57] Massopust P R 1997 Fractal functions and their applications *Chaos Solit. Fract.* **8** 171–90

[58] Massopust P R 2004 Fractal functions and wavelets: examples of multiscale theory pages *Abstract and Applied Analysis, Proceedings of the International Conference, Hanoi, Vietnam* (Singapore: World Scientific Press) pp 225–47

[59] Mehlhorn K 1984 *Data Structures and Algorithms* **vol 3** (New York: Springer)

[60] Moon F C 1992 *Chaotic and Fractal Dynamics: An Introduction for Applied Scientists and Engineers* (New York: Wiley)

[61] Moran P A P 1946 Additive functions of intervals and Hausdorff measure *Proc. Cambridge Philos* pp 15–23 Soc. 42

[62] Ngai S M and Wang Y 2001 Hausdorff dimension of overlapping self-similar sets *J. London Math. Soc.* **63** 665–72

[63] Nemytskii V V and Stepanov V V 1960 *Qualitative Theory of Differential equations* (Princeton, NJ: Princeton University Press)

[64] Ornstein D 1970 Bernoulli shifts with the same entropy are isomorphic *Adv. Math.* **4** 337–52

[65] Osikawa M and Oono Y 1981 Chaos in C^0-endomorphism of interval *Publ. RIMS* **17** 165–77

[66] Pesin Y 1997 *Dimension Theory in Dynamical Systems: Contemporary Views and Applications* (Chicago, IL: University of Chicago Press)

[67] Pesin Y and Weiss H 1996 On the dimension of deterministic and random cantor-like sets, symbolic dynamics, and the Eckmann-Ruelle conjecture *Commun. Math. Phys.* **182** 105–53

[68] Sander E and Yorke J A 2011 Period-doubling cascades galore *Ergod. Theor. Dynam. Syst.* **31** 1249–67

[69] Schöll E and Schuster H G (ed) 2008 *Handbook of Chaos Control* (Weinheim: Wiley-VCH)

[70] Singer P and Zajdler P 1999 Self-affine fractal functions and wavelet series *J. Math. Anal. Appl.* **240** 518–51

[71] Skiadas C H and Skiadas C (ed) 2016 *Handbook of Applications of Chaos Theory* (Boca Raton, FL: CRC Press)

[72] Spear D W 1992 Measures and self similarity *Adv. Math.* **91** 143–57

[73] Stella S 1992 Hausdorff dimension of recurrent net fractals *Proc. Am. Math. Soc.* **116** 389–400

[74] Tang H P, Wang J Z, Zhu J L, Ao Q B, Wang J Y, Yang B J and Li Y N 2012 Fractal dimension of pore-structure of porous metal materials made by stainless steel powder *Powder Technol.* **21** 383–7

[75] Turing A 1952 The chemical basis of morphogenesis *Phil. Trans. R. Soc. London B.* **237** 32–72

[76] Vicsek T 1992 *Fractal Growth Phenomena* (Singapore: World Scientific)

[77] Volrath H J 1977 The understanding of similarity and shape in classifying tasks *Educ. Stud. Math.* **8** 211–4

[78] Wang M C and Uhlenbeck G E 1945 On the theory of Brownian motion ii *Rev. Mod. Phys.* **17** 323–42

[79] Weierstrass K 1895 *Uber continuierliche Functionen eines reellen Arguments, die fur keinen Wert des letzteren einen bestimmten Differentialquotienten besitzen, in 'K. Weierstrass, Mathematische Werke'* (Berlin: Mayer & Miiller)

[80] Wiggins S 1988 Global Bifurcation and Chaos *Analytical Methods* (New York: Springer)

[81] Zeraoulia E and Sprott J C 2012 *Robust Chaos and Its Applications* (Singapore: World Scientific)

[82] Zmeskal O, Dzik P and Vesely M 2013 Entropy of fractal systems *Comput. Math. Appl.* **66** 135–46